H. C Godwin

Railroad Engineers' Field-Book and Explorers' Guide

H. C Godwin
Railroad Engineers' Field-Book and Explorers' Guide
ISBN/EAN: 9783743402478
Manufactured in Europe, USA, Canada, Australia, Japa
Cover: Foto ©berggeist007 / pixelio.de

Manufactured and distributed by brebook publishing software (www.brebook.com)

H. C Godwin

Railroad Engineers' Field-Book and Explorers' Guide

RAILROAD ENGINEERS' FIELD-BOOK

AND

EXPLORERS' GUIDE.

ESPECIALLY ADAPTED TO THE USE OF RAILROAD ENGINEERS

ON

LOCATION AND CONSTRUCTION,

AND TO THE NEEDS OF THE EXPLORER IN MAKING

EXPLORATORY SURVEYS.

BY

H. C. GODWIN.

SECOND, REVISED EDITION.

SECOND THOUSAND.

NEW YORK:
JOHN WILEY & SONS.
LONDON: CHAPMAN & HALL, LIMITED.
1897.

RAILROAD ENGINEERS' FIELD-BOOK.

PREFACE.

I AM publishing the following notes because I think they may possibly supply the want of a Field-book,—a want which I have often felt myself and have often heard expressed—which, while avoiding as much as possible the intricacies of mathematics, would be of more general application than any of the books of this class which I have as yet come across.

The Railroad engineer is rarely an expert mathematician: in fact it has always seemed to me that the time which must necessarily be spent by him in attaining mathematical proficiency might be very much better employed in reading up some of the more practical subjects of his profession. Bearing this in mind, I have endeavored to strip the following pages of all unnecessary mathematical deductions, making it mainly my object to give the results deduced, and yet at the same time giving sufficient explanation to enable any one possessed of the ordinary smattering of mathematics and mechanics to deduce the same results for himself.

I have avoided the insertion of Logarithmic Tables. I am well aware that to some this will appear a serious omission; but considering that this is merely a Field-book, and not a work to be consulted in cases where accuracy in the 6th figure is usually essential, I have deemed that the exclusion of the hundred pages or so which this omission permits, amply compensates for the few seconds of additional labor which the lack of them may occasionally involve. Speaking for myself, as regards Railroad work, I must say that for one time that I work by logarithms I work a hundred times by "naturals;" and I know that most engineers would bear similar testimony.

In the Astronomical problems in the latter part of the book, considerable labor may, of course, be saved by the use of

Logarithmic Tables. The method I employ myself on such work is to take with me into camp the logarithmic portion of Chambers' Mathematical Tables—which I have had bound in pocket-book form—giving the logarithms of numbers up to 108,000 and of trigonometrical functions to 7 places of decimals: in this way, high accuracy, when it *is* wanted, can be obtained much more readily and efficiently than by any table which could reasonably be inserted in a book suitable for pocket use; and as the logarithmic tables are rarely wanted outside the tent, they form a sort of stay-at-home counterpart to the Field-book itself.

Table IX is inserted solely for convenience in the reduction of indices, barometric formulæ, etc., and a few like operations, in which the use of logarithms is more or less essential.

H. C. GODWIN.

COLORADO, *January*, 1889.

INTRODUCTION.

THE Contents of this Field-book are divided mainly into four parts:

Part   I. Dealing with Railroad Location.
Part II. Dealing with Railroad Construction.
Part III. Dealing with Reconnoissance and Exploratory Surveys.
Part IV. Giving various General Information.

To these are added a Short Appendix and a Set of Tables, comprising those generally required for Field use.

Although Part III should, from its nature, take precedence over Parts I and II, since Reconnoissance is usually the first step towards Location, yet the subject of Exploratory Surveying is here treated too fully—in comparison with Parts I and II—to warrant its being regarded merely as an Introduction to them. I have therefore considered it a *special* subject, and accordingly given to it a subsequent position.

CONTENTS.

RAILROAD LOCATION.

GENERAL CONSIDERATIONS.

SEC.		PAGE
1.	Conditions of Economical Location	1
2.	Train Resistances	2
3.	Rolling Resistance	2
4.	Resistance due to Oscillation and Concussion	3
5.	Atmospheric Resistance	3
6.	Resistance due to Curvature	4
7.	Resistance due to Gravity	4
8.	Diagram of Resistance	5
	Limiting Velocity on any Grade	8
9.	Propelling Force of Locomotive	8
	Coefficient of Adhesion	8
	Sliding Friction	8
	Limiting I.H.P.	9
	Weight on Driving-wheels	9
	Grate-surface	9
10.	Diagram of Propelling Force	10
	Limiting Speed for any given I.H.P.	10
	Internal Frictional Resistances	11
	Back-pressure, Wire-drawing, etc.	11
11.	Diameters of Driving-wheels	11
12.	I.H.P. required at any given speed	12
	Most Economical Speed	12
	Limiting Grade	13
13.	Weight of Locomotives and Rolling-stock	13
14.	Resistance due to Inertia	13
	Rotative Energy of the Wheels	14
15.	Resistance caused by Application of Brakes	14
	Automatic Brakes	15
	Hand Brakes	15
16.	Initial Velocity	15
17.	Height corresponding to Velocity	16
	Table of Heights corresponding to Velocity	17
18.	Assumption of *Mean* Resistance and *Mean* Propelling Force	17
19.	Graphic Method of solving Dynamical Problems	17
20.	Examples	18

SEC.	PAGE
21. Rise and Fall....	19
Profile of Velocities..	20
22 } Effects of Rise and Fall	20
23.	
24. Maximum Grade...	21
25. Economy of Locomotive...	22
26. Compensation for Curvature...	22
27. Compensation for Brakes...	23
28. Broken Grades...	23
Momentum Grades...	23
29. Danger of breaking Train and Derailment...	24
30. Work done on Grades	24
31. Pusher-grades .	26
Table of Pusher-grades	26
32. Maximum Curvature...	26
Safe Speed on Curves	26
33. Short Tangents...	27
Location of Curves...	27
34. Table of Work done against Resistances...	28

COST OF OPERATING.

35. Cost of Work done against Resistance	28
36. Cost per Train-mile...	28
37. Economy of Construction	29
38. Cost of Operating Pusher-grades	30
39. To test Relative Cost of Various Routes..	30
40. Effect of Alterations in Alignment...	30
41. To estimate Effect of Ditto	31

RECEIPTS.

42. Deviating to catch Way-business	31

COST OF CONSTRUCTION.

43. Average Cost of Track...	32
Average Cost of General Construction...	32
Average Cost of One Mile of Track...	33
Cost of Trestle-work, Trusses, Tunnels, etc...	33

INSTRUMENTS.

44. Transit	34
Adjustments...	34
45. Remarks...	36
46. Stadia...	39
47. Compass...	42
Adjustments	42
Remarks...	42
48. Magnetic Variation...	43
Chart of Magnetic Variation...	44
49. Dumpy Level...	45
50. Y Level...	45

CONTENTS.

SEC.		PAGE
51.	Correction for Curvature and Refraction	46
52.	Hand Level	47

THE SURVEY.

53.	Reconnoissance and Preliminary Surveys	48
	Running the Line to Grade	49
	Table of Grades and Grade Angles	50
54.	Transit Work	51
55.	Latitudes and Departures	52
56.	Azimuth Observations	54
57.	A. Maximum Elongation of Polaris	55
	B. Observation of γ Cassiopeia and Polaris	56
	C. Observation of Alioth and Polaris	57
58.	Convergence of Meridians	58
59.	Simple Triangulations	60
	Offsetting the Transit-line	61
60.	Levelling	61
	Precision of a Line of Levels	62
61.	Taking Topography	62
62.	Contour Lines	64
	Locating by means of Contour Lines	64
63.	Levels and Curvature	66
64.	Equations	66
65.	Value of Topography	67
66.	Tangents and Curves	68
67.	Selection of Curves by Eye	68
68.	Balance of Cuts and Fills	69
69.	Establishing the Grades	69
	Rough Estimation of Grading	69
70.	Estimating by Centre Heights	71

CURVES.

71.	Radius and Degree of Curves	71
72.	Corrections for 50-foot Chords	72
73.	Length of Curves	74
74.	Nomenclature and Symbols	74
75.	Fundamental Formulæ	75

PROBLEMS IN SIMPLE CURVES.

76.	To lay out a curve by deflection angles	78
	To find corrected length of any sub-chord	78
	Example	78
77.	To locate a curve when the apex is inaccessible	81
78.	To locate a curve by offsets from a tangent	82
	Ditto if the apex, P.C., etc., are inaccessible	83
79.	To locate a curve by offsets from the chords produced	85
80.	To locate a curve by ordinates from a long chord	87
	Example	87
	Ditto by mid-ordinates	88

SEC.		PAGE
81.	To pass a curve through a fixed point, *I* being given...........	89
82.	To run a tangent from a curve to any fixed point................	90
83.	To connect two curves by a tangent.............................	90
84.	Given a curve joining two tangents, to change the P.C. so that the curve may end in a parallel tangent.....................	91
85.	To transfer a curve both at its P.C. and P.T. to parallel tangents.	92
86.	Given a curve joining two tangents, to change *R* and the P.C. so that the new curve may end in a parallel tangent at a point opposite to the original P.T........................	92
87.	Given a curve, to find *R* of another curve, which, from the same P.C., will end in a parallel tangent.........................	93
88.	Given a curve joining two tangents, to change *R* and the P.C. so that the curve may end in the same P.T., but with a change in direction....	93

COMPOUND CURVES.

89.	Locating compound curves..	94
90.	To locate a C.C. when the P.C.C. is inaccessible................	94
91.	Given a simple curve ending in a tangent, to connect it with a parallel tangent by means of another curve................	95
92.	To connect a curve with a tangent by means of another curve of given radius...	95
93.	Given a C.C. ending in a tangent, to change the P.C.C. so that the terminal curve may end in a given parallel tangent, without changing its radius.................................	97
94.	To connect two curves already located by means of another curve of given radius..................	98
95.	To locate any portion of a C.C. from any station on the curve..	99

TRANSITION CURVES.

96.	Advantages of Transition Curves.......................	100
97.	Method I....................	100
98.	Method II......................................	104
99.	Method III..	105
100.	Vertical Curves..	107

CONSTRUCTION.

101.	Division of the Subject	109

A. SETTING OUT WORK.

102.	Clearing Right of Way, etc............................	109
103.	Location of Culverts, etc....	109
104.	System of Drainage.—Ditches	110
105.	Checking Benchmarks and Alignment....	111
106.	Cross-sectioning.....................................	111
	Setting Slope-stakes	112
	Points at which Cross-sections should be taken..	114

SEC.		PAGE
107.	Reference Points	115
108.	Staking out Borrow-pits	115
109.	Staking out Foundation-pits for Culverts	115
110.	Setting out Bridge Foundations	116
111.	Setting out Trestlework	117
112.	Setting out Tunnels	118
113.	Giving "Grade" and centres	120
	Shrinkage and Increase	121
114.	Difference of Elevation on Curves	121
	Effect on the Dump and on Trestles	123
	Increase in Gauge on Curves	124
115.	Inspecting the Grading	124
116.	Running Track-centres and setting Ballast-stakes	125
117.	Permanent Reference-points	125
118.	Turnouts and Crossings	125
119.	Locating by Offsets	127
120.	Example	129
121.	Turnouts and Crossings on Curves	129
122.	Curving Rails	132
123.	Expansion of Rails	132

B. THE ESTIMATING OF LABOR AND MATERIAL.

124.	The Cost of Earthwork and Rockwork removed by Carts	133
125.	Ditto, by other means	136
126.	Overhaul	136
127.	The Calculation of Earthwork	137
	Areas of Cross-sections	139
128.	The Pyramid, Wedge, and Prismoid	139
129.	The Prismoidal Formula	140
130.	The Method of Average End-areas	143
	Prismoidal Corrections	143
131.	The Method of Equivalent Level Sections	146
132.	The Method of Centre-heights	147
133.	Earth-work Tables	147
134.	Correction for Curvature	148
135.	Contents of the Toe of a Dump	149
136.	General	152
137.	Timber-work	152
	Table of Board Measure	151
	Fractions of an Inch in Decimals of a Foot	153
138.	Iron-work	153
	Weight of Bolts, Nuts, and Bars	153, 154
	Railroad Spikes	154
	Angle-bars and Bolts per mile	155
	Weight of Rails per mile	155
139.	Ballast and Ties per mile	156

/ CONTENTS.

EXPLORATORY SURVEYING.

SEC.   PAGE
140. Introduction... 157

INSTRUMENTS.

141. The Sextant.—Adjustments, etc............................... 157
142.   Use of the Sextant.. 159
        Parallax 159
143.   Supplementary Arc... 161
144.   Observing Horizontal Angles........................... 161
145.   Eliminating Instrumental Errors........................ 162
146. The Artificial Horizon... 162
147. The Chronometer... 164
148. Barometers.. 165
149.   Barometric Formulæ....................................... 166
150.   Reduction of Errors of Gradient...................... 168
151.   Taking Readings... 169
152.   Diurnal and Annual Gradient........................... 169
153. The Cistern Barometer... 170
154.   To fill a barometer... 170
155.   Reading the barometer................................... 171
156.   Cleaning the barometer.................................. 171
157. The Aneroid Barometer.. 172
158.   Elevation Scales... 173

EXPLORATORY SURVEYS.

159. Division of the Subject.. 174
160. To find the distance apart, etc., of two inaccessible points....... 175
161. The "Three-point Problem"..................................... 176
162. Positions fixed by bearings..................................... 178
163.   Positions fixed by intersection....................... 178
164. Obtaining Heights of Mountains trigonometrically...... 178
165.   Refraction of the Air....................................... 180
166.   Reciprocal Angles... 180
167.   By Depression of the Sea Horizon................... 181
168. Observing Altitudes and Depressions...................... 181
        With a Sextant and Artificial Horizon................... 181
        With a Transit.. 182
169. Measurement of a Base.. 182
        Correction for Temperature.................................. 182
        Reduction to Sea-level....................................... 183
170. Example of Triangulating on Exploratory Surveys....... 183
171. To measure a horizontal angle without an instrument.... 184
172. To measure a vertical angle without an instrument...... 185
173. Measurement of Distance by Sound......................... 185
174. Measurement of Time by Vibrations......................... 185
175. Direct Measurement and Compass Courses.............. 186
        Odometers and Pedometers 186
        Estimating the Rate of Progress........................... 186

SEC.		PAGE
176.	Astronomical Observations	187
177.	Solar Time	187
178.	Equation of Time	188
179.	Sidereal Time	189
180.	Right Ascension and Declination	189
181.	Correcting for Longitude, etc	190
182.	Hour-angle	191
183.	Examples	192
184.	Refraction	194
185.	Parallax	195
186.	Correcting for Semi-diameter	197
	Augmentation	197
187.	Dip	198
188.	Summary of Corrections	198
189.	Latitude.—By Meridian Altitudes	199
190.	Remarks	201
191.	By Transits across the Prime Vertical	202
192.	By an Altitude out of the Meridian	203
193.	By Double Altitudes	205
194.	By an Altitude of Polaris at any time	206
195.	Longitude.—Local Time, by an Altitude of a Star	06
196.	Local Time, by Equal Altitudes of a Star	208
197.	Local Time, with a Transit	208
198.	By Lunar Culminations	209
199.	By Lunar Distances	210
200.	By Jupiter's Satellites	214
201.	To test the chronometer rate	214
202.	To set the transit in the meridian	214
203.	Interpolation by Successive Differences	215
204.	" Accidental Error"	216
205.	Influence of Spheroidal Form of the Earth	218
206.	Figure of the Earth	218
207.	Conversion of Angular Measure into Distance and *vice versâ*	219
208.	Given the lat. and long. of two places, to find their distance apart, etc	220
209.	To find the radius of a circle of latitude	221
210.	Offsets to a Parallel of Latitude	221
211.	Development of a Spherical Surface	221
212.	Example	222
213.	Star Map	225
	Star Tables	226, 227

MISCELLANEOUS.

214.	The Horse-power of Falling Water	228
215.	To gauge a stream roughly	228
216.	Sustaining Power of Wooden Piles	229
217.	Supporting Power of Various Materials	229

CONTENTS.

SEC.		PAGE
218.	Transverse Strength of Rectangular Beams	229
219.	Natural Slopes of Earth	230
220.	Weight of Earths, Rocks, etc., per cubic yard	230
221.	Weight of Timber and Metals per cubic foot	231
222.	Mortar, Cement, and Concrete	231
223.	Notes on Timber.—Selection of Trees	231
224.	Defects of Timber	232
225.	Felling Timber	233
226.	Seasoning and preserving Timber	233
227.	Decay of Timber	234
228.	Tests for Steel and Iron	234
229.	Strength of Rope.—Manilla, Iron and Cast Steel	235
230.	Properties of the Circle	236
231.	Trigonometry.—Plane	237
232.	General Equations	240
233.	Spherical	241
234.	Measures of Length and Surface	243
235.	Measures of Weight and Capacity	244

APPENDIX.

TABLES.

Table	I. Radii of Curves	252
"	II. Tangents and Externals to a 1° Curve	255
"	III. Tangential Offsets at 100 feet	259
"	IV. Mid-ordinates to 100-foot Chords	259
"	V. Long Chords	260
"	VI. Mid-ordinates to Long Chords	263
"	VII. Minutes in Decimals of a Degree	264
"	VIII. Squares, Cubes, Square and Cube Roots	265
"	IX. Logarithms of Numbers.—1 to 1000	282
"	X. Natural Sines and Cosines	285
"	XI. " Secants and Cosecants	294
"	XII. " Tangents and Cotangents	309
"	XIII. " Versines and Exsecants	321
"	XIV. Cubic Yards per 100 feet, in terms of Centre-height	315
"	XV. Cubic Yards per 100 feet, in terms of Sectional Area	350
"	XVI. Mutual Conversion of Feet and Inches into Meters and Centimeters	354
"	XVII. Mutual Conversion of Miles and Kilometers	355
"	XVIII. Length of 1' arcs of Latitude and Longitude	355
"	XIX. Mutual Conversion of Mean and Sidereal Time	356
"	XX. Mutual Conversion of Time and Degrees	358

PART I.

RAILROAD LOCATION.

GENERAL CONSIDERATIONS.

1. In the early days of Railroad Building, the Locating Engineer was forced to rely mainly on his individual ability, trusting principally to the correctness of his eye to detect the most suitable route, guided only by the very limited experience of others and his own common-sense. The man who worked his party the hardest, and covered most ground in the day, was in those days, unless any very obvious defects were visible in his work, too often looked upon as the best locator. But the years of experience which have followed have been years of experiment also; and the practice of Railroad Location has by degrees developed into a science, which, though yet far from perfect, forms a most important part of a Modern Engineering Education.

In a Field-book of this sort, it is impossible to do more than treat rapidly a few of the leading questions which the subject involves, and formulate, where possible, rules for guidance in the field.

A knowledge of the principles of Railroad Location must be backed up by experience in Railroad Construction. For, in order to locate well, a man must have fairly accurate ideas of the suitability and cost of the various works which his location involves. The best location for a certain road is not that which enables the traffic to be carried on with the least amount of work, or which gives the lowest Operating Expenses, but that which, in a given time, renders the

$$\text{Receipts} - \begin{pmatrix}\text{Operating}\\ \text{Expenses}\end{pmatrix} - \begin{pmatrix}\text{Interest on Capital spent on}\\ \text{Construct. Equipment, etc.}\end{pmatrix} = \text{Profits}$$

a maximum. Thus we see that more or less accurate estimates of the probable Receipts and Operating Expenses are of the utmost importance before starting the location; and it is only when these are arrived at that the amount which we are entitled to expend on construction can be fixed.

2. Before considering the Financial side of the question, however, we will glance hurriedly over some of the principal Mechanical Problems which occur in dealing with the motion of trains, for, without some slight knowledge of Railroad Dynamics, an intelligent application of the Laws of Location is impossible.

TRAIN RESISTANCES.

The Resistance due to the motion of a train on a straight level track—excluding for the present the Inertia of the train—may be regarded as being the sum of the three following components:

3. ROLLING RESISTANCE, which is composed of the frictional resistance at the journals and that at the wheels at the points of contact with the rails: these two may for ordinary purposes be classed together under the head of Rolling Resistance. Its magnitude depends largely upon the surface-bearing at the journals; the coefficient of friction decreasing as the load per unit-surface on the journals increases, so that the resistance is relatively higher in the case of Empty Cars than with Loaded ones; being at ordinary speeds about 6 lbs. per ton (2000 lbs.) of weight of train in the former case, while with Passenger Coaches or Loaded Cars it only amounts to about 4 lbs. By referring to the Diagram of Resistances, p. 6, we see that at the point of starting the Rolling Resistance is very high, being then about 20 lbs. per ton, but that at a velocity of about ten miles per hour it reaches its minimum value, and from that point increases constantly by a trifling amount through the successive higher velocities. The Initial Resistance depends largely on the length of time the train has been standing, a stop of only a few seconds causing a resistance of about one half that given in the Diagram. Since, however, there is always more or less "give" about the couplings, no two cars at the same instant offer their maximum resistance, the front end of a long train being well under way before any motion at all is transmitted to the rear. Thus the pull on the

draw-bar is not in reality so excessive as it at first appears; for if we take the whole train into consideration, the resistance at the start may be set down as about 12 lbs. instead of 20 lbs. per ton, as in the case of a single car.

The Line of Rolling Resistance starts in the Diagram from the line of the 1 p. c. grade; thus indicating that a train left standing with the brakes off on this grade, is just on the point of starting on its own account. On any grade lighter than this, a train will usually require considerable force to set it in motion. By increasing the diameters of the wheels we slightly decrease the resistance to rolling.

4. RESISTANCE DUE TO OSCILLATION AND CONCUSSION.—The amount of this we obtain approximately by assuming that it equals .005 lb. per ton at 1 mile per hour, and increases as the square of the velocity. Thus, e.g., at 40 m. p. h. it equals 8 lbs. per ton. The longer the train, however, the less this resistance amounts to per ton, for each car is more or less steadied by the force which is transmitted through it to the adjoining one; thus it is usually much more considerable in the rear than in the centre or forward end of the train. It is produced in a great measure by the inequality in elevation of the two rails on an imperfect track, and thus is often found to diminish on curves where the difference in elevation of the rails is not exactly suited to the speed at which the train is travelling, since it is then subjected to a lateral thrust which prevents the oscillations being so great as they otherwise would be.

5. ATMOSPHERIC RESISTANCE.—This is due to two causes:

(a) The opposition offered by the particles of air in the direct path of the engine, while being thrust forwards and sideways by the advancing train, together with the "suction" caused by the rear car; and—

(b) The frictional resistance of the air against the surface of the train, corresponding to the "skin resistance" in the case of ships. The former (a) amounts to about 0.3 lb. per train running through still air at a velocity of 1 mile per hour, and increases as the square of the speed: thus, e.g., at 40 m. p. h. it amounts to about 480 lbs. Probably in ordinary trains not more than one third of this resistance causes additional strain on the draw-bar, because the greater part of it is taken and overcome by the engine itself. As regards the latter resistance, (*b*) it may be ascertained with tolerable accuracy by allowing 0.03 lb. per car at a speed of 1 mile per hour, and considering it to increase as the square of the velocity. Thus, if we have a train composed of 10 loaded box-cars (see Sec. 13) hauled by an engine which, together with its tender, weighs 60 tons, the total atmospheric resistance in lbs. at 40 m. p. h. $= 480 + 480 = 960$ lbs. (assuming that the allowance already given for the engine includes the surface resistance as well); and since the weight of the train—inclusive of engine and tender—equals about 260 tons, this is equivalent to about 3.7 lbs. per ton of entire train. Suppose, in the above example, we have a **Head-wind** blowing at the rate of 20 m. p. h., we may then consider the atmospheric resistance as being that due to a train velocity of 60 m. p. h. But if this wind were blowing in the same direction in which the train is going, then the resistance caused by it would be equal to that caused by a train velocity of 20 m. p. h. in still air.

A **Side-wind** adds very considerably to the ordinary atmospheric resistances by increasing the frictional resistance at the rails, owing to the flanges of the wheels being pressed against the inner side of the leeward rail.

The above resistances are peculiar to all trains at all times; the two following, however, are accidental, and dependent on circumstances.

6. RESISTANCE TO CURVATURE.—The many causes which combine to make up this resistance, and the share which each has in forming the result as a whole, have been but vaguely determined by experiment: it is known, however, that at speeds not exceeding about 5 miles per hour, it amounts to about 2 lbs. per ton per degree of curvature, and that it decreases as the speed increases, as shown in Diagram I, till at 70 miles per hour it does not probably amount to more than $\frac{1}{4}$ lb. per ton. Thus, e.g., on a 5° curve it amounts at a velocity of 35 m. p. h. to about 2 lbs. per ton.

The use of Transition curves (page 100) is found to decrease it materially.

7. RESISTANCE DUE TO GRAVITY.—This resistance may be termed a "mathematical" one, whereas the previous ones have been based entirely on experiment; for though the coefficient of gravity is itself a quantity derived from experiment, it is merely the ratio of the inclined component AB

(Fig. 1) to the force of gravity AC, which enters into the question; or, what is the same thing, the ratio of *ab* to *ac*.

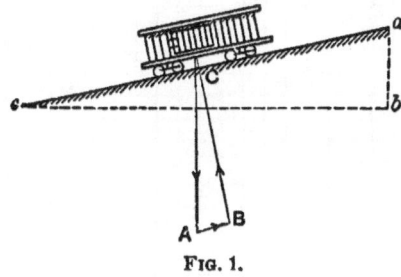

FIG. 1.

But since, in dealing with ordinary inclines, we may consider $ac = cb$, we may say that

$$\frac{AB}{AC} = \frac{ab}{cb},$$

so that the *resistance caused by gravity per ton* (2000 *lbs.*) *equals in lbs.* 20 × *rate per cent of the grade*. Thus on a 2.5 p. c. upgrade the gravity resistance equals 50 lbs. per ton.

DIAGRAM OF RESISTANCES.

8. We are now in a position to draw the **Line of Resistance** for any given train under any ordinary conditions. This line, for a train on a straight level track, is found by setting-off at the successive velocities the sum of the ordinates for the Resistances given in Sections 3, 4, and 5; and the line representing each of these component resistances can be readily plotted with the aid of the information already given. Suppose, however, that the train is running on a curve of, say, 10°, we must then measure the respective ordinates to the resistance line for the 10° curve, and add these to the ordinates already obtained. We then get the Line of Total Resistance on a 10° curve. If in addition to the 10° curve we have a $+ 0.25$ per cent grade, we have simply to add the height given on the diagram for this grade to each of the ordinates already found, in order to obtain the Line of Resistance for the train on a 10° curve and a $+0.25$ p. c. grade. If the train were *descending* the grade, it would be necessary to subtract the last ordinate instead of adding it.

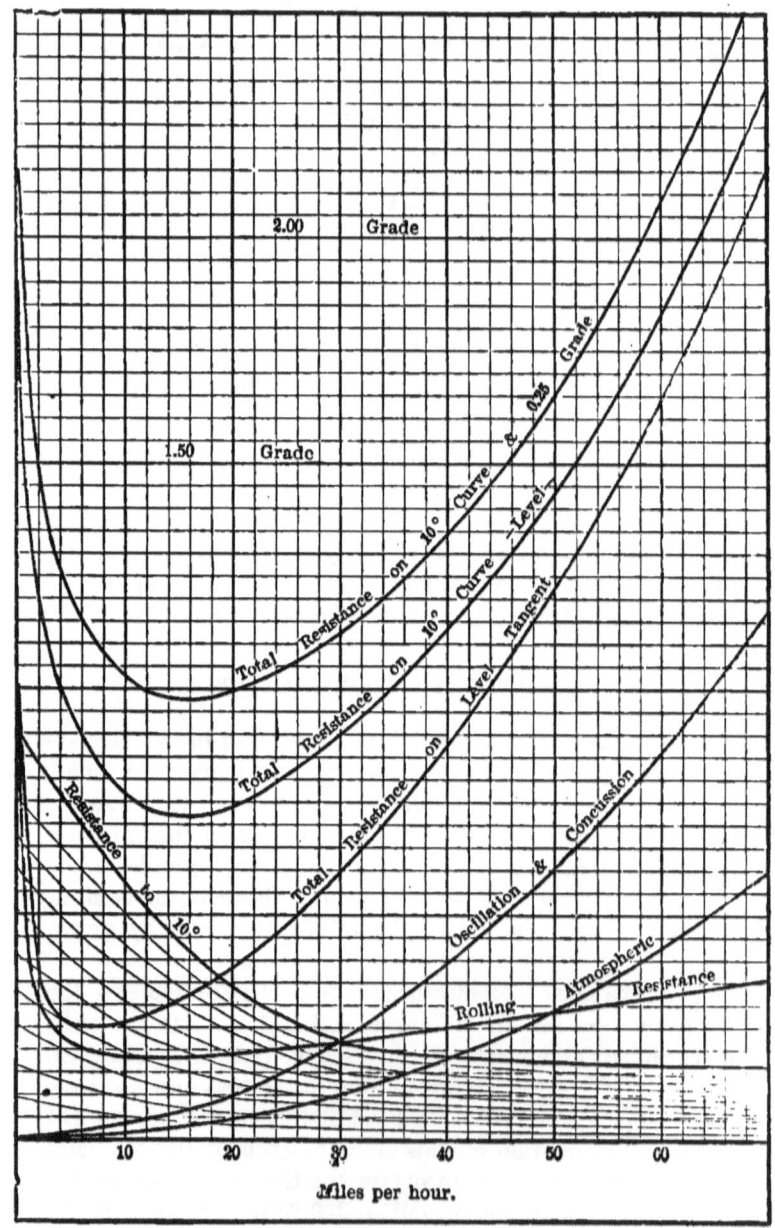

DIAGRAM I.
TRAIN RESISTANCES IN LBS. PER TON.
Engine and Tender weigh 60 tons. 10 Loaded Box-Cars, each weighing 20 tons.
SCALE, 1 inch vert. = 10 lbs.

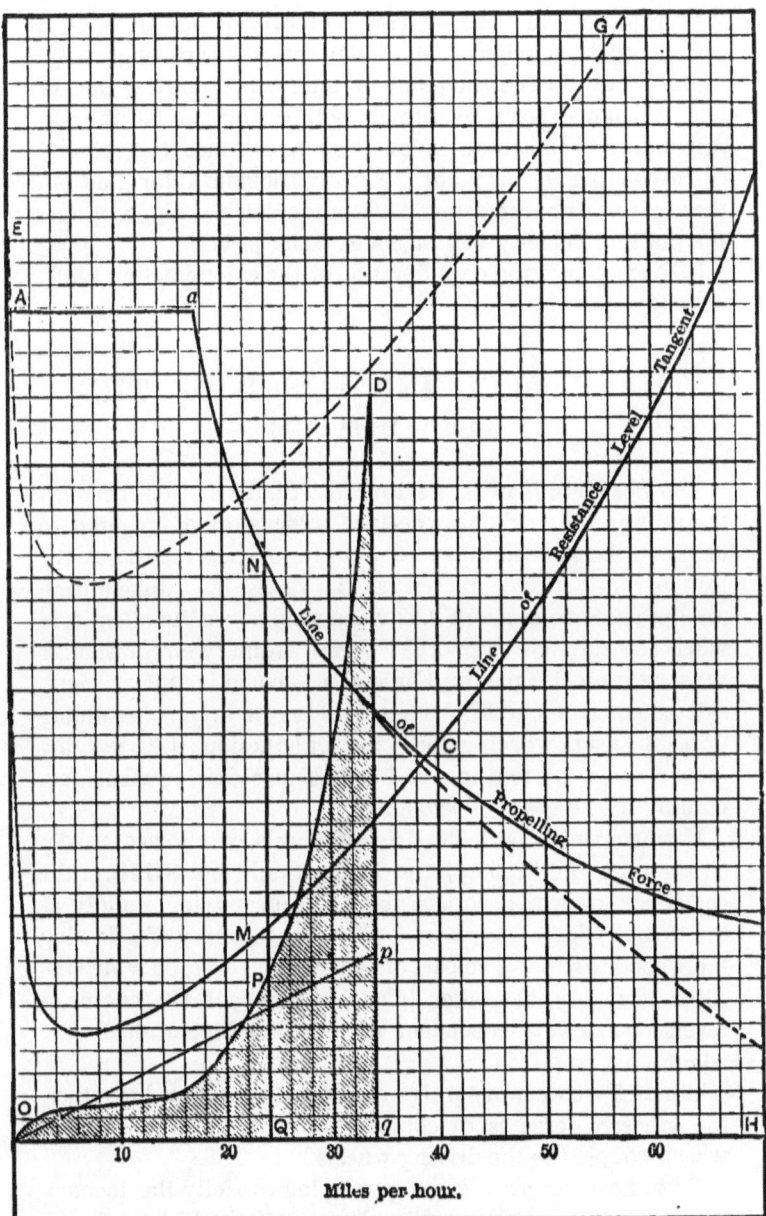

Diagram II.

PROPELLING FORCE OF LOCOMOTIVE IN LBS. PER TON.

Locomotive 500 I. H. P.   Engine and Tender = 60 tons.
$f = 0.2$ 10 Cars, 20 tons each.

SCALE, 1 inch vert. = 10 lbs.

In order to find the **Limiting Velocity** of any train on a certain grade, moving solely under the influence of gravity, we have only to find the point of intersection of the line of total resistance, for a level track, with the horizontal line corresponding to the grade in question, and notice the velocity corresponding to this point. Thus in Diagram I, for the train there given, running round a 10° curve down a 2 p. c. grade, the limiting velocity will be about 63 m. p. h.

9. Next comes the consideration of the counteracting force, namely:

THE PROPELLING FORCE OF THE LOCOMOTIVE.

The **Coefficient of Adhesion**, i.e., Static friction, between the rails and the driving wheels of a locomotive, is found to be much the same at all speeds, but to increase rapidly as the load per unit-surface increases. It varies in ordinary Railroad practice from about 0.33 when sand is used to about 0.18 when the rails are slippery. Under ordinary circumstances the maximum Propelling Force of a Locomotive may be considered equal to one fifth the weight on its drivers, assuming 0.2 as the usual working coefficient of adhesion; thus varying from about a ton to a ton and a half per driving-wheel, according to the type of locomotive.

If on starting a train the driving-wheels are allowed to slip on the rails, the friction is no longer Static but **Sliding**, the coefficient of which equals about 0.1, decreasing rapidly as the velocity increases; which shows the fallacy of allowing the wheels to slip. The part of the rail, however, on which the slipping, if any, has taken place is found, if the engine is reversed, to give a coefficient of adhesion higher than elsewhere.

Where **Two or more pairs of wheels** are coupled together, the adhesive force is, of course, due to the load on all the wheels coupled to the driving-wheels.

Now, however great steam-producing capacity the locomotive may possess, its Propelling Force is limited by the coefficient of adhesion; and though it can expend its full power in spinning the wheels around, the portion of this power which can be utilized for propelling the train is limited by the amount expressed in Indicated Horse-Power:

$$\text{I. H. P.} = 5.9\ WfV,$$

where W = total weight in tons (2000 lbs.) on the drivers,
f = coefficient of adhesion,
V = velocity in miles per hour.

This formula allows 10 p. c. for overcoming the Internal Resistances in the engine itself (see page 11). The friction at the journals of the driving-wheels, however, is not included among these, but is allowed for in the ordinary Rolling Resistance already dealt with. Thus if we take the weight on each driving-wheel as 6 tons, and $f = 0.2$, the above formula becomes

$$\text{I. H. P.} = 7NV\ \text{(nearly)},$$

where N = the number of driving-wheels.

Thus, e.g., if, in an ordinary locomotive with four driving-wheels, we have the production of steam equivalent to 400 I. H. P., we see that it is unable to utilize its full power for propelling purposes until it attains the speed of about 14 miles per hour, at which point any slight increase in pressure would cause the wheels to slip. Thus up to a certain speed the propelling power of an engine is limited by the weight on its drivers, but remains more or less constant until that speed is attained, after which, instead of being limited by the adhesion of the wheels, it is mainly a question of the steam-producing power of the boiler.

In ordinary practice, 1 square foot of **Grate-surface** is able, at ordinary speeds, to maintain the production of steam equivalent to 24 I. H. P.: so that if we know the total grate-surface of an engine and the load on its drivers,—assuming it to be tolerably well-proportioned in its various parts,—we can form a fair idea of its tractive power. The usual allowance of grate-surface varies from about 15 square feet in Passenger Engines to double this amount in some of the Heavy Freight Engines: thus the power of an ordinary Passenger Engine, when working under ordinary conditions, equals about 360 I. H. P., and in the case of a heavy Freight Engine about 720 I. H. P. Both these classes of engines can, and often do, maintain very much higher powers than these, but to work very considerably above them over a long run is a severe tax on the economy of the engine.

DIAGRAM OF PROPELLING FORCE.

10. In order to ascertain the probable effect of a given locomotive on a certain train on various grades and curves, it is best to draw the **Line of Propelling Force** of the Engine —i.e., the Line of Tractive Power exerted at the point of contact of the driving-wheels with the rails—in lbs. per ton (2000 lbs.), of the weight of the engine and train.

Suppose, as in Diagram II, we wish to find the effect of a locomotive capable of maintaining a working power of 500 I. H. P. having four drivers with 6 tons on each; and let the engine with its tender weigh 60 tons, and the train be the same as that for which the Lines of Resistance are given in Diagram I, namely, 10 loaded box cars, each weighing 20 tons—f being taken as 0.2. We then have a fair example of the working of a Light Freight Engine.

Draw the Line of Propelling Force as follows:

Make $OA = \dfrac{2000\,Wf}{\text{Tot. Weight of Train}} = 36.9$ lbs. per ton.

Then draw $Aa = \dfrac{\text{I. H. P.}}{5.9\,Wf} = 17.6$ miles per hour,

which (according to Sec. 9) gives the velocity above which slipping cannot occur. Now the theoretic curve of Propelling Force will be a hyperbola, drawn through a (AO and OH being its asymptotes). This curve may be drawn by offsets from OA thus: At a distance along OA from O equal to $\frac{1}{4}OA$, the offset equals $4Aa$; at a distance equal to $\frac{1}{2}OA$, the offset equals $2Aa$, and so on; the offset varying inversely as its perpendicular distance from O. Then C, the point of intersection of the Line of Propelling Force with the Line of Resistance, gives the **Limiting Speed** at which the engine can haul the train, under the conditions for which the line of resistance is drawn,—in this case, on a straight level track.

Then, taking any ordinate such as $NMPQ$, the part NM included between the Line of Propelling Force and the Line of Resistance gives that portion of the propelling force of the engine in lbs. per ton (2000 lbs.) which goes to overcome the Inertia of the train at the speed indicated.

But this Line of Propelling Force assumes—as we mentioned before—that 10 per cent of the I. H. P. is absorbed in overcoming the **Internal Frictional Resistances** of the engine itself—exclusive of the resistance at the journals—independent of the velocity. At low speeds this allowance is considerably too much, but at high velocities it is insufficient; for ordinary speeds, however, it will not be far from correct. The journal-friction forms probably about one third of the whole: the friction of the piston, slide-valve, valve-gear, and cross-heads also contribute considerably to the total. Very little is known as to what allowance ought to be made to cover these resistances,—in fact it is so much a matter of lubrication and mechanical detail that no general formula could be applied,—but undoubtedly they increase with the velocity, and are higher in an engine hauling a heavy train than in an engine running light.

Also we have **Back-pressure** of the steam in the cylinders, **Wire-drawing**, and various other causes entering into the question at high speeds which also tend to lessen the effective Horse-power.—**See Note A, Appendix.**

11. Now since the loss of power due to these causes depends largely on the rotary velocity of the **Driving-wheels**, in the case of two engines both developing the same I. H. P. at the same speed,—the cylinders being suitably proportioned, —the engine with the larger wheels will have a great advantage over the other at high speeds, although at low speeds the engine with the smaller wheels will have the best of it. At low speeds—since the initial pressure in the cylinders then differs but little from the boiler-pressure and the back-pressure is practically nothing—an engine with several small drivers will of course have an enormous advantage over an engine of the same I. H. P. with only a single pair of large drivers on account of its being able to utilize so much more of its power, by reason of its higher adhesive qualities. For instance, it would probably tax the engine with large drivers severely to *start* a train which the other engine could handle with ease; but when the speed reached, say, thirty miles per hour, the engine with the large drivers could work it much more easily and economically than the engine with the small ones. Thus where high velocities are required,—whether on heavy grades or not, provided the weight on the drivers is sufficient,—if the cylinders, etc., are suitably proportioned, the wheels of large diameter are decidedly the best.

Mr. Wellington states that in the case of ordinary Passenger Engines and trains of medium length, 50 per cent of the I. H. P. is consumed in the locomotive itself, overcoming its various resistances—atmospheric, rolling, internal, etc.,—so that only one half of the Horse-power produced is transmitted through the draw-bar.

From the foregoing it appears that a closer approximation to the true line of propelling force *at high velocities* may be found by drawing it as shown by the dotted line in Diagram II, somewhat below the theoretic line already drawn. The intersection of this line with OH (produced) gives the maximum speed of the engine if unopposed by any external resistances,—i.e., if running free as a stationary engine,—10 per cent only of the power developed being absorbed in overcoming internal resistances.

It must be remembered that the Line of Propelling Force shown in the Diagram is at all points the maximum which can be obtained without exceeding the I. H. P. stated; but by taking a comparatively low value of f, and a high allowance for the internal frictional resistances of the engine at low speeds, we obtain by the method given probably as correct results as can be obtained by any mathematical process.

12. If we require to know **what I. H. P. an Engine must develop** to haul a certain train at a given velocity V, we can find it at once theoretically by multiplying the total weight of the engine and train in tons (2000 lbs.) by the resistance in lbs. per ton (taken from Diagram I) and multiplying the product by $.003 V$ (V being in miles per hour). Thus with the train given in Diagram II, we should need an engine capable of developing about 950 I. H. P. in order to haul it at a speed of 50 miles per hour. The I. H. P. exerted increases nearly as V^3, and the tractive force nearly as V^2. The total amount of steam used theoretically, on a run, is nearly proportional to V^2. The **most economical speed**, as regards fuel, at which a train can be run—provided the engine is of a power suitable to the weight of the train—is found by experiment to be about 18 miles per hour, and not, as might be expected from Diagram I, at about 8 miles per hour. This is due mainly to the saving in heat owing to the engine being a shorter time on the trip, and also on account of the smaller effect produced by variations in grade at the higher velocity. To ascertain the **Limiting Grade** which it is possible to work, we find that an engine and tender weighing together 60 tons, with 24 tons on the drivers, can under ordinary conditions just make head-way up a 12-per cent grade; and that it is just all two engines of the above description can do to haul a passenger coach up a 10-per-cent grade.

13. The following may be taken as fair examples of the

WEIGHT OF AMERICAN ROLLING-STOCK:

Type.	No. of Drivers.	Weight in tons on each Driver.	Weight in tons, engine and tender, with fuel and water.
Heavy Passenger Engine...	4	5½	55
Consolidation Engine.......	8	6	75
Decapod Engine.............	10	7	95

(1 ton = 2000 lbs.)

Box car, empty, weight 10 tons. } Length 34 feet.
" " loaded, " 20 "
Flat " empty, " 8 " " 34 "
Passenger car, empty, weight 20 tons } " 50 "
" loaded, " 25 "
Drawing room car, " 35 " " 50 to 60 feet.
Sleeping-car, weight, 30 to 45 " " 50 to 70 "

RESISTANCE DUE TO INERTIA.

14. We are now able to calculate with a fair amount of precision the Propelling Force of an engine and the Total Resistance opposed to it at any given speed. The Difference between these two, such as is represented by NM, in Diagram II, gives the force in lbs. per ton which goes to overcome the inertia of the train: if the Propelling Force be the greater, increasing the velocity; but if the Resistance be the greater, decreasing it.

We will first consider the subject on the *assumption that the accelerating force remains constant at all speeds, and that there are no frictional resistances.*

It is found by experiment that a force of 1 lb. acting on a weight of 32.2 lbs. (which is perfectly free to move in the direction in which the force is acting) will, after acting on it for 1 second, give it a velocity of 1 foot per second; and that the velocity at all points increases in proportion to the interval of time during which the force acts: also, that for a given force, the velocity of a body (after it has been acted on by the force for a certain interval of time) is inversely proportional to the weight of the body. Thus the value of the Accelerating Force in lbs. per ton of train equals

$$\frac{1.518\,V}{t},$$

where $t =$ time in minutes during which force acts, and $V =$ velocity in miles per hour acquired in time t.

But this formula takes no account of the force necessary to cause the wheels to rotate; it only allows for motion in the direction in which the force acts. In order to obtain the additional force required to overcome the **Rotative Energy of the Wheels,** we may imagine the whole weight of each wheel concentrated at a point distant from its axis by an amount equal to the Radius of Gyration of the wheel. For ordinary rolling-stock we may say that this distance equals 0.75 of the radius of the wheel; and the velocity with which a point so situated rotates round the axis equals 0.75 the velocity of the train. Now the ratio of the weight of the wheels to the total weight of a train of medium length varies from about 0.1 to 0.25, according to whether the cars are loaded or empty, the proportion in the case of Passenger Cars being about the same as with Loaded Freight Cars. Therefore the Total Force necessary to overcome the entire Inertia of the train varies from about

$$F = \frac{1.6\,V}{t} \text{ to } \frac{1.7\,V}{t},$$

where $F =$ constant accelerating force in lbs. per ton (2000 lbs.) of train.

The former value is applicable to Loaded and the latter to Empty cars.

As regards the distance covered by the train from the starting-point to the point at which it attains the velocity V, it can be found by the formula

$$S = 44\,Vt,$$

where $S =$ distance in feet.

15. Now the force required to stop a train travelling with a certain velocity, in a given time, equals the force which is necessary to give it that velocity in the same time; so that the formula given above for F applies to the resistance caused by the **Application of Brakes,** as well as to the Propelling power of the engine. Now, since,—as in the case of the driving-wheels of a locomotive,—as soon as slipping begins, the adhesion at the rails decreases rapidly, therefore, in applying the brakes, the pressure should be such that the wheels will just roll on the rails; i.e., the resistance on the brakes must not be allowed to exceed the resistance at the rails, but should be as near to this limit as possible. If the pressure on the brakes could be adjusted so as to effect this in practice, we should have an efficiency for the brakes equal to the coefficient of adhesion, which we have already considered under ordinary circumstances to equal 0.2.

But it is found that with **Automatic Brakes** we cannot generally rely on a greater efficiency than 0.12, which is equal to a value of F (if the brakes are applied to the whole train) of 240 lbs. Thus the brakes may be said to offer a resistance equivalent to a 12 p. c. grade.

In the case of **Hand Brakes** it usually takes about four times as great a distance in which to stop a train when they are used, as with Automatic ones applied to the whole train.

Suppose under the above assumption we have a passenger-train running at a speed of 60 miles per hour. If steam is shut off at the same instant that the brakes are applied automatically—with an efficiency of 0.12—to three quarters of the weight of the train, the retarding value of F would equal $.75 \times 240 = 180$ lbs. per ton, and thus by our previous formula gives a value for $t = 0.53$ minutes, from which we can obtain $S = 1400$ feet. Had the train being going at only 30 m. p. h. instead of 60, it could have been pulled up in one half the time and one quarter the distance it required to stop it when running at 60 m. p. h. Thus in order to stop a train going at 60 m. p. h., we must apply four times the amount of brake-resistance which would be required to stop it if going at 30 m. p. h. in the same time.

16. So far we have dealt only with a change of velocity from Rest to V, or from V to Rest. Suppose, however, in the former case that the train, instead of being at rest, before the accelerating force F is applied, has an **Initial Velocity (v).** The formulæ given in section 14 then become changed, F in this case varying from about

$$F = \frac{1.6\,(V-v)}{t} \text{ to } \frac{1.7\,(V-v)}{t},$$

and

$$S = 44\,(V+v)t.$$

And just as the previous formulæ applied to either an accelerating or retarding force, so these apply equally well to the Propelling Force of the Locomotive or the Resistance of the Brakes.

As an **Example,** suppose we take a Passenger-train running at 50 miles per hour. The value of F necessary to reduce this speed to 30 m. p. h. in one minute $= 1.6 \times 20 = 32$ lbs. per ton, which gives a resistance equivalent to a $+ 1.6$ p. c. grade. Problems such as the above, where the value of F is assumed constant, where no account is taken of the frictional resistances, and in which the question of the time t is not directly involved, may often be solved more simply still by means of the Table of Equivalent Heights given below.

HEIGHT CORRESPONDING TO VELOCITY.

17. In the above example of the train running at 60 m. p. h. being brought to a stand-still, if the brakes had been applied to the whole train with an efficiency of 240 lbs. per ton, it would have been stopped in a distance of about 1056 ft.; or, putting it in another way, the train could have run up a 12 p. c. grade for a distance of 1056 feet before stopping, showing that it had—*stored up in it*—the Energy necessary to raise itself vertically through a height of about 127 feet. In a similar way—without going into the subjects of Kinetic and Potential Energy—every velocity may be shown to have a corresponding vertical height.

Now about 5.6 p. c. of this rise, in the case of trains, is due to the Rotative Energy of the wheels (when dealing with loaded cars) and the remainder is simply the height from which a body must fall under the influence of a force equal to its own weight,—i.e., gravity,—in order to obtain the velocity in question. But since this Rotative Energy is taken account of in the previous formulæ, we can, by finding the value of S when $F = 2000$, obtain for any given velocity the corresponding vertical height.

RAILROAD LOCATION.   17

In this way the following table has been calculated for Passenger or Loaded Freight Cars.

For a train of Empty Freight or Flat Cars, 6 p. c. should be added to the heights given.

TABLE OF HEIGHTS IN FEET CORRESPONDING TO VELOCITY IN MILES PER HOUR.

Vel.	0	1	2	3	4	5	6	7	8	9
10	3.5	4.3	5.1	5.9	6.9	7.9	9.0	10.2	11.4	12.7
20	14.1	15.5	17.0	18.6	20.2	22.0	23.8	25.7	27.6	29.6
30	31.7	33.8	36.0	38.3	40.7	43.1	45.6	48.2	50.8	53.5
40	56.3	59.2	62.1	65.1	68.2	71.3	74.5	77.8	81.2	84.6
50	88.0	91.5	95.1	98.9	102.7	106.5	110.4	114.4	118.4	122.5
60	126.7	131.0	135.3	139.7	144.2	148.7	153.3	158.0	162.8	167.6
70	172.5	177.4	182.5	187.6	192.8	198.0	203.3	208.7	214.2	219.7

Now if we have a Passenger train running at a speed of 20 m. p. h., and we wish to know what its velocity will be after descending 1000 feet of a 3 p. c. grade—ignoring as before frictional resistances—we can find it at once from the Table, thus: Its velocity at the foot of the grade will be that due to the height corresponding to a velocity of 20 m. p. h. + 30 feet = 44.1 feet, which corresponds with the velocity required, namely, 35.4 miles per hour. Or, suppose we wish to know what rate of grade would be required to decrease the speed of the above train from 40 m. p. h. to 25 m. p. h. in a distance of 1000 ft.: we have

Height corresponding to 40 m. p. h. = 56.3 feet
   "   "   " 25 " = 22.0 "

Difference = 34.3 feet.

Thus it is a 3.43 p. c. grade that would be required.

18. So far we have dealt only with the Inertia of the train on the supposition that the propelling force of the engine is constant at all speeds, and that there are no frictional resistances A method much in use in practice which partially corrects for both these fallacies is that of allowing for the *mean frictional resistance* and the *mean propelling force* of the engine, and then, by the aid of formulæ similar in effect to those given above, obtaining approximate values of S.

19. But this method of *averaging* gives very unreliable results when dealing with any but comparatively low velocities so that the following **Graphic Method,** which is extremely simple, is in most cases preferable, since the correctness of the results obtained by it depends almost solely on the care employed in working it.

Let the Lines of Resistance and Propelling Force be drawn as in Diagram II.

Take any ordinate NQ, and make $PQ = \dfrac{OQ}{NM}$.*

Similarly take other ordinates, and thus fix other positions of the point P.

Draw the curve OPD through these points. Then, if (as in Diag. II) 1 inch vertical = 10 lbs., and 1 inch horizontal = 20 miles per hour, the area (shown shaded in Diag. II) enclosed by the curve OPD, the line OH, and the ordinate corresponding to any given velocity gives the distance covered while attaining that velocity, using as a scale 1 square inch = 1 linear mile. **(See Note B, Appendix.)** And as a consequence of this, assuming, e.g., the train has an Initial velocity of 20 miles per hour, and a final velocity of 34 miles per hour, the area between the ordinates of 20 and 34 m. p. h. gives the distance traversed while the speed is being raised from the lower velocity to the higher.

By the ordinary method of averaging, at a speed of 34 m. p. h. the distance would be represented by the area Opq, instead of the shaded portion. This shows the little dependence to be placed on the averaging process, when dealing with speeds which approach the limit.

But there is a **correction** to apply to this if we wish to allow for the Rotative Energy of the wheels; and this, as we have already seen, varies from about 6 to 12 p. c. of the total energy of the train; so that in the case of Passenger or Loaded Cars 6 p. c. should be added to the distance as obtained above, and in the case of Empty Cars 12 p. c.

20. This method may be applied to a variety of problems in Railroad Dynamics: thus, **for example,** suppose we have a train travelling at 60 m. p. h., and we wish to know how far it will run if the brakes are suddenly applied, causing an additional resistance of 20 lbs. per ton—of entire train. Then the line of total resistance will be given by the dotted line EG (Diag. II), and the value of MN at any given speed will equal the entire ordinate from OH to the curve EG, for the

---

* All measured in inches on the diagram.

line of propelling force then coincides with OH—i.e., equals zero. Or, conversely, if the train be pulled up in any known distance, we can by two or three trials ascertain the efficiency of the brakes. If in dealing with such problems as these we have in the course of the distance travelled various rates of grade and curves of different "degree," we can, without serious error, draw our line of resistance for the *mean* grade and the *mean* degree of curvature.

21. We are now able to ascertain the effects of various amounts of **Rise and Fall** on the velocity of a train. In the first place, we will go back to our former assumption that the engine exerts the same tractive force at all speeds, and that there are practically no frictional resistances. Of course this is a thoroughly erroneous supposition, but by adopting it we simplify matters very considerably, and yet at the same time are able to obtain results which, for practical purposes, are sufficiently correct when we limit their application to comparatively short distances.

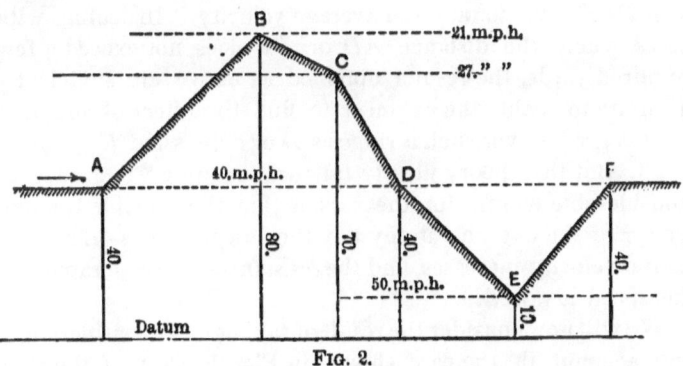

Fig. 2.

In Fig. 2 let $ABCDEF$ represent the grades on a limited portion of a certain road, then—under the assumption already made—if we have a train running along the level towards A at a uniform speed of 40 miles per hour, we obtain from the Table of Equivalent Heights in Sec. 17—

Vel. Head in ft. at $A = 56$, because $V = 40$ m. p. h.
"   " $B = 56 - 40 = 16$; ∴ V at $B = 21$ m. p. h.
"   " $C = 16 + 10 = 26$; ∴ " $C = 27$ "
"   " $D = 26 + 30 = 56$; ∴ " $D = 40$ "
"   " $E = 56 + 30 = 86$; ∴ " $E = 50$ "
"   " $F = 86 - 30 = 56$; ∴ " $F = 40$ "

By determining the speed at a few such points as these, and drawing through them the dotted lines as in Fig. 2, we have practically a **Profile of Velocities**, from which we can read approximately the speeds at different points on the grade.

22. In such a case as the above the strain on the draw-bar of the engine would at all points be constant, and the amount of work done in transporting the train from A to F would — ignoring the difference in distance, which of course in practice amounts to nothing—be the same whether the train went along the grade $ABEF$, or along a level grade ADF.

Now the effect of running over such a ridge as ABD is to lower the average speed: thus if running from A to D on the level, the train would arrive at D much sooner than by way of ABD. Again, in running over the grade DEF, its average velocity would be much higher than along the level DE. Thus the ridge ABD is detrimental to high speeds, but the depression DEF tends to raise the average velocity. In dealing with cases where the distance AD or DF does not exceed a few hundred yards, the results obtained as above are sufficiently accurate to enable the engineer to find the effect of adopting certain grades over such a ridge as D or depression E.

23. But this theory utterly fails when applied to grades of considerable length, for the reason that the possible tractive power of the engine—at any but the lower speeds—decreases as the velocity increases, and the resistances increase rapidly as the speed is raised.

We will now consider the result of taking these considerations into account in the case shown in Fig. 2. Now if the train comes on to the grade AB at a certain speed—assuming that the Effective Horse-power remains constant—it will have a velocity at B appreciably greater than that which we should obtain for it at that point by means of the Table of Equivalent Heights. So also at D it will have a velocity greater than it had at A, although by the Table the velocity at A and D should be the same. The reason of this is, that the increase in the accelerating force is more than in proportion to the increase in the total propelling force, being due to a decrease in the resistances as well as to the reduction in speed. Similar reasoning applies to the down-grades BC and CD, so that by the time the train has got to D the total amount of work done on the higher grade is relatively less than what it would have been along the level *AD*, owing to the reduced frictional resistances. Thus the train is travelling faster at *D* than it was at *A*, although it has lost time on the way. Similarly, in the case of crossing a depression such as *E*, the amount of work done will be greater by the lower route than along the level, and the train will thus have at *F* a velocity less than it had at *D*, although it will have made better time between *D* and *F* by way of *E*, than along the level *DF*.

But although the train arrives at *D* with a higher velocity than if it had proceeded along the level, yet this increase in velocity only partially makes up for the time lost between *A* and *D*. So also the decrease in speed at *F* does not entirely counteract the gain in time made along *DEF*.

The amounts by which the velocities at *D* and *F* actually differ from those obtained by the Table, depends mainly in practice on the distance between *A* and *D*, or *D* and *F*. The greater these distances are, the less reliance is to be placed on the Table; so much so in fact in dealing with long grades, as to render the energy of the train itself—considered as a store of available tractive power—practically worthless.

24. It is usual for Railroad Companies to adopt a certain rate of grade which is not—except where Pusher-grades are used—to be exceeded. This is usually termed the **Maximum or Ruling Grade**, and is selected with due consideration to the tractive power of the locomotives to be employed, the probable amount of traffic, the weight of trains to be hauled, and the speed required to be maintained. It is also selected in most cases so as to admit of a train *starting* on the grade, if by any chance it should have had to pull up. Also, it should be such that the locomotive employed can haul the train over it, altogether independent of the Momentum—or more correctly *Energy*—of the train. By means of Diagram II we can readily select the most suitable Maximum Grade by drawing the line of resistance—for a level track—and the line of propelling force suitable for the locomotives to be employed; the length of the ordinate *NM*. when scaled off, gives the equivalent resistance in lbs. per ton of the maximum grade. Thus, in the case of the **example** given in Diagram II, if the speed required to be maintained on the grade equals 24 miles per hour,—since *NM* represents to scale about 17 lbs. per ton,—the maximum grade will equal

0.85 p. c. Had the required speed been only 10 miles per hour, we might then have used a 1.6 p. c. grade. But probably in neither of these cases could the train *start* on the grade, and in order to allow for this, we must assume that the line of resistance at no point dips below 15 lbs. per ton,—i.e., 12 lbs., in accordance with Sec. 3, and a small margin of 3 lbs. to overcome the Inertia of the train.—Thus, *allowing for stoppages*, if a speed of 24 m. p. h. is to be maintained in the case shown in Diagram II, the maximum grade must not exceed 0.55 p. c.; but if 10 m. p. h. only is required, then—including allowance for stoppage—the maximum grade may be 1.1 p. c. But we must remember that where the velocity required to be maintained on the maximum grade exceeds that given by Aa, in Sec. 10, some allowance should be made for the probable increase in boiler-pressure after the train has come to a stand-still; which means that on starting, the I. H. P. of the engine may be placed considerably above its normal working power. (See Note II, Appendix.)

25. Without going into the question of the Economy of the Steam engine, we may say that a Locomotive works with its greatest efficiency when the boiler pressure remains constant and the engine is running at a uniform velocity. Thus fluctuations in speed or variations in the opposing resistances are more or less detrimental to the working of the locomotive.

As a consequence of this, if a certain elevation has to be attained, in order to make the work as easy on the engine as possible, the grade should be such as to render the sum of the resistances opposed at all points as nearly constant as possible. Thus, if the alignment be straight, the rate of grade should be uniform; but if curves or other irregularities occur, they should be compensated for, so that a constant resistance may be maintained.

26. Compensation for Curvature.—From Diagram I we see that at 10 miles per hour the resistance for each degree of curvature is about 1 lb. per ton, i.e., equivalent to a $+$ 0.5 p. c. grade, and that at about 30 m. p. h. it is about half this. The rate, however, usually adopted is .03 p. c., which is suitable to a speed of about 25 m. p. h. Thus, if the equivalent grade on a tangent is 1.5 p. c., we must reduce it on a 3° curve to 1.41 p. c. in order that the resistance may remain constant.

27. Compensation for Brakes, etc.—A point to be remembered in running a long uniform grade which does not approach the maximum is to consider at what points the train will be required to slacken or increase its speed. For example, suppose on such a grade we have a sharp curve around which the speed is not to exceed 20 miles per hour, but that on the tangent at either end of it a speed of 40 m. p. h. can be maintained. By means of the Table of Equivalent Heights we can adapt the Energy of the train so that the velocity will be reduced without the application of the brakes, and that when the curve is passed the speed of the train can be more readily increased from 20 to 40 m. p. h. But in doing this we have to be careful that at the lower end of the curve we do not increase the grade so as to tax the engine too severely. At all such points as crossings, where short stoppages are required, attention should be paid to this, for by so doing we can at times save something even in the cost of construction, besides saving considerably in fuel and in wear and tear to the Rolling stock.

28. But though the operating-expenses may be reduced to a minimum by the use of Long uniform (equivalent) grades, the amount necessarily expended on their construction may be too great to warrant adopting them. In such cases **Broken Grades** have then to be used.

Now we have already seen how to obtain the effect of undulations on the velocity and the work done, so that we can in any particular case determine for ourselves what will be the result of selecting a certain arrangement of grades. The following "pointers," however, deduced from what has already been said, may come in handy.

1. A Rise from the uniform grade is detrimental to fast traffic, and though there is a saving in actual work done on it, there is probably no saving in the consumption of fuel.

2. A Depression from the uniform grade tends to increase the mean velocity, but at the cost of a considerable amount of extra fuel.

3. Breaks in the grade which—from the point where the broken grade leaves the uniform one to the point where they next intersect—do not exceed, say, 1000 to 2000 feet, may be regarded as "*Momentum Grades,*" and accordingly are not so injurious as longer breaks where the Initial Energy of the

train is small compared with the Total Energy to be expended on them.

4. The nearer the uniform grade approaches the "Maximum grade," the more injurious do any breaks become; and the only point in connection with the "Maximum grade," where an increase in the rate is allowable, is the insertion of a "Momentum grade" at its lower end.

5. Breaks in a grade are more injurious to slow than to fast traffic—as may be seen from the Table of Equivalent Heights —e.g., an increase in elevation of 20 feet reduces the velocity from 30 to 18 miles per hour, while a velocity of 60 m. p. h. is only reduced to about 55 miles per hour.

6. Be careful in inserting Momentum grades that they will not be such as to cause the velocity at any point to exceed the safe limit. A difference in elevation of about 30 feet between the Broken and the Uniform grade should generally be taken as a limit.

29. Another point to be considered, which we have not yet referred to, is the increase in **Liability to Danger of Breaking-train and Derailment** to which an undulating grade gives rise. For, suppose in Fig. 2 we have a train running up the grade from A to B: as soon as the engine is over the summit the pull on the draw-bar becomes enormously increased, and similarly with the car-couplings throughout the entire train; so that, unless the greatest care is taken in applying the brakes, the train runs a very great risk of being broken in two. Similarly, in such a hollow as E, the cars near the centre of the train are liable to get terribly jammed together, thereby greatly increasing the chances of Derailment.

Vertical curves reduce these dangers considerably, but not entirely.

It must be remembered that it is not in the least necessary that one of the grades should be an up-grade and the other a down-grade: it is the *difference in the rate of grade* that has to be looked out for. (See Sec. 100.)

30. In Fig. 3, let ACB and ADB represent two different routes between A and B, the total Rise and Fall between the two points in each case being the same. The amount of work done in hauling the train from A to B by way of C will, supposing we are dealing with grades so long that the ques-

tion of "Momentum Grades" may be ignored, be then practically the same as by way of *D*. Similarly, if such a point as *H* in Fig. 4 has to be reached, the work done in hauling the train along the uniform grade *EH* will be practically the same as by way of *FG*. It is not the amount of work done on the *grades themselves* that has to be considered, but the amount of *extra* work which is uselessly done by a heavy engine hauling a large surplus of dead-weight (due to its own size) over

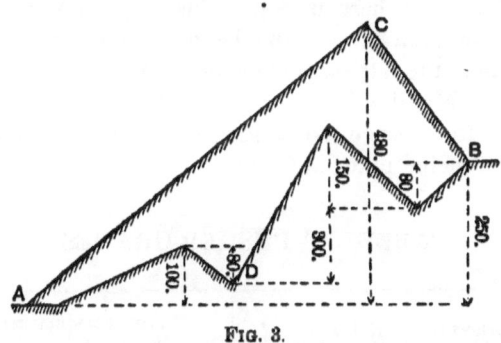

FIG. 3.

grades where a lighter engine could have hauled the train equally well. If each of the divisions *EF*, *FG*, and *GH* were a suitable length for one engine to work, the lower route would then be as economical probably as regards Operating Expenses as the higher. Besides this, we have the increased

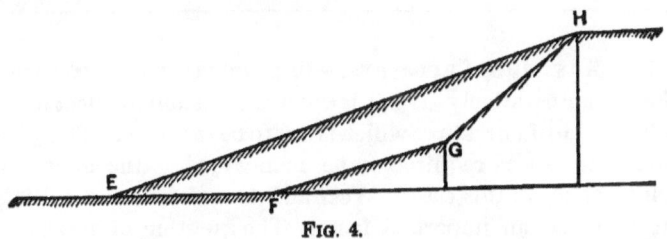

FIG. 4.

consumption of fuel, before referred to, which always accompanies variations in grade.

If we make each of the divisions along the lower route from *E* to *H* of such a length as to keep the engine employed on each fairly busy,—using a different engine on each division,—the lower route is then as economical as can be wished for, but otherwise the upper route has the advantage.

31. Now the average length of an **Engine-stage** may be considered to be about 100 miles, which is of course too long to enable us to work the lower route in the manner described above. We may often, however, by adopting a **Pusher-grade**, even at a point where at first it appears unnecessary, make a decided improvement in the economy of our grades. The length of this grade, if the Pusher is to be kept steadily employed, depends of course on the number of trains to be taken up it each day: if there are four trains a day the engine will be kept sufficiently at work if the length of the grade is only 12 miles. As to the rate of grade which may be adopted in such cases as this, Mr. Wellington gives the following Table, which is suitable for average Consolidation Engines, the coefficient of adhesion being taken at 0.25 :

TABLE OF PUSHER-GRADES.

Grade worked by one Engine.	Net Load of Train in tons.	GRADE POSSIBLE WITH—	
		1 Pusher.	2 Pushers.
Level.	2675	0.38	0.74
0.2	1758	0.75	1.26
0.5	1147	1.30	2.01
1.0	711	2.16	3.13
1.5	504	2.96	4.13
2.0	383	3.72	5.03

32. Maximum Curvature.—In countries where construction is comparatively easy, it is often the custom to select a certain degree of curvature which is not to be exceeded. The question of the speed required to be maintained is the main one which arises in this case. Wear and tear of rails and rolling-stock is also an important factor. The question of resistance —at ordinary speeds—is comparatively unimportant, since at a speed of 25 miles per hour a 10° curve only offers the resistance of about a 0.3 p. c. grade. In rough country it is impossible to fix a "maximum," for the additional cost of construction which the adoption of a limiting-grade might involve would perhaps be an inconceivably greater consideration than the loss of a few seconds—or possibly minutes—in time. As regards the question of the **Safe Speed on Curves**, it is difficult to lay down any law, but it is supposed to vary inversely as the square root of the radius. Thus if we assume that 40 miles per hour is a safe speed on a 2° curve, the speed should be limited to 20 m. p. h. on an 8° curve and to 14 m. p. h. on a 16° curve. The chances of derailment and the wear and tear of rolling-stock and rails are decreased materially by the use of Transition curves. (See Sec. 96.)

33. It is almost unnecessary to refer to the subject of Reverse Curves. In Station-yards, where the speeds are insignificant, their use is sometimes advisable; but on the Main Track an intervening tangent of at least 200 feet in length should be regarded as an absolute necessity. A fault much more frequently found is the insertion of a short tangent between two curves of the same direction. Getting on to a tangent from a curve is as hard work as getting on to a curve from a tangent; and since it is at the P. C. and P. T. that the curve gives its maximum resistance, the curves should at least be compounded so as to make the radius of curvature at all points as uniform as possible, for in each case the *total amount of curvature* will be the same. Another point to be remembered—though it is not often that it can be applied—is, that a road which has its curves at points where the speed is comparatively low has a decided advantage over one in which the curves are located at places where a high speed is required to be maintained. Thus, if a certain amount of curvature has to be got in, in such a place as DEF in Fig. 2, it should be arranged if possible so that the curvature at D and F will be sharper than at E. Curvature should also be avoided as much as possible at all points where a stoppage is required, for on starting, the resistance due to the curvature is a great consideration, and, as we saw in Sec. 6 and Diagram I, will probably make it as difficult for the train to start as a decided up-grade.

34. We have now dealt in a more or less superficial way with most of the mechanical problems which arise in connection with railroad trains; but it is convenient, for the sake of more readily comparing the value of the various resistances to passenger and freight trains at average speeds, to tabulate their mean values (as given by Prof. Jameson) as follows:

TABLE SHOWING COMPARATIVE VALUES OF RESISTANCES
AS REGARDS WORK DONE.

Items.	Distance.	Curvature.	Rise and Fall.
1 mile	5280 feet.	600°	25.0 feet.
1° Curvature	8.8 "	1°	0.041 "
1 foot Rise and Fall.	211.2 "	24°	1.0 "

"Rise and Fall" of course means *in one direction only*, and is so stated in order to take account of the Rise when running in the opposite direction. Thus in Fig. 3 the total Rise and Fall between A and B by either route equals 710 feet.

COST OF OPERATING.

35. The expense involved in overcoming the resistances referred to in Sec. 34 is not proportional to the amount of work which is performed on account of them. For instance, it is found by experience that hauling a train over one mile of level track costs on an average about the same as 150 feet of rise and fall,—not of 25 feet, as given in the last table. Similarly, with curvature, the operating of one mile of level track is found to cost the same as about 900° of curvature (not 600°); so that *as regards operating-expenses* the table given in Sec. 34 becomes—

Items.	Distance.	Curvature.	Rise and Fall.
1 mile	5280 feet.	900°	150 feet.
1° Curvature.	5.86 "	1°	0.166 "
1 foot Rise and Fall.	35.2 "	6°	1.0 "

As soon, then, as we know the expense of operating one mile of level track, we can by means of this table find the probable cost of working any certain grade or any given amount of curvature.

36. Taking $1.00—it is probably nearer 90 cts.—as the average cost of operating one mile of level track on American Railroads for each train that runs over it (and returns) each day, we can make this our unit of operating-expenses and term it the cost of one **Train-mile**. The items which go to make up the expense of the train-mile are as follow:

Motive Power..... { Oil, Fuel, Waste, Water. Driver, Fireman. Repairs.

Train Expenses.... { Train Hands. Repairs and Renewals to Cars.

Road Repairs ..... Track, Road-bed, Structures.

General........... { Stations, Terminal, Taxes. Repairs and Renewals.

Taking, then, $1.00 as the cost per train-mile, and assuming the interest on the amount capitalized at 6 p. c., we obtain the following table:

Unit.	Value per annum per daily train.	Amount Capitalized.
1 mile..................	$350	$5,833.33
1 foot..................	0.066	1.10
1° Curvature......... ...	0.39	6.50
1 foot Rise and Fall....	2.33	38.88

This assumes that each "daily" train only runs 350 days in the year, which makes a sort of allowance for Sundays, "specials," etc.

37. From the above we see that if we have ten trains making the round-trip every day, we are entitled to spend $58,333 extra on the construction of a certain route, if by so doing we can save a mile of level track; so also we should be entitled to spend $388 in the reduction of a foot of rise and fall. Thus with 10 daily trains we might safely expend $2 \times \$388 = \$776$ in lowering (only one foot) such a summit as C in Fig. 3; but if C had been the terminus of the line AC we ought only to spend $388 in lowering it one foot.

Suppose again we have two routes to select from, one of which would probably cost $40,000 more than the other, but would shorten the distance by one mile and would save a rise and fall of 100 feet. Then if there are only likely to be three trains running—including returning—each day, we are not entitled to spend more than $(\$5833 + \$3888) \times 3 = \$29,163$ to save the above distance and rise and fall; therefore it would probably be injudicious to adopt the more expensive route.

38. As regards the **cost of operating Pusher grades.** we find that a Pusher kept pretty busy costs on an average about $280 per mile of incline per annum—i.e., $140 per mile run—"all that the engine fails to do below 100 miles per day may be assumed to cost from ¼ to ⅓ as much as if it had been run, and is so much added to the cost of what is run." Thus on a 5-mile incline, with only 4 trains to be taken up it each day, the probable annual expense of the Pusher will be found thus:

$$\text{Work done,} \quad 4 \times 5 \times \$280 = \$5,600$$
$$\text{Work not done,} \quad 30 \times \frac{\$280}{4} = 2,100$$
$$\text{Total} \dots\dots\dots\dots\dots \$7,700$$

Had we been able to reach the summit without adopting a Pusher-grade—supposing the total rise and fall to be 1000 feet—the cost of "Rise and Fall" would have been for the 4 daily trains $4 \times 1000 \times \$2.33 = \$9320$, representing a difference in the operating-expenses of $1620 per annum, which at 6 p.c. would have warranted our expending $27,000 more on the route which involved the Pusher-grade, assuming curvature and distance to be the same in both cases.

39. To test the merits of different routes as regards operating-expenses, we may express them in terms of their **Equivalent Lengths** (L) in miles thus:

$$L = l + \frac{H}{150} + \frac{C}{900},$$

where
$l = $ actual length in miles,
$H = $ total rise and fall in feet,
$C = $ total curvature in degrees.

40. As regards the increase in operating-expenses caused by any slight increase in distance, such as is the result of changes in the alignment, it is not usually the case that the cost per train-mile for any small additional distance is as high as the rate already given; for many of the items, such as station and terminal expenses, which go to make up the average cost per train-mile, are not affected by an addition in distance which does not exceed 2 or 3 p.c. of the total length of the road. Thus, in selecting the choice of two routes, the engineer should not necessarily take the average cost per train mile as his standard by which to find the probable difference in the operating-expenses, but in most cases may consider about 50 cents per train-mile an amply sufficient allowance for that portion of the longer route which is in excess of the other, when that excess does not exceed the above amount.

41. In order to approximate as closely as possible to the probable cost per train-mile on any projected road, the engineer must judge by the results on other roads where the conditions are more or less similar. Where changes are to be made in the alignment of a road already in operation, the value of the proposed improvements can then be found with considerable accuracy, since the cost per train-mile is then known.

RECEIPTS.

42. The Receipts usually vary from about 1.5 to 2.0 the cost of operating; and it is not often that the locating-engineer has it in his power to affect them in any way. He may, however, by carrying the location by a slightly more circuitous route than he would otherwise have adopted, catch the traffic of some outlying village. Mr. Wellington on this subject says: "When the question comes up of lengthening the line to secure way-business, we may almost say that where there seems any room for doubt, it will almost always be policy to do so. Extra business to a railroad—the engineer will rarely err in thinking—is almost always clear profit. Of Passenger business this is literally true until the increase becomes considerable; of Freight business it is so nearly true that 80 or 90 per cent at least of the way-rate is clear profit over the usual cost of any particular shipment."

Thus, suppose we are projecting a line between two points 100 miles apart, and that half-way between them lies a small town 10 miles off the direct route. The additional distance involved in running through it is about 2 miles. Suppose, as is a reasonable estimate, the average payment per head of population is $13 per annum. Then, if there are likely to be 5 daily trains, we may put the extra cost of the two miles, including the interest on the capital spent on their construction, at about $2000 per annum. Therefore, looking at the matter only from this point of view, if the place contains, or is likely to contain before long, only about 150 people, it would probably be wise to locate the road through it.

COST OF CONSTRUCTION.

43. This is a subject which had almost better be omitted, for the range of prices is so great in different parts of the country, that values given to suit one place may be entirely misleading when applied to another place a few hundred miles off. I have, however, endeavored to strike the average prices as nearly as possible, and with these remarks they must be taken for what they are worth. They show more or less the relative cost of various works, and in this way may sometimes be of service.

First we have the following lot common to all track:

Steel rails per ton (2000 lbs.)	$25 00 to	$45 00
Angle-bars, per lb	02 "	03
Bolts, "	03 "	05
Spikes, "	02 "	04
Ties (in place), each	20 "	50
Ballast—Gravel, p. cu. yd	25 "	75
" Broken Stone, p. cu. yd	75 "	1 50
Track-laying per mile	250 00 "	500 00

Then we have the following, according to circumstances:

Solid Rock, per cu. yd	$0 75 to	$ 2 00
Loose Rock or Hard Pan, per cu. yd	35 "	75
Earth, per cu. yd	10 "	50
1st Class Masonry, per cu. yd	10 00 "	30 00
2d " " "	7 00 "	10 00
3d " " "	5 00 "	7 00
Dry rubble " "	2 00 "	5 00
Riprap, per cu. yd	1 00 "	2 00
Iron erected in bridge-work, per lb	04 "	08
Timber in Trestles, per M	25 00 "	45 00
" " Culverts, "	15 00 "	25 00
" " Log Culverts, per M	10 00 "	20 00
Piling driven, per lin. ft	25 "	75
Grubbing, per Station	12 00 "	20 00
Clearing, per acre	20 00 "	30 00
Overhaul, p. cu. yd. per Sta	01 "	02
Fencing per mile of track	300 00 "	800 00
Telegraph line—Single wire	175 00 "	250 00

By taking the mean prices of the first set, we obtain for an *average mile of standard-gauge track* (10 p. c. short rails) the following cost:

103 tons Steel rails (65 lbs. p yd.)	$3,862 00
710 Angle-bars, 20 lbs. each	355 00
1420 Bolts, 7 kegs, 200 lbs. each	56 00
5670 lbs. Spikes, 38 kegs, 150 lbs. each	171 00
2640 Ties	924 00
Ballast, 3667 cu. yds. Gravel	1,834 00
Track-laying	375 00
Total	$7,577 00

Besides these we have, of course, Right of Way, Engineering, Law, and a variety of Incidental expenses.

As regards the COST OF TRESTLEWORK, we find that for Low Pile Trestles—say 20 ft. high—assuming piling to cost 50 cents per lin. ft. driven, and the superstructure $20 per M., the cost will usually be about $6 per foot run.

For a Wooden Trestle 50 feet high at $25 per M., the cost, if resting on piles or sills, will usually be about $10 per foot run; but if 100 feet high, $20 to $25 per foot run.

The cost of Iron Trestlework varies so enormously according to the design, that it is impossible to lay down any figures which might be generally applicable. Assuming, however, that the total weight of iron in the trestle equals the total weight of wood in an equally strong wooden trestle, the cost, at 5 cents per lb., would be about double that of a wooden one. These figures are of course exclusive of Masonry foundations, and are for single-track.

As regards the COST OF TRUSSES, a Wooden Howe Truss—single-track, of 100 ft. span, Lumber at $15 per M.—costs, framed, somewhere about $2000; and an Iron Truss of the same span, at 5 cents per lb., costs about $5000. The cost in both cases varies pretty much as the square of the span. Erecting usually costs from $5 to $10 per lin. foot.

As regards the COST OF TUNNELLING, we may say it varies from $2.50 to $7.50 per cu. yd.; so that for a single-track tunnel we may consider the price per foot run to vary from about $30 to $80, including masonry. The cost of sinking a shaft or driving a heading is considerably higher in proportion than this.

For more on the subject of the **Cost of Grading**, see Sec. 124, Part II,

INSTRUMENTS.

44. The principal Instruments ordinarily used on Railroad Location are: The Transit, Compass, Level, and Hand Level; and we will consider them in the order here given. (For Instruments used on exploratory-work, see Secs. 141 to 158.)

THE TRANSIT.

Before proceeding with the adjustments of the Transit, it should be seen that the object-glass is screwed firmly home, and a short scratch made on the ring of the glass and continued on to the slide, so that, should the glass be taken out or work loose, it may be screwed up to exactly the same position it was in before. If this is not done, and the glass happens to be badly centred,—i.e., its optical axis does not lie in the centre of the telescope-tube,—if by any chance the glass is moved, the Line of Collimation will also be thrown out of adjustment.

The following are the usual adjustments for a Transit:

A. To make the vertical axis truly vertical by means of the small bubble-tubes. Turn the vernier-plate until each of the tubes is parallel to a pair of opposite plate-screws. Bring both bubbles to the centres of the tubes. Then turn the instrument through about 180°. If the bubbles are still in the centre, the adjustment of the small tubes is correct; but if not, correct for half the error in each case by means of the adjusting screws at the ends of the tubes. This adjustment should then be correct; if not, repeat the process until it is.

B. To set the cross-hairs truly vertical and horizontal.—After levelling up, test the vertical hair along its whole length on some fixed point, and if not correct, loosen the capstan-headed screws and move the diaphragm around. The horizontal hair may be tested in a similar way.

C. To make the horizontal axis of the telescope truly horizontal.—Level up the instrument and point the telescope to some object C, as in Fig. 5, at an altitude, if possible, of not less than 45°. Mark the point A where this vertical plane strikes the ground. "Reverse" the instrument, and if on pointing to C and then reducing to the ground we again strike A, this adjustment is correct. But suppose the first time the "vertical" plane had struck the ground at B, and then on reversing, instead of striking B again, it cuts through some point D. Mark a point E between D and B, distant from D by one quarter of DB. Then by means of the screws under one of the pivots of the horizontal axis bring the intersection of the cross-hairs to strike the point E. This adjustment should then be correct.

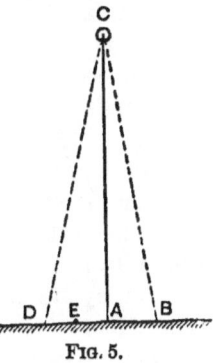

FIG. 5.

D. **To make the line of collimation perpendicular to the horizontal axis.**—Having levelled up the instrument at O, in Fig. 6, point the telescope to some object C. Turn the telescope over and mark the point A, at a distance AO

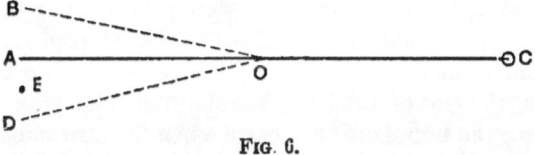

FIG. 6.

equal to about OC, where it strikes the ground in the opposite direction. By making $AO = OC$ we then obtain a correct adjustment for the line of collimation, even though the object-slide is defective; that is the only reason for making AO and OC about the same length. Reverse, and again point to C; if on turning the telescope over once more it again strikes A, this adjustment is correct. But if instead of intersecting A it cuts through some other point D, then mark a point E between D and B, distant from D by one quarter of DB, and by means of the capstan-headed screws move the diaphragm so as to bring the intersection of the cross-hairs to coincide with E. This adjustment should then be correct. This is liable to throw out adjustment B slightly, so watch that at the same time.

E. **To make the long bubble-tube parallel to the line of collimation.**—Level up the instrument and clamp the vertical arc. By means of the tangent-screw of the vertical arc bring the bubble to the centre of the tube. Then if the small bubble-tubes were sufficiently sensitive to render the vertical axis, when the instrument is levelled up, truly vertical, all points cut by the line of collimation equally distant from the instrument would have the same elevation. But it is more satisfactory to obtain a truly vertical axis by means of the long bubble-tube itself, on account of its greater sensitiveness; thus: Level up as accurately as possible by the small tubes, and then treat the long bubble-tube as if it were one of the smaller tubes, putting it into a *temporary* state of adjustment A, by means, not of the screws at the ends of the bubble-tube, but by aid of the tangent-screw of the vertical arc, and then by its means obtain a truly vertical axis. Then take the readings on two points A and B equally distant from the instrument and in opposite directions; next move the transit to a point about in the same straight line as A and B, but at as short a distance beyond either of them, say A, as the instrument can be focussed to read and level up by the small tubes. Take the reading at A, say 3.43; then if B were previously found to be 1.84 feet higher than A, the telescope should read 1.59 on B if this adjustment were correct. If we do not read this, the screws at the end of the long bubble-tube must be so altered as to bring the bubble to the centre when the instrument reads 1.59. On again pointing to A, the difference between A and B should then be *almost* 1.84. If it is not quite 1.84, proceed as before until the adjustment is correct.

By moving the instrument into the same line as A and B, as above, we avoid the necessity of levelling up this vertical axis again by means of the long bubble-tube.

Besides the above adjustments, some instruments have a means of *Centring the Eye-piece* and also of *Adjusting the Object-Slide*. (See Note C, Appendix.)

45. Remarks.—Another way of performing adjustment C is by means of an object and its reflection in still water, or even in a plate of syrup. A star at night does well for this, but it is advisable to select one as nearly east or west as possible, as its motion in azimuth is then a minimum.

If at any time adjustment C is not correct, we can obtain true results by "reversing," as in Fig. 5, and remembering that half-way between the two points so found is the correct point.

This latter remark applies also to adjustment D. It is a good plan to reverse on a back-sight every few sights, as it takes practically no extra time and at once detects if anything is wrong. By taking a point half-way between two points, as D and B in Fig. 6, we can do good work with an instrument in which this adjustment is very far from correct.

As regards adjustment E:—If we had a level handy, it is much more convenient to level two points with it; or if there is a sheet of still water at hand, two pegs driven down to its surface do equally well. To ascertain the *Index-error* of the vertical circle in instruments where it cannot be corrected for instrumentally, set the vertical axis truly vertical, as explained under adjustment E, then level up the telescope and observe the readings on the vertical arc. If they are at zero, there is no in lex-error; but if not, the difference between the readings and zero is the index-error.

If the transit has a *Striding-level* attached, adjustment C may then be more accurately performed by means of it—whether the striding-level is in adjustment itself or not, for it is only the *difference* of the readings that is required. To make adjustment C then proceed thus: Level up by the small bubble-tubes and point the telescope towards the north; take the readings of the bubble on the glass, both at its east and west end; then reverse the striding-level, end for end, and take the readings a second time: one quarter of the difference between the sum of the two east readings and the sum of the two west readings equals the number of divisions on the tube that the bubble must be moved by means of the pivot-screws in order to make the "horizontal axis" level, that end being too high the sum of whose readings is the greater. If the striding-level is in adjustment, we have only to screw up the "horizontal axis" so as to agree with it. We can, of course, *adjust the striding-level* by placing it on the pivots already levelled, and bringing the bubble to the centre of the tube.

Lighting the cross-hairs, when the instrument has no lantern attached, can be effected by fastening a piece of bright tin—or even white paper—over and partly in front of the object-glass, so as to cast the reflection of a light on the ground into the tube of the telescope; but the reflector must not obstruct more than half of the field of the object-glass. A piece of tin or paper with a $\frac{1}{2}$-inch hole in the centre of it, fastened at a suitable angle over the object-glass, answers very well.

In moving the diaphragm when the telescope has an inverting eye-piece, it has to go in the opposite direction to what appears to be the right one.

If working with an instrument the *graduation of which is faulty*, read each angle in different parts of the circle. The graduations can always be tested by reading with both verniers on various parts of the circle. In observing an angle, if we take the mean result obtained by *both verniers*, we eliminate errors due to eccentricity of the vertical axis and the graduated circle, as well as reduce the errors of graduation.

When *great accuracy is required* in reading an angle the best method to use is BORDA'S REPETITION, which slightly reduces the errors of observation, while it diminishes those of graduation in inverse order to the number of times the angle is repeated. The process is thus : Clamp the vernier-plate to zero, and read the angle by both verniers according to the usual method. Then, keeping the vernier-plate clamped, point the telescope again to the first object, and proceed as before through any number of repetitions. At the end of the final angle read the verniers, adding 360° for each complete revolution which has been made, and divide the total angular measurement by the number of times the angle was repeated. The quotient is the required angle. In this way, provided there is no play about the tangent-screws, an angle can be read with confidence to a few seconds by a very inferior instrument. **(See Note I.)**

In ordinary work, if sure of the correct centring of the vertical axis and also of the graduation itself, there is no need to read by both verniers; but it is advisable to read always by the *same* vernier if only one is used.

An instrument correct according to the adjustments given above gives correct results when dealing with objects distant from it by the amount OC in Fig. 6, but if there is *defective centring of the object-slide*—not to be confounded with eccentricity of the optical axis of the object-glass—it will not give correct results in dealing with objects at distances from it greater or less than OC. This can always be tested by ranging points in a "straight" line for a thousand feet or so, beginning as near to the instrument as the focus will permit. Then, if, on ranging the same points in again from the *other* end, they do not coincide, one half the difference between the points is the error in alignment. In this way, even with a bad instrument, a straight line can be run. We can of course also run a straight line with an instrument which has a defective object-slide, if in proper adjustment, by taking back-sights and fore-sights equal in length to OC, Fig. 6, so that then the object-slide will occupy the same position as it did when the line of collimation was adjusted.

If the object-slide works correctly, then, although the object-glass may be badly centred,—i.e., its optical axis will not coincide with the centre of the telescope tube,—if the line of collimation is in correct adjustment for *one* distance it will give correct results at *all* distances.

Parallax is caused by the focus of the object-glass and that of the eye-piece not coinciding at the cross-hairs. To correct for it, shift the eye-piece in and out until the cross-hairs are seen distinctly. Then point the telescope to some distant object and move the object-glass in and out until the image of the object is seen sharp and clear, coinciding apparently with the cross-hairs.

STADIA.

46. Transits used on location should be fitted with adjustable stadia-hairs. These are usually adjusted to read 1 foot on a rod at a distance from the centre of the instrument equal to (100 feet + distance from object-glass in its mean position to the centre of the instrument + focal length of the object-glass), usually making a distance in all of about 101.25 feet. And since the stadia-hairs should be placed so as to be equidistant from the ordinary horizontal hair, at a distance of 101.25 feet the distance read between each pair of adjacent hairs should be 0.50 feet.

If the hairs are not adjustable, but are fastened to the ordinary diaphragm, then the measurements on the rod must be regulated to suit the hairs, remembering that the apex of the angle subtended by the distance read on the rod is not at the centre of the instrument, but at a point in front of the object-glass by a distance equal to the focal distance of the object-glass, which is usually 1.25 feet in front of the centre of the instrument.

If the hairs are unadjustable, and we wish to use an ordinary levelling-rod to read on, then suppose at 101.25 feet we read 0.88 feet between the stadia-hairs, we must divide every reading in feet by 0.88 in order to obtain the distance in terms of

100 feet. Thus if at a certain point we read 4.40 on the rod, the distance will be 500 feet, or 501.25 feet from the centre of the instrument. *To find the focal-length* of the object-glass, focus it for a distant object; the distance from the cross-hairs to the object-glass then equals the focal-length.

On sloping ground, if the rodman is careful about holding the rod perpendicular to the line of sight, swaying it slowly to and fro so as to permit of the minimum reading being taken, then if the centre hair reads somewhere about 5 feet on the rod (i.e., the height of the instrument above the ground) we have only to multiply the distance as read on the incline by the cosine of the inclination, in order to obtain the true horizontal distance.

But if really correct work is wanted, it is best to have a bubble-tube attached to the rod so that it can be held vertically, and then correct for the inclination as follows:

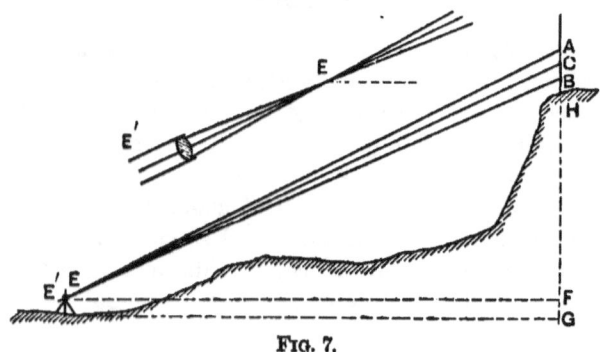

Fig. 7.

In Fig. 7 the distance

$$EF = AB \cos^2 FEC,$$

EF being in terms of 100 feet, and *FEC* being the angle of inclination as measured to *C*, the ordinary horizontal hair of the instrument, assuming that the stadia-hairs are equidistant from *C*, and that 1 foot on the rod corresponds with 100 feet in distance.

In order to reduce this to the centre of the instrument, we should of course add to *EF* the amount $1.25 \times \cos FEC$, but for ordinary inclinations we may assume this correction to equal 1 foot. Thus, if $FEC = 30°$, and $AB = 6.00$, then

RAILROAD LOCATION. 41

$E'F = 1$ ft. $+ (6 \times .75) = 451$ feet. To obtain the height HG in Fig. 7, the best way is to make CH on the rod equal to the height of the point E above the ground, say 5 feet. Then

$$HG = EF \tan FEC.$$

Thus in the above example $HG = 260$ feet. The following table gives the VALUES OF $\cos^2 FEC$, where FEC is the inclination angle:

INCLINATION.	0′	10′	20′	30′	40′	50′
0°	1.0000	1.0000	1.0000	.9999	.9999	.9998
1	.9997	.9996	.9995	.9993	.9992	.9990
2	.9988	.9986	.9983	.9981	.9978	.9976
3	.9973	.9969	.9966	.9963	.9959	.9955
4	.9951	.9947	.9943	.9938	.9934	.9929
5	.9924	.9919	.9914	.9908	.9902	.9897
6	.9891	.9885	.9878	.9872	.9865	.9858
7	.9851	.9844	.9837	.9830	.9822	.9814
8	.9806	.9798	.9790	.9782	.9773	.9764
9	.9755	.9746	.9737	.9728	.9718	.9708
10	.9698	.9688	.9678	.9668	.9657	.9647
11°	.9636	.9625	.9614	.9603	.9591	.9580
12	.9568	.9556	.9544	.9532	.9519	.9507
13	.9494	.9481	.9468	.9455	.9442	.9428
14	.9415	.9401	.9387	.9373	.9359	.9345
15	.9330	.9315	.9301	.9286	.9271	.9256
16	.9240	.9225	.9209	.9193	.9177	.9161
17	.9145	.9129	.9112	.9096	.9079	.9062
18	.9045	.9028	.9011	.8993	.8976	.8958
19	.8940	.8922	.8904	.8886	.8867	.8849
20	.8830	.8811	.8793	.8774	.8754	.8735
21°	.8716	.8696	.8677	.8657	.8637	.8617
22	.8597	.8576	.8556	.8536	.8515	.8494
23	.8473	.8452	.8431	.8410	.8389	.8367
24	.8346	.8324	.8302	.8280	.8258	.8236
25	.8214	.8192	.8169	.8147	.8124	.8101
26	.8078	.8055	.8032	.8009	.7986	.7962
27	.7939	.7915	.7892	.7868	.7844	.7820
28	.7796	.7772	.7747	.7723	.7699	.7674
29	.7650	.7625	.7600	.7575	.7550	.7525
30	.7500	.7475	.7449	.7424	.7398	.7373
31°	.7347	.7322	.7296	.7270	.7244	.7218
32	.7192	.7166	.7139	.7113	.7087	.7060
33	.7034	.7007	.6980	.6954	.6927	.6900
34	.6873	.6846	.6819	.6792	.6765	.6737
35	.6710	.6683	.6655	.6628	.6600	.6573
36	.6545	.6517	.6490	.6462	.6434	.6406
37	.6378	.6350	.6322	.6294	.6266	.6238
38	.6210	.6181	.6153	.6125	.6096	.6068
39	.6039	.6011	.5982	.5954	.5925	.5897
40	.5868	.5839	.5811	.5782	.5753	.5725

THE COMPASS.

47. The adjustments of the Compass are as follows:

A. To make the needle swing horizontally.—Level the compass, then by means of the slide-piece on the needle regulate its centre of gravity so that it will swing horizontally.

B. To straighten the needle.—See if both ends of the needle point to exactly opposite graduations while the compass is being turned completely around. If so, the needle is straight *and* the pivot is properly centred. But if not, the error will arise from either one or both of these not being correct. Turn the compass until some graduation, say 90°, comes precisely to the northern end of the needle, and bend the pivot until they do. Then turn the compass until the opposite 90° is at the north end of the needle. Mark the place where the southern end of the needle then points. Take off the needle and bend it until its southern end points half-way between 90° and the point already marked, while its northern end is kept at the opposite 90° by slightly moving the compass around. The needle will then be straight, although it will not intersect opposite degrees on account of the eccentricity of the pivot.

C. To centre the pivot.—Turn the compass around until a place is found where the opposite ends of the needle cut opposite degrees. Then turn the compass quarter-way around, or through 90°. If the needle then cuts opposite degrees, the pivot is in adjustment; but if not, bend the pivot until it does. The needle should then cut opposite degrees while being turned completely around.

Remarks.—If the magnetism of the needle gets weak, it may be renewed as follows: Cover the needle with a thin film of oil, and then with the north pole—the end marked with a line across it—of an ordinary magnet rub the south end of the needle, beginning at the centre and working outwards towards the end; similarly rub the north end of the needle with the south pole of the magnet. After doing this a few times the magnetism should be sufficiently restored.

Reading both ends of the needle corrects for eccentricity of the pivot if the needle is straight; it also of course reduces the errors of graduation.

Should the glass cover become electrified, as it will if but slightly rubbed, so that the needle sticks to the under side of it and will not "traverse" properly, touching the glass in several places with the moistened finger, or breathing on it, will remove the electricity.

A compass when left standing for any considerable time should always have its needle free, in order to prevent loss of magnetic power. Of course when carried it should always be clamped.

In taking a compass-reading, not only must all iron and steel substances be kept well away, but metal magnifiers of all sorts are liable to cause a slight deflection, owing to the possibility of impurity in the material of which they are composed. Magnifiers coated with nickel are especially bad, since nickel itself is a decidedly magnetic metal.

Since the magnetic attraction varies in different places, adjustment A, if correct in one place, will probably want looking to if the instrument is taken anywhere else.

MAGNETIC VARIATION.

48. By referring to the Chart of Magnetic Variation, we see that in North America the variation is both towards the east and the west. The "line of no variation" which separates these two divisions is found to be constantly shifting westwards at an average rate of about 4' per annum. This causes a gradual increase in all variations to the west, and a corresponding decrease in all variations to the east; and changes similar to these are going on all over the globe. Besides this *secular variation* we have *diurnal* and *annual variations*, but for practical field purposes these latter may be ignored. The former of them is such that the needle attains its extreme westerly position at about 2 P.M. each day, and its extreme easterly position at about 8 A.M.; while the latter shows itself generally by a slight increase in variations west, and decrease in variations east, during the summer.

The chart here given is more as a matter of interest than for any real use in the field. If the variation at any place is wanted accurately, usually the only satisfactory way is to take it directly by observation, as shown in Sec. 57. For very rough work, however, an *idea* of the amount of variation can be obtained from the chart by interpolating by eye.

The "lines of no variation" are shown *thicker* than the others.

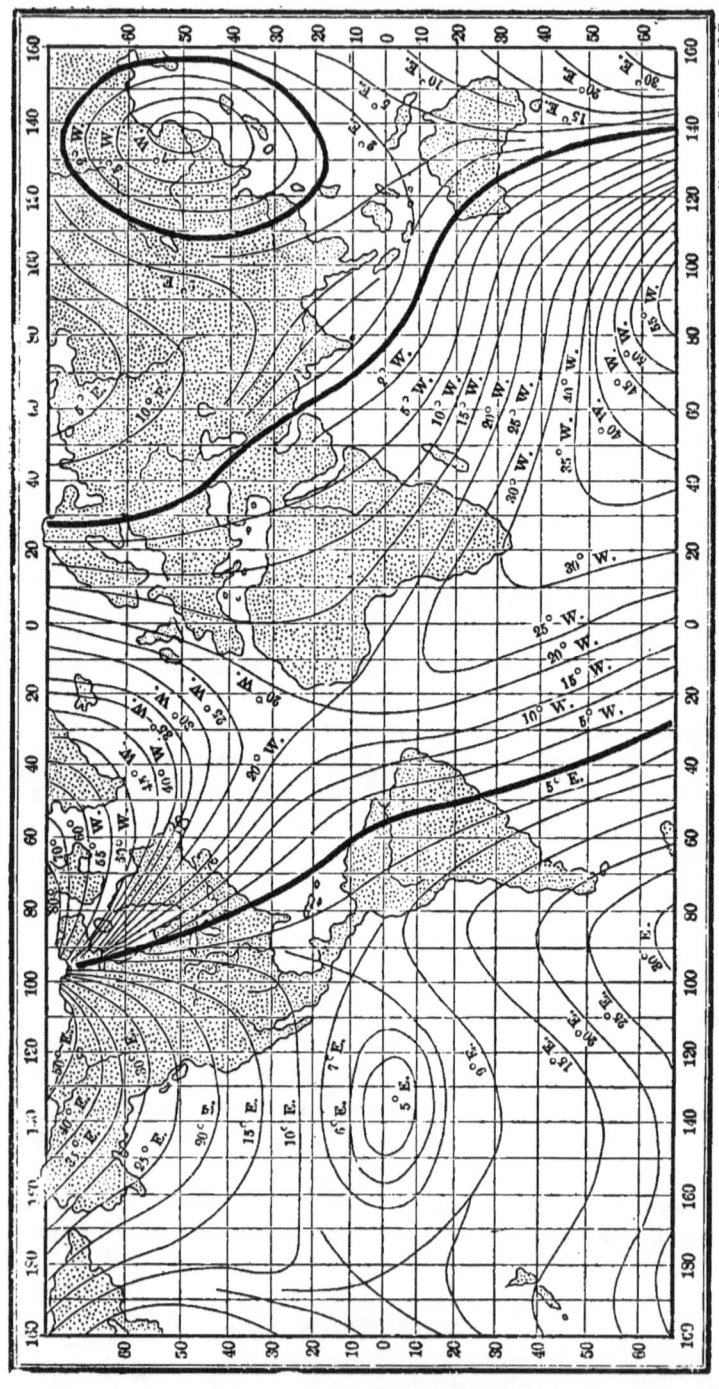

THE LEVEL.

49. We will first take the DUMPY LEVEL, which usually needs only two adjustments.

A. To make the bubble-tube perpendicular to the vertical axis.—This is done in just the same way as with one of the small bubble-tubes in adjustment A of the transit.

B. To make the line of collimation parallel to the bubble-tube.—This is done in a similar way to the adjustment of the long bubble-tube of a transit, already described, except that in this case it is the line of collimation that has to be made parallel to the bubble-tube, so that now it is the *cross-hairs* that have to be moved. In this case of course there is no necessity to set up the instrument "in about the same line as A and B" as there was in the case of the transit.

Another way of performing this adjustment is by the method of "reciprocal observations," as given for the Handlevel in Sec. 52.

The remarks which applied to the telescope of a transit apply with equal force to the telescope of a level; more especially the remark on the running of a straight line if the object-slide is badly centred. If the level has a means of **adjusting the eye-piece** and **object-slide**, see Note C, Appendix.

50. The Y Level has three adjustments as follows:

A. To make the line of collimation coincide with the axis of the telescope.—Open the clips of the Y's. To adjust the vertical hair, mark the intersection of the cross-hairs on some fixed object, and revolve the telescope in its Y's so that the level will be upside down; and then if the intersection falls to one side of the object, one half the error must be corrected for by the capstan-headed screws. To adjust the horizontal hair, turn the telescope over as before, and if the intersection of the hairs strikes above or below the object, correct as before for one half the error.

B. To make the bubble-tube parallel to the line of collimation.—This adjustment consists of two parts. First, bring the bubble to the centre and then revolve the telescope in its Y's through about 20°; if the bubble then runs to one end, half the error must be corrected for by the horizontal screws at the end of the tube, raising or lowering as may be required.

For the second part of this adjustment, place the telescope over a pair of opposite levelling-screws, open the clips, and bring the bubble to the centre of the tube. Reverse the telescope end for end in its Y's; if the bubble is not then in the centre, one half the error must be corrected for by the vertical screws at the end of the bubble-tube. On levelling-up and again reversing, this adjustment should be found to be correct.

C. To make the axis of the telescope perpendicular to the vertical axis.—Level up. Place the telescope over a pair of opposite levelling-screws. Swing the telescope half-way round on its vertical axis. If then the bubble has left the centre, bring it half-way back by means of the large capstan-headed nuts of the Y's. Then place the telescope over the other pair of levelling-screws, and if necessary proceed as before. This adjustment should then be correct.

Remarks.—As with the transit, if the object-slide of a level is defective the line of collimation when adjusted is only correct for back-sights and fore-sights of equal length with the distance of the object on which the line of collimation was adjusted.

In levelling, whenever possible, keep the fore-sights and back-sights of equal length: if so, accurate work can be done with an instrument thoroughly out of adjustment, for then the actual height of the instrument itself is of no importance. If, as in levelling uphill, it is necessary to take extremely short fore-sights, they should be counteracted by short back-sights —if not at the time, as soon afterwards as possible.

51. There is no need to allow for CURVATURE OF THE EARTH OR REFRACTION in sights under 700 feet, and then, if taking fore-sights and back-sights of about the same length, the corrections would counteract each other; so that it is only in taking an extremely long fore-sight or back-sight, which is not counteracted by a more or less equal sight in the opposite direction, that we need apply corrections for curvature or refraction.

For CURVATURE the correction in feet amounts to

$$0.67L^2,$$

where $L =$ length of sight in miles, and is to be subtracted from the reading on the rod: this being simply the tangential offset for a curve—see Sec. 78—with radius equal that of the earth.

For REFRACTION, on an average, it amounts in feet to

$$0.1L^2,$$

which is an experimental quantity, and is to be added to the reading on the rod. So that, taking the two together, we may say that the correction in feet amounts to

$$0.57L^2,$$

and is to be subtracted from the reading on the rod, the elevation as taken or given by the level being always too low. This is equivalent to about .0002 ft. at 100 feet; so that, since it increases as the square of the distance at say, 1200 feet, it will equal $.0002 \times 12^2 = .03$ foot. The following table gives the JOINT CORRECTIONS FOR CURVATURE AND REFRACTION, worked out by the above formula, and is useful in ascertaining the elevation of the surrounding country:

Distance in Miles.	Correction in Ft.	Distance in Miles.	Correction in Ft.
1	0.57	30	513
5	14.25	40	912
10	57.0	50	1,425
15	128	60	2,052
20	228	80	3,648
25	356	100	5,700

Thus from the table, if the level gives a point on a distant mountain, say 30 miles off, the elevation of that point will be equal to the elevation of the instrument + 513 feet.

52. The Hand-Level.—The only adjustment necessary as a rule with this instrument is to make the line of collimation parallel to the bubble-tube. To do this, sight from a point A to a point B, as in Fig. 8, and then back again from B to A.

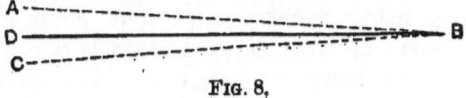

FIG. 8.

If the level is in adjustment the two sights should coincide at A. But suppose C is the point struck instead of A; then D, a point half-way between *A* and *C*, will be on a level with *B*: therefore the hair must be adjusted on the line *BD*. A handier way is of course to adjust it by means of another level, or a sheet of water.

THE SURVEY.

53. The object of the following notes is not to show the mode of conducting location,—which of course can only be picked up by actual experience in the field,—but merely to give solutions of the various mathematical and instrumental problems which arise in the course of the work.

In the case of Exploratory Surveys, the instruments used and the problems which arise being usually entirely different from those which come into question in ordinary location, they will be considered separately in Part III.

A Reconnoissance survey, as generally understood, may also be classed with the above, or it may take the form of a rough preliminary survey, a compass perhaps being substituted for the ordinary transit. As regards *Compass-surveys*, there is among engineers a strong prejudice against them, but in a country tolerably free from local attraction a compass-line is surely correct enough for preliminary work; for though by it accuracy cannot be obtained at any one point, its errors are not accumulative, but in a great measure counteract each other, so that the line as a whole should give very fair results. Another method of performing rough work is the *Stadia process*, by means of which very good results have often been obtained, the engineering staff consisting merely of an engineer and a rodman, the only instruments used being a rod and a small transit with bubble-tube and stadia attachment. Comparing compass and stadia work, the former is usually more suitable in timber and the latter in open country.

The term *Preliminary survey* is variously used, sometimes indicating a mere reconnoissance, but more generally a survey the object of which is to obtain accurate topography, in order by its means to select the final location. As regards the degree of accuracy to be employed in preliminary work, it of course depends in what way the results are to be used; but it is generally best to run the transit-line of the "preliminary" with as much accuracy as is attainable under the circumstances. If this is done we then have a line on which we can at all points depend, and which we can use as a base for other lines, knowing, if we branch off from it at one point, the exact course we must make to strike it again at any given station. We will therefore suppose in the following notes that the final location is to be selected by the aid of an accurately run preliminary line, topography having been taken on either side of the transit line to a distance of from, say, 100 to 600 feet, according to the nature of the ground.

On Preliminary Surveys, by means of a hand-level and prismatic-compass, the engineer-in-charge, keeping ahead of the party, is generally able to ascertain approximately where the line will go, and then the transit-man has merely to follow more or less the route indicated, being guided by the consideration of running the line as much as possible to a constant rate of grade. If the line, however, is being run to the maximum grade—or any other rate of grade which it is the wish of the engineer to maintain—along a continuous transverse slope, such as a mountain-side, the transit-man can choose the line tolerably well for himself, since he only has to select his stations so as to maintain the required rate, which he can do by means of the vertical arc. But in selecting these points he has to bear in mind the probable amount of curvature which there will be between the station where the instrument is standing and the place at which the front picket is to be set, and allow for it in setting the picket. (See Sec. 26.) Thus, suppose he is running the line to a 1.5 p. c. grade, and that he estimates the distance to the picket to be about 500 feet, and the probable total curvature in that distance to be 15°, then the grade-angle, instead of being 51¼′, as in the following table, will be 48½′. If he has stadia-hairs in his instrument,—as he ought to have,—he can read off the distance with sufficient accuracy on the picket itself, and in this way form his estimates more closely. The difference in distance along the straight course and along the probable location must also be allowed for where the deviation is great. The following is a

TABLE OF GRADES AND GRADE-ANGLES.

Feet per Station.	Feet per Mile.	Inclination.	Feet per Station.	Feet per Mile.	Inclination.	Feet per Station.	Feet per Mile.	Inclination.
		° ′ ″			° ′ ″			° ′ ″
.01	.528	21	.51	26.928	17 32	1.01	53.328	34 43
.02	1.056	41	.52	27.456	17 53	1.02	53.856	35 04
.03	1.584	1 02	.53	27.984	18 13	1.03	54.384	35 24
.04	2.112	1 23	.54	28.512	18 34	1.04	54.912	35 45
.05	2.640	1 43	.55	29.040	18 54	1.05	55.440	36 05
.06	3.168	2 04	.56	29.568	19 15	1.06	55.968	36 26
.07	3.696	2 24	.57	30.096	19 36	1.07	56.496	36 47
.08	4.224	2 45	.58	30.624	19 56	1.08	57.024	37 08
.09	4.752	3 06	.59	31.152	20 17	1.09	57.552	37 28
.10	5.280	3 26	.60	31.680	20 38	1.10	58.080	37 49
.11	5.808	3 47	.61	32.208	20 58	1.11	58.608	38 09
.12	6.336	4 08	.62	32.736	21 19	1.12	59.136	38 30
.13	6.864	4 28	.63	33.264	21 39	1.13	59.664	38 51
.14	7.392	4 49	.64	33.792	22 00	1.14	60.192	39 11
.15	7.920	5 09	.65	34.320	22 21	1.15	60.720	39 32
.16	8.448	5 30	.66	34.848	22 41	1.16	61.248	39 53
.17	8.976	5 51	.67	35.376	23 02	1.17	61.776	40 13
.18	9.504	6 11	.68	35.904	23 23	1.18	62.304	40 34
.19	10.032	6 32	.69	36.432	23 43	1.19	62.832	40 54
.20	10.560	6 53	.70	36.960	24 04	1.20	63.360	41 15
.21	11.088	7 13	.71	37.488	24 24	1.21	63.888	41 35
.22	11.616	7 34	.72	38.016	24 45	1.22	64.416	41 56
.23	12.144	7 54	.73	38.544	25 06	1.23	64.944	42 17
.24	12.672	8 15	.74	39.072	25 26	1.24	65.472	42 38
.25	13.200	8 36	.75	39.600	25 47	1.25	66.000	42 58
.26	13.728	8 56	.76	40.128	26 08	1.26	66.528	43 19
.27	14.256	9 17	.77	40.656	26 28	1.27	67.056	43 39
.28	14.784	9 38	.78	41.184	26 49	1.28	67.584	44 00
.29	15.312	9 58	.79	41.712	27 09	1.29	68.112	44 21
.30	15.840	10 19	.80	42.240	27 30	1.30	68.640	44 41
.31	16.368	10 39	.81	42.768	27 51	1.31	69.168	45 02
.32	16.896	11 00	.82	43.296	28 11	1.32	69.696	45 23
.33	17.424	11 21	.83	43.824	28 32	1.33	70.224	45 43
.34	17.952	11 41	.84	44.352	28 53	1.34	70.752	46 04
.35	18.480	12 02	.85	44.880	29 13	1.35	71.280	46 24
.36	19.008	12 23	.86	45.408	29 34	1.36	71.808	46 45
.37	19.536	12 43	.87	45.936	29 54	1.37	72.336	47 06
.38	20.064	13 04	.88	46.464	30 15	1.38	72.864	47 26
.39	20.592	13 24	.89	46.992	30 36	1.39	73.392	47 47
.40	21.120	13 45	.90	47.520	30 57	1.40	73.920	48 08
.41	21.648	14 06	.91	48.048	31 17	1.41	74.448	48 28
.42	22.176	14 26	.92	48.576	31 38	1.42	74.976	48 49
.43	22.704	14 47	.93	49.104	31 58	1.43	75.504	49 09
.44	23.232	15 08	.94	49.632	32 19	1.44	76.032	49 30
.45	23.760	15 28	.95	50.160	32 39	1.45	76.560	49 51
.46	24.288	15 49	.96	50.688	33 00	1.46	77.088	50 11
.47	24.816	16 09	.97	51.216	33 21	1.47	77.616	50 32
.48	25.344	16 30	.98	51.744	33 41	1.48	78.144	50 52
.49	25.872	16 51	.99	52.272	34 02	1.49	78.672	51 13
.50	26.400	17 11	1.00	52.800	34 23	1.50	79.200	51 34

TABLE OF GRADES AND GRADE-ANGLES.—Continued.

Feet per Station.	Feet per Mile.	Inclination.	Feet per Station.	Feet per Mile.	Inclination.	Feet per Station.	Feet per Mile.	Inclination.
		° ′ ″			° ′ ″			° ′ ″
1.51	79.728	51 54	1.91	100.848	1 05 39	3.55	187.440	2 01 50
1.52	80.256	52 15	1.92	101.376	1 06 00	3.60	190.080	2 03 42
1.53	80.784	52 36	1.93	101.904	1 06 20	3.65	192.720	2 05 25
1.54	81.312	52 56	1.94	102.432	1 06 41	3.70	195.360	2 07 08
1.55	81.840	53 17	1.95	102.960	1 07 02	3.75	198.000	2 08 51
1.56	82.368	53 37	1.96	103.488	1 07 22	3.80	200.640	2 10 34
1.57	82.896	53 58	1.97	104.016	1 07 43	3.85	203.280	2 12 17
1.58	83.424	54 19	1.98	104.544	1 08 04	3.90	205.920	2 14 00
1.59	83.952	54 39	1.99	105.072	1 08 24	3.95	208.560	2 15 43
1.60	84.480	55 00	2.00	105.600	1 08 45	4.00	211.200	2 17 26
1.61	85.008	55 21	2.05	108.240	1 10 28	4.10	216.480	2 20 52
1.62	85.536	55 41	2.10	110.880	1 12 11	4.20	221.760	2 24 18
1.63	86.064	56 02	2.15	113.520	1 13 54	4.30	227.040	2 27 44
1.64	86.592	56 22	2.20	116.160	1 15 37	4.40	232.320	2 31 10
1.65	87.120	56 43	2.25	118.800	1 17 20	4.50	237.600	2 34 36
1.66	87.648	57 04	2.30	121.440	1 19 03	4.60	242.880	2 38 01
1.67	88.176	57 24	2.35	124.080	1 20 46	4.70	248.160	2 41 27
1.68	88.704	57 45	2.40	126.720	1 22 29	4.80	253.440	2 44 53
1.69	89.232	58 06	2.45	129.360	1 24 12	4.90	258.720	2 48 19
1.70	89.760	58 26	2.50	132.000	1 25 56	5.00	264.000	2 51 45
1.71	90.288	58 47	2.55	134.640	1 27 39	5.10	269.280	2 55 10
1.72	90.816	59 07	2.60	137.280	1 29 22	5.20	274.560	2 58 36
1.73	91.344	59 28	2.65	139.920	1 31 05	5.30	279.840	3 02 09
1.74	91.872	59 49	2.70	142.560	1 32 48	5.40	285.120	3 05 27
1.75	92.400	1 00 09	2.75	145.200	1 34 31	5.50	290.400	3 08 53
1.76	92.928	1 00 30	2.80	147.840	1 36 14	5.60	295.680	3 12 19
1.77	93.456	1 00 51	2.85	150.480	1 37 57	5.70	300.960	3 15 44
1.78	93.984	1 01 11	2.90	153.120	1 39 40	5.80	306.240	3 19 10
1.79	94.512	1 01 32	2.95	155.760	1 41 23	5.90	311.520	3 22 36
1.80	95.040	1 01 52	3.00	158.400	1 43 06	6.00	316.800	3 26 01
1.81	95.568	1 02 13	3.05	161.040	1 44 49	6.10	322.080	3 29 27
1.82	96.096	1 02 34	3.10	163.680	1 46 32	6.20	327.360	3 32 52
1.83	96.624	1 02 54	3.15	166.320	1 48 15	6.30	332.640	3 36 18
1.84	97.152	1 03 15	3.20	168.960	1 49 58	6.40	337.920	3 39 43
1.85	97.680	1 03 35	3.25	171.600	1 51 41	6.50	343.200	3 43 08
1.86	98.208	1 03 56	3.30	174.240	1 53 24	6.60	348.480	3 46 34
1.87	98.736	1 04 17	3.35	176.880	1 55 07	6.70	353.760	3 49 59
1.88	99.264	1 04 37	3.40	179.520	1 56 50	6.80	359.040	3 53 24
1.89	99.792	1 04 58	3.45	182.160	1 58 33	6.90	364.320	3 56 50
1.90	100.320	1 05 19	3.50	184.800	2 00 16	7.00	369.600	4 00 15

When the running is tolerably easy, instead of taking a series of short courses, it is often better to insert a curve at once, selecting one which is likely—as near as can be guessed—to coincide with the probable final location; for in this way truer results can be arrived at than by a series of independent courses.

54. As regards the Instrument-work itself, the method of reading angles as so much "to the right" or "to the left" is decidedly feeble. The best way is to start with the verniers reading zero when the telescope is pointing towards the magnetic, or still better, the true north; then the first angle read is the magnetic (or true) bearing of the first course. On moving the instrument up to the front picket, the horizontal circle should be kept clamped, and the reading of the vernier again, when the instrument is next set up, constitutes a check on the former reading; for though there will probably have been some slipping of the plates, owing to the shaking while being carried from one station to the other, an error of a degree or so is easily detected. When the telescope is pointed to the back-sight the verniers should then read the same as they did at the other end of the line, and thus for the next course, on turning through the required angle, it will be its bearing—magnetic or true as the case may be—that is read. The compass-reading should also be taken for each course, at each end of the course, which thus forms an additional check on the work, and also *detects local attraction*. For if, when the instrument is set up, the needle does not on any course read the bearing corresponding with the vernier reading (if the zero corresponds with the magnetic north) or does not give the difference in the readings equal to the "variation," if the zero corresponds with the true north, if the work is correct, the cause is either the change in variation, or local attraction, or both these causes combined. If the instrument is a good one there is no need to read by more than one vernier. (See Sec. 45.) But it should usually be the same vernier that is read, and that vernier will then always be on the same side of the transit-line. If, however, the line of collimation, from some cause or other, such as a defective object-slide which cannot be remedied in the field, is unreliable, the error can be counteracted to a large extent by taking the bearings with the same vernier on opposite sides of the line at alternate stations.

55. With the bearings taken as above, or in fact taken in any way, the most satisfactory method of plotting the work is by means of LATITUDES AND DEPARTURES. This method involves a little extra work, but its advantages over the ordinary protractor method—or even the method of "chords" or "natural tangents"—are so great as to make the few minutes extra time taken in preparing the notes time well spent. The main advantage of this method is that an error

made in plotting one station is not transmitted to the next, as in the ordinary methods, for each station is plotted entirely independent of the previous one; and thus of course we can plot any one part of the location on the plan in its right position, without having to work through from the beginning. Again, if we know the position of the point we are making for, we can, without keeping a continuous plot of the work, tell at any station how much we are off our direct route, and what course we ought to steer to strike the point we are making for. The method of keeping and plotting the notes is best shown as follows:

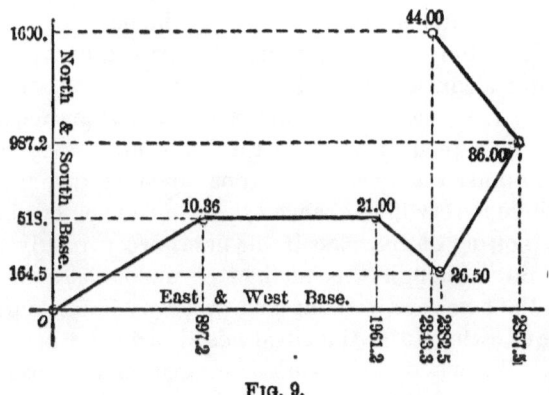

FIG. 9.

Suppose Fig. 9 represents the first five courses of a preliminary line, the notes for these courses will then be kept thus:

Sta.	Dist.	Read.	Bearing.	Lat.	Dep.	Total Lat.	Total Dep.
0	1036	60°	N. 60° E.	518	897.2		
10.36	1064	90°	E.	0	1064.	518.	897.2
21.00	550	130°	S. 50° E.	−353.5	421.3	518.	1961.2
26.50	950	30°	N. 30° E.	822.7	475.	164.5	2382.5
36.00	800	−40°	N. 40° W.	612.8	−514.2	987.2	2857.5
44.00						1600.	2343.3

Readings which give a westerly course should be considered negative; so also should latitudes *south* and departures *west*, as shown above. Then

Latitude for any Sta. = Distance × Cosine of Bearing,
Departure " " = " × Sine " "

and

Total Latitude for any Sta. =
Total Latitude for preceding Sta. + Lat. for preceding Sta.
Total Departure for any Sta. =
Total Departure for preceding Sta. + Dep. for preceding Sta.

The term "Latitude" is an abbreviation of "Difference of Latitude." The terms "Cosines" and "Sines" are more appropriate when the bearings are kept with no particular reference to the true or magnetic meridian.

By the aid of cross-section paper (if true to scale) we can plot the survey from the notes with only a straight-edge. Thus, e.g., to find the position of Sta. 26 + 50, we read off along the N. and S. base a distance to the north equivalent to 164.5 feet, and along the E. and W. base a distance to the east equivalent to 2382.5 feet; the intersection of the co-ordinates from these two points gives the position required.

On a long plan, if we have the base-lines drawn straight, and points accurately scaled off along them at, say, every 1000 feet, there is very little chance of making an appreciable error in the plotting of the plan if the notes are correctly worked out. But although this method is undoubtedly the best, unless the notes are well checked, it is very liable to give rise to errors owing to arithmetical mistakes in the notes themselves. But where good work is wanted, and in cases where probably the method of plotting by "chords" or "natural tangents" would otherwise have been used, the method of Latitudes and Departures, well checked, gives far better results, and probably takes no longer than the other ways.

56. The only way in which to feel sure that there are no appreciable mistakes in the transit-work is to check the bearing of the alignment every now and again by an observation for azimuth. This should be done, if possible, before starting the survey, or in any case as soon after as possible, and the notes then already taken reduced to their true bearings. By taking the magnetic pole as the standard of our bearings, we have no means of applying an *accurate* check to the work at a later period; but if we start with the vernier at zero, when the telescope is pointing to the *true* north, we can then check our course at any time on the survey.

Engineers generally fight rather shy of anything in connection with astronomical work; but considering that it is almost as easy to check the alignment by means of a star as by any known point on the Earth's surface,—and usually much more accurate,—it is a great pity that observations for azimuth are not used more frequently than they are. It is so much more satisfactory for the transit-man himself to know if he is doing good work; and considering that the transit-line is usually taken as the basis of all the plans to be afterwards constructed, every possible means of checking the work should be used.

57. The handiest methods of obtaining the true north are the following, one of which is applicable in most northern latitudes about every 6 hours, and can be applied without any knowledge at all of astronomical work:

A. By a Maximum Elongation.—In Fig. 10 let

Z represent the zenith,
P " the pole,
S " the Pole-star (Polaris).

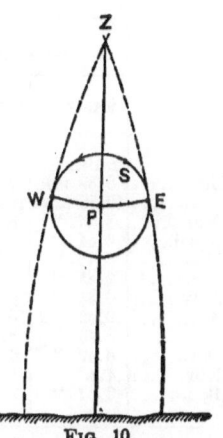

Fig. 10.

Then the small circle round the pole shows the path and direction of the star's motion, the time taken in making the circuit being nearly 24 hours. Now the radius of this small circle in angular measure is only about equal to $1\frac{1}{4}°$ (or $2\frac{1}{2}$ diameters of the sun), so that the apparent motion of the pole-star in azimuth (i.e., horizontally) will, when due east or west, be nothing at all, and for several minutes together when about east or west the motion will be inappreciable to ordinary railroad transits. Thus if we know about what time the star will be at its east or west elongation,—i.e., due east or due west,—and also the amount in azimuth by which when at those points it will be distant from the pole, we can, by setting the telescope on the star when at either of its elongations and applying the required correction in azimuth, obtain the direction of the true north. The following table shows approximately the times at which the elongations will occur. The amount of the correction in azimuth, which really equals the angle WZP (or EZP), may be found by solving the spherical right-angled triangle WPZ, the angle at W being 90°, the side WP being equal to 90°—the "declination" of the star. For Declinations of Stars see Table in Sec. 213. Thus we have

$$\sin \text{azimuth} = \cos(\text{dec.}) \sec(\text{lat.}),$$

PZ being the complement of the latitude of the place of observation. Thus suppose in latitude $50°$ N., in January 1889, we have the telescope clamped on Polaris at its eastern elongation, the vernier reading $2°.05'$; then the sine of the azimuth correction $= .0349$, which gives a value for the correction of $2°.00$, so that the telescope will be pointing due north when the vernier is set to read $0°.05'$. (*See note D, Appendix.*)

TIMES OF ELONGATIONS OF POLARIS.

Month.	1st Day.		11th Day.		21st Day.	
	Eastern.	Western.	Eastern.	Western.	Eastern.	Western.
	h. m.	h. m.	h. m.	h. m.	h. m.	h. m.
Jan......	0.39 P.M.	0.31 A.M.	11.59 A.M.	11.47 P.M.	11.20 A.M.	11.03 P.M.
Feb......	10.36 A.M.	10.25 P.M.	9.57 "	9.45 "	9.18 "	9.06 "
Mar......	8.46 "	8.34 "	8.07 "	7.55 "	7.27 "	7.16 "
April.....	6.44 "	6.32 "	6 05 "	5.53 "	5.26 "	5.14 "
May......	4.46 "	4.84 "	4.07 "	3.55 "	3.28 "	3.16 "
June.....	2.45 "	2.33 "	2.05 "	1.54 "	1.26 "	1.14 "
July.....	0.47 "	0.35 "	0.04 "	11.56 A.M.	11.25 P.M.	11.17 A.M.
Aug.	10.42 P.M.	10.34 A.M.	10.03 P.M.	9.55 "	9.23 "	9 15 "
Sep.......	8.40 "	8.32 "	8 01 "	7.53 "	7.21 "	7.13 "
Oct.......	6.42 "	6.34 "	6.03 "	5.55 "	5.24 "	5.16 "
Nov......	4.40 "	4.33 "	4.01 "	3.53 "	3.22 "	3.14 "
Dec.	2.42 "	2.34 "	2.03 "	1.55 "	1.24 "	1.16 "

Although the hour-angles from which the above times are calculated vary year by year and in different latitudes, they may be considered to be sufficiently correct between the years 1890 and 1900, and between latitudes $25°$ and $65°$ N. Where extreme accuracy is wanted, the time of observation may be calculated as in note D, Appendix. The above times increase by about 4 minutes every 10 years. But as these elongations occur only at intervals of 12 hours, more or less, it is well to have some other means of obtaining the true north, which can be used when the above method is inapplicable. The two following are similar to one another in principle, but occur about 12 hours apart, and from 5 to 7 hours from the time of the elongations given above.

B. In Fig. 11 let P be the pole and S the Pole-star, and let A

represent Alioth (ϵ Ursæ Majoris), and C represent the star "Gamma" (γ) Cassiopeia. The arrows and dotted lines show the paths and the directions of the motion of the three stars. The positions of the stars in the figure are those which they would occupy about the time of the western elongation of Polaris; but since the complete circuit occupies about 24 hours, we see that in about 6 hours C will be about vertically under S. When this occurs (i.e., when S and C are in the same vertical plane), clamp the telescope on Polaris, and wait through an interval of time which is to be found from the interval of 29 minutes 30 seconds for Jan. 1, 1889, by applying for any later date an annual correction of $+$ 19 seconds. After the lapse of this interval Polaris will be due north.

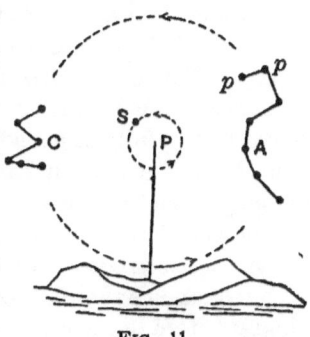

Fig. 11.

C. *The third method* consists in making use of Alioth in a similar manner to that in which we have just made use of γ Cassiopeia. But in this case, when Alioth is vertically below Polaris, Polaris will be nearly at its upper "culmination" (or "transit," as its passage across the meridian is called), but this makes no difference in the mode of procedure. The interval to wait when using Alioth was, on Jan. 1, 1889, about 27 minutes, and increases annually by 17 seconds. To calculate the above intervals, *see note E, Appendix;* but for ordinary work the figures given above are sufficiently correct as far north as 70°, and as far south as A or C are visible at their lower culminations. The altitude at which C or A will be above the horizon when due north equals about

Latitude of the place − 30°;

so that observations B and D cannot *practically* be used farther south than latitude 35° N. If, however, the instrument has a reflecting eye-piece, if either observation **B** or **C** is needed farther south than these limits, A and C can be used at their *upper* culminations, which will take place near the zenith, the intervals of time and modes of procedure will be the same as for the lower culminations.

To obtain the azimuth of Polaris at *any* time see Sec. 202.

There can be no difficulty about finding these stars if it is remembered that the altitude of the pole-star is about equal to the latitude of the place; that the " pointers" *pp*, Fig. 11, point towards it; that *A* and *C* are each about 30° from the pole-star; *C*, *A*, and *S* being all three more or less in a straight line.

The remarks made in Sec. 45 regarding the vertical axis, etc., should be carefully attended to. The times at which observations **B** and **C** will occur can be found near enough by noticing the positions of the stars themselves.

In observation **A** the instrument should be "reversed" on the star at the elongation. In observations **B** and **C**, where the star's motion in azimuth is comparatively rapid, observe, say, 2 minutes before the star is due north, and then again 2 minutes after its transit: the mean result should then be taken. An error of about 2 minutes in time in observations **B** and **C** causes an error in azimuth of about 1'. The verticality of the two stars should be also tested by a reversal of the instrument.

58. In checking the line by an azimuth observation as already described, it must be borne in mind that the **convergence of the meridians** needs a very important correction in the bearings relatively to other points east or west of the place where the observation is taken. This may be best shown by means of Fig. 12.

Let *ONEF* represent a sector of the northern hemisphere, and let *A* be the point on the earth's surface at which the survey was started, a continuous "straight" line being run which had at *A* a bearing due west. After we have traversed a difference of longitude which is represented by the angle *EOF* (or the spherical angle *N*) and have arrived at *C*, we shall be considerably south of the point *A*, our line having taken the course *AC* in the figure: so that, if at *C* we take an observation for azimuth, we shall find our line to have a bearing considerably south of west; and similarly *all* straight lines run from *A*, either towards the east or west, have a tendency to run to the south; similarly in the southern hemisphere they would have a tendency to run to the north. Thus in order to run a line from *A* to a point *B*, keep-

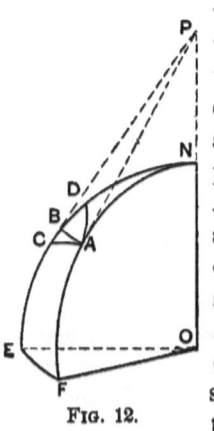

Fig. 12.

ing in the same latitude the whole way, it becomes necessary to run it as a curve. (See Sec. 209.)

Now the amount of this increase in bearing from the north is equal to the convergence of the meridians between the two places, so that in the case of A and B the difference in the bearings of the same straight line obtained by observation at each place will be represented by the angle BPA, which for ordinary work we may consider equal to the difference of longitude of the two places multiplied by the sine of their mean latitude. (*See note F, Appendix.*) Thus if in latitude 40° north we start a straight line from A due west and run it to C through 1° of longitude, the bearing obtained by observation at C should be S. 89°, 21′ W. But since it often needs some calculation to ascertain the difference of longitude, we can best proceed in ordinary work by finding from the following table the correction to be applied. Thus if in latitude 50° N. we have run a line which gives a total amount of easting or westing (i.e., Total Departure) equal to 60 miles, the amount of the correction to apply will be

$$60 \times 1' \, 02'' = 1° \, 02'.$$

TABLE OF CORRECTION FOR CONVERGENCE FOR 1 MILE OF EASTING OR WESTING.

Lat.	Correction for 1 mile.	Lat.	Correction for 1 mile.	Lat.	Correction for 1 mile.
10°	9″.18	27°	26″.52	44°	50″.19
11°	10″.13	28°	27″.66	45°	52″.00
12°	11″.07	29°	28″.85	46°	53″.83
13°	12″.02	30°	30″.03	47°	55″.67
14°	12″.98	31°	31″.26	48°	57″.67
15°	13″.96	32°	32″.49	49°	59″.83
16°	14″.93	33°	33″.83	50°	1′ 02″.00
17°	15″.92	34°	35″.17	51°	1′ 04″.17
18°	16″.91	35°	36″.50	52°	1′ 06″.67
19°	17″.93	36°	37″.83	53°	1′ 09″.17
20°	18″.94	37°	39″.17	54°	1′ 11″.67
21°	19″.98	38°	40″.67	55°	1′ 14″.33
22°	21″.02	39°	42″.17	56°	1′ 17″.17
23°	22″.10	40°	43″.67	57°	1′ 20″.00
24°	23″.17	41°	45″.17	58°	1′ 23″.00
25°	24″.30	42°	46″.85	59°	1′ 26″.25
26°	25″.38	43°	48″.52	60°	1′ 30″.00

This shows the necessity, when running a long continuous survey, of referring all bearings to an Initial Meridian, either at the point from which the survey started, or at a point near its centre. The same remarks of course apply to magnetic courses to a certain extent, but in this latter case, on account of the constantly changing variation, such corrections are hardly practical.

59. When the transit-line crosses a river or ravine or some other obstruction over which it is difficult to obtain direct measurement, the best way to proceed is by *Triangulating*, using whichever of the methods shown in Fig. 13 is most applicable to the case.

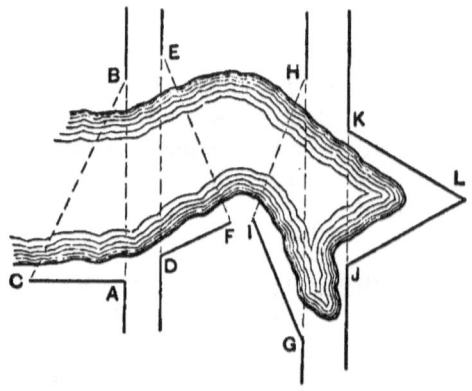

Fig. 13.

The angles at A and F each $= 90°$, and at J, K, and $L = 60°$; then

$$AB = AC \tan C,$$
$$DE = DF \sec D,$$
$$GH = IG \sin I \operatorname{cosec} H,$$
$$JK = JL = KL,$$

where $H = 180° - (I + G)$.

If the ground on which we measure our base has a tolerably uniform slope in the direction of the base, it is better to take direct measurement along the surface of the ground and multiply the distance so obtained by the cosine of the inclination to obtain the horizontal distance, than to "break-chain." Whatever difference in elevation there may be between two such points as A and B, if the base measurement is reduced to the horizontal, the distance as calculated for AB, from the angles observed with a transit, will also be the horizontal distance. If the angles were observed with a sextant, of course this would not be the case. (See Sec. 144.)

If, instead of encountering such obstructions as those given above, an obstacle which we are unable to see across presents itself, such as a huge detached rock on which we cannot set up the instrument, then perhaps as good a way as any to get round it is by offsetting the line so as to run past it on a parallel one, and then on the far side, by equal offsets, getting back on to the former line. If the obstacle, however, is too large to pass it well by this means, we can apply the equilateral triangle JKL (Fig. 13). This latter method is a good one to use whenever practicable: there is no calculation necessary in connection with it, the angles used are those most favorable to exact work, and where the obstacle *can* be seen over, a check can be applied by observing the angle at K.

After having run the line a certain distance ahead, represented by the amount L, it is often necessary to "back-up" and start the line again from the instrument so as to strike a point a certain distance d on one side of the point where the first line struck; the correction C for this may be found thus:

$$\tan C = \frac{d}{L}.$$

For more on the subject of triangulation, etc., see Part III.

60. The LEVELLER'S WORK on preliminary location consists mainly in taking the elevation at every full station, and at any intermediate points where he may consider it advisable to do so. The best form of keeping notes on such work is the following:

Sta.	B.S.	Int.	F.S.	H.I.	Elevation.
B.M.	4.25			106.60	102.35
195		4.8			101.8
+50		7.3			99.3
196	3.28		5.61		100.99

in which

Elevation in any line = H.I. − F.S. } in same line,
or "   "   "   "   = H.I. − Int.

and

H.I. in any line = Elev. + B.S. in preceding line,

The "Intermediate" column is sometimes omitted, but the insertion of it makes it easier to check each page by means of the difference of the sum of the Back-sights and Fore-sights.

To apply this check between two stations, A and B for instance, which have been used as turning-points, add together all the back-sights between A and B (including the B.S. at A, but excluding it at B); then add together all the fore-sights (excluding the F.S. at A, but including it at B): the difference of these two sums should equal the difference in elevation of A and B. If the sum of the back-sights is greater than the sum of the fore-sights, B is higher than A; but if less, then lower.

The levels should be worked out in the field whenever time permits, for reference on the work. The profile for each day's work should be made out when possible in the evening of the day on which the work was done.

As regards the **precision** of a line of levels run as above, the probable error is usually assumed to vary as the square root of the distance. The limit on the British Ordnance Survey is 0.01 foot per mile; the U. S. Coast Survey requires a limit of 0.03 per mile. If we assume a limit of 0.05 per mile for rough work, the probable error for any distance equals

$$0.05 \sqrt{\text{mile}}.$$

Thus in 100 miles the probable error $= 0.50$ ft. For more on the subject of levelling see Parts II and III.

61. The TOPOGRAPHER'S WORK consists principally in taking the ground slopes, with more or less accuracy, at every full station and at any intermediate points where he may consider it necessary, by means of which a contour plan may be constructed.

To do this he obtains from the leveller the elevation of each station and *plus* station at which he has taken levels.

There is a variety of methods in use of obtaining the slopes, and the advantage of each depends on the accuracy required, the nature of the country, and the vertical distance apart of the contour-lines.

Where the slopes are steep and accurate work is wanted, a 10-foot slope-rod with clinometer gives very good results, but is a cumbersome sort of instrument to carry about.

Where 5-foot contours are wanted, a *hand-level* is very convenient, since by considering the height of the eye above the ground to be 5 feet, the point corresponding to each contour-line is located at once by the level,—5 feet being an easy height to which to accommodate one's self,—and by pacing the distance between these points we have thus simply to enter the distances in the notes through which each contour passes. By taking the *alternate* points selected in this way, this method is of course equally applicable to 10-foot contours. Fig. 14 shows how this method is worked.

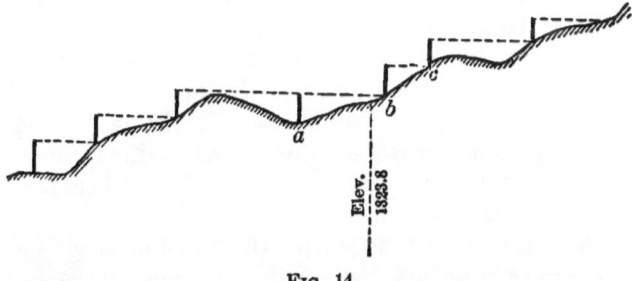

Fig. 14.

Suppose, e.g., that for a certain station the topographer obtains from the leveller the elevation of 1823.8, and that he is taking 5-foot contours. Then, if the ground is as shown in Fig. 14, he proceeds as follows: The contour-line nearest to this elevation is that of 1825 feet, the plane of which passes about 1 ft. above the ground-level at the station, so that by standing at the point a he can estimate with his eye the amount of 1.2 feet, and thus find the point b which corresponds with the contour of 1825. Similarly, standing at b he finds c, and so on up the slope as far as he considers necessary. Then returning to a, he works in the same way on the lower side. If the distances are wanted accurately, he should have a man with a tape to assist; but as a rule, *pacing*, where it is practicable, gives good enough results. The only notes to be kept in this case are the distances out (right or left) to the respective contours.

An Abney hand-level (with vertical arc) is also frequently used, and gives good results. All methods, however, which involve taking the angles of the slopes themselves necessitate extra work. One method of reducing this amount of labor is to have a set of scales for the various slopes, each made proportional to the cotangent of the inclination; but by the use of cross-section paper and a small protractor we can probably do the work equally well and equally fast.

The *stadia* method is often found very convenient for obtaining topography where the above methods would fail to give good results.

But besides taking the contours, the topographer must also take note of the courses of streams, etc., on each side of the line within a distance (usually) of a few hundred feet. The bearings of these he can take with a small prismatic-compass. He should also be constantly on the lookout for anything which may be of service in making up the preliminary estimates, such as indications of the probable classification, the flood-marks of water-courses, etc. If the topographer does his work thoroughly, he usually has difficulty in keeping up with the transit and level; but this is rarely a disadvantage, as the chances are that there will be occasional "backing-up" to be done by the party ahead.

62. The GENERAL PLAN of the "preliminary" survey showing the alignment, topography, etc., is usually plotted to a scale of 400 feet to an inch, as in Figs. 15 and 16, thus agreeing with the horizontal scale of the profile.

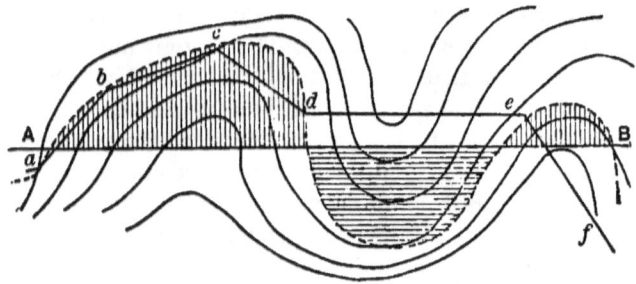

Fig. 15.

In Fig. 15 let *abcdef* represent a portion of the preliminary line as shown on the general plan, plotted to a scale of 400 feet to the inch; and let the line have been run to a $+ 1.25$ p. c. grade, and the contours be given for every 5 feet vertical. Then if each station at which the instrument was set up was at "grade," the *grade-contour* will pass through each of these points, but gradually rising from one contour to another, crossing them successively at distances of about 400 feet apart; so that if, as in Fig. 15, station *a* happens to fall on a contour-line, the grade-contour will cut the next line above, 400 feet farther on, at c; and since the next station d is only 200 feet from c, it will be situated about half-way between two of the contours.

Now this grade-contour is the line which, if adopted for the final location, would give no cuts or fills at all, so that it is the line which would render the cost of construction a minimum. The judgment of the engineer here comes in to decide how much it is advisable to deviate from this limit. So far the work has been more or less mechanical, for there are usually enough governing-points along the route to decide within two or three hundred feet the course of the preliminary line; but fitting the final location on to the plan is quite another matter. Suppose that the engineer considers that the straight line AB (Fig. 15) is about where the final line should be located. Then the shaded portions in the figure show cuts and fills alternately—shaded vertically being "cut," and horizontally "fill;" and the points where the line AB intersects the grade-contour will of course be the "grade-points." The amount of centre-cut and centre-fill can be read off at any point—not by scaling, but by counting the number of contour spaces there are between the line AB and the grade-contour. Thus, e.g., at a point in AB opposite c, there are $2\frac{1}{2}$ contour spaces, equivalent to $12\frac{1}{2}$ feet vertical, so that at this point we should have a $12\frac{1}{2}$-ft. centre-cut. By taking in this way a few points here and there, the engineer can, by means of Table XIV, form a fair idea of the number of cubic yards in each proposed cut or fill, making allowance of course where the surface-slope is steep, as shown in Sec. 69.

In this way, then, there is no great difficulty in obtaining a line which will make the cuts balance the fills, this being simply a matter of a few trials. Where curvature, however, is involved, it is not so much the question of balance as of the *total amount* of cut and fill, which needs consideration.

By having the various curves drawn on a horn protractor, or on a piece of tracing-cloth, the result of adopting any certain curve can be seen at once by sliding it up and down over the plan.

Then, again, a change of grade for a short distance may appear advisable, which necessitates altering the grade-contour. The question of overhaul, too, has to be considered, and the avoidance as much as possible of long shallow cuts. The probable classification, too, will of course affect the balance of cuts and fills. The advisability of raising the grade to avoid an expensive rock-cut also needs consideration. A little experience, however, goes a long way, and the engineer usually finds that there is little doubt to a few feet as to where the line ought to go.

63. The main features of the final location having been determined as above, and drawn on the plan, the approximate position of the points of curvature, etc., can be taken off by scale, and the line thus located on the ground; any little alterations being made, the advantages of which have become apparent when the line is seen actually staked out.

A fresh set of levels must of course be taken over the new alignment, and a profile constructed showing the rates of grade, etc., finally adopted.

As regards *compensating for curvature* where transition curves are not used, the rate of grade should be changed at the P.C. and P.T. Many engineers, however, prefer making the change at the nearest "full" station; it makes little difference, however, which way is adopted.

Bench-marks should be given at distances of a third of a mile apart or so, and guard-stakes set solidly beside the hubs. If the location is being "rushed," there is no need to fill in the *transition curves*, for that can be done equally well by the section-engineer when he takes over the work for construction. When these curves are omitted, however, it should be so shown on the plan, as in Fig. 16.

64. It often happens that after the line is located a considerable distance ahead an alteration in the alignment is deemed advisable, necessitating a shortening or lengthening of a certain portion of the line. This causes a break in the "through-chainage." Such a break as this should, wherever possible, be referred to a point where there is a change of grade, or at least to a point on a tangent, so as to simplify the running of the grades and curves as much as possible. It should be indicated *conspicuously* in the notes and on the plans and profiles in the form of an **equation**; the station on the line which comes first being read first. Thus if the left-hand side of the equation is the greater, it means that the line has been lengthened; but if the right-hand side be the greater, it has

RAILROAD LOCATION. 67

been shortened by an amount equal to the difference of the two sides.

65. The method of locating described above is of course suitable only to rolling or mountainous country; but where there is any doubt as to whether or not it is better to take contours, the engineer may generally come to the conclusion that it is better to do so. There is among some engineers an idea that the time spent in taking the topography might have been

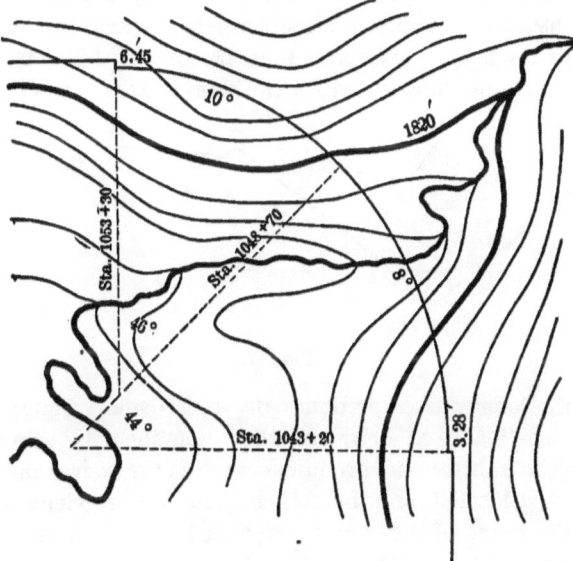

Fig. 16.

better used in running a series of trial lines. Of course in many cases this is true; but it must be remembered that a preliminary line with topography well taken to a distance on either side of, say, 500 feet (as is perfectly feasible in ordinary rolling country) covers a width of 1000 feet so completely, as to render the running of a trial-line within that area entirely needless; and that in order to settle the question absolutely as to the location through, say, a valley half a mile wide, two or at the most three lines run as above are all that can ever be required; while by the method of trial lines how many are needed before the engineer can feel satisfied that he has finally obtained as good a line as can be got? And then it is only the *best of the trial-lines* that is usually selected, which in all probability will be inferior to the line selected from the contour plan.

Besides this, if topography is taken, the engineer can at any future time show evidence as to the advisability of having adopted the route which he finally selected. It is a duty he owes to himself as well as to the Railway Company to be able to prove that the location has been good, and how is he to do this if he has simply trusted to the correctness of his eye?

66. *In country where the running is easy*, one or two trial-lines usually show pretty closely where the final line ought to go, for the long courses may then be converted into tangents, and curves be substituted for the shorter ones as in Fig. 17.

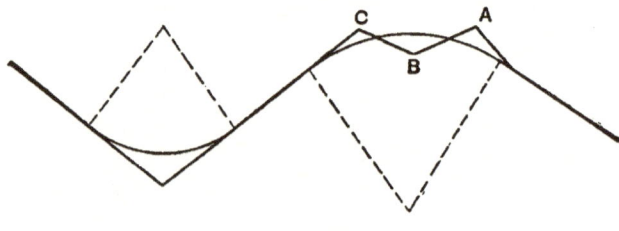

FIG. 17.

If the long courses predominate, it is usually better to get their location fixed first, and then join them by curves; but when the shorter ones are in excess, it is the curves that have to be first located, and the tangents made subservient to them.

If the notes of the courses are kept by "Latitudes and Departures," the exact curve necessary to replace such courses as ABC can be at once found according to Sec. 77.

67. An engineer with a good "eye" can often tell by merely looking over the ground what degree of curve is wanted to fit the surface, i.e., where the difference between a 3° 30′ and a 3° 45′ makes very little difference. Table II, of Tangents and Externals, is a good guide to this in many cases. For instance, by getting into position near the apex of the required curve, the engineer, with the aid of a hand-level and a prismatic compass, can often tell about how far from where he is standing the curve should pass. Thus, suppose he finds the angle of intersection to be about 40°, and that the curve should pass about 120 feet from the apex: he then finds from the Table that for an intersection-angle of 40° a 1° curve gives an external distance of 368 feet, therefore the degree of curve which he wants will be found by dividing this by 120; thus a 3° 04' curve will probably suit the case.

Where the APEX of a curve can be located without much trouble it is always better to do so; and of course this applies more especially to places where extreme accuracy in the centre-line is of importance; such as where bridge-work or trestling are required in the neighborhood of the curve.

68. The balancing of cuts and fills in comparatively *level country* is usually unadvisable, partly on account of the extra expense involved by the matter of over-haul, but mainly because, though the dump should be kept as high as possible, cuts in such country, and especially long shallow ones, generally add very considerably to the operating expenses. Thus the amount of *borrow* in such cases may often with economy be made very considerable.

69. On work of this sort the line is generally located first, and then the *grades fixed* by means of the profile. This is usually done by straining a piece of silk along the surface-line, by means of which the effect of adopting certain grades corresponding with the various positions of the thread can at once be seen; and, judging by the depth of centre-cut or fill, a fair estimate can thus be made of the amount of excavation and embankment required.

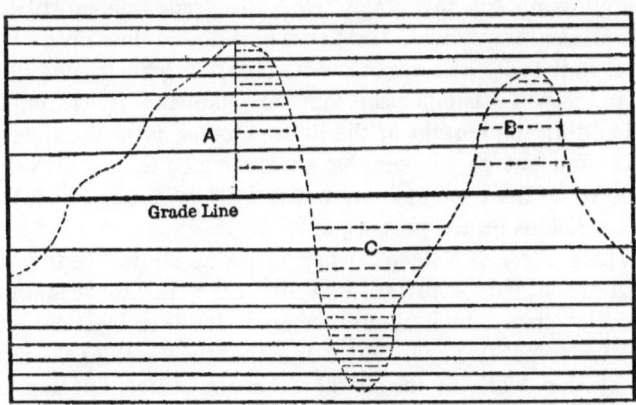

FIG. 18.

Where the work, however, is comparatively heavy, the following method will be found to give considerably better results: Suppose the dotted surface-line in Fig. 18 to be part of a profile on which we want to fix the grades so as to make the cuts and fills balance, and that in this case we wish to make a portion of cut A together with the whole of cut B sufficient to fill the hollow C. On a piece of tracing-cloth, say 10 inches long, draw a straight heavy line which is to be the grade-line; then turn to Table XIV, and see what depth of cut is required to give 1000 cubic yards contents in a length of 100 feet. Thus if the cuts are to have a 20-foot base and slopes of $1\frac{1}{2}$ to 1, as in Fig. 18, the depth of cut required will be about 8.3 feet. Then draw the parallel line above the grade-line already drawn at this distance from it, according to the vertical scale of the profile (in Fig. 18 taken as 40 feet to an inch); and again above that line draw another, distant from the grade-line by an amount corresponding to the depth of cut required to give 2000 cu. yds. in a length of 100 feet; and then draw a third for 3000 cu. yds., and so on, as many as are required. Similarly, on the lower side of the grade-line draw lines as above, suitable to the required base and slopes of the fill. Place the tracing-cloth over the profile, as in Fig. 18. If then the horizontal scale of the profile is 400 feet to an inch, take a "40" scale, and scale off along the horizontal dotted lines shown in the figure. One division of the scale then corresponds to 100 cu. yds. Thus, in order to make the cuts balance the fills (not allowing for shrinkage, etc.) the grade-line must be so placed that the sum of the horizontal dotted lines above it is equal to that of the lines below it. By sliding the tracing-cloth up and down, a balance can soon be obtained. By scaling off and adding the lengths of the lines together *mentally*, the contents of a cut or fill can be approximated to in a very few seconds; or the contents may be read off by means of the vertical divisions on the profile paper.

Where there is a steep surface-slope, an allowance must of course be added to the results as obtained by the above method. The allowance which should be made for this depends, comparatively speaking, very little on either the width of the road-bed or the depth of the cut or fill at the centre, but depends mainly on the slopes themselves; so that we may say *roughly*, that the following corrections are applicable to any ordinary depth of cut (or fill) or width of road-bed.

Thus, if by the above method we make the contents of a certain cut to amount to 20,000 cu. yds., with side slopes of

1 to 1, if the average surface-slope is about 10°, a fair estimate of the contents will be given by 21,000 cu. yds.

Slope Ratio.	Surface-slope.			
	5°	10°	15°	20°
1 to 1	1 p. c.	5 p. c.	8 p. c.	17 p. c.
1½ to 1	2 p. c.	8 p. c.	20 p. c.	45 p. c.

As to the effect of *shrinkage*, it may generally be ignored in dealing with the balancing of cuts and fills. (See Sec. 113.) A simple rule in dealing with rock-work is to assume that 100 cu. yds. of rock in excavation make 150 cu. yds. in embankment.

70. It has been assumed so far that in estimating the amount of excavation and embankment the method of centre-heights is used. In the long run the results so obtained may generally be considered to give sufficiently close results for most preliminary estimates. But when the surface-slopes are such as to necessitate continued corrections being applied, the average slopes at the different stations may be jotted down by the leveller when taking the elevations and the quantities worked out according to Mr. Trautwine's method of equivalent level sections, or some similar process.

CURVES.

71. Radius. Degree and Length of Curve.—Railroad curvature in Canada and the United States is expressed in

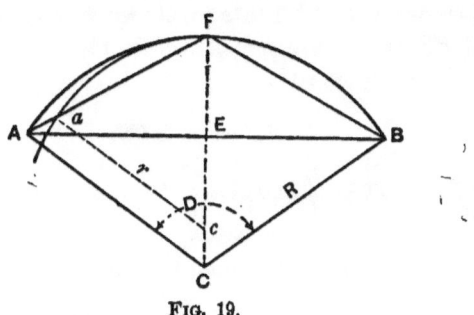

FIG. 19.

terms of the angle ACB, Fig. 19, which subtends a chord, AB,

100 feet in length; and this angle is called the *Degree* of the curve, and equals D.

In curves of small degree, i.e., of large radius, D varies very nearly inversely as the radius R.

To convert D into R, we have in the right-angled triangle AEC

$$\sin \frac{D}{2} = \frac{50}{R}; \quad \ldots \ldots \quad (1)$$

and *to convert R into D* this becomes

$$R = 50 \operatorname{cosec} \frac{D}{2}, \quad \ldots \ldots \quad (2)$$

from which formula Table I has been calculated.

From Equations 1 and 2 we see that R varies inversely as $\sin \frac{D}{2}$, and since it is only when $\frac{D}{2}$ is very small that its sine may be considered to vary as the angle itself, it follows that although we may say that the radius of a 10' curve is one tenth that of a 1' curve, by considering the radius of a 10° curve to be one tenth that of a 1° curve, we should, on accurate work, be led into an appreciable error. Thus by Equation 2,

R of a 1° curve = 5729.65 feet,
and $\qquad R$ of a 10° curve = 573.69 feet,
instead of . . . . 572.96.

72. The general practice of setting out curves on railroad construction is by means of **50-foot Subchords**, assuming that the angle subtended by any subchord at the centre C is proportional to its length. Suppose, for instance, we wish to locate a 10° curve, we see from Fig. 19 that since $AB = 100$ feet, if we wish to substitute for it two separate equal chords AF and FB, they must each exceed 50 feet in length, and the length of each must equal

$$AE \operatorname{cosec} AFC.$$

Now $AFC = 90° - \frac{D}{4}$ and $AE = 50$; therefore

$$\text{Corrected 50-ft. chord} = 50 \sec \frac{D}{4}. \quad \ldots \quad (3)$$

Thus, instead of using 50-foot chords it is the lengths given in the following table which must be used in order that two of them may give the same curve for the same deflection-angle as would be given by a 100 foot chord:

VALUES OF CORRECTED 50-FOOT CHORDS.

Deg.	Chord.	Deg.	Chord.	Deg.	Chord.	Deg.	Chord.
1°	50.000	6°	50.017	11°	50.057	16°	50.122
2°	50.001	7°	50.024	12°	50.068	17°	50.138
3°	50.004	8°	50.031	13°	50.080	18°	50.155
4°	50.007	9°	50.039	14°	50.093	19°	50.172
5°	50.012	10°	50.048	15°	50.107	20°	50.191

If the above corrections are not applied, the curve that is set out, instead of passing through the point A will pass through a at a distance from $F = 50$ feet, and its radius r will equal cF instead of CF, and

$$cF = 25 \sec CFA;$$

therefore

$$r = 25 \operatorname{cosec} \frac{D}{4}. \quad \ldots \ldots (4)$$

If we compare this equation with Equation 2, we see that the radius of a curve of any given value of D set out by 50-foot chords, according to the usual method, is exactly equal to half the radius of a curve whose degree $= \frac{D}{2}$ set out by hundred-foot chords. Thus the radius of a so called 10° curve, if set out by 50-foot chords, actually equals one half the radius of a 5° curve, i.e., 573.14 feet, not 573.69 as intended.

To find the corrected length of *any other subchord*, see Sec. 76.

The corrections which we have just seen to be necessary to accurate work, practically in a distance of 100 feet amount to nothing at all, but often in the total length of a curve they mount up considerably.

For instance, a 10° curve run in on location with a 100-foot chain, which should then of course be a true 10° curve, cannot be expected to "come out" well when tried on construction with 50-foot chords; for if the curve is 900 feet long and the instrument work and measurement absolutely correct, it will not close by 0.8 foot.

73. The **length of a curve**, in terms of 100-foot stations, as measured *along* 100-*foot* chords, may be at once found by dividing the total angle (C) at the centre, in degrees, by the degree of the curve. Thus if $L =$ *true* length of curve,

$$L = \frac{C}{D} = \frac{I}{D} \text{ (nearly)}, \quad \ldots \ldots \quad (5)$$

where $I =$ angle of intersection. (See Eq. 7.) So that if the angle subtended at the centre of a 10° curve $= 40°$, the length of the curve along the chords $= 400$ feet; and this method, on account of its simplicity, is that usually adopted on railroad work for the measurement of curves. But the *true length* of the curve will of course be greater than this in the same ratio as the arc AFB in Fig. 19 exceeds the 100-foot chord AB. Now the angle at the centre of a circle which is subtended by an arc equal to the radius equals

$$\frac{180°}{\pi} = 57°.29578,$$

so that the true length of a curve is given by the equation

$$L = \frac{CR}{57.2958} = \frac{IR}{57.2958}. \quad \ldots \ldots \quad (6)$$

Thus if $C = 40°$ and $R = 573.686$ feet (i.e., a 10° curve), $L = 400.507$ feet,—not 400 feet, as in the example above. Had this 10° curve been set out with corrected 50-foot chords, it would have measured (along the chords) 400.38 feet.

Table IV gives the length of arcs of various curves subtended by 100-foot chords, from which the true length of a curve may be at once found.

74. Before proceeding to the more practical problems in connection with the setting out of curves in the field, it will be well to consider a few of the more important equations which form the groundwork on which these problems are built up.

First, as regards the nomenclature of the various parts, as shown in Fig. 20.

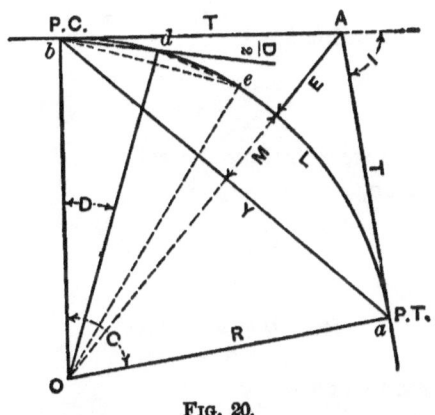

FIG. 20.

P.C. = Point of Curve.
    = Beginning of Curve.
P.T. = Point of Tangent.
    = End of Curve.
A = Apex.
I = Intersection-angle.
C = Central angle.
L = Length of Curve.

D = Degree of Curve, if bd
    = 100 feet.
T = Sub-tangent.
E = External distance.
M = Mid-ordinate to Long Chord.
Y = Long Chord.
R = Radius.

These symbols will be maintained throughout this article on curves.

75. Now because Aa and Ab are tangents to the curve at a and b, therefore OaA and ObA must each equal 90°, and the angle aAb at the apex must equal $180° - C$; therefore

$$I = C. \quad \ldots \quad \ldots \quad \ldots \quad (7)$$

Again, in the triangle bOd, since the angle at $b = 90° - \dfrac{D}{2}$, therefore the

Tangential Deflect.-Angle for a 100-foot chord $= \dfrac{D}{2}$. . (8)

In the right-angled triangle AOa

$$T = R \tan \frac{C}{2};$$

therefore, by Equation 7,

$$T = R \tan \frac{I}{2}. \quad \ldots \ldots \quad (9)$$

And if in this we substitute the value for R given in Equation 2, this becomes

$$T = 50 \tan \frac{I}{2} \operatorname{cosec} \frac{D}{2}. \quad \ldots \ldots \quad (10)$$

Again,

$$E = R \operatorname{exsec} AOa;$$

therefore, by Equation 7,

$$E = R \operatorname{exsec} \frac{I}{2}. \quad \ldots \ldots \quad (11)$$

And by combining Equations 9 and 11 we obtain

$$E = T \cot \frac{I}{2} \operatorname{exsec} \frac{I}{2};$$

therefore

$$E = T \tan \frac{I}{4}. \quad \ldots \ldots \quad (12)$$

So also

$$M = R \operatorname{vers} \frac{C}{2};$$

therefore, by Equation 7,

$$M = R \operatorname{vers} \frac{I}{2}. \quad \ldots \ldots \quad (13)$$

And by combining Equations 11 and 13, we obtain

$$M = \frac{E}{\operatorname{exsec} \frac{I}{2}} \operatorname{vers} \frac{I}{2};$$

therefore

$$M = E \cos \frac{I}{2}. \quad \ldots \ldots \quad (14)$$

RAILROAD LOCATION.

Again, by trigonometry,

$$\frac{Y}{2} = T \cos Aab;$$

therefore

$$Y = 2T \cos \frac{I}{2}. \quad \ldots \ldots \quad (15)$$

And combining this with Equation 9, we obtain

$$Y = 2R \tan \frac{I}{2} \cos \frac{I}{2};$$

therefore

$$Y = 2R \sin \frac{I}{2}. \quad \ldots \ldots \quad (16)$$

Again, by combining Equations 13 and 16, we obtain

$$Y = \frac{2M}{\text{vers } \frac{I}{2}} \sin \frac{I}{2};$$

therefore

$$Y = 2M \cot \frac{I}{4}. \quad \ldots \ldots \quad (17)$$

The above equations can readily be followed by referring to Secs. 231 and 232.

The following table may be of assistance in selecting quickly the equations required. Thus, suppose we have T and Y given, and want R; we see at once that Equation 15 will give us I; and then, by Equation 9, we can obtain R.

Given.	Required.	Use Eq.	Given.	Required.	Use Eq.
R	D	1	R, Y	I	16
I, L	D	5	M, Y	I	17
I, T	D	10	D, I	L	5
D	R	2	I, R	T	9
I, T	R	9	I, D	T	10
I, E	R	11	I, E	T	12
I, M	R	13	I, Y	T	15
I, Y	R	16	I, R	E	11
D, L	I	5	I, T	E	12
R, T	I	9	I, M	E	14
D, T	I	10	I, R	M	13
R, E	I	11	I, E	M	14
T, E	I	12	I, Y	M	17
R, M	I	13	I, T	Y	15
E, M	I	14	I, R	Y	16
T, Y	I	15	I, M	Y	17

PROBLEMS IN SIMPLE CURVES.

76. To lay out a curve by deflection-angles.—In Fig. 20 we have already seen (Eq. 8) that the angle $Abd = \frac{D}{2}$; but suppose we measure off another 100-foot chord de: then dbe also $= \frac{D}{2}$ (since $boe = 2D$, which makes $Obe = 90° - D$). Similarly, we might show that for any number of consecutive 100-foot chords the total deflection-angle would, for each one, increase by the amount $\frac{D}{2}$.

But though the Total Deflection-angle from the tangent is proportional to the number of *full* stations when these are the only points given on the curve, as we have already seen in the case of 50-foot subchords, if we insert *intermediate* stations without correcting the lengths of the subchords, the degree of the curve increases at once.

In order to find the **corrected length of any subchord** we may proceed thus: In Fig. 21 let ab represent a hundred-foot chord, then the angle $acb = D$; and let l represent any subdivision of it corresponding with the length of any uncorrected subchord; then the corrected length Y will be given by Equation 16, when

$$C : D = l : 100.$$

If we then insert this value of C in Equation 16, we obtain

$$Y = 2R \sin \frac{Dl}{200}, \quad . \quad . \quad (18)$$

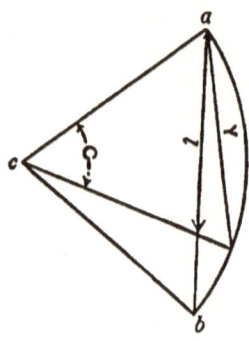

Fig 21.

Y being the corrected length of the nominal subchord l. In ordinary work, except where a sharp curve is run continuously throughout with subchords, we may ignore this correction.

Not taking the correction into account, the deflection for any subchord is to $\frac{D}{2}$ as the length of the subchord is to 100 feet; so that for any subchord we have

$$\text{Deflect. in minutes} \bigg\} = 0.3 D \times \text{Length of Subchord in feet;} \quad (19)$$

and this equation applies to a corrected subchord if we insert in it its *uncorrected* length.

Thus for a 14-foot subchord on a 3° curve the deflection-angle is 0° 12'.6.

Let us suppose that we are given a 3° Curve to the Right to locate from a P.C. at Sta. 421 + 36, I being equal to 12° 30'.

The length of the curve we find from Equation 5—since this is assumed as the standard method of measurement for railroad curves—to be 416.7 feet, therefore the P.T. will be at Sta. 425 + 52.7; then if we intend to use 50-foot subchords, our notes will be arranged as follows:

3° CURVE TO THE RIGHT.

P.C. = Sta. 421 + 36.0.
P.T. = Sta. 425 + 52.7.
Length of curve = 416.7 feet.
Intersection-angle = 12° 30'.
Subtangent = 209.2 feet.

Station.	Distance.	Deflection.	Index.	Remarks.
421 + 36			0° 0'	P.C.
+ 50	14	0° 12'.6	0° 12'.6	
422	50	0° 45'	0° 57'.6	
+ 50	"	"	1° 42'.6	
423	"	"	2° 27'.6	
+ 50	"	"	3° 12'.6	Hub.
424	"	"	3° 57'.6	
+ 50	"	"	4° 42'.6	
425	"	"	5° 27'.6	
+ 50	"	"	6° 12'.6	
+ 52.7	2.7	0° 02'.4	6° 15'	P.T.

The *Index*-reading at any station equals the sum of the *deflections* up to that station; then since the Index-reading at the P.T. is represented by the angle *Aba* in Fig. 20, and *Aba* is easily proved equal to $\frac{I}{2}$, therefore the Index-reading at the P.T. must equal half the intersection-angle, thereby giving a check on the calculations.

Having the notes worked out as above, set the transit up at the P.C. as in Fig. 22, and setting the index to zero, clamp the telescope on to a back-sight on the tangent (or on to the apex if it has been put in); then for any station the vernier must read the angle given in the index-column for that station. But suppose that when we have reached Sta. 423 + 50 we are unable to see any farther. Then set a hub (with a tack in it) at that station and a back-sight at the P.C. Set up over the hub,

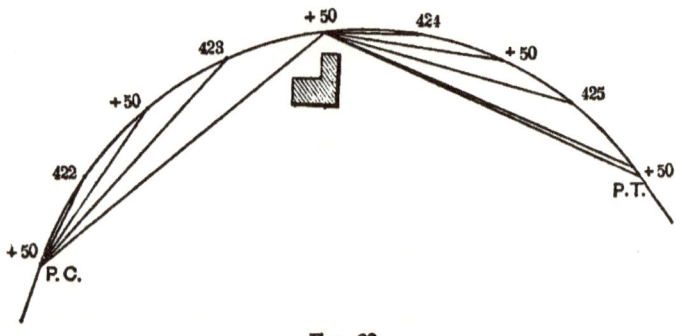

Fig. 22.

and setting the vernier back to zero, clamp the telescope on to the back-sight and turn off the remaining deflections by making the readings for the respective stations the same as those given in the Index-column. Thus:

(1) **When pointing to any station,** *the vernier must always be set to read the Index-reading for that station.*

(2) **When on the tangent at any station,** *the vernier must always be set to read the Index-reading for that station.*

By adhering to these two rules all possibility of error as regards the index-readings is avoided, and with the notes worked out as above we may locate the curve equally well from either end.

In order to find the bearing of the tangent at any station with reference to the tangent at the P.C., we have simply to multiply the index-reading at that station by two. Thus, if in the above example the tangent at the P.C. lies north and south, the bearing of the curve at Sta. 423 + 50 will be N. 6° 25'.2 E.

Usually in locating railroad curves there is no necessity to work out the deflections closer than to the nearer half-minute.

In places where accurate measurement is difficult to obtain, and great exactness is wanted, as in giving centres for piers in the middle of a river, we can often do better work by using

Two Transits, one on either side of the stream, and fixing the points by intersection. (See Sec. 163.)

77. To locate a curve when the apex is inaccessible.

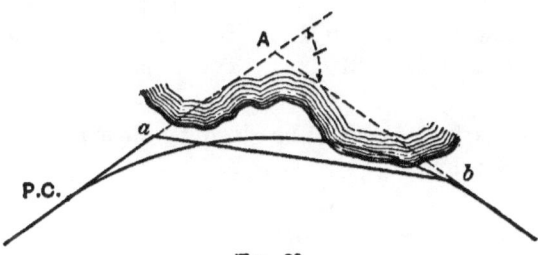

Fig. 23.

—Suppose, as in Fig. 23, we have been unable to locate the apex of a proposed curve, but have connected the two tangents at a and b by the line ab.

Then in the triangle Aab we know the distance ab and the angles at a and b; therefore we have

$$Aa = \frac{ab \sin b}{\sin A},$$

where $A = 180° - (a + b)$. We can then find the position of the P.C. For example, suppose $Aa = 320$ feet and $I = 40°$; then if we wish to connect the two tangents by a 5° curve, since the distance from A to the P.C. is given by Equation 9 (or Table II) $= 417.2$ feet, therefore the P.C. will be situated 97.2 feet back on the tangent from a.

We can then locate the curve according to Sec. 76.

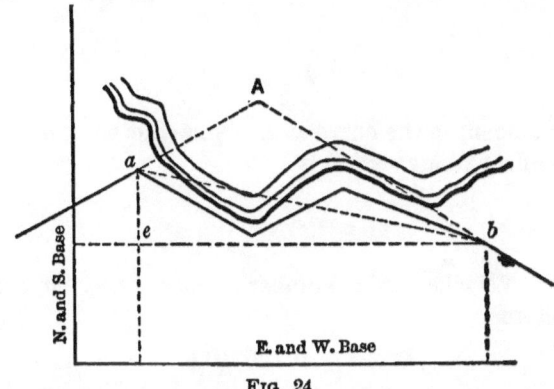

Fig. 24.

But suppose, instead of running a direct line ab, it is more convenient to run a succession of courses as in Fig. 24. Then, if the position of the stations a and b has been worked out by "Lats. and Deps." we can at once find the angles at a and b and the length ab.

For instance, let

Tot. Lat. of $a = 1020$ N.   Tot. Dep. of $a = 560$ E.
Tot. Lat. of $b = 810$ N.   Tot. Dep. of $b = 1430$ E.

Then the bearing of ab will be given by the angle at a in the triangle aeb; thus

$$\tan a = \frac{1430 - 560}{1020 - 810} = 4.143.$$

Therefore the bearing of $ab =$ S. 76° 26' E., and the length $ab = (1020 - 810)\sec a = 895.2$. Then if the bearing of the tangent at $a =$ N. 80° E., and of the tangent at $b =$ S. 60° E., we have in the triangle Aab, $a = 23°\ 34'$ and $b = 16°\ 26'$ from which we can find the position of the P.C. as above.

If the notes have not been already worked out by Lats. and Deps. the position of b with reference to a can be most easily calculated by taking the tangent at a as the N. and S. base.

78. To locate a curve by offsets from a tangent.—Let

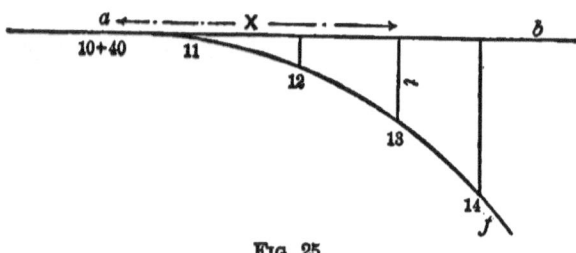

FIG. 25.

ab be a tangent to the curve at a. Now the value of the tangential offset at any station is

$$t = R \text{ vers } C.$$

But $C = ND$ where $N =$ number of Stas. *along the curve* to t, therefore

$$t = R \text{ vers } ND. \quad \ldots \ldots \quad (20)$$

Similarly, the distance *along the tangent* from a to the offset t equals

$$X = R \sin ND. \quad \ldots \quad \ldots \quad (21)$$

Thus, for *example*, suppose a falls at Sta. 10 + 40, and we wish from this point to set out a 10° curve by offsets from the tangent at a; then at Sta. 11

$$t = R \text{ vers } 6° = 3.14 \text{ feet,}$$

and the distance along the tangent at which this offset must be set off equals

$$X = R \sin 6° = 59.95 \text{ feet.}$$

The values of t at distances along the curves from a, 100 feet apart, are given in Table III, calculated by Equation 20.

A formula that often comes in handy in the field for computing tangential offsets, and which is usually true enough when X does not exceed 150 feet, is

$$t = \frac{X^2}{2R} \text{ (nearly).}$$

Tangential offsets may often be made use of when, on account of some obstacle or other, the method given in Sec. 76 cannot be used. By offsetting the tangent itself occasionally, as in Fig. 26, we can with ease run a curve past a succession of obstacles, and at the same time keep the offsets comparatively short.

Fig. 26.

Another occasion on which this method can be used to advantage is when the **apex, P.C. and P.T. are inaccessible.**

Suppose, by way of *example*, that we have to locate a 10° curve in a position such as is represented in Fig. 27, the angle of intersection having been found according to Sec. 77 to be,

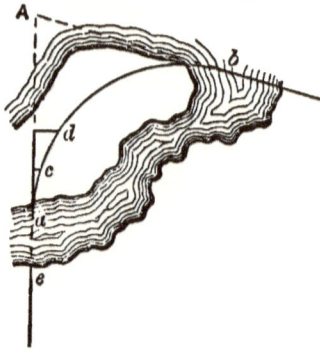

FIG. 27.

say, 90°, and the distance from A to some fixed accessible point e to be 723.7 feet: then ae will equal 150 feet. Suppose we are able to begin running in the curve at c, a point 200 feet along the curve from a: then the offset at c will equal 34.6 feet, at a distance from a along the tangent of 196.2 feet or from $e = 346.2$ feet; and the offset at d, 300 feet along the curve from a, equals 76.9 feet at a tangential distance of 286.8 feet from a, or from $e = 436.8$ feet. Thus we have two points c and d fixed on the curve, by means of which we can locate any other part of the curve accessible to them, as shown in Sec. 76.

Or, suppose we have such a case as that shown in Fig. 28, where we have run the curve ab round as far as d, but find that the P.T. is inaccessible, and yet wish to get on to the tangent without adopting the method given in Sec. 77. A convenient method of doing this is to locate the apex A, if accessible, by setting off from e, the middle point of the curve, the external E, found by Equation 11; then we have *one* point on the tangent Ab.

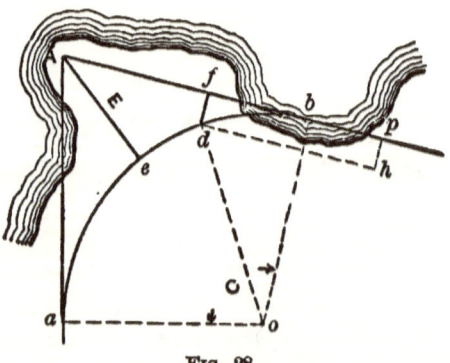

FIG. 28.

Again, by running on the curve as far as is possible to d, we can there set the vernier to read the (Index-reading for b + Diff. of Index between d and b) − 90° and set off $df = t$, found by Equation 20 : thus if the Index-reading for $d = 40°$ and for $b = 60°$, the vernier must read $60° + 20° - 90° = -10°$.

The angle ND in Equation 20 equals of course the angle dob. We thus have a *second* point f on the tangent Ab, and therefore we have its direction. Then by Equation 9, since $Ab = T$, we can, by triangulation, find the distance of A from some accessible point p on the tangent; then $bp = Ap - T$. Or, since by Equation 21, $fb = R \sin dob$ we can triangulate from p to f instead.

If A is inaccessible also, instead of proceeding as in Fig. 27, we might when at d set the vernier to the (Index-reading for b + Diff. in Index between d and b), which will give a line dh parallel to the tangent at b. Thus the vernier must read 80°. We can then set off $ph = df$, and thus obtain two points f and p in the direction of the tangent Ab; and since we know $dh = fp$ by direct measurement, and fb by calculation, we thus have the distance bp.

Again, if we have an obstacle on the curve itself, we can run a tangent from some point on the curve which will clear it, and so connect the curve at the further side in a similar way to that shown in Fig. 27; or we might run a Long Chord past it and lay it off by ordinates as in Sec. 80.

79. To locate a curve by offsets from the chords produced.—Let it be required to locate a 10° curve an by offsets

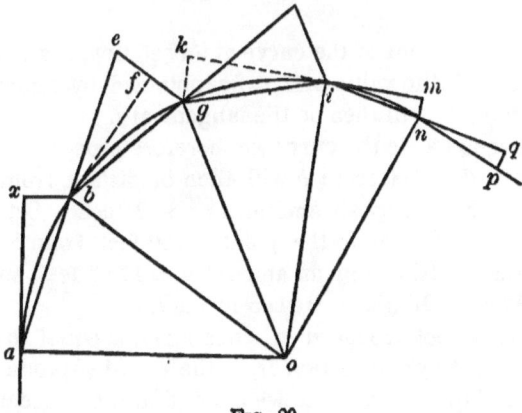

Fig. 29.

from the chords produced, and let, for example, the length of the curve = 360 feet. In Fig. 29—exaggerated for the sake of clearness—let *ab*, *bg*, and *gi* be 100-foot chords. then if *eb* is in the same straight line as *ab* and is equal to *bg*, the triangle *beg* is similar to the triangle *obg*; therefore

$$bg : R = eg : bg.$$

So that, calling the chord $bg = c$, and the chord deflection $eg = d$, we have

$$d = \frac{c^2}{R}. \quad \ldots \ldots \ldots (22)$$

but this value of d of course only holds good when the length of the preceding chord (as *ab*) is equal to c.

Again, if $fg = \frac{1}{2}eg$, then the triangle $bfg =$ the triangle axb, therefore $xb = \frac{1}{2}eg$. Therefore, if $t =$ the tangential offset,

$$t = \frac{c^2}{2R}, \quad \ldots \ldots \ldots (23)$$

a formula (already given in the last section in other terms) which holds good for any lengths of chord, provided the angle at $x = 90°$.

When $c = 100$ feet, we also have the formula

$$t = 100 \sin \frac{D}{2}.$$

To find a tangent to the curve at any station, say *i*, we have only to set off the value of $t = kg$, obtained by Equation 23, at station *g*; *ki* will then be the tangent at *i*.

In order to locate the curve we therefore proceed as follows:

Measure $ab = 100$ feet; *b* will then be distant from *ax*, the tangent produced, by an amount $t = 8.72$ feet. Set pickets at *a* and *b*, and range in the point *e*, 100 feet from *b*; then *g* will be distant from *e* by an amount $d = 17.43$ feet, and from *b* by 100 feet. Similarly we can locate *i*.

But the 60-foot subchord *in*, since it is not equal to *gi*, cannot be located by a deflection from the chord *gi* produced, according to Equation 22. So we must find the tangent at *i* by setting off at *g* the amount $kg = 8.72$ feet; then, having obtained the tangent at *i*, we can calculate the offset *mn* for the 60-foot chord *in* (by Equation 23), which equals 3.14 feet, and this brings us to the P.T. of the curve. In order to find the direction of the tangent at n we may either set off at i the value of t for the chord in, or we may produce the chord in to q, making $nq = in$, and then from q set off an offset $qp = mn$; np will then be the direction of the tangent.

Theoretically, we ought always to make the angle between a tangent and its offset $= 90°$, and between a chord produced and its offset $= 90° - \frac{1}{2}$ angle subtended by the chord at the centre; but in ordinary work there is no need to be particular about this.

80. To locate a curve by ordinates from a long chord.—Suppose, as in Fig. 30, we have two stations a and b given, we then have the length of the arc adb, and so we can find C by Equation 5.

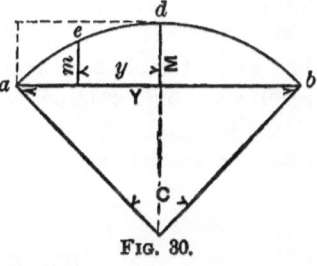

Fig. 30.

Now if d is the middle point on the curve the deflection-offset t from the tangent at d to $a = M$, the ordinate at d; therefore, by Equation 20,

$$M = R \text{ vers } \frac{C}{2}, \quad \ldots \quad (24)$$

M being the *mid-ordinate* to the long chord Y. The length of an ordinate from the chord to any other station e will be given by the equation

$$m = M - R \text{ vers } ND, \quad \ldots \quad (25)$$

where $N =$ the number of stations measured *along the curve* from d to e; and the distance from the centre of the long chord at which m must be set off is given by

$$y = R \sin ND, \quad \ldots \quad (26)$$

which is the same as Equation 21 for the value of X in Sec. 78.

To take **an example**: Suppose a is at Station $2 + 20$ and b at Station $6 + 40$, then d will fall at Station $4 + 30$. Let $D = 10°$, then $C = 42°$; and we can find Y either by direct measurement or by Equation $16 = 2R \sin 21° = 411.2$ feet. Similarly by Equation 23 we find $M = 38.1$ feet.

If we then wish to set off an ordinate to Sta. 3.00, we have $N = 1.3$; therefore $y = R \sin 13° = 129.1$ feet, and m, by Equation 24, $= 38.1 - R \text{ vers } 13° = 23.4$ feet.

It is usually unnecessary to calculate the values of y, except perhaps when near the ends of the chord. Thus, in the above example, had we assumed $y = 100 N = 130$ feet, it would practically have made no difference in the position of Sta. 300.

If we have the length of the chord Y given, we may obtain C directly from it by means of Equation 16; or, conversely, when we know C we can obtain Y.

The lengths of Long Chords subtending arcs up to 6 stations are given in Table V; also the length of arcs subtended by 100-foot chords. Thus, if $C = 20°$ and $D = 10°$; Y instead of being equal to 200 feet, really equals 200.254 feet, which is the result we should obtain if we used Equation 6 instead of Equation 5 to obtain the value of L. The middle ordinate may *also* be correctly found thus:

$$M = R - \sqrt{R^2 - \frac{Y^2}{4}}, \quad \ldots \ldots (27)$$

and any other ordinate

$$m = M - R + \sqrt{R^2 - y^2}. \quad \ldots \ldots (28)$$

An approximate formula, which is really a corruption of Equation 27, is

$$M = \frac{Y^2}{8R} \text{ (nearly)}. \quad \ldots \ldots (29)$$

It is sufficiently true, however, when Y is small, the error on a 20° curve, in the case of a 50-foot chord, only amounting to .002 foot. By comparing Equation 29 with Equation 23, we see that the mid-ordinate to a short chord may be considered equal to one quarter the tangential offset at a distance along the tangent equal to the chord.

A convenient method of locating small arcs is that shown

in Fig. 31, where, having found M by Equation 24 or 29, the mid-ordinate for the subchord ac may be considered equal to

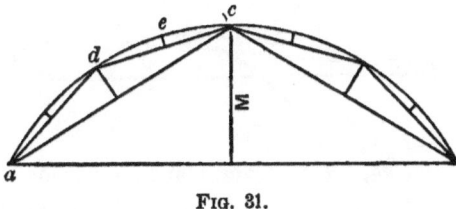

FIG. 31.

$\frac{1}{4}M$, and the ordinate e of the sub-subchord dc similarly equal to one quarter the ordinate at d.

81. To pass a curve through a fixed point, the angle of intersection being given.—Suppose we first find the

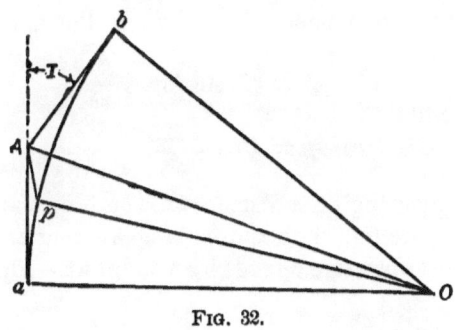

FIG. 32.

position of the fixed point p (Fig. 32) with reference to Aa in terms of the distance Ap and the angle aAp: then

$$pAO = 90° - \left(aAp + \frac{I}{2}\right),$$

and

$$\sin ApO = \sin pAO \sec \frac{I}{2};$$

therefore in the triangle ApO we have pO equal

$$R = Ap \sin pAO \operatorname{cosec} (pAO + ApO).$$

ApO always exceeds 90°.

82. To run a tangent from a curve to any fixed point.
—Let p (in Fig. 33) be the fixed point, and a and b be any two

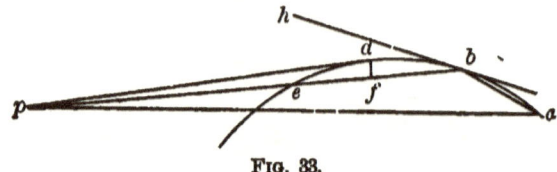

Fig. 33.

points on the curve,—b, however, being on the side remote from p, yet as near to the probable situation of the tangent-point d as is possible. Then taking the chord ab as a base, the length of which is given by Equation 16, observe the angles at a and b in the triangle abp; then

$$bp = ab \sin a \operatorname{cosec} apb,$$

when $apb = 180° - (a + b)$.

Now if bh is the tangent at b, we know the angle hbp, and can thus find

$$eb = 2R \sin hbp.$$

But by Euclid

$$dp = \sqrt{bp \times ep}.$$

Thus by measuring off a distance $bf = bp - dp$ and offsetting to the curve, we find the required tangent-point d.

83. To connect two curves by a tangent.—First suppose,

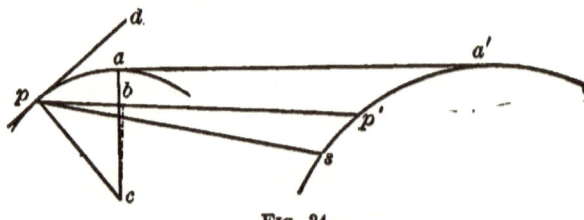

Fig. 34.

as in Fig. 34, that *both curves are of the same direction*. On the curve of smaller radius R select a point p slightly more remote from the other curve than the tangent-point at a probably is. On the curve of larger radius R' find a point p' which has its tangent parallel to the tangent at p. This may be done by running a trial-line to some station s; and then, by comparing the direction of the tangents at p and s, we find how far along the curve from s, p' will be situated.

RAILROAD LOCATION.   91

Now if pd is the tangent at p, and cb is perpendicular to pp', we have $pca = dpp' - acb$, and

$$\sin acb = \frac{(R' - R) \text{ vers } dpp'}{pp' + (R' - R) \sin dpp'} \text{ (nearly)},$$

pp' being obtained by direct measurement; and

$$aa' = pp' + (R' - R) \sin dpp' - (R' - R) \sin acb,$$

from which we can find the position of a'.

But suppose, as in Fig. 35, the *two curves are of opposite direction*.

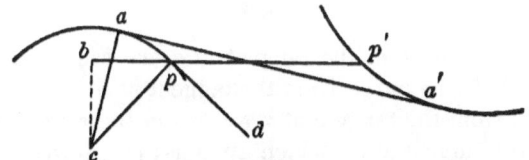

Fig. 35.

Then select p on the side of a towards the other curve. Then, as before, $pca = dpp' - acb$; but in this case

$$\sin acb = \frac{(R' + R) \text{ vers } dpp'}{pp' + (R' + R) \sin dpp'} \text{ (nearly)},$$

and

$$aa' = pp' + (R' + R) \sin dpp' - (R' + R) \sin acb.$$

The distance ap should never exceed 100 feet when the curves are of the same direction, or 75 feet when of opposite direction, and should always be taken *as small as possible*.

84. Given a curve joining two tangents, to change the P.C. so that the curve may end in a parallel tangent.

Let it be required to move the P.C. at a (in Fig. 36) so that the curve ab, instead of ending at b, will end in a parallel tangent, distant from the tangent at b by the amount e.

Then, since it is simply a case of shifting the curve bodily in the direction of the tangent aa', we have

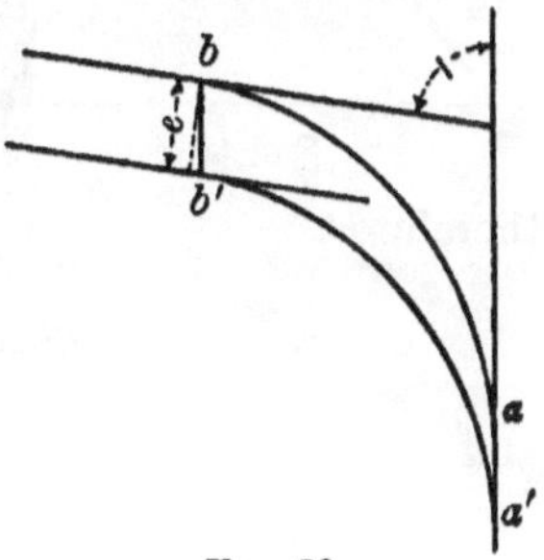

$$aa' = e \text{ cosec } I.$$

Fig. 36.

Had $a'b'$ been the given curve, and it were required to shift it outwards to the parallel tangent at *b*, the same equation of course applies.

85. Suppose we have such a case as that shown in Fig. 37, where *ab* is the given curve, and it is required to shift it to **parallel tangents at each end,** as at *a'* and *b'*.

Fig. 37.

Then starting from the tangent at *a*, we can, as above described, shift the curve from the tangent at *b* to the tangent at *b'*, and from the tangent at *a* we can in the same way shift it on to the tangent at *a'*, which gives us the required positions of *a'* and *b'*.

86. Given a curve joining two tangents, to change the radius and the P.C. so that the new curve may end in a parallel tangent at a point opposite to the original P.T.

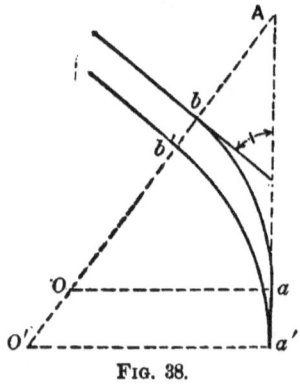

Fig. 38.

In Fig. 38 let it be required to change the radius of the curve *ab* and also the position of *a*, so that the curve, instead of ending in *b*, will end in a parallel tangent at *b'* (*b'* being directly opposite to *b*). Then if O is the centre of the curve *ab* and R its radius, and O' the centre of the curve *a'b'* and R' its radius, by Equation 11,

$$Ab = R \operatorname{exsec} I,$$
and
$$Ab' = R' \operatorname{exsec} I;$$

therefore

$$R' - R = \frac{bb'}{\operatorname{exsec} I},$$

and

$$aa' = bb' \cot \frac{I}{2}.$$

Had *a'b'* been the given curve, and it were required to shift it outwards to the parallel tangent at b, the same equations of course apply.

87. Given a curve joining two tangents, to find the radius of another curve which, from the same P.C., will end in a parallel tangent.

Let it be required to change the radius of the curve ab, so that it will end in a parallel tangent at b'.

Let O be the centre of the curve ab and R its radius, and O' be the centre of the curve ab' and R' its radius. Then $R - R' = OO'$; therefore

$$R - R' = \frac{e}{\operatorname{vers} I}.$$

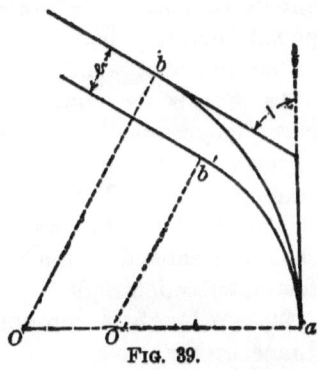

Fig. 39.

Had ab' been the given curve, and it were required to shift it outwards to the parallel tangent at b, the same equation of course applies.

88. Given a curve joining two tangents, to change the radius and position of the P.C. so that the curve may end in the same P.T., but with a given change in direction.

In Fig. 40 let it be required to change the radius and P.C. of the curve ab, so that at b it will have a difference in direction equal to $I' - I$. Then if O is the centre of the curve ab and R its radius, and O' and R' are the centre and radius of the curve $a'b$,

$$R \operatorname{vers} I = R' \operatorname{vers} I';$$

therefore

$$R' = \frac{R \operatorname{vers} I}{\operatorname{vers} I'},$$

and $aa' = R \sin I - R' \sin I'$.

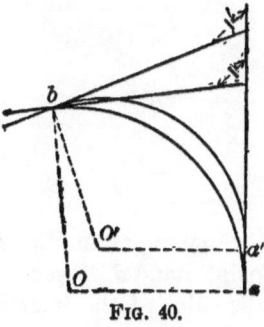

Fig. 40.

COMPOUND CURVES.

89. A compound curve, being merely a series of two or more simple curves, the manner in which it is located is by setting out its components separately, each P.C.C. (Point of Compound Curvature) being treated as a P.C. or P.T., the direction of the tangent at each P.C.C. being given by its Index-reading.

As regards the notes, instead of keeping them for each curve independently, it is better to carry the Index-reading through continuously from the P.C. to the P.T., so that the reading for the P.T. equals half the *total* intersection-angle.

The length and intersection-angle of *each* component curve should be entered in the notes, and also the total length and total intersection-angle.

90. To locate a compound curve when the P.C.C. is inaccessible.

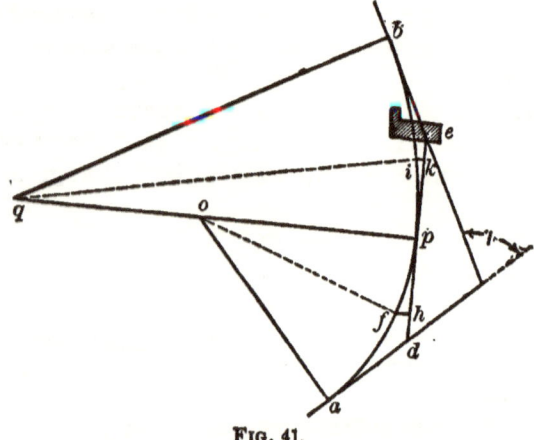

FIG. 41.

Suppose, as in Fig. 41, *p* (the P.C.C.) is inaccessible. The points *e* and *d*, if accessible, may then be found by inserting the value of the intersection-angle, in the case of each curve separately, in Equation 9, and thus obtaining for *T* the distances *ad* and *be*.

Then from the tangent *de* the curve can be located by offsets, as already shown.

If the points *d* and *e* are also inaccessible, select in the curve some convenient point *f*, and from it set off the offset *fh* =

of vers *fop* (by Equation 20). Similarly, from a point in the other branch of the curve lay off an offset $ik = qi$ vers iqp. We can then find the position of p by Equation 21; thus:

$$hp = of \sin fop.$$

91. Given a simple curve ending in a tangent, to connect it with a parallel tangent by means of another curve.

1. Let *ac* in Fig. 42 be the given curve, and *bc* the required curve: then we have

$$\cos C = 1 - \frac{e}{R-r},$$

from which we can at once find the P.C.C.

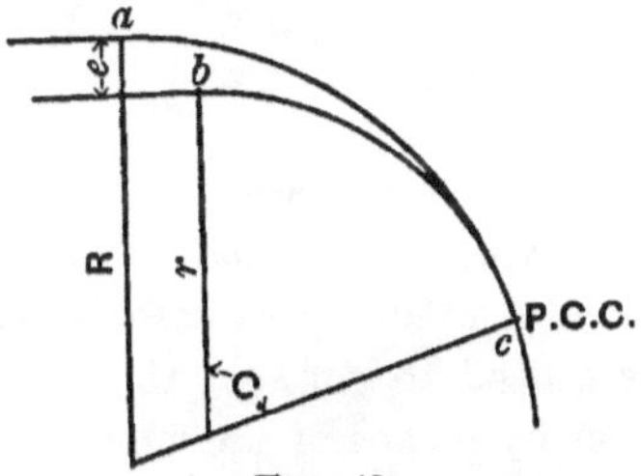

2. Let *bc* be the given curve, and *ac* the required curve: then since C, the central angle, is the same for both curves, the above equation holds good also in this case.

Fig. 42.

92. To connect a curve with a tangent by means of another curve of given radius.

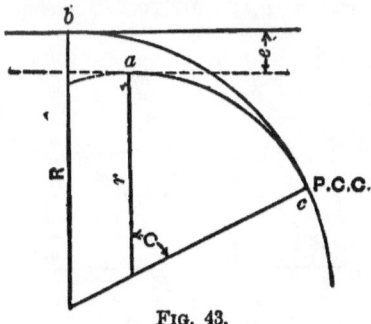

Fig. 43.

1. Let *ac* in Fig. 43 be the given curve which it is required to connect with a given tangent at *b*. Find the point *a* on the given curve which has its tangent parallel to the given tangent, and measure *e*: then, since

$$\cos C = 1 - \frac{e}{R-r},$$

we can thus find the position of the P.C.C.

2. But if the radius of the required curve is less than that of the other curve, then, as in Fig. 44, find the point d at the intersection of the tangent at b with the given curve ac, and observe the angle of intersection at $d = aod$; then

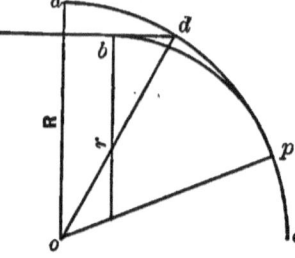

Fig. 44.

$$\cos aop = \frac{R \cos (aod) - r}{R - r}.$$

Thus p, the P.C.C., will be situated at a distance along the curve from d represented by the curvature $aop - aod$.

3. An analogous case is that shown in Fig. 45, where it is required to connect the curve ac with a *tangent on the convex side* by means of the curve pb.

Then, as before, find d and observe the angle of intersection at $d = aod$; then

$$\cos (aop) = \frac{R \cos (aod) - r}{R + r},$$

from which we can find p as above.

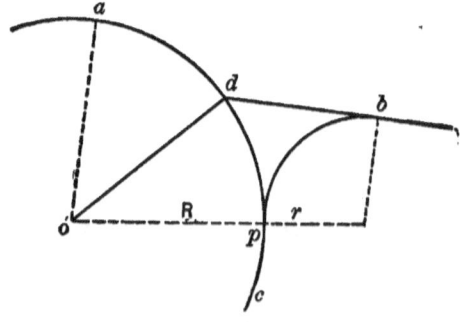

Fig. 45.

Suppose in case 3 the point d were found to coincide with a; then we merely have the case of a Y located on the tangent db, in which case the above formula becomes

$$\cos (aop) = \frac{R - r}{R + r}.$$

93. Given a compound curve ending in a tangent, to change the P.C.C. so that the terminal curve may end in a given parallel tangent without changing its radius.

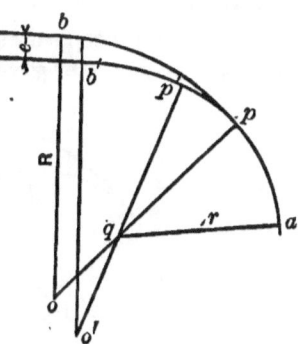

FIG. 46.

1. In Fig. 46 let the radius of the terminal curve *pb* be greater than the radius of the other curve *pa*; then,

A. *If we want to shift the curve inwards* to b', then to find p', the new position of the P.C.C., we have

$$\cos o' = \cos o + \frac{e}{R-r};$$

but,

B. If abp' were the given curve, and it were *required to shift it outwards* to b, then

$$\cos o = \cos o' - \frac{e}{R-r};$$

and since in both cases

$$pqp' = o - o',$$

we can thus find the position of p or p', as the case may be.

2. Suppose, however, the radius of the terminal curve *bp* is less than the radius of the other curve *pa* as in Fig. 46, and that it is required to shift the tangent (A) *inwards* to b: then

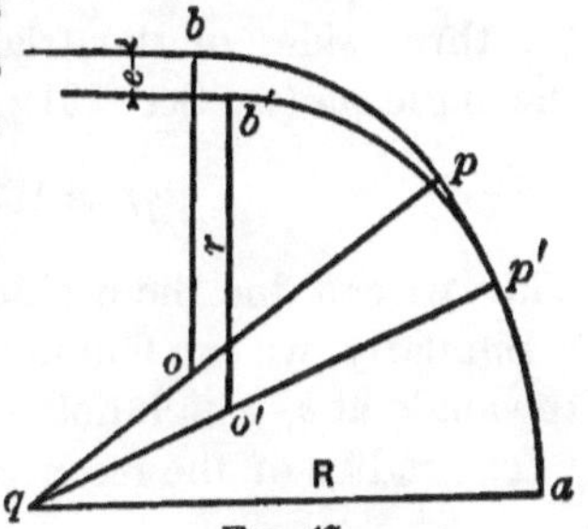

$$\cos o' = \cos o - \frac{e}{R-r}.$$

FIG. 47.

But (B) if $ap'b$ were the given compound curve, and it were required to shift it *outwards*, then

$$\cos o = \cos o' + \frac{e}{R-r}.$$

Then since in both cases (A) and (B) $pqp' = o' - o$, we can find the position of p or p' as the case may be.

94. To connect two curves, already located, by means of another curve of given radius.

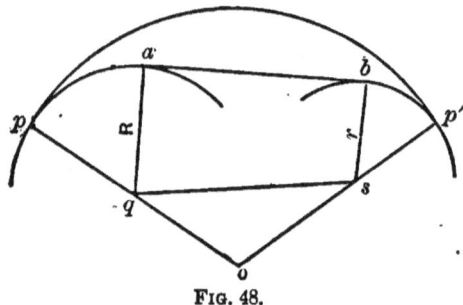

Fig. 48.

As in Fig. 48, let R be the radius of the easier curve, and r the radius of the sharper curve. Find the tangent ab as shown in Sec. 83, and also the distance ab by direct measurement or calculation; then

$$\tan (aqs) = \frac{ab}{R-r},$$

and

$$qs = ab \operatorname{cosec} (aqs).$$

Then, since $oq = op - R$ and $os = op' - r$, where op and op' are each equal to the radius of the required curve, we have the three sides of the triangle oqs, from which we can find the angle oqs (see Sec. 231); and

$$aqp = 180° - (oqs + aqs).$$

Thus we can find the position of p.

Similarly, we can find the position of p'; or we can calculate the angle at o, which does equally well.

The radius of the required curve must exceed

$$\frac{qs + R + r}{2}.$$

If $R = r$, then

$$\sin (aqp) = \frac{ab}{2(op - R)}.$$

95. To locate any portion of a compound curve from any station on the curve.

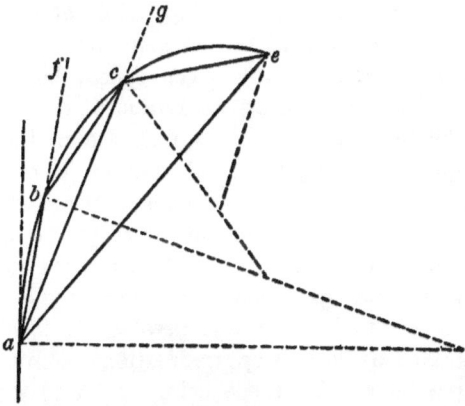

FIG. 49.

Let *abce* in Fig. 49 be a compound curve, and *a* any station on the curve, and let it be required to establish the point *e*; the P.C.C.'s at *b* and *c* being inaccessible.

Assume, for the sake of simplicity, that the chords *ab*, *bc*, and *ce* are equal, and let the curvature of *bc* equal twice the curvature of *ab*, and that of *ce* three times the curvature of *ab*.

Now if $d = $ the deflection from the tangent at *a* for Sta. *b*, then, if *ab* be produced to *f*, the angle $fbc = d + 2d = 3d$. Again, if the chord *bc* be produced to *g*, the angle $ecg = 2d + 3d = 5d$. Then in the triangle *abc*, the angle at $b = 180° - 3d$; and since the length of the chords can be found by Equation 16 (Sec. 74), we can find the side *ac* and the angles at *a* and *c*. Again, in the triangle *ace*, the angle at $c = 180° - (bca + 5d)$; thus we can find the angle at *a*. Similarly we can find the angle subtended at *a* by the chord *bc*, and thus we have the total deflections to *b*, *c*, and *e*. When the chords are of different lengths, as is of course usually the case in practice, and the curvature varies irregularly, we can by plotting the curves and drawing the tangent at each P.C.C. see at once in each case what the deflection-angle at any P.C.C. will be from the chord produced. The principle will be just the same as in the case above described.

Sec. 96 is an application of this problem.

TRANSITION CURVES.

96. Since the elevation and depression of the outer and inner rails, respectively, at the entrance to a curve must be made gradually, and for any given speed the difference in elevation varies inversely as the radius of curvature, it follows that the curvature should also decrease gradually, having a radius equal to infinity at the P.C. and a minimum at the centre of the curve. If we assume, as is usual, that the difference in elevation of the two rails increase at a uniform rate until the maximum curvature is attained, then the theoretic curve which should be adopted is a form of the elastic curve, which, on account of the trouble involved in locating it, has been supplanted by various approximations, such as the curve of sines, parabolæ, etc.; these being easier to locate in the field.

The use of Transition Curves is found not only to cause less resistance to the passage of trains than a similar curve whose ends are not eased off, but also generally to enable the curves to be fitted better to the ground than in the case of plain circular ones.

That Transition Curves *are* of advantage in actual practice is shown by the fact that all Simple Curves at their P.C.'s and P.T.'s have a decided tendency to assume the form of the Elastic Curve; and since this lateral creeping is caused by the pressure of the flanges of the wheels, increased wear and tear to rails and rolling-stock is the result.

It is to be noticed that the easing of curves in many cases involves an increase in curvature at the centre of the curve, but this is usually so slight as to be practically inappreciable, and is much more than compensated for by the reduction of curvature at the ends of the curve. Thus, for example, where a 9° simple curve defines the limit of curvature in the case of uneased curves on any road, by inserting transition curves a 10° curve would be perfectly allowable

The three following methods of inserting transition curves are simple and easily applied:

97. Method I.—Suppose, as in Fig. 50, that we have a 5° 30′ curve *ab*, which it is required to ease off by means of a transition curve.

Now if we do not wish to shift the main curve inwards from the tangent at a, it becomes necessary to shift the tangent at a

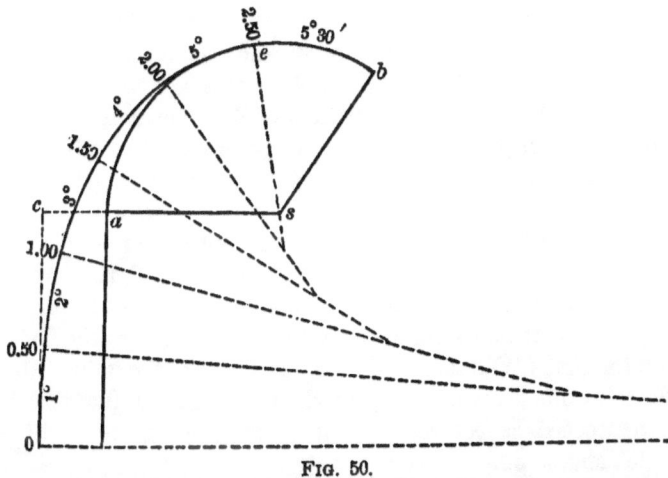

Fig. 50.

itself outwards by the amount ac, and also to throw the P.C. at a backwards by the amount oc, so that the point o becomes the new P.C.

Now
$$ac = Y \sin d - R \operatorname{vers} C,$$
and
$$oc = Y \cos d - R \sin C,$$

where $Y =$ the long chord to the end of the transition curve; $d =$ the total deflection-angle from Sta. o to the end of the transition curve (given in top line of Tables A and B in this section); $C =$ the total curvature of the transition curve, as represented by the angle esa (values of which are given in Tables A and B); and $R =$ Radius of the main curve.

The values of the first term in each of these equations are also given (i.e., $Y \sin d$ and $Y \cos d$) in Tables A and B.

Suppose we consider that a transition curve which increases its curvature by 1° in every 50 feet (as in Table A) will suit the case in question, then we want 250 feet of such a curve in order that the increase in curvature at no point may exceed 1°, and in that case we find from the above formula that $oc = 113.40$ feet and $ac = 3.06$ feet; so that the tangent must be offsetted to the left a distance of 3.06 feet, and the new P.C. will be situated 113.40 feet back from the original one.

Set the transit up at the point *o* and locate the curve in the usual manner, the zero of the instrument coinciding with the direction of the tangent, the index-readings being taken from the top line in Table A. The point *e* at Sta. 2.50 from *o* will then be the P.C.C. of the 5° branch of the transition curve and the 5° 30′ main curve. Should the point *e* not be visible from *o*, the transit may be moved up to any of the intermediate stations, and the total deflection for the other stations from the tangent at any station are given in the tables; so that, suppose we had found it necessary to move up to Sta. 1.50, then we can get the zero of the instrument to coincide with the direction of the tangent at that station, by setting the vernier to the deflection for Sta. 1.50 (taken from the top line in the table) when the telescope is clamped on to the back-sight at Sta. *o*. We then proceed as before; e.g., our index-reading for *e* will be 3° 25′, and so on.

Had a change of 1° in every 50 feet extended the transition curve too much, we might have adopted the curve given in Table B.

TABLE A.—CHANGING 1° IN EVERY 50 FEET.

TOTAL DEFLECTIONS FROM THE TANGENT AT ANY STATION, AND THE VALUES OF C, $Y \sin d$, AND $Y \cos d$.

0	.50	1.00	1.50	2.00	2.50	3.00
Transit.	0° 15′	0° 37½′	1° 10′	1° 52½′	2° 45′	3° 47½′
0° 15′	Transit.	0° 30′	1° 07½	1° 55′	2° 52½	4° 00′
0° 52½′	0° 30′	Transit.	0° 45′	1° 37½′	2° 40′	3° 52½′
1° 50′	1° 22½′	0° 45′	Transit.	1° 00′	2° 07½′	3° 25′
3° 07½′	2° 35′	1° 52½′	1° 00′	Transit.	1° 15′	2° 37½′
4° 45′	4° 07½′	3° 20′	2° 22½′	1° 15′	Transit.	1° 30′
6° 42½′	6° 00′	5° 07½′	4° 05½′	2° 52½′	1° 30′	Transit.
C	0° 30′	1° 30′	3° 00′	5° 00′	7° 30′	10° 30′
$Y \sin d$ in feet.	0.22	1.09	3.05	6.54	11.98	19.80
$Y \cos d$ in feet.	50.00	99.99	149.95	199.81	249.41	298.74

TABLE B.—CHANGING 2° IN EVERY 50 FEET.

0	.50	1.00	1.50	2.00	2.50	3.00
Transit.	0° 30′	1° 15′	2° 20′	3° 45	5° 30′	7° 35′
0° 30′	Transit.	1° 00′	2° 15′	3° 50	5° 45′	8° 00′
1° 45′	1° 00′	Transit.	1° 30′	3° 15′	5° 20′	7° 45′
3° 40′	2° 45′	1° 30′	Transit.	2° 00′	4° 15′	6° 50′
6° 15′	5° 10′	3° 45′	2° 00′	Transit.	2° 30′	5° 15′
9° 30′	8° 15′	6° 40′	4° 45′	2° 30′	Transit.	3° 00′
13° 25′	12° 00′	10° 15′	8° 10′	5° 45′	3° 00′	Transit
C	1° 00′	3° 00′	6° 00′	10° 00′	15° 00′	21° 00′
$Y \sin d$ in feet.	0.44	2.18	6.10	13.06	23.89	39.37
$Y \cos d$ in feet.	50.00	99.98	149.80	199.32	248.12	295.70

The stations located as above need only be considered as *temporary* ones, by means of which the true stations may be located. These may be best obtained as follows: Suppose Sta. *o* falls really at Sta. 304 + 34, then Sta. 304 + 50 can be located by stretching a tape between temporary Stations *o* and 0.50 and setting off the ordinate M (Equation 24, Sec. 80) 16 feet along it from *o*, and so on between the different stations. Values of M are given in the following table for a 1° curve. The value of M for any other curve may be considered to vary as the curvature, so that, for example, for a 9° curve the ordinate at any point will be 9 times that given in the table for the corresponding distance.

VALUES OF M FOR 1° CURVE, 50-FT. CHORDS.

Dist. from Temp. Sta.	M in feet.	Dist. from Temp. Sta.	M in feet.	Dist. from Temp. Sta.	M in feet.
2 ft.	.011	10 ft.	.035	18 ft.	.050
4 "	.016	12 "	.040	20 "	.052
6 "	.022	14 "	.044	22 "	.054
8 "	.030	16 "	.048	24 "	.054

The principal objection which can be urged against this curve is its rigidity; this is in a great measure overcome by having the option of the two sets of curves given above, one changing by 1° every 50 feet, and the other by 2°. Generally speaking, the former is adapted to curves not exceeding 7°, and the latter to curves of from 6° to 14° curvature; while for curves of from 5° to 8° either set may be employed.

Another objection which may be brought against it, and one which is often brought against transition curves generally, is that it is not worth the trouble taken in locating it. As regards this, the use of transition curves, not only theoretically but practically, is found to reduce the resistance of the curve very materially, to lessen the cost of maintenance of way, to reduce the chances of derailment, and considerably to ease the motion of the cars.

There is no need to set out the transition curves during the location, but the tangent in any instance should be run to *c* (Fig. 50) and the transit then offsetted to *a*, from which point the main curve can be located. The amount of the offset *ac*, and the distance *oc*, should be added to the notes of the curve, and also the distance *ae*, which represents *C*. The general plan of the location then shows the curves as in Fig. 16. Then when the engineer takes charge of the work for construction he has simply to "reference" the points *o* and *e*, and run in the curve by means of the above table, as easily as he would any simple curve.

98. Method II.—Another form of transition curve is that shown in Fig. 51. It is especially suitable in cases where it is more convenient to offset the curve than the tangent itself. It practically converts the original simple curve into a 3-centre one, but where the curvature of the main curve is light, it answers the purpose of easing off the curvature at its ends sufficiently in ordinary cases.

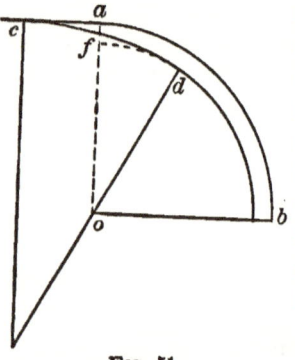

FIG. 51.

In Fig. 51, let r = radius of the original main curve *ab*.

Offset *ab* inwards by an amount $af = e$; then if R = radius of the terminal curve *cd*, we have

$$\cos fod = 1 - \frac{e}{R - (r - e)},$$

from which we can find the position of *d*; and

$$ca = R - (r - e) \sin fod,$$

from which we can find the position of c. The curve cd can then be best located with a transit from the point c.

A convenient method of applying this principle in practice is to make $e = 0.2$ foot for every degree of curvature of ab, and to make $R = 3(r - e)$; then if we make $fd = 33.9$ feet, d is the P.C.C., and

$$ca = 2(r - e) \sin fod,$$

fod being found from the formula

$$\cos fod = 1 - \frac{e}{2(r - e)}.$$

For ordinary curves ca then varies from 75 to 100 feet.

99. Method III.—Another method of substituting a 3-centre curve for a simple one, when we do not wish to change the original tangent-points, is as follows:

In Fig. 52 let o be the centre of the original simple curve afb, the radius of which $= R$; and let o_1 be the centre of the new main curve ced, whose radius $= R_1$. And let o_2, o_2 be the centre of the terminal curves ac and db, whose radii $= R_2$.

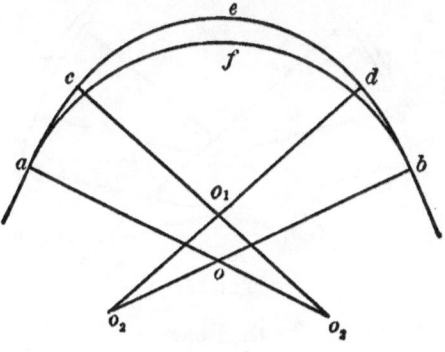

FIG. 52.

1. *Given R_1 and R_2.*
Then

$$\sin \frac{co_1 d}{2} = \frac{(R_2 - R) \sin \frac{aob}{2}}{R_2 - R_1},$$

and

$$ao_2 c = \frac{aob - co_1 d}{2}.$$

Thus we obtain the position of the points c and d.

2. *Given R_1 and $ao_2c = bo_2d$.*

Then

$$R_2 = \frac{R \sin \frac{aob}{2} - R_1 \sin \frac{co_1d}{2}}{\sin \frac{aob}{2} - \sin \frac{co_1d}{2}}.$$

The curvature of the arc ced should never exceed that of ab by more than 1° (about 50′ excess is usually a suitable amount), and R_2 should equal about $3R$.

The distance

$$fe = (R_2 - R_1) \sin ao_2c \operatorname{cosec} \frac{aob}{2} - (R - R_1).$$

Suppose, however, in substituting the 3-centre curve for the simple one, it is advisable for the points e and f to coincide as in Fig. 53.

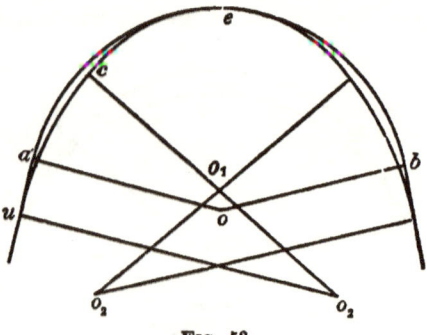

Fig. 53.

1. *Given R_1 and R_2,* we then have

$$\operatorname{vers} uo_2c = \frac{(R - R_1) \operatorname{vers} \frac{aob}{2}}{R_2 - R_1}.$$

Then a must be put back on the tangent to u, and

$$au = (R - R_1) \operatorname{vers} \frac{aob}{2} \left(\cot \frac{uo_2c}{2} - \cot \frac{aob}{4} \right).$$

RAILROAD LOCATION.

2. *Given R_1 and uo_2c*, we then have

$$R_2 = R_1 + \frac{(R - R_1) \text{ vers } \frac{aob}{2}}{\text{vers } uo_2c},$$

au being found as above.

VERTICAL CURVES.

100. We have already considered the dangers which arise from sudden changes of grade (see Sec. 29). Where these changes are considerable, amounting to, say, 0.5 p. c. in the difference of grade, it is advisable to round off the angle at the junction of the two grades by means of vertical curves. On bridge-work this should be more especially attended to. Theoretically, the curve which should be applied is a parabola, and this happens also to be the simplest form of curve to insert in practice.

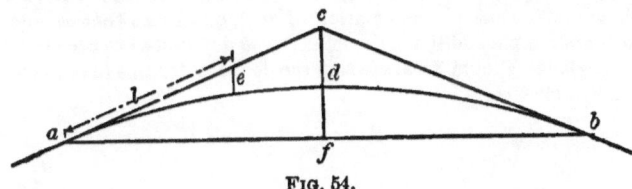

FIG. 54.

In Fig. 54 let ac and cb be two grades between which it is required to insert a vertical curve.

Now $cf = 2cd$; therefore, if the letters a, b, and c stand respectively for the elevations at those points,

$$cd = \frac{c}{2} - \frac{a+b}{4},$$

and the correction e at any other point is given by the equation

$$e = \frac{cd \cdot l^2}{(ac)^2}.$$

ac and cb are usually made about 200 feet each.

Vertical curves are not usually inserted during location, or even shown on the location profile; but the corrections for them should be worked out before the cross-sectioning begins, and the grade as shown on the construction profile should be the corrected grade.

Note.—In dealing with deflection-angles and offsets of curves, the engineer—entirely ignorant of the Differential Calculus—may often save himself a considerable amount of labor by making use of the principle of *Successive Differences*, an application of which is given in Sec. 203, Part III. Thus, e.g., the deflection-angles given in Tables A and B, Sec. 97, may be calculated up to 300 feet merely by the application of the 2d differences, and may be extended considerably beyond that amount by using the 3d differences. More especially is this method applicable in calculating offsets to a curve which may be considered to vary as the *Square* of the tangential distance, for then their 2d differences will be constant. As an example of this, the values of $(H - H')^2 \frac{sL}{27 \times 6}$, given in Sec. 130,—varying as the *square* of $(H - H')$,—have for their 2d difference 1.852, which does not change; therefore the differences of the differences of the values in the table increase *regularly*, the difference between any two values being greater than the preceding difference by this amount; thus the calculation of such a table as that is merely a matter of simple addition as soon as the 2d difference has been obtained. The engineer should be always on the lookout for this in the construction of tables, etc.

Part II.

CONSTRUCTION.

101. The Field-work of engineering during Construction may be divided into two parts, the first (A) dealing with the setting out of the work, and the second (B) with the estimating of the labor and material employed in its execution; and in this order it will be well to consider the subject.

A. THE SETTING OUT OF WORK.

102. An engineer, when given a subdivision of a road to look after during its construction, often finds merely the centre-line staked out at every 100 feet,—with hubs indicated by Guard-stakes at the transit stations,—and bench-marks every half-mile or so apart. He is provided with a copy of the location profile and of the transit-notes and bench-marks, and with the notes and plans connected with any special features in the construction on his subdivision for which he will be held responsible—such as plans of bridge-sites, culverts, etc.

If in a timber country, the first thing he has to do is to see to the **Clearing of the Right of Way**, which he does by marking out the limits—if the clearing is to be carried to the full width—by blazing the trees at distances of a hundred feet or so apart on either side of the centre-line, and inscribing the letter C.

While the clearing is being done, he usually has time to examine the country along the line with an eye to the location of culverts and the size of openings necessary, and to make a closer examination of the probable classification of the cuts than the location party probably had the opportunity of doing.

103. In order to obtain a correct idea as to what size of openings may be necessary, he is guided by the flood-marks along the water-courses; and if there is any doubt about these in the neighborhood of the line, he must follow them up until he finds some definite indication of the amount of flow, or else forms a more or less accurate estimate of it for himself, by an examination of its source.

In selecting the points for culverts and the sizes required, the engineer must bear in mind the effect of drainage upon the natural well-defined water-courses: for instance, water that before the construction of ditches ran more or less broadcast over the country,—as is frequently the case in low marshy land,—thereby perhaps in a dry season showing no indications of its existence at another time of the year, or which in a wet season may be simply indicated by a saturation of the soil, may, when conducted by ditches to the mouth of a culvert, present a very decided reality.

Often too, by cutting a small ditch, two streams can be brought together at a less cost than would be involved by the construction of two separate culverts. For a masonry culvert is an expensive article in the first place, and the usual substitute —a timber one—a still more expensive article in the long run. When the dump is low, open wooden culverts are the best to use as temporary expedients, for any defects in them are readily visible, and masonry culverts can be built to replace them with very little trouble. For small openings piping does admirably, but should be well bedded; as a temporary substitute for pipes, small plank culverts may be inserted, which may afterwards serve as a means of inserting the pipes themselves.

104. A thorough system of drainage along each side of the road-bed should be one of the first points to which the attention of the engineer should be given, for it is often possible to greatly decrease the cost of construction by constructing ditches some little time before the commencement of the work.

As regards the form and size of such ditches, it is usually sufficient to make them with slopes of 1 to 1, but with plenty of width in the base: as a rule, for each foot of water likely to be in the ditch there should not be less than three feet of base; and the rate of fall should be made as uniform as is compatible with the cost of construction. For small ditches, the rate of fall should not be less than 0.2 p. c. if possible; but a large ditch which is likely to have a depth of water of not less than one foot will draw tolerably well with a fall of only 0.1 p. c. Neither should the fall be so great as to permit scouring to any large extent.

Small extra ditches are usually staked out with centre-stakes only, and the amount of excavation calculated from the centre-heights. But for larger ones slope-stakes should be set, and if the surface is irregular it must be properly cross-sectioned.

105. It is often the case that the cross-sectioning of the work has been done by a party detached from the main location party: if so, the engineer usually has time to *check the bench-marks* and insert new ones for himself at points which he may consider suitable. These B.M.'s should not be less than 10 stations apart; their positions should be such as to do away as much as possible with turning-points. They should be marked B.M., and the elevation of each inscribed on it. At each bridge-site there should be a bench-mark close at hand. It is a good plan also, if there is time, to *check the alignment* from the transit-notes. Any error discovered, either in the levels or the alignment, should be at once reported. For discrepancies arising in the checking of the alignment by using short chords, see Part I.

106. When, however, the subdivision engineer has the *cross-sectioning* to do himself, if the construction is being started at various points on his work almost simultaneously with his taking charge, he then has his time from the very first fully occupied in taking cross sections.

The amount of work which this involves depends a good deal on the manner in which the grading is to be measured. If measured in excavation only, then it is merely the cuts that have usually to be cross-sectioned; but if measured in cut and fill, both must receive equal attention. In the former case, where borrowing has to be done, it is often necessary, however, to have the fills also cross-sectioned, for, owing to the impossibility of measuring the borrow-pits correctly, the work may have to be measured in the fills, and this must be borne in mind at the time of cross-sectioning. Also, to obtain a *correct* estimate of the over-haul it is necessary to have the fill connected with it cross-sectioned. At all points, too, where the question of the distribution of material is likely to arise, cross-sections of the fills are useful, but these need not be taken with the same accuracy as those required for the measurement of the work.

To cross-section *properly*, five men are wanted besides the engineer,—namely, a rodman, a man to carry stakes, another to drive them and another to mark them, and a tapeman,—for though the *setting of slope-stakes* is sometimes done separately from the cross-sectioning, it usually saves both time and expense to do both at once.

Before starting to cross-section, the engineer will do well to construct a small table for each different width of road-bed and set of slopes which he is likely to use, giving the "distances out" to the slope-stakes for various amounts of side-heights. For though he rapidly acquire these after a little practice,—and should be checked in his calculations of them by the rodman,—still, by having a table before him, he saves considerable mental work and insures greater accuracy. He should also be provided with a small scratch-block.

The best way to explain the method of cross-sectioning is by means of an example.

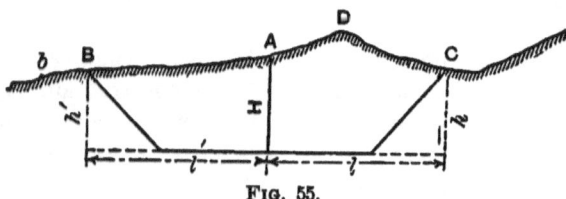

Fig. 55.

Let $bBAC$, in Fig. 55, represent a surface which we wish to cross-section. We first take the elevation at the centre A, which should correspond within a tenth or so with that given on the location profile. By subtracting the grade at the station from this elevation we thus have H, the centre cut at A. The rodman then goes to the left and holds the rod at some point b near where he judges the slope-stake will come. If on obtaining the side-height for b it is found that the proper distance out from A for this height does not agree with the distance out as actually measured, other points must be tried until a point is obtained, such as B, where these two correspond. An error of only a few tenths in distance can be estimated for by eye without taking a separate reading to correct for it, so that two or three trials are usually all that are required to fix the position for the slope-stake; and on comparatively level ground the point can be usually hit off by a good rodman at the first trial.

Similarly on the right the point C must be fixed.

If there are any decided irregularities in the surface, such as is represented at D, the elevations of such points must also be taken.

The following rules give all that is required as regards the actual levelling:

1. **When H.I. is above grade.**—*If the rod-reading exceed the difference in elevation of the H.I. and Grade, the excess = the fill; but if it is less, the deficiency = the cut. Consequently, when the rod-reading = the difference of H.I. and Grade that point is a* **Grade-point.**

2. **When H.I. is below Grade,** *the rod-reading + the difference of H.I. and Grade = the fill.*

Cut is always indicated by a *positive*, and **Fill** by a *negative* sign.

The following is a good form for keeping the notes:

Sta.	L.	C.	R.		B.S.	F S.	H.I.	Elev.	Grade.	Remarks.
1020	$\frac{0.0}{7.0}$	+1.0	$\frac{+3.0}{14.5}$		1.3		102 30	101.0	100.00	+1 p.c. grade Roadbed 14′ in fill, 20′ in cut. Slopes 1½ to 1.
1021	$\frac{-1.0}{8.5}$	0.0	$\frac{+3.3}{6.0}$	$\frac{+1.0}{11.5}$		1.3		101.0	101.00	
1022	$\frac{-3.0}{11.5}$	-2.0	$\frac{0.0}{7.0}$			2.3		100.0	102.00	

There is no need to work out the elevations in the field, but so doing in the office afterwards forms a useful check on the work, since H.I. − F.S. (which of course *is* the elevation) should agree within a tenth or so with the sum of grade ± centre-height, F.S. representing the rod-reading at the centre. We see from the above that it is the *Difference of H.I. and Grade* which is the foundation of the calculation at each station, and this, when worked out for the next station after a turning-point, can be modified for the succeeding stations by merely adding or subtracting the difference in grade. Thus the calculation is simpler than it at first appears from the above rules.

The **slope-stakes** should be marked S.S. on the outer sides and the numbers of the stations on the inner. The **centre-stakes** should have the cut or fill marked on them.

As to the **points at which cross-sections should be taken**, the rodman in selecting them should bear in mind that it is not necessarily the highest or lowest points that are required, but those points which, when joined by straight lines, will give the contents as nearly as possible equal to the true volume. It is impossible as well as unnecessary to take account of many of the small irregularities which occur, but by a judicious selection of points these may to a considerable extent be made to counteract each other. Where the contents are calculated by "average areas"—as is usually the case—we can easily find from Sec. 130 what limit should be adopted as regards the difference in centre-heights and widths between the slope-stakes of two cross-sections, in order that the error in the volume as calculated shall not exceed a certain amount. For *exact* work a difference of two feet between the centre-heights of two adjoining cross-sections is about the limit which should be allowed; but in ordinary practice we may say that a cross-section should be taken every 50 feet when the difference in centre-height amounts to about 5 feet. This is, of course, mainly to reduce the errors which arise from using an approximate method of calculating the quantities, and not to take into consideration the irregularities of surface. To counteract as much as possible these latter, judgment in the selection of the cross-sections has a better effect than labor spent in obtaining a large number of cross-sections a few feet apart. They should also be taken whenever "grade" occurs on either the edge of the road-bed or in the centre; and whenever a cross-section is taken where a grade-point falls in the road-bed its position must be obtained. For if a grade-point is the only point obtained at any station, it necessitates assuming centre- and side-heights afterwards in working out the contents, in order to make use of that grade-point, so that it is much more satisfactory—and in the end involves no more work—to obtain these heights by direct measurement.

There is of course no need to take cross-sections any closer together on a curve than on a tangent, as may be easily seen from Sec. 134.

When in doubt as to the material in a certain cut, i.e., as to whether it is earth or rock, etc., it is best to cross-section it for the usual earth-slopes and have it stripped to that width in one or two places; if then rock is encountered in a solid bed, the rest of the cut may be cross-sectioned for rock, and as soon as the rock is reached the earth trimmed off to its proper slopes before the rock is worked. This of course necessitates a cross-sectioning of the rock surface as well as of the original ground-surface, and these cross-sections should be taken *at the same stations*, so as to facilitate the calculation of the respective volumes of earth and rock.

107. The referencing of the P.C.'s and P.T.'s is a part of the engineer's work which must also be attended to before construction begins. **Reference-points** should be placed, two on each side of the alignment, at angles of about 45° with it, and sufficiently distant to be free from all chance of disturbance during construction; the point referenced thus lies at the intersection of the two lines joining the opposite points. Sometimes, however, especially on side-hill work, it is necessary to place all the reference points on one side of the track, in which case the apex of the angle formed by the lines passing through each pair of reference points is the point referenced. Each reference-point should be marked R.P. on a guard-stake set beside it, and the magnetic bearings and distances of the points entered in the notes.

108. **The Staking out of Borrow-pits** consists in driving stakes at the corners of the proposed pits, and obtaining elevations of the ground-surface so as to form the upper line of a set of parallel cross-sections of the pit, the lower line being obtained by taking levels immediately under those taken on the surface, when the excavation is completed. In order that the bottom levels may be properly connected with those taken on the surface, reference-points must be established. The simplest way of doing this is by driving hubs, say 10 feet back from the edge of the pit, in the line of each cross-section. By taking the cross-sections 27 feet apart, as is often done, there is some little labor saved in calculating the contents, since the mean of any two cross-sections in square feet equals the volume between them in cubic yards.

A sketch plan of each pit should be made in the note-book, and properly lettered to accord with the notes.

109. **Staking out Foundation-pits for Culverts**, either masonry or timber, consists of setting stakes at the corners as

given by the foundation plan and marking on each stake the cut necessary. A sketch of each pit should be made in the note-book, and of course the amount of cut at each stake recorded. When the foundation consists of timber, the pit should be low enough to insure the timber being at all times, if possible, kept under water, or at any rate moist; about 18 inches is the average depth for foundation-pits for wooden culverts on Railroad work. In staking out, it should also be remembered that the culverts should not have a fall of more than, say, 1 in 10, so that when the ground slopes transversely to a greater extent than this the culvert must be put on the skew so that its inclination will not exceed this amount. If the depth of the foundation-pit exceeds 4 or 5 feet, it should be staked out a foot wide all round to allow room for working.

110. Setting out Bridge-foundations.—When a bridge is *on a tangent* there is no difficulty about staking out the foundation-pits, that needs particular mention. The work is usually best done with a transit and tape from the centre-line,—an optical square comes in very handy for this,—the offsets being obtained by scale or otherwise from the foundation plan. In this way there is less liability to make an error than in any other, since each point is set out independently of the previous ones. When the material is not likely to stand vertically, it should be given a slope sufficient to warrant its stability. If there is not room to admit of this, then of course the sides must be shored-up in some way.

When, however, the bridge is *on a curve*, if the span is short, it is from the tangent at the centre of the bridge that the offsets must be set off. In dealing, however, with bridges of comparatively long spans, the centre of the curve on the bridge will by no means coincide with the centre of the structure, as is shown by Fig. 56.

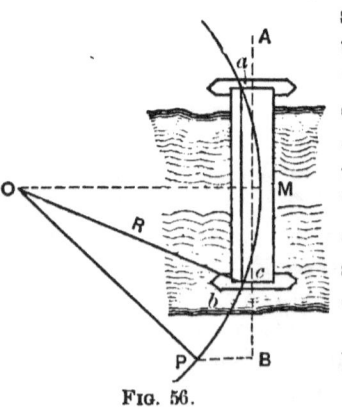

FIG. 56.

Now AB will be the centre-line of the bridge, where

$cb = \frac{1}{2}$ ordinate at M to ab

(see Equation 23, Sec. 80); so that the true centres of the piers lie considerably outside the centre-line at those points. If any pier, as c, is inaccessible, c (its centre) may be located as follows:

In the centre-line of the track take some accessible point P, and set off PB perpendicular to AB, making

$$PB = R \text{ (vers } POM - \tfrac{1}{2} \text{ vers } bOM);$$

then will

$$Bc = R\left(\sin POM - \frac{ab}{2}\right).$$

C may then be located either by direct measurement from B, or by intersection.

In setting out bridge-foundations great care should be given to a thorough system of referencing all important points, and the reference-points must be so selected as not to be obstructed by staging or scaffolding during the progress of the work.

111. Setting out Trestlework.—In locating the position for the piles in low pile-bents, it is sufficient to locate the centre of each bent and then set off the positions for the piles by measuring out from the tangent at the centre, finding the angle by eye; if possible, the position of each pile should be marked with a stake.

When piles are being driven on a curve by a floating pile-driver, in water too deep to drive stakes, the centre of each bent must be given by the intersection of the lines given by two transits, as in Sec. 76.

If, however, the trestle is on a tangent, by placing pickets on either bank in line with each row of piles the centre for any pile can be given without the aid of an instrument; or pickets can be so set that the pile-driver can line itself in without the assistance of any one on the bank; the distances between the bents may be taken by measurement from one bent to the next. In the case of framed bents resting on sills, it is advisable to have the sills brought to a solid foundation at *about* an indicated elevation before the framing-bill is made out: in this way a firmer foundation is often obtained at a cost of less labor than if the exact elevation for the sills was prescribed. The sills for each bent should then be accurately levelled and centred.

In dealing with high trestles, the transverse centre-line of each bent should be referenced, the reference-points being at a considerable distance from the bent itself, so as the better to permit the line being carried to a high elevation in the structure if required. The length of the chords should be corrected according to Sec. 76.

Where pony-bents are used they should be so skewed around as to conform with the contour of the ground; they must be accurately levelled before the sills are laid on.

In giving points for "cut-offs" in piling out of reach, the pile should be blazed and a tack driven into it, the distance above the tack—which should be in full feet—being inscribed. The position of the tack is best found as follows: For example, let the difference of H.I. and grade = 6.11 feet; then if the point of cut-off is 2 feet below grade, and it is wished to put in the tack so as to read "5 feet below cut-off," we must read on the rod 0.89 foot. The position of the tack is then at the foot of the rod.

112. Setting out Tunnels.—This is work which often needs considerable time and care, in order that the results obtained may be satisfactory.

Let Fig. 57 represent the section of a tunnel in course of construction.

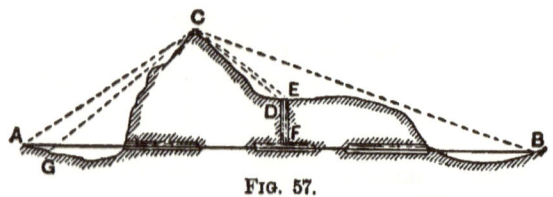

FIG. 57.

The first thing to do is to establish some point C in the alignment from which a good view—if possible—may be had of the mouths of any shafts which it may be required to sink, and also of two distant points A and B, also in the same straight line. If the instrument is then set up at C and the telescope clamped on to A, on reversing it the point B should be intersected. By repeated trials the three points A, B, and C are then established in the same straight line, and these points should be permanently marked.

In order to obtain the centre-line of the tunnel, say at the left end, another point G in the same line as AB must be given, and the centre-line is then obtained by the production of AG.

But suppose the work is to be carried on also from one or more shafts as EF, then the alignment has to be "dropped" from ED to the elevation of the tunnel at F, and in this operation the greatest care is necessary. There are three or four ways in which this can be done, but the following is that usually adopted for tunnel-work, as it admits of greater accuracy than the others, which are more suitable for simpler mining operations:

Two instruments such as that shown in Fig. 58 should be firmly bolted on either side of the shaft as D and E, and near to its edge, both being lined in vertically over the centre-line of the tunnel.

Each instrument consists of a plate p—with a narrow vertical slit in it and scale s attached—which can be moved sideways by means of the screws a and b, so that it can be set to any desired reading on the scale—the scale being read by a vernier v attached to the main body of the instrument. Having set these two instruments approximately in line, then, by a series of observations taken at different times,—so as to counteract as much as possible the varying conditions which affect each separate sight,—ascertain for each instrument the *mean* of the readings. Having then set the plates to give that reading, the centres of the vertical slits coincide with the mean alignment.

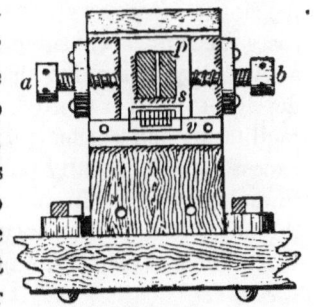

FIG. 58.

Two fine steel wires must then be carried from one slit to the other, each being placed against the vertical edge, so that they form two parallel lines, close together, across the shaft, one on each side of the alignment. Midway between these two wires, and as near to the edge of the shaft as possible, but on opposite sides of it, two fine copper wires should be passed, long enough to reach down to the tunnel at F, and to the ends of these two *heavy* plumb-bobs should be attached. The wires should be enclosed in wooden tubes to protect them from currents of air, falling water, etc. The plumb-bobs themselves should be immersed in buckets of water to lessen their oscillations. Scales should then be placed so as to read these oscillations slightly above the plumb-bobs. The mean of these *sets* of readings then gives a point on the alignment, and from the two points so obtained the centre-line of the tunnel may be extended in either direction by first establishing a point in one direction, and then in the other; and these points can then be checked by observing whether all four are in the same straight line: if found to be correct, they should be permanently established. The *levels* may be dropped by means of a steel tape, with which the levelling-rod used has been previously compared.

The length of the tunnel may be found either by direct measurement (breaking-chain) or by triangulating.

In *locating a tunnel*, it should be remembered that it is usually cheaper to open a cut at depths under 60 feet than to bore. In many clays, however, a cut of this depth would be barely practicable owing to the increase in the inclination of the slopes necessary on account of the depth itself, and in such cases the limit is considerably less than this. As regards the advisability of sinking shafts, it is mainly a question of the depth of shaft required, the need of ventilation, and the facilitating the transport of material. Where the depth is not excessive it is usually policy to sink several shafts in a long tunnel, and work from each independently, for the work is thereby considerably hastened, and after its completion the shafts themselves form admirable means of ventilation.

Side-drifts, where they are possible, accomplish the same results as shafts, and are usually to be preferred to them on account of less risk to life and property during construction, and their convenience afterwards.

Where the alignment has not to be carried to any great distance from the points dropped to the bottom of a shaft as above described, it is better to sink the shaft a few feet on one side of the centre-line, and to reach the tunnel from it by means of a cross-heading.

The centre line in the tunnel is best given by points on the roof from which plumb-lines can be hung when required.

113 Giving Grade and Centres forms a very large portion of the work to be done by the engineer during construction. The giving of "grade" may be greatly facilitated by having stakes driven to grade, from which at any future time the levels may be given with a hand-level –an instrument highly useful during railroad construction. To have to carry a heavy level for several miles just to give grade at two or three stations, as is frequently done, is absurd. By having a bubble-tube attached to the telescope of the transit a considerable amount of trouble may also be saved, and with it the elevations can be given quite as correctly as are ever required on a railroad dump.

In setting grade-stakes, allowance must be made in dealing with material which is likely to shrink in order to allow for it. The amount of the **Shrinkage** depends considerably on the pressure to which the material is subjected, consequently on the height of the fill: as an average, however, in earthy soils the linear contraction is about 10 p. c., so that a 10-foot fill should be "put up" 1 foot above grade. In dealing with wet or frozen soils greater allowance should be made, but with dry sandy material, less.

The allowance also depends very largely on the manner in which the dump is constructed. A dump well trodden by horses usually shrinks very little, and in many such cases there is no need to allow for shrinkage at all; but where the work is put up by tipping or shovelling, double the allowance may in some cases be none too much.

The **increase in bulk in rock**, as well as the shrinkage of earth, necessitates an allowance being made when arranging for the distribution of material. A good general rule for this is, that 10 yards of earth in excavation make 9 yards in embankment, and 10 yards of rock in excavation make 17 yards in embankment.

As regards "giving centres" during construction, it should be seen that the slope-stakes are intact, and then by their means the centres for a cut or fill may be usually obtained from the cross-section notes, without the trouble of setting up the transit, with accuracy quite sufficient to enable the contractor to proceed with his work.

114. Difference of Elevation on Curves.—The centrifugal force brought into play by the inertia of the train when going round a curve must be counterbalanced by a more or less equal and opposite force in order to prevent the flanges of the outer wheels being pressed too severely against the rails The simplest way of bringing a counteracting force into play is to make use of a component of the weight itself, which may be done by canting the track as in Fig. 59.

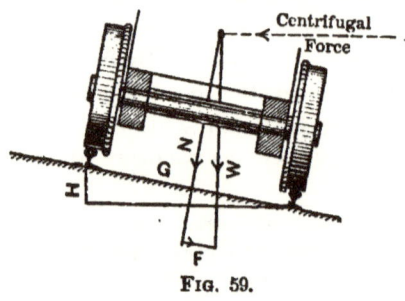

Fig. 59.

Thus, if the force W, representing the weight of a car, be resolved into its rectangular components N (normal to the track) and F (parallel to the track), we see from Sec. 7 that F is proportional to $\frac{H}{G}$, H being the difference in elevation of the rails, and G the gauge—or more strictly, the distance from centre to centre of rails. Now the value of the centrifugal force in pounds equals $\frac{v^2}{32R}$, where $v =$ velocity in feet per second, and R the radius of the curve; so that when there is no tendency to tip over on either side—if we assume, as we may well do in practice, that F is the component parallel to the centrifugal force—we have

$$\frac{H}{G} = \frac{v^2}{32R}; \text{ therefore } H = \frac{Gv^2}{32R}.$$

So that, substituting for R the value given in Sec. 71, and substituting V, velocity in miles per hour, for v, we have

$$H = .00067 \, GV^2 \sin D;$$

or, as an approximate formula, easy to remember, we have

$$H = \frac{GV^2}{15R} \text{ (nearly)}.$$

If we take $G = 4'\ 8\frac{1}{2}''$, we then have

$$H = .0032 \, V^2 \sin D.$$

The following table, abbreviated from that given by Mr. Searles, calculated for the value of *F parallel to the centrifu-*

gal force, and for a distance from centre to centre of rail = 4′ 10¾″ (suitable to the 4′ 8½″ gauge), gives the difference in elevation of the two rails in feet, at various speeds for different degrees of curvature.

Vel. in m. p. h.	Degree of Curve.								
	1°	2°	3°	4°	5°	6°	7°	9°	12°	16°
10	.006	.011	.017	.023	.029	.034	.040	.051	.069	.091
20	.023	.046	.069	.091	.114	.137	.160	.206	.274	.365
30	.051	.103	.154	.206	.257	.308	.359	.460	.611	.809
40	.091	.183	.274	.365	.455	.545	.634	.811	1.069	—
50	.143	.285	.427	.568	.707	.844	.979	—	—	—
60	.206	.410	.612	.811	1.006	1.196	—	—	—	—

A convenient rule, much used in practice for a gauge of 4′ 8½″, is, that the difference in elevation equals one half inch for every degree of curvature.

In order to allow for the difference in elevation *on the dump*, the road-bed should have its outer edge higher, and its inner edge lower, than grade. To allow for it *on trestles*, whether in pile-bents or framed bents, the posts must be cut so as to give the required inclination to the cap on which the stringers rest: the batter of the batter-posts and the verticality of the upright posts remain unchanged.

It is usual to adopt a difference in elevation in the rails suitable to the mean speed of the trains which pass over them: the consequence of which is, that the rails on *both* sides get worn, but in different ways—the outer ones by the fast trains and the inner ones by the slow trains. The coning of wheels, which was at one time largely resorted to, is rarely used now on account of the increased oscillation and concussion (see Sec. 4) to which it gave rise, so that the flanges of the wheels, by means of their pressure against the inner sides of the rails, have themselves to keep the balance between the centrifugal force and the component of gravity which is set to counteract it, more or less. In curves uneased by transition curves, the difference in elevation at the P.C. and P.T. must be at least equal to what it is at any other part of the curve, so that it must begin some little distance back on the tangent and increase gradually until it reaches its maximum at the P.C. or

P. T., as the case may be. For a 3° curve it is usually sufficient to begin the difference in elevation about 100 feet back, and for a 10° curve about 200 feet back on the tangent. When transition curves are used, they must be treated with a difference in elevation at all points more or less suitable to their curvature; but where the transition curve is merely a simple curve inserted to ease the approach to a sharper one, the difference in elevation for the terminal curve must begin back on the tangent as above, and for the main curve some little distance back on the terminal curve, so as to admit of its reaching its maximum at the P.C.C.

It is usual to slightly increase the *gauge on curves*, generally by about ⅛" for every degree of curvature up to 5°.

115. *Inspecting the Grading.*—The engineer should, if possible, pass over every portion of his subdivision at least twice a week, and the oftener the better. In open country there is comparatively little chance of having the dump badly put up owing to lack of supervision, except perhaps through the use of a superabundance of "sods;" but in timber country where there is plenty of grubbing to be done, and the work is largely let as "station-work," the engineer must be constantly on the lookout for the presence of roots and stumps in the dump. In winter too, snow, frozen moss, etc., at the bottom of a fill serve admirably as a *temporary* means of bringing it up to grade. He should see that there is a fair line of stumps at the side of the track after the completion of the work in places where grubbing has occurred, or that they have really been burnt; and when there is snow on the ground he must have it swept well to the side before the filling is begun. He must see that the ditches on either side of the embankments, etc., as well as those in the cuts themselves, are taken out properly, and thoroughly cleared of all obstructions, that the slopes are neatly dressed off and well out to the slope-stakes. For the final inspection of the road-bed, grades and centres must be carefully run, and the width tested wherever it appears lacking. All litter along the side of the track must be cleared away or burnt, and anything in danger of falling on to the road-bed removed. About this latter injunction the engineer cannot be too careful, and when in doubt as to the stability of a piece of rock or an overhanging tree, he should have it removed at any cost. He must also remember that a rock or tree which at the time of inspection looks tolerably firm, may be a considerable source of danger after the disintegrating effects of a hard winter, or a season of heavy rains, and that it costs very much less to have it removed during construction than at a later period.

116. *Running Track-centres and setting Ballast-stakes.*—Where the ballasting is done before the track is laid, ballast stakes must be driven every 50 feet, so that their tops indicate the elevation of the top of the ballast. They should be placed on either side of the centre-line at the foot of the ballast-slopes. Centre-stakes should also be set every 100 feet apart on tangents and every 50 feet apart on curves, to guide the track-layers; tacks should be inserted in them.

When the track is laid without first ballasting, a line of centres must be given before the track is laid, and usually afterwards as well, to guide the surfacing gang, for the centres previously put in are almost sure to have been knocked out in laying the track.

It sometimes happens in hasty work that the engineer who has the track-centres to run cannot get *his* centres to coincide with the centre of the dump or with the centres of the bridges. As regards the centres on the dump, he must use his own judgment as to what is best to do: if it is clear that the dump is out of line, he must stand by his own centres; but if otherwise, it is usually better for him to increase or ease his curvature a little, so as to make it conform with the centre of the road-bed. On bridges or open culverts he must *make* his own centres fit the centres of the structures, and if this cannot be done without seriously affecting the adjacent track, the case must be reported at once.

117. *Permanent Reference-points.*—After the track is laid large hardwood stakes—or better still, stone monuments—should be set to mark the P.C.'s, P.C.C.'s, and P.T.'s. They should be placed on the outer side of the curves, at right angles to the track, usually about 5 or 6 feet from the centre.

TURNOUTS AND CROSSINGS.

118. In dealing with the subject of turnouts and crossings, we will assume that the Common Stub Switch is used, since it is the simplest, and the formulæ for it are readily applied to any other form of switch.

Let Fig. 60 represent a turnout from a straight track, A and a forming the "heel" and B and b the "toe" of the switch.

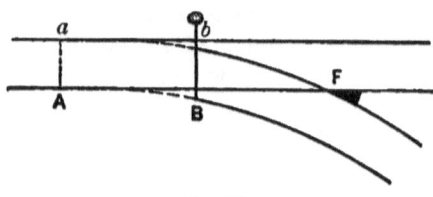

FIG. 60.

Then if
G = gauge,
N = number of the frog,
F = "Frog angle,"
   = Angle of Intersection at F,

R = radius of turnout curve,
AF = frog distance,
AB = length of switch-rail,
D = degree of curve,

we have

$$N = \frac{\cot \frac{F}{2}}{2}, \qquad \tan \frac{F}{2} = \frac{G}{AF},$$

$$AF = 2GN, \qquad AF = \left(R + \frac{G}{2}\right) \sin F,$$

$$R = 2GN^2, \qquad R = \left(AF \csc F\right) - \frac{G}{2},$$

$$AB = \sqrt{4GN^2 \times \text{Throw}}.$$

The *throw* according to Sec. 78 $= \dfrac{AB^2}{2R}$.

The *number of a frog* may of course always be found by measuring the tongue; thus if at a certain point we find its width to be 5 inches, this divided into the distance of that point from the theoretic point of the tongue gives the number of the frog; thus if that distance were 4' 2", it would be a No. 10 frog.

RAILROAD CONSTRUCTION.

The following table gives these values for a gauge of 4 8½" and a throw of 5".

N	F	AF in feet.	R in feet.	D	AB in ft.
4	14° 15'	37.66	150.66	38° 46'	11.2
5	11° 25'	47.08	235.40	24° 32'	14.0
6	9° 32'	56.50	338.98	16° 58'	16.8
7	8° 10'	65.91	461.58	12° 27'	19.6
8	7° 09'	75.33	602.62	9° 31'	22.4
9	6° 22'	84.74	762.70	7° 31'	25.2
10	5° 43'	94.16	941.60	6° 05'	28.0
11	5° 12'	103.58	1139.84	5° 02'	30.8
12	4° 46'	112.99	1355.90	4° 14'	33.6

This table may be applied to *other gauges;* F of course remaining unchanged, AF and R will vary directly as the gauge; D will, of course, vary inversely as R. Thus for a 3-foot gauge and a No. 9 Frog we must multiply the above values of AF and R by $\frac{3.000}{4.708} = .637$; and the above value of D must be multiplied by $\frac{4.708}{3} = 1.57$. AB is of course dependent on the value of the throw adopted.

119. Suppose, however, that the turnout instead of starting from a straight track, as in Fig. 60, starts from a curve as in Figs. 61 and 62; then we may assume that when the main curve and the turnout curve are both in the *same direction,* that the case, as regards the position of the frog, etc., is equivalent to a turnout from a straight track, the curvature of the turnout curve being equal to the *difference* of the curvature of the main and of the turnout curve; and if in *opposite directions,* then the curvature of the turnout curve may be taken as being equal to the *sum* of the curvatures.

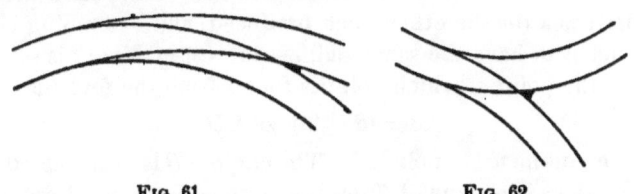

FIG. 61.   FIG. 62.

Suppose we have two parallel tracks AD and CB, as in Fig. 63, which we wish to join by a crossing; or, having the track AD only, we wish to insert a turnout AB which shall connect the side track B with the main track AD. Since the former case differs only from the latter in the fact that the dotted portion C, with the accompanying frog, is omitted, the two cases may be treated together as follows:

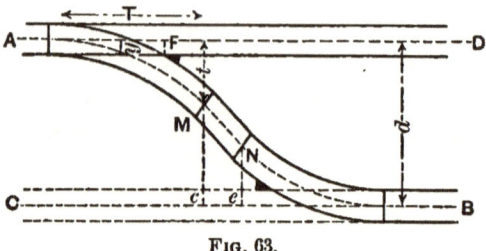

Fig. 63.

Starting from the centre-line AD with a given frog number, we select a certain length n, expressing the length of the branch AM in terms of 100-foot stations. The length of the offset t at M is then given, according to Sec. 78, by the formula

$$t = R \text{ vers } nD,$$

and the distance along the track AD to this offset equals

$$T = R \sin nD.$$

Thus by setting off the offset t at a distance T along the tangent from A, we locate the point M. The position of the frog at F is found by taking from the above table the value of AF, and measuring it off along AD, offsetting F by an amount equal to half the gauge.

Another offset $y = \frac{1}{4}$ gauge may also be set off at a tangential distance $= \frac{1}{2}AF$. These points, together with the toe of the switch, are usually all that are wanted in the curve AM. The length of any other offset, if required, may be found from Sec. 78.

The offset t is then produced across to the centre of the other track (or the other track produced) and—assuming both branches to have the same radius—the offset $Ne = t$ is set off from the point e, which point is found from the formula

$$ce = (d - 2t) \cot nD.$$

We thus have the point N. The curve NB is then located by using the same value of T, and the same offsets as before, only of course in reverse order.

By obtaining n from the formula

$$\text{vers } nD = \frac{d}{2R},$$

which gives its limiting value, we have a simple reverse curve without the intervening tangent MN : but this is bad practice when it can be avoided.

Should the radius of NB be required different from that of AM, the tangential distance for NB must then be calculated afresh.

The advantages of this method are, that any length of intervening tangent can be used,—provided that the curves are carried up to the frogs,—so that the engineer can select any value of n for himself; and with simply a tape, he can locate the crossing in a manner a good deal simpler than the ways ordinarily in use.

120. As an example, let $d = 40$ feet and let No. 8 frogs be used; and suppose we select 1.3 as a value for n. Then from the table, $AF = 75.33$, $R = 602.62$, and $D = 9° 31'$,—the gauge being 4' 8¼".

Then from the above formulæ we have
$$nD = 1.3 \times 9° 31' = 12° 22',$$
$$t = 602.6 \times \text{vers } 12° 22' = 14 \text{ feet},$$
$$T = 602.6 \times \sin 12° 22' = 129 \text{ feet},$$
$$ce = 12 \times \cot 12° 22' = 54.7 \text{ feet},$$
and $y = 1.2$ feet.

The notes for the setting out of the crossing may then be arranged as follows:

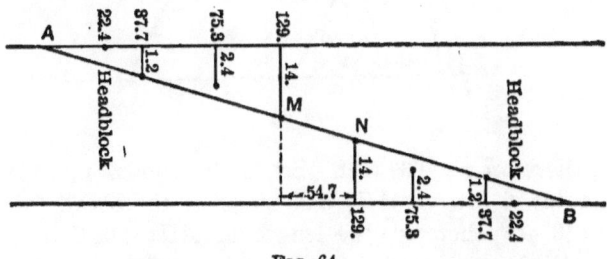

Fig. 64.

When the distance between the two tracks is great, the crossing should be run in with a transit.

121. If the turnout or crossing falls on a curve, it is best to locate it with a transit according to one of the two following methods:

1. If the curvature of the main track is tolerably sharp and the distance d between the centres of the two parallel tracks comparatively small, we can avoid the insertion of a reverse curve without materially lengthening the crossing as follows:

In Fig. 65 let $D =$ the degree of the turnout curve AC,
$R =$ radius of the outer track A,
and $r =$ radius of the turnout curve AC

The length of AC may then be found in terms of nD, thus:

$$\text{vers } nD = \frac{d}{R-r};$$

and the length of the tangent equals

$$CB = (R - r) \sin nD.$$

For example, let the outer track A be on a 4° curve; then $R = 1433$, and let $d = 40$ feet, and the given frog number for the main curve $= 11$.

Then, according to Sec. 119, D for the turnout curve must be that value which is required to make the difference in curvature of the track A and the curve AC equal about 5°, both curves being in the same direction; and since this value

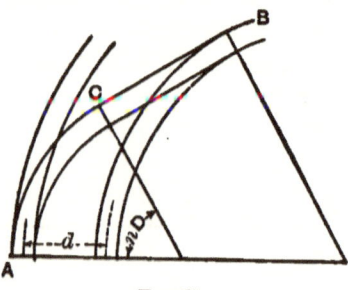

Fig. 65.

is 9°, therefore $r = 637$ feet. Set the instrument up at A and locate the 9° curve AC; and since by the above formula $nD = 18° \ 15'$, therefore the length of $AC = 202.7$ feet, and similarly the length of $CB = 249.2$ feet. Thus we find the point B.

To run from B to A would be simply a reversal of the above.

The frog for the track B will of course be that suitable to a turnout radius equal to the radius of the track B.

But suppose this method would in any particular case cover too much ground, or be unsuitable in some other respect, we can then use the following one, which, though involving the use of a reverse curve, is well enough for station-yards, etc., where no high speeds are attained.

2. In Fig. 66 let R = radius of the inner track B,
   r = radius of branch CB,
   r_1 = radius of branch AC.

Then
$$\text{vers } BHC = \frac{d\left(R - r_1 + \dfrac{d}{2}\right)}{(R + r)(r + r_1)},$$

from which we can find the length of the branch BC; and

$$\text{vers } BOA = \frac{d\left(r - r_1 - \dfrac{d}{2}\right)}{(R + r)(R + d - r_1)};$$

and since the angle
$$AEC = BOA + BHC,$$

we can thus find the length of the arc AC, and locate the crossing with the transit, starting from either end A or B.

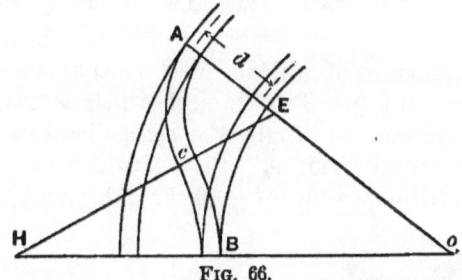

FIG. 66.

In order to use frogs of the same number for tracks A and B, we must have the change of curvature at A equal to that at B. The positions of the frogs may be found according to Sec. 119.

The positions of the frogs may be found according to Sec. 119.

In the case of a *Double Turnout* the engineer can, by applying the formulæ given above, always locate it with accuracy sufficient for ordinary purposes, without the aid of special formulæ. The length of switch-rails given in Table in Sec. 118 are the *proper* lengths for a 5" throw, but in practice a difference of 5 feet or so in the length of the rail will be of very little importance. In the same way there is no necessity for the frog to have exactly the number which it should have according to the table. The laxity which is allowable in these matters depends on the speeds at which the trains are likely to pass over the switch.

122. Curving Rails.—The following table gives the mid-ordinates in inches for curves of various lengths. Rails should also be tested for *Uniformity of Curvature* by testing one half of their length for ¼ of the mid-ordinate. (See Sec. 80.)

Deg. of Curve.	Length of Rails in Feet.						
	30	28	26	20	18	14	10
	In.	In.	In.	In.	In.	In.	In.
1°	.240	.192	.156	.096	.072	.048	.024
2°	.456	.408	.348	.204	.168	.096	.048
3°	.696	.612	.528	.312	.264	.144	.072
4°	.948	.828	.720	.420	.348	.216	.108
5°	1.19	1.08	.888	.528	.420	.264	.132
6°	1.40	1.22	1.06	.624	.504	.312	.156
7°	1.64	1.44	1.25	.732	.588	.360	.180
8°	1.90	1.64	1.43	.840	.672	.408	.204
10°	2.35	2.05	1.78	1.04	.852	.540	.264
12°	2.83	2.47	2.15	1.26	1.02	.636	.312
14°	3.30	2.87	2.48	1.46	1.19	.732	.360
16°	3.76	3.28	2.83	1.67	1.36	.840	.420

123. Expansion of Rails.—Steel expands about 1 part in 150,000 for each degree Fah. through which its temperature is raised; so that for 30-ft. rails the spaces between their ends should vary from about $\frac{1}{16}''$ at a temperature of 120° F. to about $\frac{5}{16}''$ at a temperature of − 40° F. This must be *carefully* attended to.

B. THE ESTIMATING OF LABOR AND MATERIAL.

124. The Expense of Grading is of course almost entirely dependent on the cost of the labor expended on it, the value of the material not entering into the question; so that estimating the cost of it is simply a matter of ascertaining the time and wages which are absorbed in its execution.

The following notes on the subject of handling earth and rock, which are taken from Trautwine on Excavations and Embankments,—than whom possibly no better authority could be quoted,—serve to show the relative cost of the different processes through which the material has to pass before being finally disposed of in the embankment; and, consequently, from them the aggregate cost may be obtained with a greater or less amount of precision. These processes we will consider in the order in which they occur, taking as the standard of *wages* $1.00 *per working day* of 10 hours, and the expense of a horse as $0.75 (including Sundays).

A. THE COST OF EARTHWORK REMOVED BY CARTS.

1. Loosening the Earth ready for the Shovellers.— A *two-horse plough*, with two men to manage it, will loosen about 250 yards per day of strong heavy soil, about 500 yards of common loam, or about 1000 yards of light sandy soil; thus the cost of loosening these materials per cubic yard will respectively be about 1.5 cents, 0.8 cent, and 0.4 cent—i.e., assuming the total cost of the plough and men and horses connected with it to be about $3.87 per day. When *a four-horse plough* is needed, as in dealing with stiff clays or cemented gravel, the cost runs up to about 2.5 cents per cubic yard.

Loosening by *picks* costs about three times as much as by ploughs, where the latter can work to advantage. The amount which a man can loosen with a pick in a day varies from about 14 to 60 yards, according to the material.

2. Shovelling the loosened earth into carts.— The shovellers are usually actually at work from 5 to 7 hours out of the day. If we assume that each cart carries, as a working load, $\frac{1}{3}$ cu. yd., a shoveller can load it in from 5 to 7 minutes, according to the nature of the material; and suppose he is actually shovelling for 6 hours out of the day, then in the course of the 10 hours he handles about 24 yards of light sandy soil, 20 yards of loam, and 17 of heavy soil at the cost of 4.2 cents, 5 cents, and 5.8 cents, respectively.

3. Hauling away the earth, dumping and returning. —The average speed of horses when hauling is about 200 feet per minute, so that every 100 feet of *lead* occupies about one minute; dumping and turning occupies about another 4 minutes; so that the number of trips per cart per day equals

$$N = \frac{M}{4+L},$$

where $M =$ number of minutes in the working day (here 600) and $L =$ length of the lead in terms of 100 feet. Then $\frac{1}{3}N$ equals the number of cubic yards moved by each cart per day; and $\frac{1}{3}N$, divided into the total expense of the cart per day, gives the cost of hauling per cubic yard. Assuming that one driver attends to four carts (doing nothing else), the total cost per cart may be set at $1.25 per day.

4. **Spreading on the embankment.**—The cost of this varies considerably, but may be said to average about 1½ cents per cu. yd. When the earth is dumped over the end of the embankment, or is "wasted," ¼ cent per cu. yd. should be allowed for keeping the dumping-places clear.

Keeping the hauling road in good order.—This is an item highly expensive if neglected, but if well looked after, $\frac{1}{10}$ cent per cu. yd. per 100 feet of lead is usually sufficient to cover it.

Wear and tear of tools.—"Experience shows that ¼ of a cent per cubic yard will cover this item." This also includes the interest on the cost of the tools.

Besides the above, 1½ cents per cubic yard should be added to cover the cost of *superintendence and water-carriers*, and about ⅛ cent for extra trouble in ditching and trimming up.

As regards the profit to the contractor, it may be set down as from about 6 to 15 per cent, according to the magnitude of the work and the risks incurred; out of this he usually has to pay the clerks, store-keepers, cost of shanties, etc., but these as a rule cover their own expenses.

The following table gives the cost, exclusive of profit to the

Length of Lead in feet.	Cu. yds. hauled per day per cart.	TOTAL COST, PLOUGHED AND SPREAD, IN CENTS.			
		Light sandy soil.	Common loam.	Strong heavy soil	Stiff clay or cemented gravel.
50	44.4	10.4	12.2	13.7	14.7
100	40.0	10.8	12.5	14.0	15.0
200	33.3	11.5	13.2	14.8	15.8
300	28.6	12.2	14.0	15.5	16.5
400	25.0	12.5	14.7	16.2	17.2
600	20.0	14.4	16.1	17.7	18.7
800	16.7	15.8	17.6	19.1	20.1
1000	14.3	17.3	19.0	20.6	21.6
1200	12.5	18.8	20.5	22.0	23.0
1400	11.1	20.2	21.9	23.4	24.4
1600	10.0	21.7	23.4	24.9	25.9
1800	9.1	23.1	24.8	26.3	27.3
2000	8.3	24.6	26.3	27.8	28.8
2500	6.9	28.2	29.9	31.4	32.4
3000	5.9	31.8	33.5	35.0	36.0
4000	4.5	39.0	40.8	42.3	43.3
5000	3.7	46.4	48.1	49.6	50.6

contractor, of *earth when ploughed and spread in the embankment*. When loosened with picks, from 1.3 to 4.5 cents per
cu. yd. should be added to the values given, according as to
whether the material is of a light sandy nature or a stiff clay.
If merely dumped over the embankment, then the values
given may be reduced by about 1 cent per cubic yard.

B. THE COST OF ROCK REMOVED BY CARTS.

The total cost of loosening hard rock—with wages at $1.00
per day—is usually covered by 45 cents per yard in place ; in
dealing with soft shales which can be loosened by pick, being
sometimes as low as 20 cents, while in shallow cuttings of
tough rock, in which the strata lie unfavorably, $1.00 may be
insufficient.

A good churn-driller will drill from 8 to 12 feet of 2-inch
holes, about 2¼ feet deep, per day, at a cost of about 12 to 18
cents per foot.

A cart suitable for ⅛ cu. yd. of earth as a working load
will take about ⅛ cu. yd. of rock. Rock takes longer to
shovel into the carts than earth, so that we may say the equation given above for earth becomes in the case of rock

$$N = \frac{M}{6 + L},$$

and the number of yards hauled per day is given by ⅛N.
Loading costs about 8 cents per cu. yd., and the repair of the
hauling-road about ⅛ cent per cu. yd. per 100 feet of lead.
Thus we have, exclusive of the profit to the contractor—

Length of Lead in feet.	No. of cu. yds. per cart per day.	Cost per cu. yd. for hauling and emptying.	Total cost per cu. yd.
50	18.5	6.8	60.0
100	17.1	7.3	60.5
200	15.0	8.3	61.7
300	13.3	9.4	63.0
500	10.9	11.5	65.5
700	9.2	13.6	68.0
1000	7.5	16.7	71.7
1500	5.7	21.9	77.9
2000	4.6	27.1	84.1
2500	3.9	32.3	90.3
3000	3.3	37.5	96.5
4000	2.6	47.9	108.9

"Loose Rock" usually costs about 30 cents per yard less than the above cost for hard rock.

125. Both rock and earth can generally be moved at about the same cost by wheelbarrows as by carts when the lead is equal to about 200 feet; for shorter hauls the wheelbarrows have the advantage, but for longer, the carts.

As regards the cost of removal by scrapers or any other form of vehicle, it may be approximated to in the same manner as the removal by carts in Sec. 124. A scraper generally moves from 30 to 60 cubic yards per day with a short haul. A medium-size steam-shovel, if kept tolerably busy, should, under ordinary conditions, load the cars at a cost of from 2 to 3 cents per cu. yd. Grading-machines, 8 or 12 horse, in light soil and with low fills, can generally turn over from 500 to 1000 cu. yds. per day.

126. Estimating Overhaul.—It is common to allow an extra price, usually from 1 to 2 cents for every cubic yard of material, either earth or rock, for each 100 feet that it is hauled beyond what is termed the *limit of free haul*, represented by l in Fig. 67.

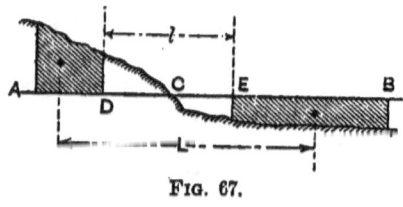

FIG. 67.

Let us suppose that the material in the cut AC is just sufficient to make the fill CB, then the material on which *overhaul* must be charged is that lying between A and D (or B and E), and the distance which that material is hauled is represented by L, the distance between the centres of gravity of the two solids AD and EB; consequently the length of overhaul $= L - l$, and if S represents the contents of AD (or EB), then the amount of overhaul $= S(L - l)$.

Thus, for example, if $L = 1000$ ft., $l = 600$ ft., and $S = 4000$ cu. yds., the cost of overhaul at 1 cent per cu. yd. per 100 ft. will be $160.

But though the distance l is always given, in order to locate it on the profile we must find the points D and E, such that the material in $DC =$ the material in EC. This may usually be done by inspection of the profile; and in the same way the points A and B may be fixed. In cases where the centre-heights are not fair indications of volume, these points may be quickly found to within a few feet, by means of the cross-section note-book. The positions of the centres of gravity of the two solids AD and EB may also usually be fixed by inspection. On this subject the *Engineering News* says: "As quick a way as any is to plot the *volumes* of each solid as ordinates, as one would plot a profile, on stiff card-board, cut out the area thus drawn, and balance it on a knife-edge; but a way which we can recommend as much the best and fairest of any, in competent hands, is to guess at it, throwing the benefit of a doubt for or against the contractor according to the character of the haul, and to some extent of the material excavated. The actual haul cannot fairly be taken at times as the crow flies, nor is it exactly fair that haul over good solid gravel should have the same allowance as haul from a shallow cut through muck. As a contract is a contract, and must be general, no considerable deviations on account of such contingencies as these are admissible, but no considerable ones are necessary, the limits of error in guessing at the 'centre of mass' being very small, and having reference to a small item of price, whereas the limits of error in one unavoidable kind of guessing which is usually going on at the same time, that of classification, are very large, and have reference to a very large item. This consideration alone ought to show the folly of any great hair-splitting in mathematical computations of the precise overhaul; but there is a certain class of minds who are never happy unless they can find some hair to split, and who will split it with just as much care although there may be a log of wood alongside which they can't split, to which the right half of the hair is to be added."

THE CALCULATION OF EARTHWORK.

127. The three solids with which engineers have mainly to deal in the calculation of earthwork are the pyramid, the wedge, and the "prismoid;" for though, owing to the irregularities of surface, these figures, mathematically speaking, are never actually met with in practice where the surface of the ground forms one or more sides of the figure, yet the contents as given by them are sufficiently accurate under ordinary circumstances, when the work has been properly cross-sectioned. But before dealing with the calculation of the contents of these solids, it will be well to consider the methods of obtaining the **areas of the cross-sections** themselves, on which the computations are based.

1. When the cross-section is of *triangular form*, as in Fig. 69, its area of course—taking for instance the triangle ABC—equals $AB \times \frac{1}{2}$ the perpendicular distance from C to AB, or AB produced.

2. When the cross-section is an *ordinary 3-level one*, as in Figs. 71 and 72, then if $B =$ width of road-bed and $H, h, h', l,$ and l' are as shown in Fig. 55,

$$\text{Area} = \frac{H}{2}(l + l') + \frac{B}{4}(h + h'),$$

which is the formula most generally in use.

3. If the *surface is horizontal*, then this becomes

$$\text{Area} = H\left(\frac{B}{2} + l\right).$$

4. Or, if *regularly inclined*,

$$\text{Area} = \frac{B \cdot h}{2} + lh',$$

where h is the greater side-height, and l its corresponding distance out from the centre, h' being the smaller side-height.

5. But it frequently happens that we have such a section as that shown in Fig. 68. Such an area may be best calculated

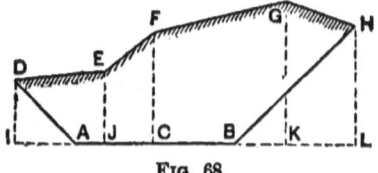

FIG. 68.

by first finding the contents of the figure $IDHL$, and then deducting from it the areas DIA and HLB; thus the area of this cross-section equals

$$\frac{ID + EJ}{2}(IJ) + \frac{EJ + FC}{2}(JC) + \frac{FC + GK}{2}(CK) +$$

$$\frac{GK + HL}{2}(KL) - \frac{ID \cdot IA}{2} - \frac{BL \cdot HL}{2}.$$

The above forms of cross sections are really all that are required in practice, 1, 2, and 5 being those most generally in use. Neither of these forms requires plotting, but it is usually advisable to plot cross-sections of large area which are very irregular even though calculated as above, for by so doing mistakes are much more readily apparent. Where the work consists largely of irregular cross-sections, a good and rapid method of obtaining the areas is to plot the cross-sections and use a planimeter. The error in ordinary cross-sections, plotted on cross-section paper to a scale of 10 feet to an inch, should never—where the planimeter is carefully adjusted so as to allow for the shrinkage of the paper, etc.—exceed 1 p. c.; and considering that these errors to a large extent cancel each other and are free from errors of calculation, which are usually much more probable than errors in reading the planimeter scale, the result in the long run is at least equally likely to be as near the truth as that obtained by the more laborious process of calculation.

128. The areas of the cross-sections having been obtained, the calculation of the contents of the solids which they bound is the next point to deal with, and we will consider them in the order given above.

A. The Pyramid.—The usual cases in which pyramids occur are those shown in Fig. 69, which need no explanation.

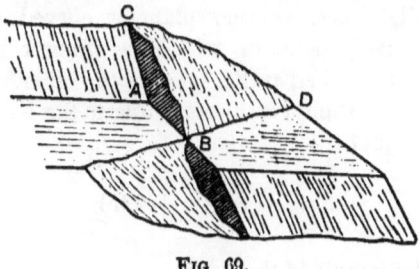

Fig. 69.

The contents of such a pyramid as $ABCD$ are found by the formula

$$S = ABC \times \frac{AD}{3},$$

and this rule applies to *any* form of base.

B. The Wedge.—The various forms of wedge which present themselves in calculating the contents of earthwork, of which that represented in Fig. 70 is the usual type, can only be estimated *correctly* by the application of the **Prismoidal**

Formula. But since at the points where the wedge form of solid occurs the cut or fill is always small, the error involved

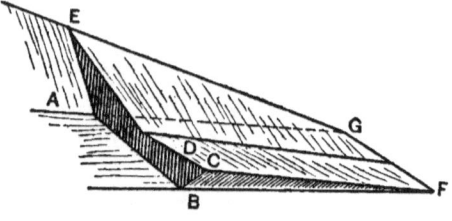

Fig. 70.

by using the formula for the *rectangular wedge* is immaterial; thus we may say that the contents

$$S = \text{area } ABCDE \times \frac{AG}{2}.$$

C. The Prismoid.—Though the term "prismoid" strictly applies only to such solids as are contained by 6 plane surfaces, the two end-faces being parallel, and two of the other faces being *not* parallel, the extended application of the ' prismoidal formula" has corrupted its true meaning, so that it is now applied very generally in Railroad work to all solids having two parallel faces, whether plane or curved, upon which, and through every point of which, a straight line may be drawn from one of the parallel faces to the other.

The contents of such a solid according to the PRISMOIDAL FORMULA equal

$$S = \frac{L}{6}(A + a + 4M),$$

where $L =$ the length of the solid,
A and $a =$ the areas of its two parallel faces,
 and $M =$ the cross-section parallel to A and a, and half-way between them.

This formula at first looks simple enough, but the calculation of M is the difficulty.

129. To explain the application of this formula, suppose we have two end-areas A and a as in Fig. 71.

Now in order to obtain the mid-section, we must know the points in A and a from which the straight lines joining them start, and at which they end; thus in Fig. 71, if the cross-section notes simply give the elevations for the 3-level sections A and a, we assume that the upper surface between them is

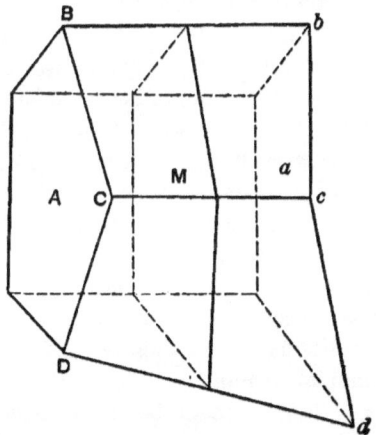

FIG. 71.

composed of two warped surfaces, $BCcb$ and $CDdc$, which is what follows from supposing that the centre and side heights of M are the averages of the corresponding heights of A and a. So that if the surface were actually as shown in Fig. 72,

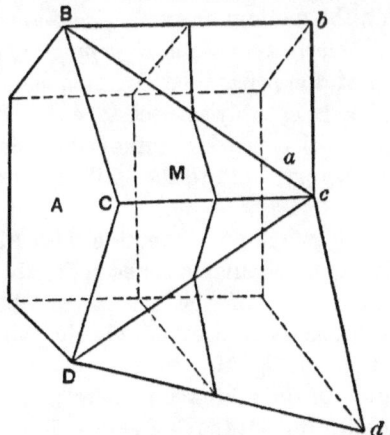

FIG. 72.

we should obtain entirely erroneous results by taking the value of M given by Fig. 71. Thus when the surface is such that points in A and a, other than those directly corresponding, are to be considered as being joined by straight lines, it becomes necessary to indicate in the notes between what points in A and a the straight lines *are* assumed to be drawn; and then the surface, instead of being made up of two or more warped surfaces, will be composed entirely of a series of plane surfaces as in Fig. 72. This is best done, where required, by drawing, in the cross-section note-book, lines connecting the notes of the points to be joined. This would also have to be done between two cross-sections A and a which did not happen to have the same number of points taken in each. At times cases occur in which it is advisable to fill in slope-lines in this way, but they are very few and very far between; for the labor involved in the calculation of M in such cases would usually have been very much better expended in actually taking a cross-section between A and a. Therefore, as a rule, where the prismoidal formula is to be used in the calculation of the contents, it is very much better to cross-section a little more closely, where necessary, and to omit the filling-in of the slope-lines, than to take cross-sections a little farther apart and fill in the slope-lines by inspection.

The value of the prismoidal formula, as applied in the case of Fig. 71, is not so much to rectify irregularities in surface as to make suitable allowance for the difference in the heights of A and a, which the method of average end-areas does not do. In practice, however, *where the work is properly cross-sectioned*, the application of the prismoidal formula is a mathematical refinement which is entirely unnecessary, for the method of average end-areas—that usually employed—then gives results sufficiently satisfactory, both to the Railway Company and the Contractor.

It is an interesting fact in connection with Figs. 71 and 72, that if the contents be calculated for each possible arrangement of slope-lines, the mean of the results so obtained will be equal to the result as derived by merely the joining of corresponding points, as in Fig. 71.

The calculation of the mid-area is merely a matter of simple proportion. In dealing with such a case as Fig. 72, by plotting A and a on a sheet of cross-section paper, the drawing of the mid-sections may be done by simply drawing parallel lines; so that this should be done as a check to the calculations and also as a means of facilitating them.

130. The method used nowadays almost entirely for the calculation of grading, is that of **Average End-areas,** which assumes that

$$S = \frac{A+a}{2} L.$$

Now this method, which is the simplest of any to work, unfortunately has a considerable tendency to *excess;* the results obtained by it are, however, the same as those given by the prismoidal formula—applied as in Fig. 71,—therefore presumably correct, under the following circumstances:

1. Whenever the centre-heights of A and a are the same, whatever the difference in side heights may be.

2 Whenever the entire widths between the slope stakes at A and a are the same, whatever the difference in centre-heights may be.

When, however, the smaller centre-height is at the same end of the solid as the greater width between the slope-stakes, the volume as given by average end-areas will be actually *deficient.*

But since these cases are the exceptions, the results as given by this method are in the long run considerably too high, unless care is taken in cross-sectioning to limit the excess. To correct for this tendency a **Prismoidal Correction** may be used, found by deducting the prismoidal formula from the formula for average end-areas; and this correction, when the *surface of each end-section is horizontal,* equals in cubic yards

$$C = (H - H')^2 \frac{sL}{27 \times 6},$$

where H and H' are the end centre-heights in feet, s the slope-rates, and L the lengths of the solid in feet.

Taking $s = 1\frac{1}{2}$ and $L = 100$, we obtain the following values for C, which serve in making up preliminary estimates to show the errors involved by a rough system of cross-sectioning when the contents are calculated by average end-areas.

TABLE OF PRISMOIDAL CORRECTION FOR 100 FEET IN CU. YDS. FOR HORIZONTAL SURFACES WHERE $s = 1\frac{1}{2}$.

$H - H'$	0	1	2	3	4	5	6	7	8	9
0	0	1	4	8	15	23	33	45	59	75
10	93	112	133	156	181	208	237	268	300	334
20	370	408	448	490	533	578	626	675	726	779

This value of C is altogether independent of the width of the road-bed; so that, for example, suppose on ground sloping in the direction of the length of the solid we have, between two sections 100 feet apart, a difference in centre-heights of 23 feet, if $s = 1\frac{1}{2}$ and there is no slope transversely, the contents as given by average end-areas will be 490 cubic yards too much, even with a 14-foot road-bed; or, if the fill at one end is 2 feet and at the other end 25 feet, the prismoidal formula gives 1957 cubic yards as the volume, while the method of average end-areas gives 2447 cubic yards, or 25 p. c. too much.

But the above values of the prismoidal correction only apply when the surfaces of the sections are horizontal. If, however, in dealing with 3-level sections we call W and W' the entire width between the slope-stakes at each end, then the prismoidal correction equals, in cubic yards,

$$C = (H - H')(W - W') \frac{L}{27 \times 12},$$

which is independent of the side-slopes and width of the road-bed. So that, having calculated the contents according to the formula for average end-areas, we have simply to find for each cross-section the value of $(H - H')$ and $(W - W')$, and take out from the following table, which gives the values of C, the amount in cubic yards which is to be added to the contents already obtained in order to obtain the result which would be given by the prismoidal formula. Should, however, the smaller centre-height be at the same end of the solid as the greater width between the slope-stakes, then C must be subtracted.

RAILROAD CONSTRUCTION.

TABLE OF THE VALUES OF C, WHEN $L = 100$ FEET.

$W - W'$ in feet.	$H - H'$ in feet.									
	1	2	3	4	5	6	7	8	9	10
1	.3	.6	.9	1.2	1.5	1.8	2.1	2.4	2.7	3.1
2	.6	1.2	1.8	2.4	3.0	3.6	4.3	4.9	5.5	6.2
3	.9	1.8	2.7	3.6	4.6	5.5	6.5	7.4	8.3	9.3
4	1.2	2.4	3.6	4.9	6.2	7.4	8.6	9.8	11.1	12.3
5	1.5	3.1	4.6	6.2	7.7	9.2	10.8	12.3	13.8	15.4
6	1.8	3.6	5.5	7.4	9.2	11.1	12.9	14.8	16.6	18.5
7	2.1	4.3	6.5	8.6	10.8	12.9	15.1	17.3	19.4	21.5
8	2.4	4.9	7.4	9.8	12.3	14.8	17.3	19.7	22.2	24.6
9	2.7	5.5	8.3	11.1	13.8	16.6	19.4	22.2	25.0	27.7
10	3.1	6.2	9.3	12.3	15 4	18.5	21.5	24.6	27.8	30.8
11	3.4	6.8	10.2	13.6	17.0	20.3	23.7	27.1	30.6	33.9
12	3.7	7.4	11.1	14.8	18.5	22.2	25 8	29.5	33.3	37.0
13	4.0	8.0	12.0	16.0	20.0	24.0	28.0	32.0	36.0	40.1
14	4.3	8.6	12.9	17.3	21.5	25.8	30.1	34.5	38.8	43.2
15	4.6	9.2	13.8	18.5	23.1	27.7	32.3	37.0	41.6	46.3
16	4.9	9.8	14.8	19.7	24.6	29.5	34.5	39.4	44.3	49.3
17	5.2	10.4	15.7	20.9	26.2	31.4	36 6	41.9	47.1	52.4
18	5.5	11.1	16.7	22.2	27.8	33.3	38.8	44.4	49.9	55.5
19	5.8	11.7	17.6	23.4	29.3	35.1	41.0	46.9	52.7	58.6
20	6.2	12.3	18.5	24.6	30.8	37.0	43.2	49.4	55.6	61.8
21	6.5	12.9	19.4	25.8	32.3	38.8	45.3	51.8	58.3	64.8
22	6.8	13.5	20.3	27.1	33.9	40.6	47.4	54.3	61.1	67.9
23	7.1	14.2	21.3	28.4	35.4	42.5	49 6	56.8	63.9	71.0
24	7.4	14.8	22.2	29.6	37.0	44.4	51.8	59.2	66.7	74.1
25	7.7	15.4	23.1	30.8	38.5	46.2	54.0	61.7	69.4	77.1
26	8.0	16.0	24.0	32.0	40 0	48.1	56.1	64.1	72.1	80.2
27	8.3	16.6	24.9	33.2	41.5	49.9	58.3	66.6	74.9	83.3
28	8.6	17.2	25.8	34.5	43.1	51.8	60.5	69.1	77.7	86.4
29	8.9	17.8	26.8	35.7	44.7	53.7	62.7	71.6	80.5	89.5
30	9.3	18.5	27.7	37.0	46.3	55.6	64.9	74.1	83.3	92.6

There is no need to apply these corrections at the time when the quantities are worked out by average end-areas, as generally the engineer is then too much occupied in obtaining *rough* estimates of the work; but they can subsequently be applied, with very little trouble, to such solids as in his opinion need correcting.

The application of this method undoubtedly reduces the final estimate of the grading very considerably, rarely by less than 1 p. c., and in some cases, where the cross-sectioning has been carelessly done, by as much as 4 or 5 p. c. But it must be remembered that in this way the true volume is obtained more nearly than by any other of the approximate processes, and that the results are slightly higher than those obtained by the use of such tables as "Trautwine," "Rice," etc., founded on the principle of Equivalent Level Sections. Without the application of the prismoidal correction the contractor is entirely at the mercy of the engineer who does the cross-sectioning (if the method of average end-areas is used), who has it, often unconsciously, in his power to make a difference in the final estimate of 3 or 4 per cent, by not paying attention to the differences in centre-heights and widths of the cross-sections he is taking. And though the errors in any given piece of work are in favor of the contractor, still the uncertainty to which they give rise, in the long run do him considerably more harm than good. If a correction is not used, some limiting value for $(H - H') \times (W - W')$ should be established.

Some standard system of measuring grading is much wanted. As it is now, a contractor on one piece of work gets the benefit, possibly of 3 p. c. due to the use of average end-areas, uncorrected; while on the next contract he takes very likely he has the quantities actually *cut down*, owing to the use of tables of equivalent level sections. It is true that if the work is properly cross-sectioned the excess as given by the method of average end-areas should not exceed 1 or 2 p. c., but in the ordinary way in which cross-sectioning is done, a considerable amount of trouble is taken in order to correct for small surface irregularities, while the great errors which are involved by the difference in centre-heights are barely considered so long as the slopes between the sections are tolerably uniform.

When the cross-sections are irregular, the prismoidal correction can usually be applied with sufficient accuracy by treating them as 3-level sections, and thus applying the value of C as given above.

131. The Method of Equivalent Level Sections is an incorrect means of applying the prismoidal formula by reducing the end-sections to sections equivalent in area but with their surfaces horizontal, and then taking as the area of the mid-section that which is given by the mean of the corrected centre-heights. But unfortunately the results so obtained are only correct—

1. When the two end-areas are "similar"—i.e., the corresponding surface-slopes from the centre to the slope-stakes are the same at both ends, provided the road-bed is not intersected between them;

2. When the surface is regularly warped from one end to

the other, provided that no two of the straight lines connecting corresponding points, such as A, a, etc., in Fig. 71 are inclined to grade in opposite direction (as they *are* in Fig. 71).

In cases where these conditions do not hold, then, assuming that the true result is given by the prismoidal formula if merely the corresponding points A, a, etc., are joined by straight lines, the method of equivalent level sections gives results *too small*. But if the surface is intersected by undulations, running obliquely, necessitating the use of "slope-lines" as in Fig. 72, then the results may either be too small or too great, according to circumstances. But since this latter method of applying the prismoidal formula is the exception, and the results as obtained by applying it in the manner shown in Fig. 71 more generally correct, the general tendency of the method of equivalent level sections is to deficiency, but not by an amount usually sufficient to warrant the use of a correction. The real objection to this method is the labor involved in applying it when dealing with cross-sections in the slightest degree "irregular," and even in dealing with 3-level sections the work involved is greater than that by the method of average end-areas, corrected; while the result in the former case is an approximation, in the latter it is presumably correct.

132. The method of centre-heights, which is very useful in making preliminary estimates, simply assumes that the contents between any two cross-sections are given according to the method of average end-areas, the area at each end being taken as the area of a horizontal section with a height equal to the actual centre-height. The results so obtained naturally err, sometimes in excess and sometimes in deficiency—the tendency in the former direction being, however, the more common. But since there is no decided tendency to cumulative error, the result obtained as a whole for several stations where the direction of the surface slope is varied, agrees tolerably well with the true volume, though for any one station the error may be very considerable. *In the long run* more accurate results are usually given by this method than by that of average end-areas. (See Secs. 69 and 70.)

133. By the use of Table XIV the labor of applying the method of Centre-heights is greatly reduced.

Table XV saves considerable labor in reducing areas to cubic yards, by avoiding the necessity of multiplying by 100 and dividing by 27. There is no need to take the quantities out closer than to the nearest yard. In using the table for lengths other than 100 feet a good deal of trouble may be saved in the way of multiplication and division by reducing each time the simpler of the two values with which the table is entered; thus if we have an average area of 634 square feet for 50 feet, the amount opposite 317 gives the quantity required, instead of dividing 2348.2 by 2.

134. Correction for Curvature.—We have hitherto assumed that the cross-sections are parallel to each other—i.e., that the track is straight. Suppose, however, that in Fig. 73, exaggerated for the sake of clearness, o represents the centre of a certain curve whose radius $= R$, the cross-section $ACaB$ representing any cross-section on the curve.

Now it is clear that if we have two cross-sections whose centres are 100 feet apart (along the curve) and take in each a point b, situated outside the centre by a distance y, the distance between these two selected points, measured along a line parallel to the centre-line, is to 100 feet as $R + y$ is to R, arcs

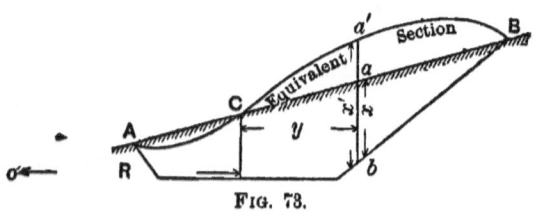

Fig. 73.

subtended by equal angles at the centre being proportional to their radii. But instead of calculating the contents for the *varying distance*, it is simpler to assume that the track is straight, and to correct the *sections themselves* so as to allow for it: so that, instead of using the above proportion, we may consider that the area of a section at any distance y from the centre must be increased or decreased in the proportion

$$x' = \frac{x(R \pm y)}{R},$$

where x' represents the corrected area and x the original area; y being positive if falling, as in Fig. 73, on the outside of the curve, and negative if falling inside. So that if at any point as a we measure the ordinate x and its distance from the centre y, the above equation gives us x', the corrected length of x, which, being measured upwards from the point b, gives us a', the new position of a. Similarly by finding other positions of a', the curved line $ACa'B$ being drawn through them, gives the *equivalent section* on a straight track.

In curves of 8° and upwards, where the slope is comparatively steep in one direction, this correction should be applied. It is best to assume an average section for two or three stations together, and to divide the radius by 10, so as to make R a distance easily scaled, and then to divide the correction so obtained by 10. Thus, if the section is taken as an average one for 300 feet on a 10° curve, we plot $R = 57$ feet, and the correction so obtained—which is of course equal to the difference between the contents given by the actual section and the equivalent section—must itself be divided by 10; or, what is the same thing, be considered to apply only to a length of 30 feet. Two or three ordinates are usually sufficient to locate with sufficient accuracy the surface of the equivalent section. Where the surface is level there will of course be no correction necessary, for then the excess on one side of the centre-line balances the deficiency on the other.

This method is equally easy to apply to *any form of cross-section*, however irregular it may be.

135. The contents of the toe of a dump are commonly calculated according to the formula given in Sec. 128 for a wedge, but the result so obtained is always considerably too small; neither can the prismoidal formula be directly applied.

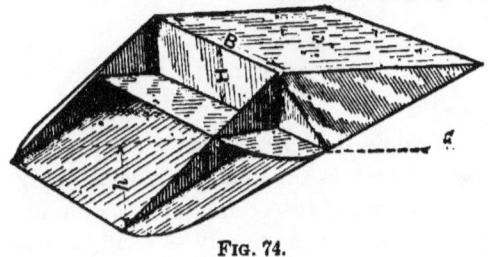

Fig. 74.

First, let us assume the surface of the ground to be level; then the simplest way to obtain correctly the contents of the toe is to consider each corner as a quarter of a cone; then if H equals the height of the fill in feet, and s the slope ratio, the contents of the two corners together equal

$$.523H^3s^2;$$

so that the entire contents of the toe are given by the formula

$$S = .523H^3s^2 + .5BH^2s;$$

B being the width of the road-bed in feet. This formula is easily worked out by means of Table VIII. S must then be divided by 27 to reduce it to cubic yards.

If $s = 1\frac{1}{2}$, then the above equation becomes

$$S = .75BH^2 + 1.178H^3.$$

But when the ground slopes downward in the direction of the toe, as is the more common case, then we may consider the toe to be divided into two portions, as shown in Fig. 74; the upper one, which we have just dealt with, having a vertical height equal H, and the lower one with a vertical height $= h$. Then, omitting for a moment the consideration of the circular corners, the contents of the upper portion are to the contents of the lower portion as H is to h. Now, though this does not quite hold good when taking the corners into account, the error involved by assuming it to do so is immaterial; so that we may say, that when the ground slopes forward as in Fig. 74, the *total contents* equal

$$S' = S\left(1 + \frac{H}{h}\right),$$

the value of S being obtained as above.

The value of h may be obtained quite well enough by plotting H and the slopes of the ground and the dump.

If the ground slopes transversely as well, the case becomes decidedly complicated, and the engineer must then assume such values, as will when inserted in the above formulæ, give what he considers fair results.

In dealing with the toe of a dump less than 10 feet in height the wedge formula is sufficiently accurate, but where the fill

TABLE OF BOARD MEASURE.

Thickness in Inches.

Width in In.	1	1¼	2	2½	3	3½	4	4½	5	5½	6	7	8	9	10	11	12	14	16
1	.0833	.1250	.1667	.2083	.2500	.2917	.3333	.3750	.4167	.4583	.5000	.5833	.6667	.7500	.8333	.9167	1.000	1.167	1.333
1¼	.1250	.1875	.2500	.3125	.3750	.4063	.5000	.5625	.6250	.6875	.7500	.8750	1.000	1.125	1.250	1.375	1.500	1.750	2.000
2	.1667	.2500	.3333	.4688	.5625	.5833	.6667	.7500	.8333	.9167	1.000	1.167	1.333	1.500	1.667	1.833	2.000	2.333	2.667
2½	.2083	.3125	.4167	.5208	.6250	.7292	.8333	.9375	1.042	1.146	1.250	1.458	1.667	1.875	2.083	2.292	2.500	2.917	3.333
3	.2500	.3750	.5000	.6250	.7500	.8750	1.000	1.125	1.250	1.375	1.500	1.750	2.000	2.250	2.500	2.750	3.000	3.500	4.000
3½	.2917	.4375	.5833	.7292	.8750	1.021	1.167	1.313	1.458	1.604	1.750	2.042	2.333	2.625	2.917	3.208	3.500	4.083	4.667
4	.3333	.5000	.6667	.8333	1.000	1.167	1.333	1.500	1.667	1.833	2.000	2.333	2.667	3.000	3.333	3.667	4.000	4.667	5.333
4½	.3750	.5625	.7500	.9375	1.125	1.313	1.500	1.688	1.875	2.063	2.250	2.625	3.000	3.375	3.750	4.125	4.500	5.250	6.000
5	.4167	.6250	.8333	1.042	1.250	1.457	1.666	1.875	2.083	2.292	2.500	2.917	3.333	3.750	4.167	4.583	5.000	5.833	6.667
5½	.4583	.6875	.9167	1.146	1.375	1.603	1.833	2.063	2.292	2.521	2.750	3.204	3.667	4.125	4.583	5.042	5.500	6.417	7.333
6	.5000	.7500	1.000	1.250	1.500	1.750	2.000	2.250	2.500	2.750	3.000	3.500	4.000	4.500	5.000	5.500	6.000	7.000	8.000
7	.5833	.8750	1.167	1.458	1.750	2.042	2.333	2.625	2.917	3.208	3.500	4.083	4.667	5.250	5.833	6.417	7.000	8.167	9.333
8	.6667	1.000	1.333	1.667	2.000	2.333	2.667	3.000	3.333	3.667	4.000	4.667	5.333	6.000	6.667	7.333	8.000	9.333	10.67
9	.7500	1.125	1.500	1.875	2.250	2.625	3.000	3.375	3.750	4.125	4.500	5.249	6.000	6.750	7.500	8.250	9.000	10.50	12.00
10	.8333	1.250	1.667	2.083	2.500	2.917	3.333	3.750	4.167	4.583	5.000	5.833	6.667	7.500	8.333	9.167	10.000	11.67	13.33
11	.9167	1.375	1.833	2.292	2.750	3.208	3.664	4.125	4.583	5.042	5.500	6.417	7.333	8.250	9.167	10.08	11.000	12.83	14.67
12	1.000	1.500	2.000	2.500	3.000	3.500	4.000	4.500	5.000	5.500	6.000	7.000	8.000	9.000	10.00	11.00	12.00	14.00	16.00
13	1.083	1.625	2.167	2.708	3.250	3.792	4.333	4.875	5.417	5.958	6.500	7.583	8.666	9.750	10.83	11.92	13.00	15.17	17.33
14	1.167	1.750	2.333	2.917	3.500	4.083	4.667	5.250	5.833	6.417	7.000	8.167	9.333	10.50	11.67	12.83	14.00	16.33	18.67
15	1.250	1.875	2.500	3.125	3.750	4.375	5.000	5.625	6.250	6.875	7.500	8.750	10.00	11.25	12.50	13.75	15.00	17.50	20.00
16	1.333	2.000	2.667	3.333	4.000	4.667	5.333	6.000	6.667	7.333	8.000	9.333	10.67	12.00	13.33	14.67	16.00	18.67	21.33
18	1.500	2.250	3.000	3.750	4.500	5.250	6.000	6.750	7.500	8.250	9.000	10.50	12.00	13.50	15.00	16.50	18.00	21.00	24.00
20	1.667	2.500	3.333	4.167	5.000	5.833	6.667	7.500	8.333	9.167	10.000	11.67	13.33	15.00	16.67	18.33	20.00	23.33	26.67

amounts to about 20 feet the difference in the results by the two methods is very considerable.

136. The original notes of the cross-sections should be copied on the left-hand pages of another note-book, and opposite them, on the right-hand pages, the sectional areas, contents, etc., should be entered as soon as worked out. A "Record" should also be kept, into which each separate item should be entered as soon as completed,—not in detail, but simply the total amounts; these notes then form the groundwork of the final estimate. The *details* are entered separately in note-books apportioned to each class of work.

As regards taking notes for the monthly estimates, the simplest way is to walk over the work and sketch on the progress profile the state of construction at the time. Another way, possibly more convenient in light work, is to note the percentage of the total amount which is done up to date.

The classification is often a matter of considerable difference of opinion, especially in the allowance for "loose rock." All boulders, etc., exceeding the limit for loose rock must be carefully measured. When there is much of this to do, a good plan is to have a man especially to look after it on two or three subdivisions, who can also take the Force Account and give to the contractors any simple information they may require concerning the work. The subdivision engineers and their men are thus saved a very considerable amount of time and work.

TIMBER-WORK.

137. Timber is usually measured in railroad structures in B. M. (Board Measure), the contract for culverts, etc., being let by the 1000 feet B. M. One foot B. M. = 144 cubic inches, so that the B. M. of any given stick is found by multiplying together the width and thickness in inches and the length in feet, and dividing the result by 12.

The first portion of this calculation and the division by 12 is accomplished by means of the table on page 151.

In altering the length of trestle-posts, etc., to make allowance for the difference in elevation of the two rails, the following table will be found useful, as well as in many similar operations:

RAILROAD CONSTRUCTION.

FRACTIONS OF AN INCH IN DECIMALS OF A FOOT.

In.	0	1	2	3	4	5	6	7	8	9	10	11
0	Foot	.0833	.1667	.2500	.3333	.4167	.5000	.5833	.6667	.7500	.8333	.9167
1/32	.0026	.0859	.1693	.2526	.3359	.4193	.5026	.5859	.6693	.7526	.8359	.9193
1/16	.0052	.0885	.1719	.2552	.3385	.4219	.5052	.5885	.6719	.7552	.8385	.9219
3/32	.0078	.0911	.1745	.2578	.3411	.4245	.5078	.5911	.6745	.7578	.8411	.9245
1/8	.0104	.0938	.1771	.2604	.3438	.4271	.5104	.5938	.6771	.7604	.8438	.9271
5/32	.0130	.0964	.1797	.2630	.3464	.4297	.5130	.5964	.6797	.7630	.8464	.9297
3/16	.0156	.0990	.1823	.2656	.3490	.4323	.5156	.5990	.6823	.7656	.8490	.9323
7/32	.0182	.1016	.1849	.2682	.3516	.4349	.5182	.6016	.6849	.7682	.8516	.9349
1/4	.0208	.1042	.1875	.2708	.3542	.4375	.5208	.6042	.6875	.7708	.8542	.9375
9/32	.0234	.1068	.1901	.2734	.3568	.4401	.5234	.6068	.6901	.7734	.8568	.9401
5/16	.0260	.1094	.1927	.2760	.3594	.4427	.5260	.6094	.6927	.7760	.8594	.9427
11/32	.0286	.1120	.1953	.2786	.3620	.4453	.5286	.6120	.6953	.7786	.8620	.9453
3/8	.0313	.1146	.1979	.2813	.3646	.4479	.5313	.6146	.6979	.7813	.8646	.9479
13/32	.0339	.1172	.2005	.2839	.3672	.4505	.5339	.6172	.7005	.7839	.8672	.9505
7/16	.0365	.1198	.2031	.2865	.3698	.4531	.5365	.6198	.7031	.7865	.8698	.9531
15/32	.0391	.1224	.2057	.2891	.3724	.4557	.5391	.6224	.7057	.7891	.8724	.9557
1/2	.0417	.1250	.2083	.2917	.3750	.4583	.5417	.6250	.7083	.7917	.8750	.9583
17/32	.0443	.1276	.2109	.2943	.3776	.4609	.5443	.6276	.7109	.7943	.8776	.9609
9/16	.0469	.1302	.2135	.2969	.3802	.4635	.5469	.6302	.7135	.7969	.8802	.9635
19/32	.0495	.1328	.2161	.2995	.3828	.4661	.5495	.6328	.7161	.7995	.8828	.9661
5/8	.0521	.1354	.2188	.3021	.3854	.4688	.5521	.6354	.7188	.8021	.8854	.9688
21/32	.0547	.1380	.2214	.3047	.3880	.4714	.5547	.6380	.7214	.8047	.8880	.9714
11/16	.0573	.1406	.2240	.3073	.3906	.4740	.5573	.6406	.7240	.8073	.8906	.9740
23/32	.0599	.1432	.2266	.3099	.3932	.4766	.5599	.6432	.7266	.8099	.8932	.9766
3/4	.0625	.1458	.2292	.3125	.3958	.4792	.5625	.6458	.7292	.8125	.8958	.9792
25/32	.0651	.1484	.2318	.3151	.3984	.4818	.5651	.6484	.7318	.8151	.8984	.9818
13/16	.0677	.1510	.2344	.3177	.4010	.4844	.5677	.6510	.7344	.8177	.9010	.9844
27/32	.0703	.1536	.2370	.3203	.4036	.4870	.5703	.6536	.7370	.8203	.9036	.9870
7/8	.0729	.1563	.2396	.3229	.4063	.4896	.5729	.6563	.7396	.8229	.9063	.9896
29/32	.0755	.1589	.2422	.3255	.4089	.4922	.5755	.6589	.7422	.8255	.9089	.9922
15/16	.0781	.1615	.2448	.3281	.4115	.4948	.5781	.6615	.7448	.8281	.9115	.9948
31/32	.0807	.1641	.2474	.3307	.4141	.4974	.5807	.6641	.7474	.8307	.9141	.9974
	0	1	2	3	4	5	6	7	8	9	10	11

For notes on the strength, etc., of timber, see Part IV.

IRON-WORK.

138. In estimating the weight of **Bolts and Nuts** the weight of the heads and nuts themselves may be taken from the following table, assuming them to be of ordinary proportion:

Diameter of Bolt.	1/4	3/8	1/2	5/8	3/4	7/8	1	1¼	1½	1¾	2	2¼
	lbs.	lbs	lbs.	lbs.	lbs.	lbs.	lbs.	lbs.	lbs.	lbs.	lbs.	lbs.
Hex. Head and Nut............	.017	.057	.128	.27	.43	.73	1.1	2.2	3.8	5.6	8.8	17
Sq. Head and Nut............	.021	.069	.164	.32	.55	.88	1.3	2.6	4.4	7.0	10.5	21

The weight of the shanks of the bolts may be found from the following table of the weight and strength of iron rods. If, however, the screw end is *upset*, with a consequent enlargement of the nut and head, the usual allowance for the weight due to upsetting, and square head and nut, will be equal to about 13 diameters of additional length of the shank of the bolt. If the nut and head are hexagonal, 11 diameters are then sufficient. This allowance is suitable when the length of the upsetting equals about 6 diameters of the shank. Thus if we have a 1-inch bolt upset for 6″, if 36″ long and the head and nut square, its weight will be given by the weight of a 1-inch bar 49″ long.

WEIGHT AND STRENGTH OF ROUND WROUGHT-IRON BARS.

Diam. in inches.	Weight in lbs. per foot run.	Breaking Strain in lbs.	Diam. in inches.	Weight in lbs. per foot run.	Breaking Strain in lbs.
$\tfrac{1}{8}$	.0414	550	$1\tfrac{1}{8}$	3.35	42340
$\tfrac{3}{16}$	.093	1240	$1\tfrac{1}{4}$	4.13	52200
$\tfrac{1}{4}$	.165	2200	$1\tfrac{3}{8}$	5.00	63170
$\tfrac{5}{16}$	.258	3430	$1\tfrac{1}{2}$	5.95	75260
$\tfrac{3}{8}$	.372	4950	$1\tfrac{5}{8}$	6.99	88260
$\tfrac{7}{16}$	.506	6720	$1\tfrac{3}{4}$	8.10	102370
$\tfrac{1}{2}$	.661	8800	$1\tfrac{7}{8}$	9.30	117600
$\tfrac{9}{16}$	.837	11130	2	10.6	133700
$\tfrac{5}{8}$	1.03	18750	$2\tfrac{1}{8}$	12.0	142900
$\tfrac{11}{16}$	1.25	16620	$2\tfrac{1}{4}$	13.4	160400
$\tfrac{3}{4}$	1.49	19780	$2\tfrac{3}{8}$	14.9	178500
$\tfrac{13}{16}$	1.75	23300	$2\tfrac{1}{2}$	16.5	198000
$\tfrac{7}{8}$	2.03	26880	$2\tfrac{5}{8}$	18.2	218200
$\tfrac{15}{16}$	2.33	30910	$2\tfrac{3}{4}$	20.0	239400
1	2.65	35170	3	23.8	285000

As a *safe working strain* one fifth of the above breaking strains may usually be taken.

The two *washers* generally used to each bolt weigh together about the same as a length of shank = 14 diameters; but if the bolt is upset, they then weigh about the same as a length = 22 diameters.

Railroad Spikes.—The following table gives the weight, etc., of the spikes commonly used for fastening the rails to the ties:

Length in inches.	Thickness in inches	No. per keg of 150 lbs.	No. per lb.	Length in inches.	Thickness in inches.	No. per keg of 150 lbs.	No. per lb.
4½	½	400	2.66	5½	½	350	2.33
5	⅝	705	4.70	5½	9/16	289	1.93
5	7/16	488	3.25	5½	⅝	218	1.46
5	½	390	2.60	6	½	310	2.07
5	9/16	295	1.97	6	9/16	262	1.75
5	⅝	257	1.71	6	⅝	196	1.30

The following table gives the **angle-bars** and **bolts** necessary for 1 mile of track:

Length of Rails in feet.	No. of Angle-bars.	No. of Bolts.	Length of Rails in feet.	No. of Angle-bars.	No. of Bolts.
24	880	1760	27	782	1564
25	844	1688	28	754	1508
26	812	1624	30	704	1408

The following table gives the **weight of Rails required** for 1 mile of track:

Weight of Rail per yard.	Weight per mile.		Weight of Rail per yard.	Weight per mile.		Weight of Rail per yard	Weight per mile.	
lbs.	tons.	lbs.	lbs.	tons.	lbs.	lbs.	tons.	lbs.
40	62	1920	56	88	0	65	102	320
45	70	1600	57	89	1280	68	106	1920
48	75	960	60	94	640	70	110	0
50	78	1280	62	97	960	72	113	820
52	81	1600	64	100	1280	76	119	960

The weight of iron required per mile is very nearly given by the rule: Multiply the weight in lbs. per yard by 1¾; the product is the weight required in tons of 2000 lbs. (the tons in the table = 2240 lbs.)

The weight of iron in lbs. per yard is given by multiplying its sectional area in inches by 10, assuming the iron to weigh 480 lbs. per cubic foot. Steel rails usually weigh about 490 lbs. per cubic foot.

139. BALLAST AND TIES.—The following table gives the amount of **ballast required per mile of road**:

Depth in inches.	Top Width, Single Track.			Top Width, Double Track.		
	10 Ft.	11 Ft.	12 Ft.	21 Ft.	22 Ft.	23 Ft.
	cu. yds.	cu. yds.	cu. yds	cu. yds.	cu. yds.	cu. yds.
12	2152	2347	2543	4308	4499	4695
18	3374	3667	3960	6600	6894	7188
24	4694	5085	5474	8096	9388	9780
30	6111	6600	7087	11490	11980	12470

This table assumes that the side-slopes of the ballast are at the rate of 1 to 1, and that there is a space of 6 feet clear between the tracks.

The following table gives the number of **Ties required per mile of track**:

Centre to Centre in inches.	No. of Ties.	Centre to Centre in inches.	No. of Ties.
18	3520	27	2347
20	3168	30	2112
22	2899	33	1920
24	2640	36	1760

For useful information in connection with Construction, see Part IV.

PART III.

EXPLORATORY SURVEYING.

140. IN Part I we have already considered the subject of "Preliminary Surveys," made principally with the object of obtaining topography by means of which the final location for a railroad may be selected. We will *here* deal with the subject of rough Reconnoissance and Exploratory Surveys, in which accuracy—such as it is generally understood—is not essential, and in which the general bearings of rivers and streams, and the elevations of mountain passes, etc., plotted to a scale of a mile or so to an inch, are the main points to be established.

But before dealing with the problems which arise in exploratory surveying it will be well to consider the Instruments usually employed in this class of work.

INSTRUMENTS.

141. The Instruments generally used in Reconnoissance and Exploratory Surveys are the following: The Sextant, Chronometer, Artificial Horizon, and the Cistern and Aneroid Barometers. To these may be added with advantage, a light portable Transit.

We will treat each separately in the order here given.

The Sextant.

There are in common use two forms of sextant—the Nautical and the Box sextant; but since the latter is nothing more than the former reduced into a small portable shape, we can consider them both under one head. For astronomical work the box-sextant may be considered almost worthless, but for taking ordinary topography it is an extremely handy instrument, and in more extensive work it is a very useful support to a nautical sextant in many ways. The ADJUSTMENTS of the sextant are as follows:

A. To place the index-glass perpendicular to the plane of the instrument.—Set the index to about 60°, and then, looking at the image of the limb of the instrument as reflected in the index-glass, the real limb and the image should appear to form one continuous arc. If they do not do so, the index-glass must be moved by means of the screws at its back (see Fig. 75) until it does.

B. To place the horizon-glass perpendicular to the plane of the instrument.—Clamp the index near to zero, and then, looking at some well-defined object, turn the tangent screw of the index until the object, as seen directly, and its reflected image are brought, if possible, to coincide. If they cannot be made to coincide the horizon-glass is out of adjustment and must be corrected by means of the adjusting screws with which it is fitted.

C. To obtain the index-error.—For the purpose of measuring the index-error when it is *negative*, i.e., when the correction for it is to be added, the graduations of the limb are carried a short distance back from zero through what is termed the ARC OF EXCESS. The index-error is obtained by noticing the reading when the coincidence mentioned in Adjust. B is obtained. But in this case the object must be a far distant one, so that the reading may not be affected by instrumental parallax. Had the index been set exactly at zero when the above-mentioned coincidence was made, there would of course be no index-error, but it is usually better to apply an index-error than to attempt to obtain an exact coincidence at zero.

A very accurate method of obtaining the index-error is to measure the diameter of the sun several times " on and off the arc "—i.e., on the positive and negative side of zero: the mean of the readings will then be the correction, positive if on the main arc, and negative if on the arc of excess. Thus, for example, if the diameter of the sun measured on the main arc = 32' 20", and on the arc of excess 30' 40", the mean being 0' 50" on the main arc, shows that 50" has to be *subtracted* from all angles as read from zero on the main arc, i.e., that the coincidence mentioned in Adjust. B occurs when the reading is 50" on the main arc.

D. To correct for eccentricity.—A common error to which all sextants are liable is eccentricity of the centre of motion of the index-arm and the centre of the graduated arc. It unfortunately admits of no adjustment, but corrections for it may be obtained as follows: "As it has no appreciable effect on small angles, it is advisable—using the artificial horizon—to take a set of altitudes, say 10, which will form a mean of about 100° on the arc, noting the time of each accurately by a trustworthy chronometer; should the time so found coincide with the known rate of the chronometer there is no error. Should the results differ by several seconds of time, it may be assumed that the error of the instrument, combined with personal error, has caused it. By the rate at which the sun was rising or going down during the observations, the amount of angle due to those seconds is easily found (see Sec. 195). Half that amount will be the error of the sextant upon that angle. As an EXAMPLE, suppose by a morning observation the true reflected altitude = 100°, while the instrument made it 100° 01', the calculation would make it about 3 seconds later than the truth. In the afternoon a similar error would make it 3 seconds earlier. Thus a disagreement of about 6 seconds arises for about 1' of altitude. By 4 or 5 such sets of altitudes at different parts of the arc sufficient data will be procured from which to form a table of corrections for all altitudes."

142. The sextant, unlike the transit, has the apex of the angle which it measures not coincident with any particular part of the instrument, but varying its position according to the magnitude of the angle observed. This is due to what is usually called **Instrumental Parallax**, and arises from the fact that the index-glass is not situated in the direct line of sight. This may be best shown by means of Fig. 75.

Suppose S and R are two objects, the angle between which we wish to measure. When the index-arm has been so placed that the image of S is reflected from the index-glass I, so as to coincide with R as seen directly through the horizon-glass H, the angle which is given by the sextant is the angle SAR, where A is a point in the line of sight, found by producing SI to its intersection. But suppose S' and R were the two objects between which the angle is to be observed, then a will be the apex of the angle measured. Finally, if S is situated at s, so that sI is parallel to RA, then the angle given by the sextant between s and $R = 0°$ (i.e., if there were no index-error the reading should be zero), and if the reflection of R were brought to coincide with R as directly seen, then the angle observed would be negative, and would thus be read on the "arc of excess," and be equivalent to IRA. If R is at a distance from the instrument so great that RI and RA are sensibly parallel,— as was assumed in Adjustment C,—the question of instrumental

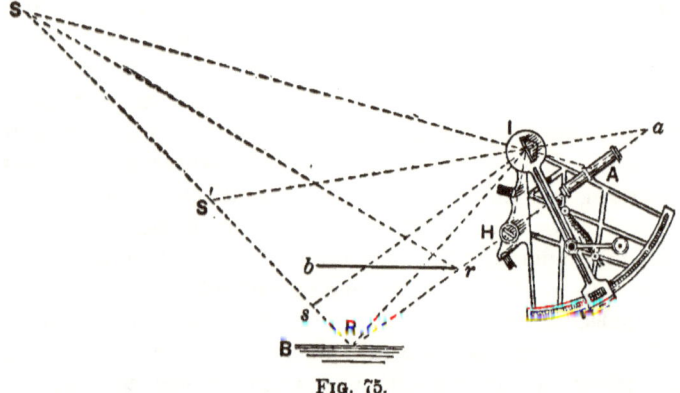

Fig. 75.

parallax may be ignored; but in measuring angles between two objects when the object directly looked at is *near at hand*, the instrument must be either so placed that the apex will coincide with the position at which the angle is to be observed, or else a correction applied, the angle as given by the sextant—taking, say, the index-glass as the constant apex of the angle—being always too small.

In using an artificial horizon there is another form of parallax which sometimes needs consideration due to the apex A of the angle observed not coinciding with the artificial horizon. Let R be the image of a star S reflected in the artificial horizon Then if SA is parallel to SR, as is sensibly the case when dealing with objects at a considerable distance from the instrument, the angle SAR may be considered equal to twice the angle SRB; i.e., the altitude read on the sextant is the "double-altitude" of the star, which needs dividing by two in order to obtain the altitude; but where S is comparatively close at hand, then we cannot consider $SAR = 2SRB$, and consequently by dividing the reading on the arc by two, it is not the altitude as reflected *from the horizon* which is observed, but from a point r so situated that the angle ASr is equal to the angle RSr. Suppose we select this point r in the line of sight, as in Fig. 75, then it may be easily proved that if rb is parallel to RB (the surface of the artificial horizon) $Srb = \frac{1}{2}SAR$. And since the sines of small angles may be assumed to be proportional to the angles themselves, we may consider the point r to be situated half way between A and R. Thus in observing an altitude with the artificial horizon, where the distance RA is appreciable compared with the distance SA, it becomes necessary either to apply a correction, or to arrange the positions of the horizon and the instrument so that the point r may coincide with the apex of the angle which it is wished to observe.

143. A sextant is usually only graduated up to about 140°. For nautical work this is amply sufficient, but where an artificial horizon is used—since the angle read is double the real altitude—the altitude will be limited to about 70°. To obviate this difficulty, sextants are often supplied with a contrivance which consists of a small mirror below the index-glass, fixed in such a position that when the index is at the mark numbered 180° upon what is called the SUPPLEMENTARY ARC, those two mirrors are at right angles to each other, and the objects whose images appear to coincide in direction really lie in diametrically opposite directions.

144. In observing angles with the sextant, when the two objects and the observer's eye are not in the same horizontal plane, in order that the angle measured may be a horizontal one, it becomes necessary either to arrange matters in such a way that the angle observed between the objects may be the horizontal angle, or to apply a correction to the angle observed.

In the former case two vertical rods may be ranged in line with the objects and the observer's eye, and the angle between them then measured with the plane of the sextant horizontal. But the most accurate method is to observe the angle between the objects themselves, and then to observe the angle of altitude, or depression of each.

Thus, in Fig. 76, let A and B be the two objects, O the position of the observer. Then if Z be the zenith and a and b points where the vertical planes through A and B respectively intersect the horizontal plane abO, then Aa and Bb represent respectively the altitudes of A and B, and the complement of

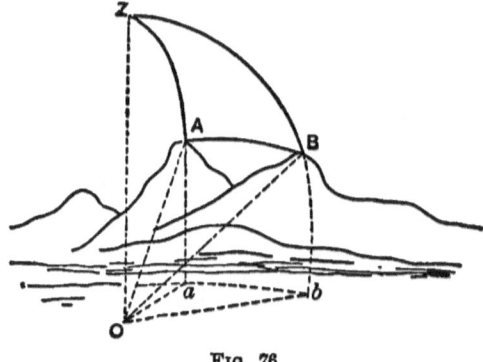

Fig. 76.

the altitude of each equals its "zenith distance," AZ or BZ. Then in the spherical triangle ABZ, since we know all three sides, therefore (since $ab = Z$)

$$\cos \frac{ab}{2} = \sqrt{\frac{\sin S \sin (S - AB)}{\sin AZ \sin BZ}},$$

where $S = \dfrac{AZ + BZ + AB}{2}$.

145. Every possible means should be taken in observing angles with a sextant to *eliminate instrumental errors*. In order to do this all careful observations should be in "doubles:" thus if the observation is for latitude, a star north and a star south should be observed; the errors of the instrument will then affect the result in opposite directions, and taking the mean of the results will eliminate the errors. So also an observation for time should be taken in "doubles;" namely, a star east and a star west. Also in taking Lunar Distances the sets should be taken in "doubles," one set of distances to a star east of the moon and one to a star west.

The Artificial Horizon.

146. The best substance to use for an artificial horizon is mercury, mainly on account of its bright reflecting surface. In a wind, however, syrup is better than mercury, being more viscous and consequently less liable to be affected by currents of air, but its reflecting surface is decidedly inferior. Oil, too, is frequently made use of. A sheet of water on a still night makes a fairly good horizon.

Black glass horizons, which can be levelled up by means of adjusting screws, are sometimes used, but though at times more convenient than a liquid surface they are considerably less reliable. The best way to carry mercury is in an iron bottle, which can be made by any blacksmith out of a piece of iron pipe, fitted with a screw stopper in the cap. Mercury must be kept carefully away from all greasy substances, and also from lead, gold, or silver, with which it amalgamates. A glass cover in the form of a triangular prism is often of use in shielding the horizon from the wind; but owing to the increased probability of error, due to refraction in the cover itself, it is to be avoided when possible. The mercury can usually be protected from the wind by placing it in a hole slightly below the general surface of the ground, or by building up a sort of protection around it. A wooden trough makes the best form of saucer to hold it in; copper also does well. It should have an outlet at one corner to facilitate the pouring back of the mercury into the bottle. About 5 inches by 3 inches is a good size for the trough. It should also be of about uniform depth, which need not exceed half an inch.

To PREPARE THE HORIZON, pour the mercury into a small chamois-leather bag, leaving, however, a little behind in the bottle as "scum," and then squeeze it out gently into the trough. The surface so obtained is usually as clear as could be wished for, but if the trough or the leather happens to have been a little dirty, a film of dust will sometimes be found on the surface. This can easily be cleared away by sweeping it lightly with a feather. The horizon is then ready for use.

If a class cover is used over it, the observation should be taken twice, the cover being turned around for the second observation, and the mean of the results taken; in this way the error arising from the refraction of the glass is more or less eliminated.

The mercury should always be carried as steadily as possible, the bottle being kept "end up."

Altitudes less than about 6° cannot be read with the artificial horizon on account of the obliquity of the rays.

An artificial horizon is almost always to be preferred to a natural horizon, such as is given at sea, on account of the refraction of the air, as regards the horizon itself, not entering appreciably into the question.

The Chronometer.

147. Chronometers have been found by experience, when subjected to the shakings and joltings which necessarily more or less accompany their transportation on land, to be very unreliable instruments. A small pocket-chronometer is usually almost as reliable for land work as one of larger and finer make, being less liable to derangement.

As regards the care of chronometers, they should always be kept as much as possible in the same position, and be always wound at the same time of day, and *wound to the butt.* Also, they must be kept away from all magnetic influence, such as is often caused by their proximity to iron. They should, of course, be rated before starting out, but if they are new chronometers they will probably gain on their "rate." The "shop-rate" is almost always different from the field-rate, so that really very little dependence is to be placed on them compared with that placed on chronometers at sea. But though the rate when out on the work may be entirely different from what it was before starting, yet the rate in the field will be more or less constant; and though no great dependence can be placed on the actual position as given by a chronometer after considerable jogging and jolting, yet it serves to connect the various stations observed, relatively to each other, with a fair amount of accuracy when the intervals of time between the observations are not great. These positions can then be finally corrected after the general field-rate of the chronometer has been ascertained.

As regards allowing for temperature, that can only be done by an actual testing at different temperatures. Every chronometer goes fastest in some certain temperature which has to be calculated from the rates that it makes at three fixed temperatures; then as the temperature varies from that at which the chronometer goes fastest, so its rates vary in the ratio of the square of the distance in degrees of temperature from its maximum gaining temperature. A fair test for a pocket-chronometer is to place it in four extreme positions and let it stay in each for 24 hours; if the rate for any position does not vary by more than five seconds from the rate in any other position, the watch is as good as can generally be found.

BAROMETERS.

148. There are two kinds of barometers used in exploratory surveying—the "CISTERN" form of the mercurial barometer, and the "ANEROID."

The Cistern barometer, owing to its size, is mainly suitable for use in camp as a standard with which the Aneroids may be compared.

The nature of the difficulties involved in observing the difference of elevation between any two points may be best shown as follows:

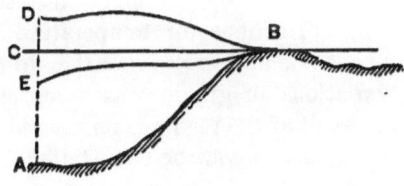

FIG. 77.

In Fig. 77, suppose we have two stations, A and B, whose difference in elevation we wish to determine. If the atmosphere were in a state of rest there would be no difficulty in devising formulæ which should give correct results, supposing the instruments themselves recorded correctly, for then the barometric reading along the horizontal line CB would at all points be the same, and we should simply have to obtain a formula founded on Boyle and Mariotte's law for the pressure of gases, to obtain the difference in the heights of A and C which should correspond with the observed difference in pressure. But since the atmosphere is always more or less subject to disturbing influences, such as temperature, humidity, etc., which cause the *barometric gradient* at B to assume such forms as BD or BE, no formula founded on statical principles can possibly be expected to give correct results; yet any formula which attempts to take account of the fluctuations in gradient necessitates a knowledge of the temperature, humidity, and general state of the atmosphere between A and B, which it is impossible to obtain. By taking observations at points immediately between A and B some allowance may be made for these various disturbances, but as a rule very little is gained by so doing compared with the time and labor which it involves.

Since the variations in gradient are generally too rapid to allow of the state of the atmosphere at one hour being of much service in indicating its probable condition a few hours—or even minutes—later, it follows that labor spent in reducing barometric readings between two such stations as A and B, by applying corrections for latitude and various other requirements which are often employed, simply results in a mathematical illusion which is possibly erroneous to the extent of 50 or 100 p. c.

The best way to proceed in ordinary practice is to make use of formulæ which assume the air to be in a state of equilibrium—applying corrections for temperature which experience has shown to be necessary—and then to eliminate the errors due to variations in gradient as much as possible by taking the mean result of the readings on several occasions, or by observing simultaneously at the two stations, as described in Sec. 150.

149. The first information necessary in devising a formula for the reduction of barometric readings is the relative weight of mercury and air. This ratio amounts to about 1050, depending upon which values of the densities are employed. The barometer at the time is supposed to be at sea-level in latitude 45° at a temperature of 32° F. This ratio, if multiplied by 5.74—which is a factor obtained from Boyle and Mariotte's law that the density of a gas varies directly as the pressure to which it is subjected—gives a product known as the *barometric coefficient*. Various values are given for this coefficient, but probably that given by Regnault is the most accurate, namely, 60,884; from this, taking no account of the effects of temperature or latitude, we find that the difference in elevation in feet equals

$$X = 60384 \log \frac{H}{h},$$

where H is the barometric reading at the lower station and h

is the barometric reading at the upper station. The correction for *temperature*, as usually applied, assumes that the mean temperature of the air *between* A and B is the mean temperature of the air *at* the two stations. If we then take .004 as the coefficient of expansion of air for 1° Centigrade, the above formula needs multiplying by $1 + .002(T + t)$, where T and t are the temperatures on the Centigrade scale at the lower and the upper station, respectively; and if we take T and t as the temperatures on the Fahrenheit scale, then this factor becomes

$$1 + \frac{T + t - 64}{900},$$

and this is usually called the "temperature term."

Another factor is often employed to correct for the different effects of gravity, due to difference of *latitude*. According to Laplace, this "latitude term" equals

$$1 + .0026 \cos 2L,$$

where $L =$ the latitude. He also applied a correction for the effect of altitude above sea-level on the force of gravity; but this may be altogether neglected. A correction is also sometimes applied to allow for the effect of temperature on the barometers themselves—which is ascertained by having thermometers attached to them. And since changes of temperature affect both the mercury and the scales in opposite directions, if we take .0001 as the *relative* expansion of mercury for 1° F. to the expansion of the scales, in order to correct the barometers themselves for temperature, the above value of X should be multiplied by

$$\frac{1}{1 - .0001(T' - t')},$$

where T' and t' are the temperatures as recorded by the "attached" thermometer at the lower and the upper station, respectively.

Thus the *complete formula* becomes

$$X = 60384 \log \frac{H}{h} \left(1 + \frac{T + t - 64}{900}\right) \times$$
$$\left(1 + .0026 \cos 2L\right) \left(\frac{1}{1 - .0001(T' - t')}\right).$$

A correction for humidity is sometimes applied, but it necessitates observations of the state of the air being taken with a hygrometer; and since it is doubtful, even then, whether any material advantage is derived by so doing, we may ignore this correction entirely. We may simplify the above equation considerably by dispensing with the latitude term, which in ordinary practice is never required. In aneroid barometers the last term of course does not enter into the question at all; so that the formula generally applicable to *aneroid barometers* is

$$X = 60384 \log \frac{H}{h}\left(1 + \frac{T+t-64}{900}\right).$$

If H and h do not differ by more than about 3000 feet we may do away with the logarithms in the above equation, which thus becomes, approximately,

$$X = 52450 \frac{H-h}{H+h}\left(1 + \frac{T+t-64}{900}\right).$$

The error involved by this formula is inappreciable within the limits stated.

By assuming $(T+t)$ to equal 108° this formula becomes

$$X = 55000 \frac{H-h}{H+h},$$

which is generally known as Belville's Formula and is convenient for rough work.

The table opposite gives the VALUES OF $\left(\frac{T+t-64}{900}\right)$.

150. The results which are obtained by using only one barometer, carrying it from station to station, are of course subject to all the errors of gradient; and these errors usually increase with the distance between the two stations; but by taking the mean of *several* results, the probable error becomes greatly reduced. (See Sec. 204.) Errors of gradient may be more or less eliminated by using TWO BAROMETERS, and observing simultaneously at each station, the barometers being

EXPLORATORY SURVEYING.

$T+t$	$\dfrac{T+t-64}{900}$	$T+t$	$\dfrac{T+t-64}{900}$	$T+t$	$\dfrac{T+t-64}{900}$	$T+t$	$\dfrac{T+t-64}{900}$
20°	−.0489°	66°	+.0022°	112°	+.0533°	158°	+.1044°
22	.0467	68	.0044	114	.0556	160	.1067
24	.0444	70	.0067	116	.0578	162	.1089
26	.0422	72	.0089	118	.0600	164	.1111
28	.0400	74	.0111	120	.0622	166	.1133
30	−.0378	76	+.0133	122	+.0644	168	+.1156
32	.0356	78	.0156	124	.0667	170	.1178
34	.0333	80	.0178	126	.0689	172	.1200
36	.0311	82	.0200	128	.0711	174	.1222
38	.0289	84	.0222	130	.0733	176	.1244
40	−.0267	86	+.0244	132	+.0756	178	+.1267
42	.0244	88	.0267	134	.0778	180	.1289
44	.0222	90	.0289	136	.0800	182	.1311
46	.0200	92	.0311	138	.0822	184	.1333
48	.0178	94	.0333	140	.0844	186	.1356
50	−.0156	96	+.0356	142	+.0867	188	+.1378
52	.0133	98	.0378	144	.0878	190	.1400
54	.0111	100	.0400	146	.0911	192	.1422
56	.0089	102	.0422	148	.0933	194	.1444
58	.0067	104	.0444	150	.0956	196	.1467
60	−.0044	106	+.0467	152	+.0978	198	+.1489
62	.0022	108	.0489	154	.1000	200	.1511
64	.0000	110	.0511	156	.1022	202	.1533

compared before and after the observations: and these errors may of course be still further reduced by taking the mean of several simultaneous observations; and in this way the best results can probably be obtained. But between two stations there is usually a *permanent gradient* dependent on local causes, such as the topography and nature of the ground, which no number of observations would tend to eliminate, and for which allowance can rarely be made. It is largely due to this cause that the heights of mountains, calculated from the mean of a large number of observations which differ but little from each other, are often found, when obtained by more accurate means, to be very largely in error.

151. There are two or three points in connection with the READING OF BAROMETERS that are worth remembering. For instance, readings should never be taken in the immediate vicinity of any body which obstructs the wind. "If the barometer is observed on the windward side of a mountain the reading will be too high; if on the leeward side, too low." Neither should readings ever be taken directly before or after a storm of wind or shower of rain, as the atmosphere is then usually in an unsettled state.

152. "The pressure of the air everywhere undergoes a

daily oscillation. The gradient introduced by this daily change is called the DIURNAL GRADIENT. The pressure has two maxima and two minima which are easily distinguishable. Near the sea-level the barometer attains its maximum about 9 or 10 A.M. In the afternoon there is a minimum about 3 to 5 P.M.; it then rises until 10 to midnight, when it falls again until about 4 A.M., and again rises to attain its forenoon maximum. The day fluctuations are the larger."

"The annual progress of the sun from tropic to tropic throws a preponderance of heat first on one side of the equator and then on the other, which produces an annual cycle of changes in the pressure, and gives rise to what has been called the ANNUAL GRADIENT. The amount of this variation is quite small, but increases rapidly toward the poles; at the equator it rarely exceeds one quarter of an inch per year, while in the polar regions it is often as much as two or three inches in a few days."

We will now consider the barometers themselves.

A. The Cistern Barometer.

153. This is an awkward instrument to carry about, but its usefulness on exploratory work usually fully makes up for the inconvenience which it causes. It is found by experience to be absolutely necessary in carrying forward an extended system of barometric observations to have at hand a standard barometer with which the aneroids may be from time to time compared.

A supply of tubes and mercury should accompany the barometer in case of accident, and it should be provided with a wooden and leather case. When moved from one place to another, even across the room, it should be screwed up so that the tube and cistern will be perfectly full, and gently turned over, end for end, so that the cistern will be uppermost. In wheeled vehicles it should be carried by hand, and on horseback strapped across the rider's shoulder. By carrying it with the cistern uppermost any particles of air which may be contained in the mercury become disengaged by the jolting, and escape at the end where they do no harm.

154. TO FILL A BAROMETER, should it become necessary to do so in the field, proceed as follows: Warm both the mercury and tube and filter in through a paper funnel—the hole of which does not exceed $\frac{1}{50}$ of an inch—to about $\frac{1}{4}$ of an inch from the top. Close the end and turn the tube on its side; the mercury will then form a bubble which can be made to travel from end to end and gather all the small air-bubbles visible that adhered to the inside of the tube while filling. Let the bubble pass to the open end, fill up with mercury and close the tube. Reverse the tube over a basin, when, by slightly relieving the pressure against the end, some of the mercury will be forced out, forming a vacuum above, which ought not to exceed half an inch. Close up again tightly and let this vacuum-bubble traverse the length of the tube as before, on the several sides, absorbing the minute portions of air still left, now greatly expanded by the reduction in pressure. Perfect freedom from air can be detected by the sharp concussion with which the mercury beats against the sealed end, when, with a large vacuum-bubble, the horizontally held tube is slightly moved. Any air which may still be left—which will probably not affect the reading by more than a few thousandths of an inch—will soon escape if the barometer is carried about cistern uppermost.

Filling by boiling is a slightly more efficient method, but it is a much more difficult proceeding.

155. IN READING THE BAROMETER, first of all note the temperature on the attached thermometer, then screw up the mercury in the cistern so that its surface just touches the ivory point, being careful that the barometer hangs vertically. Give a gentle tap near the top of the mercurial column to destroy the adhesion of the mercury. Set the vernier by bringing its front and back edges into the same horizontal plane with the top of the mercury; then read.

156. Should the mercury in the cistern become so dirty that neither the ivory point nor its reflection in the mercury can be seen, the instrument must be taken apart and cleaned. To do this "screw up the adjusting screw at the bottom until the mercury entirely fills the tube, carefully invert, place the instrument firmly in an upright position, unscrew and take off the brass casing which encloses the wooden and leather parts of the cistern. Remove the screws and lift off the upper wooden piece to which the bag is attached; the mercury will then be exposed. By then inclining the instrument a little, a portion of the mercury in the cistern may be poured out into a clean vessel at hand to receive it, when the end of the tube will be exposed. This is to be closed by the gloved hand, when the instrument can be inverted, the cistern emptied, and the tube brought again to the upright position. Great care must be taken not to permit any mercury to pass out of the tube. The long screws which fasten the glass portion of the cistern to the other parts can then be taken off, the various parts wiped with a clean cloth and restored to their former position." Everything used in the operation must be clean and dry, and all breathing on the parts avoided as much as possible.

If the mercury is dusty or dimmed by oxide it may be cleaned by filtering through chamois leather, but if chemically impure it must be rejected and fresh mercury substituted. The cistern should then be filled as nearly as possible and the wooden portion put together and fastened. The screw at the bottom of the instrument should then be screwed up. "The instrument can then be inverted, hung up and readjusted. The tube and its contents having been undisturbed, the instrument should read the same as before."

B. The Aneroid Barometer.

157. The "Aneroid" is a valuable instrument for engineering and exploratory purposes on account of its portability, and though not to be compared in accuracy with the mercurial barometer, the results given by it will often not differ from those given by the latter sufficiently to be of importance. It is in such cases as these that the aneroid is eminently useful. But it is too liable to derangement, and subject to too many defects, to warrant its being used in any other way than to supplement some more accurate form of obtaining elevations. In dealing with the mercurial barometer, after the correction for temperature has been applied, the instrumental errors which need correcting are very small; but with an aneroid the same cannot be said. Most of the better class of aneroids are supposed to compensate automatically for changes in temperature. This compensation should be tested by comparison at different temperatures with a standard barometer, and the errors tabulated and kept for future reference.

While reading, the aneroid should always be held horizontally, for the weight of the parts themselves has a very considerable influence on the readings: a difference corresponding to fifty feet being not uncommon when held in different positions. The aneroid may be adjusted by means of the small screw at its back, so as to agree with the reading of a standard barometer, but when the difference is only slight it is better to regard it as an "index error," and correct in that way, than to alter the reading.

158. Cheap aneroids commonly have the SCALE of inches subdivided so as to read the elevations above sea-level. This would be very convenient if only the corresponding pressure at the sea-level were always the same as given on the index and the atmosphere always in a state of equilibrium. The pressure at the sea-level is generally assumed as being equivalent to 30 inches.

Another method which is convenient, though "unscientific and inaccurate," is that of having a movable scale of elevations which can be set to agree with the barometer reading at any known elevation. But the best way to obtain a reading is to observe the reading in inches, and then to reduce it by one of the formulæ already given.

BAROMETRIC AND ATMOSPHERIC HEIGHTS.

Bar. in.	Alt'de feet.	Bar. in.	Alt'de. feet.	Bar. in.	Alt'de. feet.	Bar. in.	Alt'de. feet.	Bar. in.	Alt'de. feet.
21.	9900.1	**23.**	7375.1	**25.**	5060.6	**27.**	2924.4	**29.**	940.9
.1	9768.3	.1	7254.7	.1	4949.8	.1	2821.8	.1	845.4
.2	9637.1	.2	7134.7	.2	4839.5	.2	2719.6	.2	750.2
.3	9506.5	.3	7015.3	.3	4729.6	.3	2617.8	.3	655.3
.4	9376.4	.4	6896.5	.4	4620.1	.4	2516.3	.4	560.7
.5	9247.0	.5	6778.1	.5	4511.0	.5	2415.2	.5	466.5
.6	9118.3	.6	6660.2	.6	4402.3	.6	2314.4	.6	372.6
.7	8990.0	.7	6542.8	.7	4294.0	.7	2214.0	.7	279.0
.8	8862.4	.8	6426.0	.8	4186.3	.8	2114.0	.8	185.7
.9	8735.3	.9	6309.6	.9	4078.9	.9	2014.3	.9	92.7
22.	8608.9	**24.**	6193.8	**26.**	3971.9	**28.**	1915.0	**30.**	0.0000
.1	8483.0	.1	6078.3	.1	3865.4	.1	1816.0	.1	— 92.5
.2	8357.7	.2	5963.4	.2	3759.3	.2	1717.4	.2	— 184.7
.3	8233.0	.3	5848.9	.3	3653.6	.3	1619.2	.3	— 276.6
.4	8108.7	.4	5734.9	.4	3548.3	.4	1521.3	.4	— 368.2
.5	7985.1	.5	5621.4	.5	3443.4	.5	1423.7	.5	— 459.5
.6	7862.0	.6	5508.3	.6	3338.8	.6	1326.5	.6	— 550.6
.7	7739.4	.7	5395.7	.7	3234.6	.7	1229.6	.7	— 641.4
.8	7617.5	.8	5283.6	.8	3130.8	.8	1133.0	.8	— 731.9
.9	7495.9	.9	5171.9	.9	3027.4	.9	1036.8	.9	— 822.2

No advantage seems to be gained by the use of large aneroids; in fact experience shows that when the barometer is subjected to much shaking, the best work is usually done by instruments not exceeding 3 inches in diameter. The elevations according to which the elevation-scales on aneroids are usually divided are as given on the preceding page, and are obtained by a formula similar to those already given, assuming the temperature to be 60° Fahr.

Many scales, however, adopt a temperature of 32° F., in which case the corresponding elevations will be reduced in the proportion of 1.058 to 1.

The uncertainty which is connected with barometric observations is greatly dependent on the latitude; the barometric pressure being very much more regular in the tropics than in the polar regions.

EXPLORATORY SURVEYS.

159. There are three distinct ways in which exploratory surveys may be carried on:
    A. By a series of triangulations.
    B. By direct measurement and compass courses.
    C. By astronomical observations.

And though usually an explorer makes use more or less of all three methods, it will be better for the sake of clearness to consider each separately.

A. By a Series of Triangulations.

The method of triangulating is mainly suitable to mountainous country, or at any rate to country where a view of distant mountain-peaks is to be had.

Before, however, considering the practical working of this system, it will be well to deal with a few of the principal trigonometrical problems which arise in work of this sort.

In Sec. 59 we have already dealt with some of the simpler forms of triangulation, suitable in cases where a straight line has to be continued over an inaccessible surface; but we will here consider the cases of obtaining distances and directions of points relatively to each other.

160. Given two inaccessible points A and B, to find their distance apart and bearing relatively to each other.—In Fig. 78 let CD be a line the length and bearing of

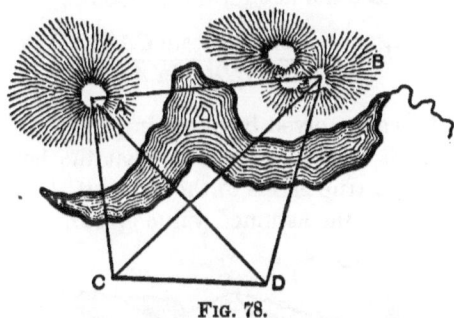

Fig. 78.

which are known. Observe the angles ACD, BCD, ADC, and BDC. Then in the triangle CDA we have the angles at C and D and the length CD, and can thus find CA. Similarly in the triangle CBD we can find CB. Then in the triangle CAB we have the side CA and CB and the angle at C, from which we can obtain the distance AB and its bearing relatively to CD.

The following equations, however, reduce the work which the direct solution given above involves. Find an angle K such that

$$\tan K = \frac{\sin ADC \sin CBD}{\sin CAD \sin BDC};$$

then

$$\tan\left(\frac{CAB - ABC}{2}\right) = \tan(45° - K)\cot\frac{ACB}{2};$$

then

$$CAB = \frac{CAB + ABC}{2} + \frac{CAB - ABC}{2},$$

and

$$AB = CD\,\frac{\sin BDC \sin ACB}{\sin CBD \sin CAB}.$$

If C can be ranged in line with A and B we can then find the position of A and B separately, as shown in Sec. 59; the difference of the distances so obtained gives the length of AB, and the bearing is obtained by direct observation.

176   EXPLORATORY SURVEYING.

Suppose, however, that in Fig. 78 the length and direction of AB is known, and it is the distance CD which is required.

Then observe the angles at C and D and obtain CAB as before, but in this case the last formula becomes

$$CD = AB \frac{\sin CBD \sin CAB}{\sin BDC \sin ACB}.$$

This might be also solved by assuming a certain length for CD, and from it finding as above what the length of AB must be; then the true AB is to the value of AB so obtained as the true CD is to the assumed value of CD.

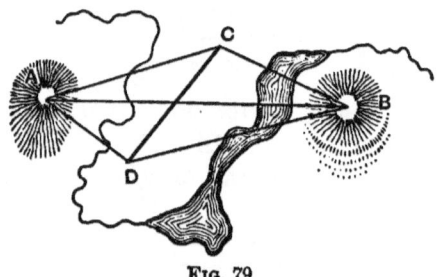

Fig. 79.

If, as in Fig. 79, the lines AB and CD cross each other, the above formulæ apply equally well.

161. The problem known commonly as the "**Three-point Problem**" is probably the most useful method there is of establishing the position of any given point; it is as follows:

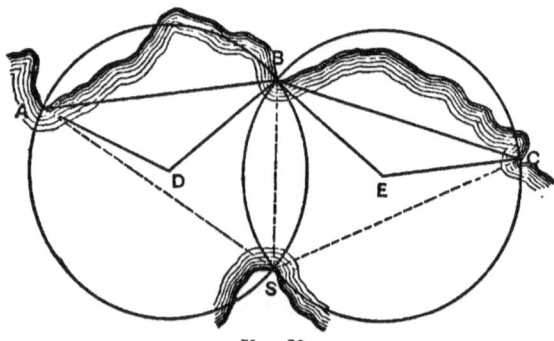

Fig. 80.

Suppose, as in Fig. 80, we know the position of three points A, B, and C and wish to fix the position of the point S; we can do it by simply observing the angle ASB and BSC.

EXPLORATORY SURVEYING. 177

Then, in order to obtain the position of S *geometrically*, proceed as follows:

Find D, the centre of the circle ABS (by setting off at A and B angles equal to $90° - ASB$). Then draw the circle through the points A, B, and S. Similarly find the centre E and draw the circle BCS. Then S, the point of intersection opposite B, is the position required.

When one of the angles is obtuse, set off its difference from $90°$ on the opposite side of the line joining the two objects to that on which the point of observation lies.

When the angle $ABC =$ the supplement of the sum of the two angles, the position of S will be indeterminate by this method.

S may often be obtained with sufficient accuracy *instrumentally* by plotting the angles ASB and BSC on a piece of tracing-cloth, and sliding it over the plan until the required position is obtained. The "station-pointer" is an instrument much used for this purpose, especially in hydrographers' offices, where soundings are usually plotted in this way.

If accuracy is required the position of S may be found *analytically* thus, as given by Prof. Gillespie:

Let $AB = c$; $BC = a$; $ABC = B$; $ASB = S$; and $BSC = S'$.
Also make $T = 360° - S - S' - B$,
and let $BAS = U$, and $BCS = V$.
Then

$$\cot U = \cot T \left(\frac{c \sin S'}{a \sin S \cos T} + 1 \right);$$

$$V = T - U;$$

$$SB = \frac{c \sin U}{\sin S}, \quad \text{or} \quad SB = \frac{a \sin V}{\sin S'};$$

$$SA = \frac{c \sin ABS}{\sin S}, \quad \text{and} \quad SC = \frac{a \sin CBS}{\sin S'}.$$

Thus if $ASB = 33° 45'$, $BSC = 22° 30'$,
$AB = 6000$ ft. and $BC = 4000$ ft.,
we find $ABC = 104° 28' 39''$.
Then $U = 105° 08' 10''$;
whence $V = 94° 08' 11''$.
$SB = 10425.1$ ft., $SA = 7101.9$ ft., and $SC = 9342.9$ ft.

162. The position of a point may also be fixed by observing the bearings from it of **two known points**, and may be found on the plan by drawing through those points the bearings so obtained; their intersection gives the point required.

163. Another common method of fixing the positions of

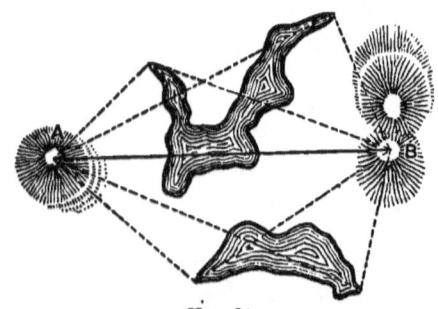

FIG. 81.

outlying points is by **intersection**, as in Fig. 81, the position of the two points of observation A and B being known.

164. While on the subject of triangulation, it will be as well to consider the methods of obtaining **the heights of mountains trigonometrically**.

In the first place, suppose we are able, as in Fig. 82, to ob-

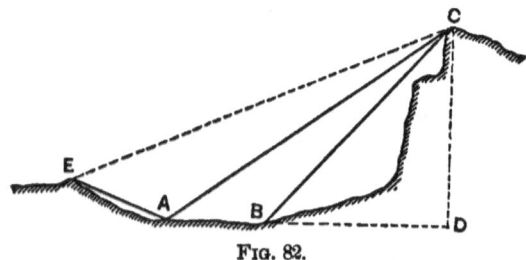

FIG. 82.

tain two points A and B in the direction of C (a point the elevation of which we wish to obtain) both at the same elevation, and to measure the distance between them; then

$$CD = \frac{AB}{\cot CAD - \cot CBD}.$$

If, however, the two points cannot be taken at the same level, but have to be taken such as E and A, observe the angle CEA,

EXPLORATORY SURVEYING.

and at A the altitudes of C and E, either with an artificial horizon or with the vertical arc of a transit. Then in the triangle EAC

$$AC = EA \sin E \operatorname{cosec} C,$$

where the angle at $C =$ the sum of the altitudes of E and C (taken at A) — the angle at E. Then $CD = AC \sin CAD$.

This would of course hold equally good if EA sloped the other way, but then $C =$ alt. of C from A — alt. of A from E — angle at E. The correction for curvature and refraction given in Sec. 51 must be added to the height as obtained above.

But suppose it is not convenient to obtain a base as above in the same direction as C. Then, as in Fig. 83, measure a

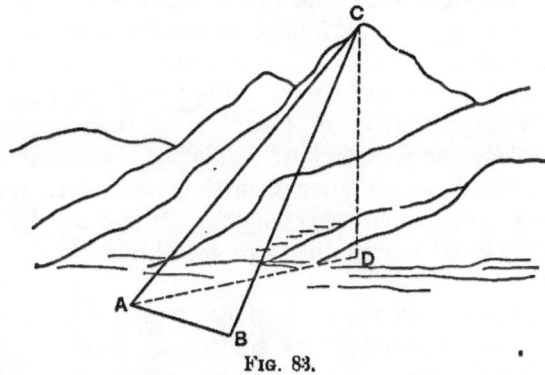

FIG. 83.

base AB (not necessarily level) and observe the angles CAB and CBA. Then in the triangle ABC

$$AC = AB \sin B \operatorname{cosec} C.$$

Next observe the altitude of C from A, i.e., the angle CAD; then

$$CD = AC \sin CAD.$$

To the height so obtained, the correction for curvature and refraction given in Sec. 51 should be added.

Suppose it is required to find the *difference in elevation* of two inaccessible points, the simplest way is to find the elevation of each separately, as above, and subtract the one from the other.

165. In observing altitudes, the **refraction of the air** enters so largely into the question and varies so enormously according to the condition of the atmosphere, that every precaution must be taken to eliminate the errors due to it, where accurate work is wanted.

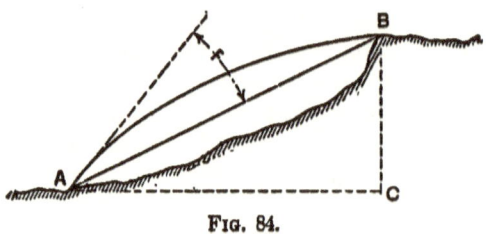

FIG. 84.

Its nature is such that suppose A and B are two stations visible from each other, the line of sight between A and B, instead of being straight, follows a curved course as shown in Fig. 84, making the altitude as observed at A too great, by the amount F, which is termed the "angle of refraction." Similarly the *depression* of A as observed from B will be too small. Thus the tendency of refraction is to make objects appear at a higher elevation than they really are; so that in observing altitudes a correction for refraction should be always subtracted from the *apparent* altitude to obtain the *true* altitude.

In ordinary work the corrections given in Sec. 51 for both curvature and refraction are sufficiently correct. But for highly accurate work—on which this article does not treat—various allowances and corrections must be made.

Refraction diminishes with altitude and is slightly greater over water than land. It is generally at its maximum during the night, and at its minimum about noon; but it is *steadier* in the night than in the day time, and for this reason night work is usually as reliable as work done during the day. About sunrise and sunset are the worst times to observe altitudes, for not only is refraction then high in quantity, but also extremely variable. A day with the sky overcast is a good day on which to take an observation. Clear days are more subject to rapid changes than dull ones. (For Astronomical Refraction, see Sec. 184.)

166. A method of eliminating to a great extent the effect of refraction in observing the difference of elevation of two stations A and B, is that of observing **Reciprocal Angles.** Thus in Fig. 84, at A, the altitude of B should be observed, and at B (when practicable) the depression of A. Half the difference of these angles will be the combined correction, and the tangent of half their sum, multiplied by the horizontal distance between them, will give the difference of level, after adding the correction for *curvature of the earth* given in Sec. 51. This method assumes that the coefficient of refraction is the same at both A and B; therefore the angles should, if possible, be observed simultaneously, lest the refracting power of the air should change in the interval. (For the correction for Refraction, see Sec. 51.)

167. To obtain the height of a mountain by the observed depression of the sea horizon.—The depression of the horizon, or as it is commonly called at sea the "Dip," taking $R =$ the earth's mean radius of curvature in feet, equals in seconds

$$D = 206265 \sqrt{\frac{2H}{R}}$$
$$= 63.8 \sqrt{H};$$

therefore

$$\sqrt{H} = \frac{D}{63.8},$$

where $H =$ Height in feet.

Thus, were it not for refraction, we could find the elevation of A (Fig. 85) by merely observing the dip D. But D' is the dip actually observed; so that, taking refraction into account, the above formula becomes

$$\sqrt{H} = \frac{D'}{55} \text{ (nearly),}$$

which can only be depended on to give approximate results.

168. *In observing altitudes with a sextant and artificial horizon,* as for instance in Fig. 84, the altitude of B will be one half the altitude read on the arc, since it is the

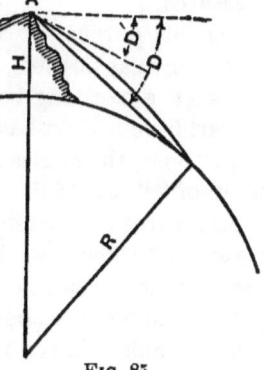

Fig. 85.

"double altitude" that is actually observed. To find a point C on the same level as the instrument the altitude can then be measured down from AB. *To observe the depression* of A from B with a sextant and artificial horizon, we must establish some point—as far off as possible so as to reduce parallax—the altitude of which exceeds about 6°, and observe its altitude correctly, and then obtain the angle between it and the object whose depression we wish to find. At night a star may often be made use of for this purpose, allowance being made for its motion. This method may also be employed in reading altitudes which would otherwise need the use of a supplementary arc. (See Parallax, Sec. 142.)

To read an altitude or depression with a transit, observe the altitude first in the usual way, then "reverse" and point the telescope to the object and read its supplement; the mean altitude so obtained is free from error due to the "horizontal axis" not being truly perpendicular to the "vertical axis" of the instrument. The errors of graduation and observation are also somewhat reduced.*

169. It is essential that a survey which consists of a series of triangulations should have an accurate base to start from. Sometimes in exploratory surveys the distance between two mountain peaks, or some prominent objects near the point at which the survey starts, is already known with sufficient accuracy to warrant the line joining them being accepted as a base, but more usually it is necessary to obtain the distance between such points from a base more or less accurately measured.

For this purpose of course as level a piece of ground must be obtained as possible, and as there is often difficulty in finding such a site long enough for a base, it becomes necessary to start from a short base and then extend it by a series of triangulations, the angles of which fall, if possible, between the limits of 30° and 120°.

As regards the MEASUREMENT OF A BASE for *ordinary* work we can consider a steel tape, properly tested at a given temperature, to be sufficiently accurate. The *correction for temperature* amounts to about .000007 of the length of the tape for every 1° Fah. Thus a 100-foot tape, tested at a temperature of 50° F., would give a result too long by about 3 feet in 2 miles at a temperature of 90° F.

---

* Adjustment E, page 35, is also corrected for.

EXPLORATORY SURVEYING. 183

Since all maps are made on the assumption that the linear measurements are *reduced to the sea-level*, in dealing with high altitudes the length of the base may be multiplied by

$$1 - \frac{h}{r} \text{ (nearly)},$$

where h = elevation above sea-level, r = radius of the earth (see Sec. 206), in order to reduce it to sea-level. But this is a refinement which is usually only needed in work requiring great accuracy.

170. In making a regular triangulation survey, the angles of the main triangles are of course themselves observed; but in such work as exploratory surveys, where mountain peaks are selected as "stations," such a method of procedure would, on account of the time and difficulty involved, be out of the question. A readier method of proceeding may be best shown by an example as in Fig. 86. It depends upon *always having in view at any station at least two points whose positions are known.*

Suppose we have obtained, by triangulation or otherwise, the distance between and bearing of two conspicuous points A and B, and suppose our route lies along the dotted line *abcd*.

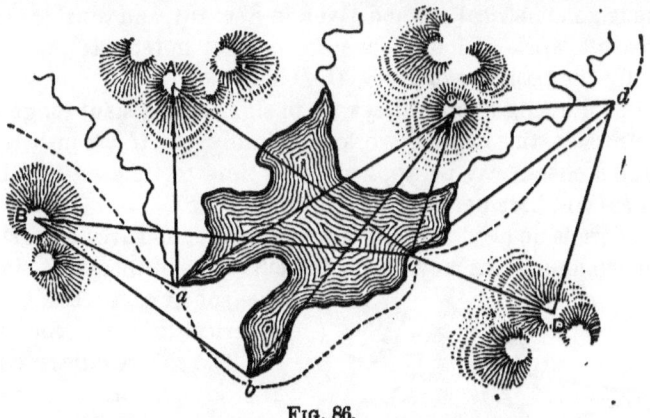

Fig. 86.

At a, a point from which A and B are visible, we observe the bearings of A and B, and thus fix the position of a. Suppose that from a a distant mountain peak C is visible, we take the bearing of it also; then if we wish to fix the position of such a point as b, from it we observe the bearings of B and C. When we get to *c* we locate its position by bearings from *A* and *B*; but suppose we can see *A* and *B* no farther, it then becomes necessary to establish two other points which we may use as we have already used *A* and *B*. A bearing to *C* will then locate *it*. We also observe the bearing of *D*. When *d* is reached, we observe the bearings of *C*, *c*, and *D*, which fix *its* position and also the position of *D*.

No simpler way of keeping a course can be had than this; and it has the enormous advantage over many of the methods in use, that it fixes the main topographical features bordering along the route at the same time as positions on the route itself. The explorer must be constantly on the lookout for points ahead on his probable route and in the neighborhood. The drawback to the method is its inaccuracy when worked by magnetic bearings alone. But if the points are well selected, an error of a degree or so in the bearings is really immaterial in work of this class, and the errors usually more or less counteract each other. Besides, from time to time the courses and distances can be easily checked by the establishment of another base, and the work already done more or less corrected, and a fresh start made.

If we keep *three* or more points in view we are able to apply the trigonometrical method given in Sec. 161, and thus do very *accurate* work so long as we are careful in establishing correctly the positions of *A*, *B*, *C*, *D*, etc.

In following along valleys, or in sight of a distant range of mountains, this method works admirably, and if a transit is at hand a check may be applied from time to time on the distances and bearings with very little trouble.

There is no need to apply any correction, however extensive the triangulations may be, for the curvature of the earth, since the spherical excess of a spherical triangle containing 75.5 square miles is only 1"; so that in a triangle containing 4530 square miles the sum of the three angles only exceeds 180° by 1'.

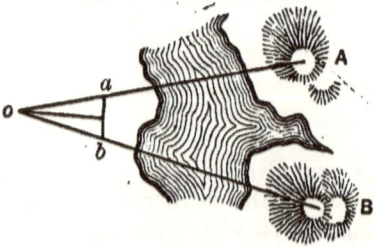

FIG. 87.

171. To measure a horizontal angle without an instrument between two such points as A and B from O, as in Fig. 87. Range in a and b with A and B, each distant from O by, say, 50 feet. Measure ab, then

$$\sin \frac{AOB}{2} = \frac{ab}{100}.$$

172. To measure a vertical angle without an instrument, probably the simplest way is to hold a pencil vertically out at arm's length and note the length subtended on it. Then if the distance from the eye to the pencil $= l$ and p is the length subtended on the pencil,

$$\tan A = \frac{p}{l},$$

where A is the angle required. Similarly if L were the distance of some object whose height H we wish to obtain,

$$H = \frac{Lp}{l}.$$

173. Distance across an open stretch of water can often be taken with sufficient accuracy by observing the time occupied by the passage of the **report of a gun** from one point to the other. This may be done in the day-time if there is a telescope handy to watch for the smoke, but otherwise the flash of course can be best seen at night. The velocity v, in feet per second, with which sound travels, depends greatly on the temperature; thus at 32° F., $v = 1090$; at 60° F., $v = 1125$; and at 100° F., $v = 1175$.

By taking the mean of 3 or 4 shots, the distance may be obtained with confidence to a quarter of a mile. If the wind is blowing hard in the direction from which the sound comes, the velocity of the wind may be added to v.

174. We can observe an **interval of time when a watch is not at hand** by counting the vibrations of a stone tied to the end of a string. If from the centre of gravity of the stone (and the string) to the point of suspension is 39.1 inches, each vibration occupies one second. For any other length L, each vibration occupies

$$\sqrt{\frac{L}{39.1}} \text{ seconds.}$$

The vibrations should be kept as small as possible so as to reduce the resistance of the atmosphere. In this way a tolerably long interval may be measured with a fair amount of confidence. The *best way*, however, is to compare the vibrations with a watch subsequently.

B. BY DIRECT MEASUREMENT AND COMPASS COURSES.

175. By far the most convenient and accurate method of obtaining direct measurement on exploratory surveys is by means of an **odometer**, which answers the same purpose as the patent log at sea, only more efficiently; but unfortunately it necessitates the use of some wheeled vehicle, which is not always a convenient appendage to an exploring outfit.

Pedometers answer well in country where the condition of the ground is comparatively regular and walking easy, but where the surface is much broken they are worse than useless, being misleading as well. The best means of then ascertaining the distance travelled is by estimating the rate of progress and keeping track of the time. The approximate rate may always be found by noting the time occupied in covering, say, 100 yards; then if $t=$ the time occupied in seconds, the velocity in miles per hour equals

$$v = \frac{200}{t} \text{ (nearly)};$$

so that we have the following values of v for various values of t:

t secs.	v m. p. h.	t secs.	v m. p. h.	t secs.	v m. p. h.	t secs.	v m. p. h.
200	1	80	2.5	40	5	25	8
133	1.5	66	3	33	6	22	9
100	2	50	4	28	7	20	10

As regards keeping the courses by compass, in open country, it is best to establish the bearing of some point ahead on the probable route and then to correct it by estimation, if, when abreast of that point, it should be found to be considerably to one side of the route taken. In timber country, the bearing of the sun being taken from time to time, it forms a highly useful guide when no distant landmarks are visible. At night the pole-star forms as good a guide as could be wished for.

C. BY ASTRONOMICAL OBSERVATIONS.

176. Before attempting the solution of astronomical problems in connection with the establishment of positions on the earth's surface, it will be well to give a few explanations as briefly as possible regarding the fundamental principles involved, and definitions of the terms used.

TIME.

177. Civil or Common Time is really what is termed in astronomical language **Mean Solar Time**, with this difference, that a civil day being reckoned from midnight to midnight, the corresponding astronomical day is reckoned from the noon of that day to the following noon, and is also counted continuously up to 24 hours. Thus 4 A.M. on Jan. 10 would be stated in mean solar time as $16^h 0^m$ Jan. 9. Now the velocity with which the earth travels round the sun varies in different parts of its orbit. Owing to this cause and also to the obliquity of the ecliptic (see Sec. 180) the sun's apparent motion is irregular. Thus we find that the sun is apparently travelling faster in winter than its average rate, and in summer slower. It is simpler to consider the earth as stationary and the celestial bodies as revolving round it. In speaking of the velocity of the sun's motion, then, it is its motion among the stars—or on the star sphere—that is referred to, not its actual motion in the sky; the average rate of this motion is about 59' per day and in a direction opposite to that in which the whole star sphere is apparently revolving, so that the motion of the sun *in the sky* is really slower than that of any given star, the result of which is that the star apparently revolves round the earth 366 times while the sun only makes 365 revolutions (nearly).

Now, owing to the irregularity in the sun's motion, it is more convenient to substitute for the real sun a *fictitious* one, termed the "Mean Sun," which is imagined to make the same number of revolutions in the course of the year as the real sun, but always to maintain the same rate of motion. Thus it follows that the mean sun sometimes crosses the meridian—i.e., is due south—before, and sometimes after, the real or, as it is termed in the Nautical Almanac, the *apparent* sun.

178. The interval of time between the passage of these two suns across the meridian is called the **Equation of Time**, which when the mean sun is ahead of the apparent sun is considered *positive*, and when the apparent sun is ahead, *negative*. Thus, since the mean sun is always south at mean noon, by adding or subtracting (as the case may be) the equation of time to or from 24 hours—subtracting 24 hours if necessary—we obtain the mean solar time at which the apparent sun is on the meridian, i.e., apparent noon. Thus, if for a certain day the equation of time is given as $+ 12^m 04^s$, the apparent sun will be on the meridian $12^m 04^s$ after mean noon, or at $0^h 12^m 04^s$ astronomical mean time. Had the equation been negative, apparent noon would have occurred at $23^h 47^m 56^s$ mean astronomical time.

Expressing the relative positions of the two suns in the form of an equation, we have

   Mean Time = Apparent Time ± Equation of Time.

The mean time of that sun is the greater whose R.A. is the less. (See Sec. 180.)

Day of Month.	Jan.	Feb.	March.	April.	May.	June.
1	$+ 4^m\ 0^s$	$+13^m\ 54^s$	$+12^m\ 28^s$	$+ 3^m\ 50^s$	$- 3^m\ 03^s$	$- 2^m\ 24^s$
11	$+ 8\ \ 21$	$+14\ \ 29$	$+10\ \ 06$	$+ 0\ \ 58$	$- 3\ \ 48$	$- 0\ \ 36$
21	$+11\ \ 41$	$+13\ \ 47$	$+ 7\ \ 12$	$- 1\ \ 25$	$- 3\ \ 37$	$+1\ \ 31$

	July.	August.	Sept.	Oct.	Nov.	Dec.
1	$+ 3^m\ 36^s$	$+ 6^m\ 04^s$	$- 0^m\ 13^s$	$-10^m\ 27^s$	$-16^m\ 19^s$	$-10^m\ 39^s$
11	$+ 5\ \ 15$	$+ 4\ \ 56$	$- 3\ \ 35$	$-13\ \ 19$	$-15\ \ 49$	$- 6\ \ 23$
21	$+ 6\ \ 05$	$+ 2\ \ 53$	$- 7\ \ 06$	$-15\ \ 22$	$-13\ \ 53$	$- 1\ \ 31$

The above values of the Equation of Time show approximately the positions of the two suns relatively to each other throughout the year. These values change but little from year to year; and are sufficiently accurate to enable an engineer

EXPLORATORY SURVEYING. 189

to find mean time to a few seconds whenever he may not have a Nautical Almanac at hand; or to correct the reading of a sun-dial, which of course gives apparent solar time, in order to reduce it to mean time.

179. Now the interval of time between the passage of a star across the meridian one day and its passage on the following day is equal to one **Sidereal day**; and since the sun makes only 365.242 revolutions to 366.242 of the stars, we have

A sidereal day = $23^h\ 56^m\ 4^s.09$ mean solar time,
or, A mean day = $24^h\ 03^m\ 56^s.55$ sidereal time;
or, in other words,

To convert a sidereal interval of time into mean solar units, it has to be reduced at the rate of 9.830 seconds per hour;—while

To convert a mean solar interval into sidereal units, it has to be increased at the rate of 9.856 seconds per hour.

Sidereal time is reckoned from the " vernal equinox," or the moment at which the sun crosses from the southern to the northern hemisphere, and is thus, in a way, altogether independent of mean solar time; but if we know the moment at which the vernal equinox occurs in mean time, we thus have a means of connecting sidereal with mean time. But instead of having to start our calculations from the vernal equinox each time, the sidereal time of mean noon is given for every day in the year in the Nautical Almanac; so that

To convert sidereal time into mean time, we have this rule: From the sidereal time given (increased if necessary by 24 hours) subtract the sidereal time at the preceding noon, and then reduce the result at the rate of 9.830 seconds per hour;—and,

To convert mean time into sidereal time: Increase the mean time at the rate of 9.856 seconds per hour; the time thus obtained, added to the sidereal time at the preceding noon (subtracting 24 hours if necessary), gives the corresponding sidereal time.

The *Conversion of the Intervals* may be greatly facilitated by means of Table XIX.

DECLINATION AND RIGHT ASCENSION.

180. These are terms used to denote the positions of celestial bodies in the star sphere relatively to the equinoctial (which is really its "equator") and a plane perpendicular to it passing through the vernal equinox; in the same way as terrestrial Latitudes and Longitudes give the positions of places on the earth's surface, relatively to the equator and the meridian of Greenwich.

The plane of the earth's equator produced to the star sphere gives what is called the *Equinoctial;* and the *Ecliptic*, which is really the plane occupied by the earth's orbit, is inclined to the equinoctial at an angle of about 23° 27′ (slightly varying), which is termed the *Obliquity of the Ecliptic.*

Instead, however, of expressing the Right Ascension of bodies as so many degrees E. or W. of the vernal equinox, it is more convenient to adopt the phraseology of sidereal time and denote the positions of bodies according to the interval of time at which they cross the meridian after the zero of sidereal time, i.e., the vernal equinox. Thus it follows that the sidereal time at which a body is on the meridian is given by its Right Ascension (R.A.), so that instead of speaking of the "sidereal time at preceding noon" as in the rules given in Sec. 179, we might have said "the R.A. of the mean sun at preceding noon," for the sidereal time at noon is often so stated in almanacs. And if we know the sidereal time at mean noon, say at Greenwich, we can, by adding or subtracting the equation of time (as the case may be) obtain the R.A. of the apparent sun at mean noon at Greenwich, and by correcting the sidereal time at mean noon at the hourly rate of $+9.856$ seconds, and also correcting the equation of time, we can find the sun's R.A. at any *later* hour.

The **Declination** of a body, which is really its angular measure on the star sphere, north or south of the equinoctial, is considered *positive* when north, and *negative* when south.

181. But so far we have assumed, except in the case just mentioned above, that it has been unnecessary to correct either the equation of time, R.A. or Dec., as given in the almanac; but since these quantities are always varying, and they are only given for a certain hour at a certain place, when required for any other hour the values as given in the tables must be corrected—usually with sufficient accuracy by simple interpolation—to reduce them to the time for which they may be required. And since every 15° of longitude *west* is equivalent to 1 hour *later* and 15° *east* to 1 hour *earlier*, if in longitude 90° west of Greenwich we want the declination of the sun at

4 P.M., and for noon on that day it was given in the almanac as $+17°\ 40'$, and at noon on the following day as $+18°\ 00'$, the declination at 4 P.M. in longitude 90° west (which is equivalent to 10 hours later) will be $17°\ 48'.3$; and in the same way the R.A. and Equation of time must be corrected.

In dealing with stars, daily and hourly corrections are unnecessary, since their Decs. and R.A.'s change but little in the course of the year (see Sec. 213); but in dealing with the moon, the change is so rapid as to necessitate a more accurate interpolation than would be given by simple proportion as above.

HOUR-ANGLE, ETC.

182. The "hour-angle" is a term which may best be explained by means of Fig. 87.

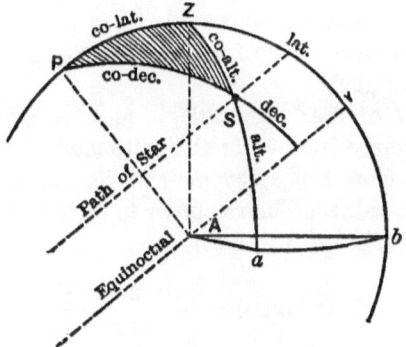

Fig. 87.

Suppose a person stationed at A, on the earth's surface, observes a star S at an altitude Sa above the horizon ab. Then if P is the celestial pole and Z the zenith, since he knows the declination of the star, if he *also* knows his latitude, he has the three sides of the spherical triangle PZS given by the complements of these values; and this triangle, if PZb is the meridian of A, is generally known as the **astronomical triangle**, and the angle ZPS is the hour-angle, which, if expressed in time, is really the difference in R.A. of the star S and of a point on the meridian at the moment of the observation; or, in other words, it equals the difference between the R.A. of the star and the sidereal time at the moment. Thus if the hour-angle in sidereal time $= H$ and the local sidereal time $= T$, we have, *to convert the hour-angle into sidereal local time,*

$$T = H + \text{R.A.} \; (- 24 \text{ hours if necessary});$$

and *conversely,*

$$H = T \, (+ 24 \text{ hours if necessary}) - \text{R.A.},$$

which is the formula for obtaining the hour-angle when the body observed is either the *moon*, a *planet* or *star;* the R.A. being the R.A. of the body observed at the moment of observation. In the case of the *sun*, in order to convert the hour-angle into mean local time, we have simply to reduce it to apparent time by dividing by 15 (as given below), and then apply the equation of time (corrected for the time of observation) to reduce the apparent time to mean time; and the converse of this—to find the hour-angle when given the mean local time—is simply a reversal of the process, for the sun's apparent time *is* its hour-angle.

The value h of the hour-angle in angular measure, as obtained for instance by solving the astronomical triangle, must be subtracted from 360° *when the star lies in east* in order to give it its true value. Then in order to convert h into H, since 1 hour is equivalent to 15°, we have

$$H \text{ (in hours)} = \frac{h \text{ (in degrees)}}{15};$$

and this equation of course holds good if for the words "hours" and "degrees" we substitute on both sides either the word "minutes" or "seconds." So that, for instance, if we obtain by an observation of a star in the *east* a value for the hour-angle—as obtained from the astronomical triangle—of 40°, we have $h = 320°$; therefore $H = 21^h\,20^m$.

Table XX greatly facilitates the conversion of H into h, or *vice versâ.*

183. The following *examples* serve to illustrate what has already been said.

1. *At what hour will Arcturus culminate* (i.e., *be on the meridian*) *on Sept. 18, 1889, at Greenwich?* From the Nautical Almanac we find that the sun's mean R.A. at mean noon at Greenwich on Sept. 18 $= 11^h\,50^m\,22^s.8$, and also that the R.A.

EXPLORATORY SURVEYING. 193

of Arcturus will then $= 14^h\ 10^m\ 37^s.8$; and since the R.A. of the star is really the sidereal time at which it culminates, we have merely to convert its R.A. into mean time according to Sec. 182. Thus Arcturus will be on the meridian at $2^h\ 20^m\ 15^s$ mean astronomical time, i.e., at $2^h\ 20^m\ 15^s$ P.M.

2. *What will be the R.A. of the apparent sun on Nov.* 15, 1889, *in longitude* 90° *W. at* 4 P.M.? Since 4 P.M. in 90° W. occurs 10 hours after mean noon at Greenwich, and from the Nautical Almanac we find the Sun's mean R.A. at mean noon on Nov. $15 = 15^h\ 39^m\ 03^s.0$. Since the correction for 10 hours $= +10 \times 9^s.856 = 1^m\ 38^s.5$, the Sun's mean R.A. corrected to date $= 15^h\ 40^m\ 41^s.5$. Similarly the equation of time corrected to date $= 15^m\ 08^s.3$; and since the apparent sun is then ahead of the mean sun, the R.A. of the apparent sun for the date required $= 15^h\ 39^m\ 40^s.5 - 0^h\ 15^m\ 08^s.3 = 15^h\ 24^m\ 32^s.2$.

3. *Find the Sun's declination at* 8 A.M. *July* 22, 1889, *in longitude* 30° *E.* Now 8 A.M. at 30° E. occurs 6 hours before mean noon at Greenwich; and from the Nautical Almanac the declination at Greenwich at mean noon on July $22d = +20°\ 12'\ 16''$, which, corrected to **6 hours earlier**, $= +20°\ 15'\ 15''$, which is the declination required.

4. *Given* $10^h\ 24^m\ 08^s$ *as the local astronomical mean time on Feb.* 1, 1889, *in longitude* 60° *W. to convert it into local sidereal time.* According to Sec. 179, we must first convert this time into a sidereal interval by increasing it at the rate of 9.856 secs. per hour, which gives $10^h\ 25^m\ 50^s.5$, and the sidereal time at mean noon 4 hours later than Greenwich mean noon $= 20^h\ 48^m\ 11^s.2$, thus the local sidereal time (deducting 24 hours) $= 7^h\ 14^m\ 01^s.7$.

5. *Suppose on June* 1, 1889, *we observe Castor at* $2^h\ 30^m\ 04^s$ A.M. *local time, in longitude* 105° *W. what is the hour-angle in angular measure?*

This in mean astron. time equals, May 31........$14^h\ 30^m\ 04^s$
Increase at rate of $9^s.856$ per hour............... $2^m\ 22^s.9$

Sidereal interval in sidereal time...............$14^h\ 32^m\ 26^s.9$
Sidereal time at mean noon in 105° W. May 31.. $4^h\ 37^m\ 50^s.7$

Sidereal local time of obs. $= T$.................$19^h\ 10^m\ 17^s.6$
R.A. of Castor................................... $7^h\ 27^m\ 32^s.7$

Hour-angle H (subtracting 24 hours).... .... $2^h\ 37^m\ 50^s.3$
Therefore Angular equivalent $h =$.............$39°\ 27'\ 35''$

194   EXPLORATORY SURVEYING.

6. *Given the hour-angle of the apparent sun in the east, as obtained from the astronomical triangle, as* 14° 29' 10" *on June* 14, 1889, *in longitude* 90° *E., find the mean local time.* Since the observation is in the east, $h = 345°\ 30'\ 50''$, which corresponds with $23^h\ 02^m\ 03^s$; therefore the observation occurred $23^h\ 02^m\ 03^s$ apparent time after apparent noon on June 14; and at that moment the mean sun was ahead of the apparent sun by $0^m\ 10^s$, therefore the mean local time of observation $= 23^h\ 02^m\ 13^s$ June 14.

REFRACTION, PARALLAX, SEMI-DIAMETER, AND DIP.

184. In Secs. 51 and 165 we have already considered the effect of Refraction when dealing with objects on the earth's surface. The same uncertainty exists in dealing with celestial objects as to the amount of the correction necessary to counter-

Alt.	Ref.	Alt.	Ref.	Alt.	Ref.	Alt.	Ref.	Alt.	Ref.	Alt.	Ref.
° '	' "	° '	' "	° '	' "	° '	' "	°	' "	°	' "
0 00	33 00	2 30	16 23	6 30	7 52	12 20	4 16	30	1 38	60	0 33
0 05	32 11	2 35	16 04	6 40	7 41	12 40	4 09	31	1 35	61	0 32
0 10	31 22	2 40	15 45	6 50	7 31	13 00	4 03	32	1 31	62	0 30
0 15	30 36	2 45	15 27	7 00	7 21	13 20	3 57	33	1 28	63	0 29
0 20	29 50	2 50	15 09	7 10	7 12	13 40	3 51	34	1 24	64	0 28
0 25	29 06	2 55	14 52	7 20	7 03	14 00	3 46	35	1 21	65	0 27
0 30	28 23	3 00	14 35	7 30	6 54	14 20	3 40	36	1 18	66	0 25
0 35	27 41	3 05	14 19	7 40	6 46	14 40	3 35	37	1 16	67	0 24
0 40	27 00	3 10	14 03	7 50	6 38	15 00	3 30	38	1 13	68	0 23
0 45	26 20	3 15	13 48	8 00	6 30	15 30	3 23	39	1 10	69	0 22
0 50	25 42	3 20	13 33	8 10	6 22	16 00	3 17	40	1 08	70	0 21
0 55	25 05	3 25	13 19	8 20	6 15	16 30	3 11	41	1 05	71	0 20
1 00	24 29	3 30	13 05	8 30	6 08	17 00	3 05	42	1 03	72	0 19
1 05	23 54	3 40	12 39	8 40	6 01	17 30	2 59	43	1 01	73	0 17
1 10	23 20	3 50	12 14	8 50	5 55	18 00	2 54	44	0 59	74	0 16
1 15	22 47	4 00	11 50	9 00	5 49	18 30	2 49	45	0 57	75	0 15
1 20	22 15	4 10	11 28	9 10	5 43	19 00	2 44	46	0 55	76	0 14
1 25	21 44	4 20	11 07	9 20	5 37	19 30	2 40	47	0 53	77	0 13
1 30	21 15	4 30	10 47	9 30	5 31	20 00	2 36	48	0 51	78	0 12
1 35	20 46	4 40	10 28	9 40	5 26	20 30	2 32	49	0 50	79	0 11
1 40	20 18	4 50	10 10	9 50	5 20	21 00	2 28	50	0 48	80	0 10
1 45	19 51	5 00	9 53	10 00	5 15	21 30	2 24	51	0 46	81	0 09
1 50	19 25	5 10	9 37	10 15	5 08	22 00	2 20	52	0 45	82	0 08
1 55	18 59	5 20	9 21	10 30	5 00	23 00	2 14	53	0 43	83	0 07
2 00	18 35	5 30	9 07	10 45	4 54	24 00	2 07	54	0 41	84	0 06
2 05	18 11	5 40	8 53	11 00	4 47	25 00	2 02	55	0 40	85	0 05
2 10	17 48	5 50	8 39	11 15	4 41	26 00	1 56	56	0 38	86	0 04
2 15	17 26	6 00	8 27	11 30	4 35	27 00	1 51	57	0 37	87	0 03
2 20	17 04	6 10	8 15	11 45	4 29	28 00	1 47	58	0 36	88	0 02
2 25	16 44	6 20	8 03	12 00	4 23	29 00	1 43	59	0 34	89	0 01

act the refractory power of the air, as we found to exist when the objects observed were near at hand; but in the case of *Astronomical Refraction* the altitude of the object is a much more important factor than in the previous case; for the lower the altitude not only the more obliquely do the rays pass through the successive layers of air, but the extent of atmosphere which they have to traverse is greater than at a higher altitude. The preceding table of *Mean Refractions*, calculated for a barometer pressure of 29.6 inches and a temperature of 50° F., may be used at all times under ordinary circumstances, when dealing with celestial objects whose altitudes exceed 30°.

At *low altitudes* the corrections given in the table should be corrected by multiplying them by the factors B and T, which make allowance respectively for the height of the Barometer and the Temperature of the air: thus

True Refraction = Mean Refraction $\times B \times T$.

VALUES OF B.

Bar. In.	28	28.5	29	29.5	30	30.5	31
B	0.946	0.963	0.980	0.997	1.014	1.031	1.047

VALUES OF T.

Temp.	− 30° F.	− 10° F.	+ 10° F.	+ 30° F.	+ 50° F.	+ 70° F.	+ 90° F.
T	1.180	1.130	1.082	1.038	1.000	0.960	0.925

The correction for refraction must of course be *subtracted* from the observed altitude.

185. The positions of all celestial bodies as given in the Nautical Almanac are calculated with reference to the Centre of the Earth; thus if, as in Fig. 88, an observer at A observes the altitude of the sun S to be the angle SAH, in order to reduce this angle to the centre of the earth, i.e., to the angle SOh, he must add to it the angle ASO, which is termed the *Parallactic angle*.

Now if S were just on the horizon, i.e , at H, then

$$\sin AHO = \frac{AO}{HO} = \frac{\text{Radius of Earth}}{\text{Distance of Sun}},$$

where AHO is termed the **Horizontal Parallax**, and is given in the Nautical Almanac. In the case of the sun it varies

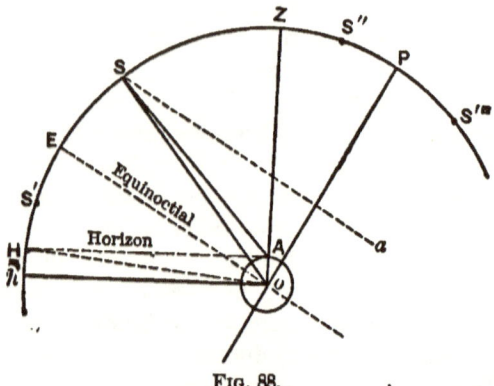

Fig. 88.

from about $8''.7$ to $9''.0$. In order to reduce this to **Parallax in Altitude**, we have from the above figure

$$\sin ASO = \sin AHO \sin SAZ;$$

therefore

$$\sin (\text{Par. in alt.}) = \sin (\text{Hor. Par.}) \cos (\text{alt.});$$

or, assuming the sines of small angles to be proportional to the angles themselves,

$$\text{Par. in alt.} = \text{Hor. Par.} \times \cos (\text{alt.}).$$

Thus, at an altitude of $45°$, Parallax in altitude $= 6''$, and at $60° = 4''$.

In the case of the *moon*, since its distance from the earth compared with the radius of the latter makes it important what value of the radius is used, the Hor. Par. is given in the Nautical Almanac as *Equatorial horizontal parallax*, meaning that the value of the radius used is that at the Equator; thus for other latitudes the correction taken from the following table should be subtracted from it before applying the correction for altitude, in order to obtain the value of the Horizontal parallax suitable for the latitude in question:

Eq. Hor. Par.	Latitude.								
	10°	20°	30°	40°	50°	60°	70°	80°	90°
53'	0''.3	1''.2	2''.7	4''.4	6''.2	8''.0	9''.4	10''.3	10''.6
61'	0''.4	1''.4	3''.1	5''.1	7''.2	9''.2	10''.8	11''.9	12''.2

186. Correcting for Semi-diameter.—In taking an altitude of the *sun*, the upper or lower "limb" is generally observed, and the altitude so obtained corrected by the subtraction or addition of the semi-diameter—obtained from the Nautical Almanac—to reduce it to the sun's centre. In observing with an artificial horizon, the application of the correction for semi-diameter can be avoided by bringing the reflections to coincide. With either a transit or sextant a good way is to observe one limb and note the time, and immediately after observe the other limb and note the time; the mean altitude may then be considered to give the altitude of the sun's centre at the mean time.

Similarly in observing the transit of the sun across any vertical plane we take the mean time of the passage of its east and west limbs.

In observing the *moon*, we usually can only observe one limb; and in this case, on account of its proximity to the earth, it is necessary to apply a correction to the semi-diameter as given in the Nautical Almanac, which assumes the observer to be at the centre of the earth, in order to allow for the increase in its semi-diameter on account of his being nearer to it than the centre of the earth. This is termed correcting for the **Augmentation of the Semi-diameter.** The corrections are given in the following table:

Semi-diam.	Apparent Altitude.								
	10°	20°	30°	40°	50°	60°	70°	80°	90°
14′ 30″	2″.4	4″.7	6″.9	8″.8	10″.5	11″.8	12″.9	13″.5	13″.7
17′ 0″	3″.4	6″.5	9″.5	12″.1	14″.4	16″.3	17″.7	18″.6	18″.8

In finding the time occupied by the semi-diameter of the sun or moon in crossing the meridian, it must be remembered that it is only when the declination $= 0°$ that (if the R.A. is not changing) the semi-diameter will travel across the plane at the rate of 15° to one sidereal hour (or 15° 2′ 24″ to one mean hour). At any other declination we have, as the rate of travel,

$$15° = 1 \text{ sid. hour} \times \cos(\text{dec.}),$$

on just the same principle as the length of a degree of longitude decreases as the cosine of the latitude. In the same way, it is only when the body is on the horizon that its semi-diameter can be measured, without correction, by the horizontal circle of a transit, for as the altitude of the body increases, so also does the horizontal circle increase its reading in proportion to the secant of the altitude.

The change in R.A. during the passage of the semi-diameter must of course be added to the time which it would have occupied had its R.A. been constant.

187. Dip.—This is a correction only necessary when the sea-level is taken as the horizon, and is practically the same as that given in Sec. 167. It is to be subtracted from the observed altitude. The following are its approximate values, but refraction enters too largely into the question to enable accuracy to be obtained by the use of a sea-horizon:

Height above Sea-level in feet,	5	10	20	30	40	50	60	75
Dip, ...	2' 5"	3' 0"	4' 10"	5' 10"	6' 0"	6' 40"	7' 20"	8' 10"

Other values may be found from the values of H, calculated according to Sec. 167.

188. We will now sum up the corrections (which we have already considered) necessary to apply in taking ordinary observations.

1. **Observation for Altitude.**
   A. *Using a sea horizon or level.*
   If a *Star*. Observed Altitude (− Dip) ± Index-error − Refraction = True Altitude.
   If the *Sun*, or a *Planet*. Observed Altitude (− Dip) ± Index-error − Refraction ± Semi-diameter + (Hor. Parallax × cos alt.) = True Altitude.
   If the *Moon*. Observed Altitude (− Dip) ± Index-error − Refraction + (Hor. Eq. Parallax corrected for latitude and converted into Par. in alt.) ± Semi-diameter, reduced for Augmentation = True Altitude.
   B. *Using an artificial horizon.*
   In this case the double-altitude as read on the arc + or − the Index-error must be divided by 2 in order to obtain the observed altitude, and then the other corrections—except of

course for Dip, which only comes in when using a sea-horizon—applied as above. If the two reflections are brought to coincide, there will be no correction needed for semi-diameter; but a more perfect observation can usually be obtained by bringing the limb of one reflection in contact with the opposite limb of the other, in which case the semi-diameter must be corrected for as above.

"Index-error" includes errors of any sort in connection with the instrument for which allowance must be made.

2. **Observation for Azimuth.**

If a *Star*. Observed Azimuth = True Azimuth.

If the *Sun* or a *Planet*. Observed Azimuth ± (Semi-diameter × sec alt.) = True Azimuth.

If the *Moon*. Observed Azimuth ± Semi-diameter (reduced for Augmentation) × Sec. alt. = True Azimuth.

Having now considered all the corrections which need be applied in the case of ordinary field observations when using either a sextant or small portable transit, we will next consider the methods by which the latitude and longitude of a place may be established by astronomical observations.

LATITUDE.

189. A. By a Meridian Altitude.—In Fig. 88, if for the moment we assume the observer to be at the centre of the earth, so as to do away with the idea of parallax, if *PSH* is the meridian and *S* the Sun, *SE* represents the Sun's Dec. N.; and if its declination did not change, since *Sa* indicates its path, we can easily see that its altitude would be greatest when on the meridian. But since its declination *is* always changing, the Sun attains its maximum altitude in the northern hemisphere when its declination is changing towards the north, *after* it has passed the meridian, and when changing towards the south, *before* it reaches the meridian. The difference between its meridian altitude and its maximum altitude does not exceed at any season 1", so that in ordinary work the maximum altitude is assumed as being equal to the meridian altitude.

In taking an observation of the moon with a sextant it is necessary to allow for this, especially about the time of the equinoxes, the difference between its meridian and maximum altitudes sometimes amounting to as much as 2' 15".

When a transit is used to observe the meridian altitude, it is usually set in the meridian, so that no correction is then required.

For the amount of the correction, see Note G, Appendix.

Now in Fig. 88, if Oh were the observer's horizon, the altitude of the Sun is represented by the angle SOh, Z is the Zenith, and the latitude of the place of observation is given by the angle ZE. Therefore the latitude of the place equals

$$ES + SZ = \text{Dec. N.} + \text{Zenith distance.}$$

And since the Zenith distance is the complement of the altitude, we are thus able, by means merely of the meridian altitude, to obtain the latitude; and this applies equally well to *all* celestial bodies, so that in the northern hemisphere, if, as S in Fig. 88, the Dec. is N., then

$$\text{Lat.} = \text{Dec. N.} + \text{Zenith distance.} \quad \ldots \quad (a)$$

If declination is south, as S',

$$\text{Lat.} = \text{Zenith distance} - \text{Dec. S.} \quad \ldots \quad (b)$$

If the Star is above the Zenith, as S'',

$$\text{Lat.} = \text{Dec. N.} - \text{Zenith distance.} \quad \ldots \quad (c)$$

If the Star is below the pole, as S''',

$$\text{Lat.} = \text{Altitude} + \text{Co-declination.} \quad \ldots \quad (d)$$

In the Southern Hemisphere the same formulæ apply, bearing in mind that what is South in the southern hemisphere is equivalent to what is North in the northern.

The altitude taken "below the pole" is of course the *minimum* altitude. The altitudes of S'' and S''' are observed in the north.

Suppose, for instance, we observe the meridian altitude of Regulus on Mar. 17, 1889, to be 40° 16′ 40″.

Now the declination of Regulus at that date = 12° 30′ 30″; so that we have

EXPLORATORY SURVEYING.   201

Observed altitude of Regulus..............	40° 16′ 40″
Correction for refraction.................	− 1′ 07″
True altitude	40° 15′ 33″
Therefore, zenith distance...............	= 49° 44′ 27″
Declination of Regulus...................	12° 30′ 30″
Therefore, Latitude by Eq. (*a*)...........	= 62° 14′ 57″ N.

Again, suppose on Feb. 8, 1889, in longitude 105° W., the meridian altitude of the sun's upper limb is observed to be

	48° 27′ 20″
Correction for refraction.................	− 50″
" " parallax.................	+ 5″
" " semi-diameter.............	− 16′ 15″
True altitude of sun's centre.............	48° 10′ 20″
Therefore, zenith distance...............	= 41° 49′ 40″
Now the sun's declination S. at Greenwich at app. noon on Feb. 8.................	= 14° 49′ 30″
Correction for 7 hours later..............	− 5′ 36″
Sun's declination at date.................	14° 43′ 54″
Therefore, Latitude by Eq. (*b*)...........	= 27° 05′ 46″ N.

190. It is always preferable to use a star instead of the sun or moon for a meridian altitude. The moon should only be used in thick weather, when the stars are invisible. In selecting a star for the observation, the altitude should not be less than 30° if possible, on account of refraction. In order for a star to appear above the horizon on the meridian, the sum of the declination and co-latitude must exceed 0°, and the excess equals the true altitude, remembering that declination north is + and south −; this gives a check before the observation is taken, preventing the wrong star being used. For stars below the pole as S''' in Fig. 88, in order that the star may be visible above the horizon at its minimum altitude the latitude must exceed the co-declination, the excess being the true altitude.

When using a transit, we may proceed in two ways:

1. By observing the maximum altitude and correcting according to Sec. 189, and Note G, Appendix.

2. By setting the transit in the meridian, and then observing the altitude of the passage.

The meridian may best be obtained by an Elongation of Polaris as described in Sec. 57, or by the other methods described in Secs. 57 and 202.

In taking meridian altitude it is well to observe a star in the north as well as a star in the south; the mean result is then tolerably free from instrumental errors.

Polaris, either at its upper or lower transit, is a good star to use on account of its slow motion admitting of several altitudes being taken.

B. By Transits across the Prime Vertical.

191. This is the most accurate method of obtaining the latitude, but necessitates the use of a transit.

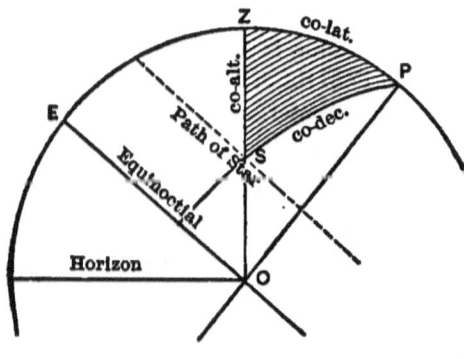

FIG. 89.

In Fig. 89 let PZE represent the meridian, Z the zenith, P the celestial pole, and S the body, the time of whose transit across the prime vertical—i.e., the vertical plane ZO, lying due east and west—we wish to observe, in order by it to obtain the latitude. Now in the spherical triangle ZPS the angle at $P =$ the hour-angle h (see Sec. 182), and $ZS =$ the co-alt. of the body when on the prime vertical, ZP the co-latitude, and PS the co-declination.

Therefore, since $Z = 90°$,

$$\tan (\text{lat.}) = \tan (\text{dec.}) \times \sec h.$$

But in order to obtain h, we must know the exact local time of the observation, which may be obtained according to Secs.

195, etc. The longitude we need only know with sufficient accuracy to admit of correcting the sidereal time at mean noon, i.e., for ordinary work, to about 20 miles.

This method of determining the latitude of a place admits of high precision, since an error of 1 second in the local time only causes an error of about 1¼ seconds in latitude, or about 170 feet.

The passage of the star across the prime vertical should be observed both in the east and the west (or else another star used), and the mean result taken to eliminate errors.

The altitude of a body when on the prime vertical is given by the equation

$$\sin (\text{alt.}) = \sin (\text{dec.}) \operatorname{cosec} (\text{lat.});$$

and the hour at which the observation occurs is given by the equation

$$\sec h = \tan (\text{lat.}) \cot (\text{dec.}).$$

If the transit has three vertical hairs, which it should *at least* have for astronomical work, the star may be observed at, say, its eastern transit on the north side of the prime vertical upon the hair which is to the left of the collimation centre; then after reversing the instrument, the star may be observed again on the same hair. If the telescope is left in the last position until the star comes to its western transit, it is observed again on the same hair to the south of the prime vertical, and then reversing the telescope the star again crosses the same hair on the north side. Thus a latitude determination is arrived at free from instrumental errors and with the errors of observation greatly reduced. It is best to select a star with as small a declination as possible, as its motion in azimuth will then be more rapid.

C. By an Altitude out of the Meridian.

192. It often happens that just about the time when the sun or star is on the meridian suitable for obtaining the latitude according to method A, it becomes obscured by passing

clouds. If, however, the local time is known approximately, the latitude can still be obtained in the following way:

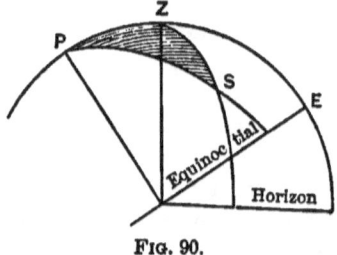

Fig. 90.

Suppose in Fig. 90 *PZE* is the meridian and *S* a star which has only a short time before crossed the meridian. Then in the "astronomical triangle" *PZS*, if we know *ZS* = co-alt., *PS* = co-dec. and the hour-angle *ZPS*, we can at once, by solving the spherical triangle, find the side *PZ* = co-lat. But instead of using the common formulæ (as given in Sec. 233), the following will be found simpler:

Make
$$\tan A = \cos ZPS \times \tan PS,$$
and
$$\cos B = \cos A \times \cos ZS \times \sec PS.$$

Then, if the six-o'clock circle and the prime vertical lie on the same side of *S*, as will always be the case when *S* is near the meridian,

$$\text{co-latitude} = A - B;$$

but if *S* lies between them, we have

$$\text{co-latitude} = A + B.$$

But since this method is really only suitable for use within an hour or two of the meridian circle, it is the former of these two equations which is almost exclusively used.

When the latitude and declination are of contrary signs, we then have simply

$$\text{Lat.} = (A + B) - 90°.$$

To use this method, it is necessary to know the value of the hour-angle with tolerable accuracy. This can be obtained by one of the methods given in Secs. 195, etc.; or in the case of a star it can easily be obtained by observing its altitude before reaching the meridian,—assuming that it is only cloudy about the time of the meridian passage,—noting the time by an ordinary watch; then on the other side of the meridian, if the moment is observed at which it again reaches the same altitude, half the interval (converted into a sidereal interval) = hour-angle H (see Sec. 182). With the sun this is only applicable when its declination is changing but little, or when near the zenith.

D. By double Altitudes.

193. The following are very convenient methods of obtaining the latitude when the local time is not known.

A. *By two altitudes and the interval of time between them.*— In Fig. 91 let Z be the zenith, P the celestial pole, S and S' the two positions of the star at the moments at which the altitudes and times are observed.

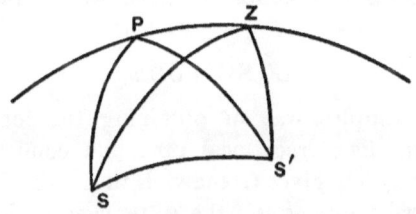

FIG. 91.

Then the interval between the two observations in sidereal time = the hour-angle, which converted into angular measure = SPS'. Then in the triangle PSS', $SP = S'P$ = co-declination; thus we can find SS' and $PS'S$. Then in the triangle $ZS'S$, since we have the three sides we can find the angle $ZS'S$, which, subtracted from $PS'S$, gives the angle $PS'Z$. Then in the triangle $PS'Z$ we have $S'P$, $S'Z$, and the angle $PS'Z$, from which we can find PZ = co-latitude.

A good common watch is all that is required to observe the intervals.

But instead of taking two altitudes of the *same* star, it is better to observe—

B. *By simultaneous altitudes of different stars.*—The hour-angle is given by the difference in R.A. of the two stars, and the rest of the working is the same as above. When, however, there is but *one observer*, so that the altitudes must be taken in succession, he must proceed thus: The altitude of one star must be taken, and the time noted by the watch; the altitude of the other star must then be taken, and the time again noted. After a short interval the altitude of the second star must again be taken, and the time noted. He thus finds the motion in altitude of the second star in a given time, from which, by proportion, he can find what its altitude was when the first star was observed.

In both A and B the altitudes as observed must of course be reduced to the true altitudes in order to obtain SZ and $S'Z$.

194. On the last page of the Nautical Almanac for each year is given a Table for computing the latitude from an observed **Altitude of Polaris** at any time, the hour-angle being approximately known; and as full instructions accompany the table, these need not be repeated here. The local time being known, the hour-angle H is of course obtained as in Example 5, Sec. 183.

LONGITUDE.

195. The simplest way of obtaining the longitude of a place is to find its correct local time, and compare it with a chronometer which gives Greenwich time; the difference between the two times equals the difference of longitude: so that if we have a chronometer at hand keeping Greenwich time, obtaining the longitude is simply a matter of obtaining the local time.

A. To obtain Local Time by an altitude of a star.

If it were not for the slowness of the motion of a star when near the meridian, a convenient method of obtaining the local time would be to reduce its R.A. to mean time at the moment of its maximum altitude, which would then be the mean local time of its transit. But in order to obtain a well-defined moment of observation, it is necessary for the motion in altitude to be as rapid as possible, and for this reason a star should be selected as near the prime vertical as possible. Suppose at a certain moment by the chronometer we observe the altitude of a star S (see Fig. 90); then if the latitude is known, in the triangle PZS, since $PZ =$ co-lat. $= l$, $PS =$ co-dec. $= d$, and $SZ =$ co-alt. $= a$, we have, by spherical trigonometry,

$$\cos \frac{h}{2} = \sqrt{\frac{\sin s \sin (s-a)}{\sin d \sin l}},$$

where $s = \dfrac{a+d+l}{2}$ and $h =$ the hour-angle ZPS; if the declination and the latitude are of opposite signs, $d =$ dec. $+ 90°$.

Now the nearer S is to the prime vertical, the less is an accurate knowledge of the latitude essential, and the less does an error in altitude affect the result. Thus the body should be observed as nearly east or west as possible, and certainly not within an hour or two of its transit.

The following table shows the errors in longitude in minutes of arc involved by an error of 1 minute in latitude, when S is observed at different bearings in different latitudes.

Bearing.	LATITUDE.						
	10°	20°	30°	40°	50°	60°	70°
10°	5′.67	5′.76	6′.55	7′.40	8′.82	11′.38	6′.03
20°	2′.75	2′.79	3′.17	3′.59	4′.27	5′.49	2′.92
40°	1′.19	1′.21	1′.38	1′.55	1′.85	2′.38	1′.27
60°	0′.58	0′.59	0′.67	0′.75	0′.90	1′.15	0′.62
80°	0′.18	0′.18	0′.20	0′.23	0′.27	0′.35	0′.19

Thus in latitude 30° if the bearing of a star when observed is 80° an error in latitude of 5 miles would only cause an error of about half a mile in longitude.

An error in the altitude is of much more importance, as the following table, giving the errors in longitude in minutes caused by an error of one minute in altitude, shows:

Bearing.	LATITUDE.						
	10°	20°	30°	40°	50°	60°	70°
10°	5′.91	6′.25	6′.65	7′.50	8′.96	12′.17	16′.87
30°	2′.03	2′.17	2′.30	2′.64	3′.14	3′.98	5′.84
50°	1′.32	1′.39	1′.51	1′.71	2′.03	2′.63	3′.78
90°	1′.01	1′.06	1′.15	1′.31	1′.55	2′.00	2′.90

Since the accuracy of the altitude is of great importance, it is well to take several sights, say 3 or 5, within a minute or so of each other, and note the corresponding chronometer readings; the mean altitude may then be considered to correspond with the mean time. If the local time which was used in order to correct the sidereal time at noon for the assumed longitude is found to have been appreciably in error, allowance must be made for this.

In observing altitude for time, if great accuracy is desirable, it is well to observe both in the east and the west; the mean result of the two sets is thus practically free from instrumental errors. This method of course applies equally well to the sun as to a star; and since the co-declination is always a large arc, whatever error there may be in it, there will only be half that error in the half sum; and since the errors in these altitudes oppose one another, an error in the co-declination such as might arise from an error of two or three degrees in the longitude assumed to correct the sun's declination will not seriously affect the result.

B. To obtain local time by equal altitudes of a star.

196. All that we have to do in this case is to observe the altitude of a star in the east and note the time, then note the time when in the west it again descends to the same altitude. Half the interval between the two observations is the "middle-time," which corresponds with the local sidereal time given by the star's R.A. Thus we have simply to convert the star's R.A. into mean local time and compare it with the middle-time by the watch to obtain the watch-error.

By taking a set in the east and a set in the west, since index or instrumental errors do not enter into the question at all, the mean altitude for the mean time should give a really good result. There is no necessity to apply a correction for refraction, unless the barometric pressure or temperature has changed considerably between the observations.

C. To obtain local time with a transit.

197. The best way to proceed with a transit is to set it in the meridian and observe the time of transit of the sun or one or more stars; the correct local time is then found by merely converting the R.A. of the body at the time of its transit into mean time.

198. But so far in obtaining the longitude we have assumed that we have had at hand a chronometer rated to Greenwich time. But since little reliance can be placed on chronometers when travelling across country, one of the following methods should be adopted as a check on the chronometer from time to time.

TO OBTAIN THE LONGITUDE BY LUNAR CULMINATIONS.

The principle on which this method of obtaining Greenwich time is based is as follows:

In the Nautical Almanac the moon's R.A. is given for every hour during the year at Greenwich. If then in any other longitude we find the moon's R.A. at a certain moment, that moment will correspond with the time at Greenwich at which the moon would have the same R.A. as that which we observed. Thus, if the moon's R.A. in the Almanac at 6 P.M. were given as 8^h, if in a certain longitude we find at exactly 10 P.M. local time the moon's R.A. to be 8^h, we know we are in a longitude 4 hours ahead of Greenwich, i.e., 60° E. To obtain the R.A. of any body by observation, we have only to find the mean local time of its transit across the meridian and convert it into sidereal time, which is the R.A. required. Thus we proceed as follows:

Find the correct local time by the watch. Set the transit in the meridian. Observe the moment of transit of the moon's bright limb. Again find the correct local time by the watch. The moon's semi-diameter, which is given for every 12 hours in the Almanac, must then be found and divided by 15 to reduce it to equivalent time, which would then be the sidereal time occupied by its passage if its declination $= 0°$ and its R.A. were unchanging. But since its R.A. is always increasing, the passage of the semi-diameter will occupy a time longer than this by an amount which may be obtained from the Almanac by simple proportion, by seeing what the increase in R.A. is at the assumed Greenwich time of the observation; the total time of the passage so obtained multiplied by the secant of the declination (see Sec. 186) then gives the time actually occupied in the passage; and this added to, or deducted from, the observed time of transit of the limb, gives the time of transit of the moon's centre, which, converted into sidereal time, gives the moon's R.A. at the moment of observation.

It is well to take a set of observations for time before and after the moon's passage; and the instrument, if possible, should not have less than 3 vertical hairs, the passage across each of which may be observed and reduced to the centre hair.*

Every possible precaution should be taken in this observation, for the error of a second of time in observing the moon's limb, compared with the corrected watch time, —i.e., an error of 1 second in R.A.,—may easily cause an error in longitude of 5 miles. Thus by a single observation with a small transit we cannot depend on our longitude to within about 10 miles. But if the observer is stationed for 3 or 4 days at any one place, by taking the mean result of 3 or 4 observations he should be able to obtain the longitude with a probable error, say, not exceeding 4 or 5 miles, corresponding with an error in Greenwich time (in ordinary latitudes) of from 20 to 30 seconds.

Having now obtained the moon's R.A., the next thing to do is to find the hour at Greenwich with which it corresponds.

Since the moon's change in R.A. is usually rapid, and great accuracy is necessary, the ordinary method of simple interpolation will not apply here. The following formula may therefore be used instead:

$$T - t = \frac{60(A - a)}{D + \frac{d}{2}\left(\frac{T-t}{3600}\right)};$$

where $T =$ the hour required;

$t =$ the hour for which R.A. is given in the Almanac, previous to T;

$A =$ R.A. corresponding with T;

$a =$ R.A. corresponding with t;

$D =$ Increase in R.A. in 1 mean minute at time t;

$d =$ Increase in D in 1 mean hour at time t.

If D is decreasing, d is of course negative. In the term involving the unknown value $(T - t)$, the *probable* value must be used, which is correct enough. We thus have the value of the Greenwich time corresponding with the observed local time of the transit of the moon's centre, the difference of which, divided by 15, gives the difference of longitude.

199. TO OBTAIN THE LONGITUDE BY LUNAR DISTANCES.—This method is similar in principle to the preced-

---

* See Sec. 197.

ing one, the difference being that here it is the distance from the moon to some star which is observed instead of its R.A. The present case, since it does not involve the use of a transit and admits of several observations being taken on one night, is more suitable for exploratory work, and is the method altogether used for checking the chronometers at sea. The distances between the moon's centre and certain stars of the first and second magnitude are given in the Nautical Almanac for every three hours at Greenwich, so that it is simply a case of measuring the distance from the moon's limb to a star, and correcting for refraction, semi-diameter, etc., noting the local time of the observation, and then finding from the Almanac what hour at Greenwich corresponds with the corrected distance.

In Fig. 92 let M' and S' be the positions of the moon and star at the moment of observation, and Z the zenith; then $M'S'$, corrected for semi-diameter, equals the apparent Lunar distance, and $M'Z$ and $S'Z$ the co-altitudes. The true positions will differ from these by the differences in altitude MM' and SS': the moon, on account of the correction for parallax exceeding that for refraction, will be elevated above its apparent position; whilst the star, on account of refraction only, will be depressed below its observed position.

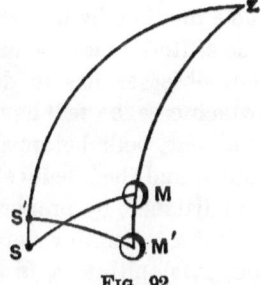

Fig. 92.

Now, if the apparent altitudes are observed at the time of observing the lunar distance $S'M'$, we have the three sides of the triangle $S'ZM'$, so that the angle at Z may be found trigonometrically. Then the two sides $S'Z$ and $M'Z$, being corrected for refraction and parallax, give the sides of the corrected triangle SZM; and since we thus have two sides and the included angle Z, we can calculate the true lunar distance SM. This operation is termed **"Clearing the lunar distance."**

The following formula, by Borda, is probably the most convenient to use for effecting this:

$$\sin \frac{D}{2} = \cos \frac{H + H'}{2} \cos C,$$

where
$$\sin^2 C = \frac{\cos s \cos (s \sim d) \cos H \cos H'}{\cos h \cos h' \cos^2 \frac{H+H'}{2}},$$

where $s = \dfrac{h+h'+d}{2}$,

and $h =$ app. alt. of moon's centre, $h' =$ app. alt. of star;
$H =$ true alt. of moon's centre, $H' =$ true alt. of star;
$d =$ app. distance $S'M'$, $D =$ true distance SM.

An error of a minute or two in the altitude makes no appreciable difference in the distance.

The vernier should be set to a division easily read off, and at the moment when the distance agrees with this reading the observer should call "stop," at which signal the assistant should note the time by the watch, and at the same instant, if possible, the altitudes may be observed by two assistants. But usually one observer has to do the whole work with the sextant, in which case he will have to observe the altitudes of the moon and star, both before and after the observation, and note the times, and then deduce the altitudes at the time of measuring the distance, by proportion.

But a better way is to spend the time otherwise occupied in observing altitudes, in obtaining a large number of lunar distances and then to *compute the altitudes* as follows:

Since we know the time of each observation, we can obtain the hour-angle at that moment, which, in either the case of the moon or a star, is merely the difference in R.A. of the body and the sidereal time at the moment $+ 24$ hours if necessary, the R.A. in the case of the moon being corrected for the time of observation by assuming a probable value for the longitude. Then if $L =$ latitude and $d =$ co-declination,

$$\sin (\text{alt.}) = \frac{\sin L \sin (E+d)}{\sin E},$$

where
$$\cot E = \cot L \cos h,$$

and $h =$ the hour-angle. If h *exceeds* $90°$ $\cos h$ is negative, which will make $\cot E$ also negative; so that to avoid the use of supplements, it is simpler to say

$$\sin(\text{alt.}) = \frac{\sin L \sin (E - d)}{\sin E}.$$

These are of course the *true* altitudes.

In selecting stars from which to measure the distance, it should be remembered that the mean of two distances, one measured to a star on the right and the other on the left, will be practically free from instrumental errors; so that this plan of observing should always be adopted when possible. It is well, too, to select stars the distances between which and the moon are varying most rapidly,—for there is a considerable difference sometimes between the rates,—and yet at the same time the altitudes should not be less than, say, 10°.

A complete lunar observation should consist of 6 "sets," each set including 3 simple distances; 3 of these sets should be taken to the left of the moon and 3 to the right; also two observations for latitude, one in the north and one in the south, to eliminate instrumental errors; and two sets of observations for time, one to a star in the east and another in the west, one *before* and the other *after* the measuring of the distances.

Having thus obtained the *mean* lunar distance for the *mean* local time, the corresponding Greenwich time may best be deduced according to the instructions and data given in the Nautical Almanac with sufficient clearness to render any further explanation superfluous, as that work must of necessity be an accompaniment to the observations. Since, however, the Nautical Almanac assumes that the computer has at hand a table of **Ternary Proportional Logarithms,** such as is given in Chambers' Mathematical Tables or Bowditch's Navigator, it will be well to see how these may be calculated, in the event of such not being the case.

A Proportional Logarithm for any portion of a certain period is merely the difference of the logarithms of the period and of the portion. Thus, taking the period as 3 hours, since lunar distances are given in the Almanac at intervals of every 3 hours, or 10,800 seconds, the logarithm for it $= 4.0334$; then since the logarithm for 1 hour ($= 3600$ seconds) $= 3.5563$, the proportional logarithm for 1 hour $= 0.4771$. The explorer, however, should provide himself with some portable form of logarithmic tables if likely to have much of this sort of work to do.

200. Another method of obtaining Greenwich time is by observing with a powerful telescope the local time of the **Eclipses of Jupiter's Satellites.** But this method, for a variety of reasons, is considerably less reliable than those given above. The Nautical Almanac gives instructions and data as to the manner of obtaining Greenwich time by this method.

TO TEST THE CHRONOMETER RATE.

201. Whenever a halt is made for over 24 hours, it is a very simple matter to check the rate of the chronometer. With a transit this can best be done by setting it in a vertical plane lying fairly north and south, and noting the moments of the passages of 3 or 4 stars. The interval of time before the respective passage of each on the following evening $= 23^h\ 56^m\ 04^s.9$. With a sextant this may best be done by observing the altitudes of 3 or 4 stars lying fairly east or west—their motion being greater in altitude when near the prime vertical—and noting the chronometer times; after the lapse of the above interval, each will again be at the same altitude on the following night.

TO SET THE TRANSIT IN THE MERIDIAN.

202. Three methods of obtaining a north and south line have already been given in Sec. 57; the method by Maximum Elongations of Polaris is the best, for it admits of plenty of time to reverse the instrument and establish a true north and south line. When Polaris is not convenient for this purpose, any other star (which has an elongation) may be used as shown in Note D, Appendix. In the same way, if neither Alioth nor γ Cassiopeia is convenient for observation, other stars may be used as shown in Note E, Appendix. When, however, neither of these methods is exactly suitable, the azimuth of Polaris out of the meridian may be found at any moment by solving the astronomical triangle PZS in Fig. 87, and thus obtaining the angle at Z, which is the azimuth.

To do this we have given the declination, and we must also have two of the following three: latitude, altitude, and hour-angle. Since the latitude is most easily obtained, and the

EXPLORATORY SURVEYING.

altitude gives the best result if *near the elongations*, these two should *then* be used. If, however, the star is *near the meridian*, the latitude and the hour-angle should be employed.

In the *former* case we have

$$\cos \frac{Z}{2} = \sqrt{\frac{\sin s \sin (s-d)}{\sin a \sin l}},$$

a, d, and l being the complement of the altitude, declination and latitude respectively, and s the half sum of a, d, and l.

In the *latter* case we have

$$\cos a = \cos d \cos l + \sin d \sin l \cos h,$$

from which we obtain

$$\sin Z = \sin h \sin d \operatorname{cosec} a.$$

$h =$ hour-angle. (See Sec. 182.)

When the latitude and declination are of opposite signs, $d = \text{dec.} + 90°$.

203. In observing the altitude of the moon for time or latitude, as is often practicable in thick weather when the stars are invisible, and more accurate interpolation of its declination is necessary than is obtained by simple proportion, the method usually adopted for this purpose is that known as INTERPOLATION BY SUCCESSIVE DIFFERENCES. The interpolation formula is

$$F^n = F + \frac{nd_1}{1} + \frac{n(n-1)}{1 \times 2} d_2 + \frac{n(n-1)(n-2)}{1 \times 2 \times 3} d_3 +, \text{etc.}$$

For example, suppose we wish to find the moon's declination at Greenwich at 2^h 15^m on Nov. 15, 1889.

From the Nautical Almanac we find the declination given for every hour. We select the declination at the hour before the one for which we wish to interpolate ($= F$), and put it in the first column as below; beneath it we put in order the declinations for, say, 3 or 4 following hours, as given in the Almanac. In the second column we put down the first differences of these (d_1) obtained by subtracting *downwards* and prefixing the proper algebraic sign. In the third column we place the second difference (d_2) (i.e., the differences of the first differences), and so on.

Now n is the ratio of the fractional period for which we wish to interpolate, to the interval between which the values are given; in this case 15 minutes to 1 hour, therefore $n = \frac{1}{4}$: so that now we have merely to insert the upper values in the columns for d_1, d_2, etc., and the above value of n, in order to find the declination at $2^h\ 15^m$.

	F	d_1	d_2	d_3
Dec. at $2^h =$	$18°\ 17'\ 4''$	$- 7'\ 59''$	$- 6''$	$+ 1''$
" $3^h =$	$18°\ 09'\ 5''$	$- 8'\ 05''$	$- 5''$	
" $4^h =$	$18°\ 01'\ 0''$	$- 8'\ 10''$		
" $5^h =$	$17°\ 52'\ 50''$			

Thus,
$$F^n = 18°\ 17'\ 4'' - 1'\ 59''.8 + .56'' - .07'';$$
therefore,
$$\text{Dec. at } 2^h\ 15^m = 18°\ 15'\ 04''.75.$$

In such a case as the above, as it happens, the simple method of interpolation would have given $F^n = 18°\ 15'\ 04''.2$, which of course would have been amply near enough for anything in the way of ordinary work. But where the explorer is desirous of obtaining a really accurate observation this method is often of high value.

204. Adjustment of Observations.—It is a well-recognized fact in practice, when making a series of measurements of any quantity, that after every possible means of eliminating and correcting for instrumental errors have been employed, there still remain certain *accidental* errors which no experience or skill on the part of the observer can rectify, since the causes to which they are due are themselves unknown. Thus it happens that each measurement in the set may be different, although, judging from the care taken in observing each and the apparent similarity of the conditions under which they were taken, no such differences should exist. The question then arises as to what is to be taken as the most probable result.

Now according to the Theory of Least Squares, the method usually adopted for the solution of these problems, the most probable value of any number of measurements of the same quantity, each measurement being considered to be equally reliable, is that which makes the sum of the squares of the

EXPLORATORY SURVEYING.

"errors" a minimum; and the value which does so is the arithmetical mean of all the measurements. The "error" in the case of each measurement being its difference from the mean.

But it often happens that the circumstances under which the several measurements are made are such as to warrant greater "weight" being given to some of them than to others. These weights are often deduced from the observations themselves, or from them in connection with a special series of observations; but in ordinary field practice, weights assigned arbitrarily after a thoughtful perusal of all the attendant circumstances are more likely to be of value than those found by a strict application of the formulas of Least Squares. Weights being thus assigned, the most probable value of the results will be found by multiplying each observed value by its weight, and dividing the sum of the products by the sum of the weights, the result being that value which renders the sum of the products of the squares of the errors and the respective weights a minimum. And this value is termed the Weighted Mean. This may be best illustrated by an example.

Suppose that we have, as several corrected measurements of a base, the following numerators, and that, considering all the attendant circumstances, we have assigned to each the weight shown as its denominator, assuming, for the sake of simplicity, that the weight of the least reliable is expressed by unity:

$$\frac{2056.32 \text{ feet}}{1}, \frac{2056.20 \text{ feet}}{4}, \frac{2056.16 \text{ feet}}{3}.$$

Then the most probable value of the result is given by

$$\frac{2056.32 + (2056.20 \times 4) + (2056.16 \times 3)}{1 + 4 + 3} = 2056.20.$$

A fair test of precision in dealing with a set of measurements is afforded by means of the "probable error" of a single determination, which is found by taking the difference between each individual result and the mean, squaring these quantities, and dividing their sum by $(n-1)$ where n represents the number of individual results; then, on extracting the square root of this quotient and multiplying by 0.674, we obtain the so-called Probable Error. But this term does not mean that that error is more probable than any other, but merely that in a future observation the probability of committing an error greater than the probable error is equal to the probability of committing an error less than the probable error.

The probable error of the arithmetical mean may be similarly found, the value $n(n-1)$ being substituted for $(n-1)$ in the rule given above for a single determination.

Errors in excess are considered positive; those in defect, negative.

205. Having now examined the various methods of obtaining positions on exploratory surveys, we next come to the subject of ascertaining the bearings and distances of these positions relatively to each other or to other points, when taking into consideration the curvature of the earth's surface.

From what has already been said in Sec. 58 on the subject of the Convergence of the Meridians, we can see what form the corrections will have to take in order to allow for the spherical —or more correctly spheroidal form of the earth; and now, by means of 3 or 4 simple problems, we can obtain all the formulæ necessary for the construction of the groundwork of a map, or the calculation of courses, which are ever likely to be needed in connection with exploratory surveys.

In Engineering Geodesy it is usually sufficiently accurate to assume the earth to be a sphere, the radius of which equals the mean radius of curvature of the spheroid; but it may be as well here to examine the subject roughly, in order that the engineer may have an idea of the extent of the errors which this assumption involves.

206. THE FIGURE OF THE EARTH.—According to Col. Clarke,

the mean Equatorial semi-axis = 20926202 feet,
and the Polar Semi-axis = 20854895 feet.

Also the radius of curvature *in the direction of the meridian* in any latitude L equals in feet

$$R = 20890564 - 106960 \cos 2L + 228 \cos 4L;$$

EXPLORATORY SURVEYING.

and the radius of curvature *in a direction perpendicular to the meridian* equals in feet

$$r = 20961932 - 35775 \cos 2L + 46 \cos 4L.$$

Thus at the Equator

$$R = 20783832 \text{ feet}, \quad r = 20926203 \text{ feet};$$

and at the poles

$$R = 20890564 \text{ feet}, \quad r = 20961932 \text{ feet}.$$

So that for engineering purposes we may take 20,890,000 feet as the **mean radius of curvature**. Again, according to the same authority, the length of a **degree of latitude** equals in feet

$$D = 364609.1 - 1866.7 \cos 2L + 4 \cos 4L,$$

and the length of a **degree of longitude** equals in feet

$$d = 365542.5 \cos L - 311.8 \cos 3L + 0.4 \cos 5L.$$

The value of the foot taken above is the English standard, which is less than the American standard in the ratio of 1 mile to 1 mile and 3.677 inches.

For rough work we may consider

$$D = 364000 \text{ feet} \quad \text{and} \quad d = D \cos \text{Lat}.$$

Table XVIII gives the true values of 1 minute of arc, to the nearest foot.

207. Now from the formula for the length of a circular arc given in Sec. 73, if we take the above value of the mean radius of curvature, we find the length of an arc on the earth's surface in feet equals

$$l = 6076n \text{ (nearly)},$$

where $n =$ the number of minutes in the arc; and the converse of this,

$$n = \frac{l}{6076} \text{ (nearly)},$$

enables us to convert any given distance into its equivalent in angular measure.

If it is desirable to obtain the value of l more accurately than by this means, we can do so by obtaining first the value of l in the direction of the meridian, either from Table XVIII, or more correctly by dividing the value of D, given in Sec. 206, by 60. Also the length of a 1' arc perpendicular to the meridian is needed, which may be obtained by means of the value of r, given in Sec. 206. Then if we call this latter value l', the length of an arc subtending 1' at the earth's centre, which makes an angle A with the meridian, equals

$$l \cos^2 A + l' \sin^2 A.$$

208. Given the latitude and longitude of two places to obtain their distance apart, and the bearing of the course joining them.—Suppose A and D in Fig. 12 are the two given places, then the arc AF and the arc ED represent their latitudes. Then in the spherical triangle AND, since N = difference of longitude, and AN and ND are equal to the co-latitudes of A and D, we can find AD thus:

$$\cos AD = \sin a \sin d + \cos a \cos d \cos AND,$$

where a and d are the latitudes of A and D. And the bearing of the arc AD, which at A is represented by the angle NAD, is then given by the equation

$$\sin A = \cos d \operatorname{cosec} AD \sin AND.$$

Or, if A and D are in the same latitude, we have

$$\tan A = \cot \tfrac{1}{2} AND \operatorname{cosec} \text{lat.}$$

The arc so obtained can be converted into feet as shown in Sec. 207; and this is the distance along the arc of the great circle passing through A and D, i.e., the shortest distance between them on the earth's surface.

Conversely, given the latitude and longitude of A, and the bearing and distance of another place D, to find the latitude and longitude of D.—First convert AD into angular measure according to Sec. 207; then we have the sides

AD, AN, and the included angle A. Then to find d we have

$$\sin d = \cos AD \sin a + \sin AD \cos a \cos A.$$

Then AND, the difference of longitude, is given by

$$\sin AND = \sin A \sin AD \sec d.$$

The bearing of AD at D may be obtained from the equation

$$\sin D = \sin AND \cos a \operatorname{cosec} AD.$$

The formulæ given in this section are simply those ordinarily used for the solution of spherical triangles. (See Sec. 233.)

209. To find the radius of a Circle of Latitude.—In Fig. 93 let C be the centre of the earth, N the pole, and L any given latitude; then, considering the earth to be a sphere, the angle $LPC =$ the latitude of L, so that

$$PL = LC \text{ cot latitude,}$$

where $PL =$ radius of the circle of latitude. LC may be taken as equal to 20,890,000 feet.

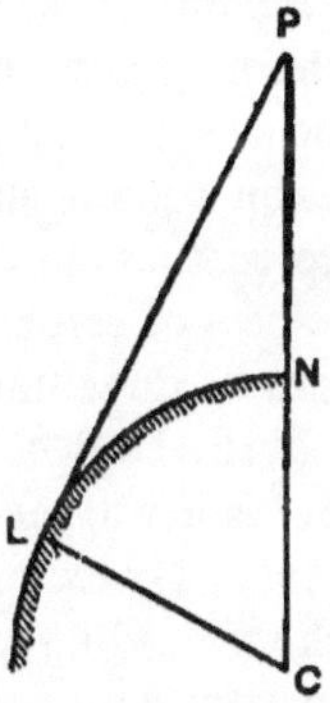

210. To calculate the offset at any point C to a parallel of latitude AC from a straight line AB, tangent to AC at A.—We can do this by treating the parallel of latitude AC in Fig. 94 as a curve to which the arc of a great circle AB is tangent at A, and thus obtain the offset CB according to Sec. 78; or, we can solve the right-angled spherical triangle ANB, and so find the latitude of B, if we know the difference of longitude N, thus:

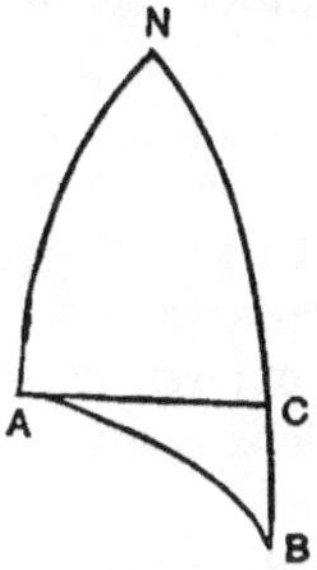

$$\tan (\text{lat. } B) = \tan (\text{lat. } A) \cos N.$$

CB then equals the difference of latitude of A and B.

211. We are now in a position to consider the influence of the spherical form of the earth, assuming for the moment the earth to be a sphere, on a map the linear measurements of which have been computed on the supposition that the surface of the earth is a plane,

Fig. 93.

Fig. 94.

Now a spherical surface cannot be developed on a plane surface, but can only be developed on a sphere of equal radius. Thus no map can, theoretically even, be correct to the same scale in all its parts. In nautical charts, which are generally made on Mercator's Projection, this difficulty is overcome by the use of a scale of meridional parts, the scale at all points being proportional to the secant of the latitude. And this is a very convenient method, where all positions are obtained astronomically and where the error involved by calculating the courses according to "Middle Latitude Sailing" is of no importance. But in constructing a map this method is inconvenient; for if the same scale is used throughout, it assumes that parallels of latitude are right lines, and that there is no convergence of the meridians. In plotting exploratory surveys, simplicity is an important factor; also, the map must be adapted to the same scale throughout, and be so arranged as to be suitable to the plotting of topography as on a plane surface. To approximate as near as possible to correctness in the more important portions, and to throw the excess of error into the less important parts, is the best that can be done under any circumstances.

212. In Sec. 58 we referred to the corrections which it was necessary to make on account of the convergence of the meridians. By extending this method we are able, with the aid of the preceding problems, to construct the groundwork of our map without any other principles than those already explained. The best way is to take an **example** and work it out as if in actual practice.

Suppose from A in Latitude 60° N. and Longitude 120° W. we intend starting off straight across country for B, a place which, from the maps, we find to be situated in about Lat. 59° N. and Long. 110° W., and wish before starting to lay out the groundwork of a map to be constructed from the knowledge of the topography which we intend to obtain on the way—that we may have some reliable means of plotting our results as soon as obtained, and also of determining positions relatively to each other by means of bearings and distances.

At A we draw, as in Fig. 95, the base-lines AS and AD. Then find the length of AC from Table XVIII, calculating as if it were in the mean latitude of A and B, i.e., 59° 30′ N.; thus $AC =$ about $10 \times 60 \times 3095 =$ say 1,857,000 feet. If great accuracy were required, we could find the value of d in latitude 59° 30′ according to Sec. 206, then $AC = 10d$.

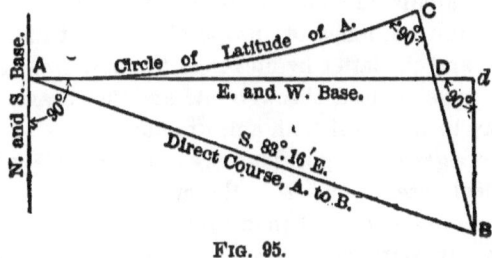

Fig. 95.

Next we make $AD = AC$, and through D draw the meridian CB, the bearing of which on the map, relatively to A, = the convergence between A and $B = 8° 36'$. Therefore the angle $CDA = 81° 24'$.

The length of the offset CD may be found according to Sec. 78, and is equal to about 140,000 feet; and since B lies 1° to the south of C, and on the meridian passing through D, we have $DB =$ about 225,400 feet. Then by solving the plane triangle ADB, we obtain $AB = 1,903,800$ feet, and the angle $BAD = 6° 44'$. Thus the direct course from A to B is S. 83° 16′ E., and $Ad =$ "Total departure" $= AB \cos 6° 44' = 1,890,700$ feet, and $Bd =$ "Total latitude" $= DB \cos 8° 36' = 222,800$ feet.

We have thus the groundwork of our map ready for the plotting of the courses, and if we use sheets of cross-section paper, with 10 divisions to the inch, and plot to a scale of 10,000 feet to an inch, we then have a map of tolerably convenient size, plotted to a scale sufficiently large to show the main features of the country, since any important parts which may have been made the subjects of special survey can be best shown separately.

In order to connect the Astronomical work with that which is plotted by Latitudes and Departures, or by protractor, and which we may call our "dead-reckoning," we must draw meridians and curves of latitude at about every 30′. To fill in these meridians, divide AC equally into 20 parts, and draw the meridians perpendicular to the curve at each of these points, i.e., dividing up the convergence equally among them. The curve of latitude AC, since we know the distance CD, can be drawn by assuming that the offset half-way between A and $D = \frac{1}{4}CD$, and so on, according to Sec. 78.

The advantages of this method of plotting are, that we can readily connect positions taken by astronomical observations with those calculated from dead-reckoning, the former being plotted by the guidance of the parallels of latitude and the meridians, and the latter by means of the base *Ad*. Also, that the same scale is used throughout, and the bearings of all points may be taken off with a protractor.

If the topographical positions are obtained solely by direct astronomical observations, *then* the method of Mercator's Projection is more convenient than that given above.

To plot our route we proceed as follows: Suppose we take rough compass courses; these we plot lightly on the map, having worked them out, say, by Latitudes and Departures, correcting the "latitudes" *absolutely* according to any latitude observations we may take, the "departures" being guided to a reasonable extent by the observations for longitude. Thus our course is constantly being broken, involving a new "total latitude" for each fresh start. This we can best find by scaling from *Ad*, after having plotted the position astronomically. At the end of our journey, whatever error in longitude we may have, may usually be divided up proportionally along the whole route, if the trip has been made at a tolerably uniform pace. The error in latitude should be inappreciable.

The above example shows what must be considered in plotting an extensive survey; and though a more rough and ready method is usually correct enough, yet where the field-work is run in such a way as to warrant a tolerably accurate plot of it being made, the little extra time involved in making a good map is time well spent.

As regards the mode of procedure in keeping a course astronomically, Col. Frome says: "It is probably inconvenient always to obtain latitude at noon, but we can generally do so, and more correctly, at night by the meridian altitude of one or more of the stars. The local time can immediately before or after be ascertained by a single altitude of any other star out of the meridian—the nearer the prime vertical the better; and if a pocket-chronometer is carried, upon which any dependence can be placed, the explorer has thus the means, by comparison with his local time, of obtaining his approximate longitude, and laying down his position on paper. The longitude should also be obtained occasionally by Lunar Dis-

EXPLORATORY SURVEYING. 225

tances, or some other method. The latitude he should always get correct to half a mile, and the longitude to 8 or 10 miles."

213. The Star Map given below will be found convenient in selecting suitable stars for observations. The stars are plotted from their R.A.'s and Decs. in the same way that a map of the earth is plotted by longitudes and latitudes, i.e., *looking down on it.*

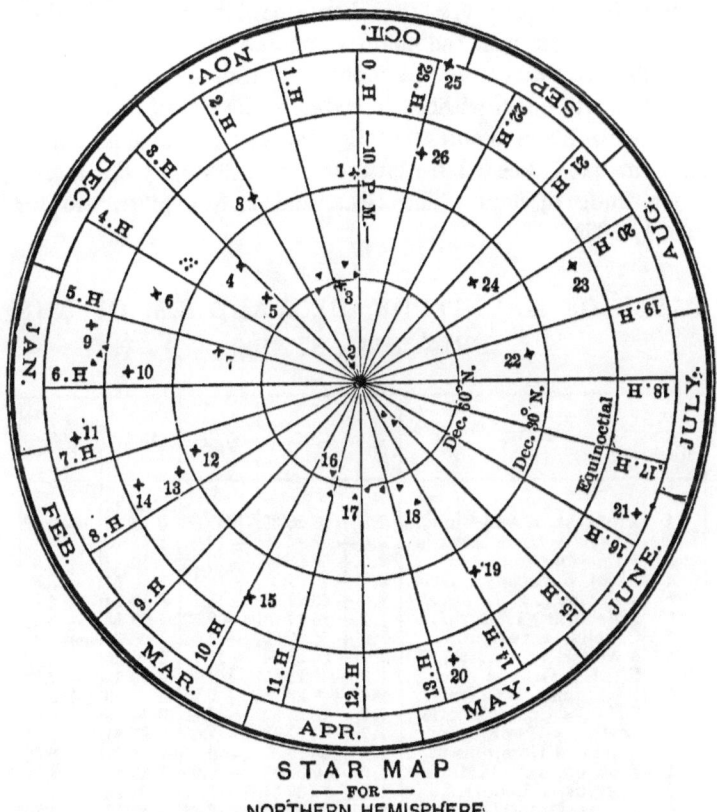

STAR MAP
—FOR—
NORTHERN HEMISPHERE.

The centre is the celestial pole, and the 24 radiating lines divide the 24 hours of R.A. Now the initial point for R.A. being on the meridian at 10 P.M. about Oct. 21, we can divide the circle into 12 divisions, and arrange them so that the radiating line marked 0 *Hours* will cut the 10-o'clock division about two thirds along it. Thus we read off that about Oct. 21 the star marked 1 will be on the meridian, i.e., due south, at

10 P.M. Similarly the star marked 23 will be on the meridian at 10 P.M. about Aug. 17.

But suppose we want to know what star will be near the meridian about 8 P.M. on Jan. 10. Imagine the margin of the map, with the months marked on it, to be stationary, and the interior portion to rotate in the same direction as the hands of a watch, once in $23^h 56^m$; then, since the map shows the position *at 10 p.m.*, at 8 P.M. (two hours earlier) the star marked 5 will have been near the meridian on Jan. 10.

In this way we can tell at about what time any meridian observation will occur without referring to the Nautical Almanac. Thus with this map and the following key and table no Nautical Almanac is needed for latitude observations, by the meridian altitudes of stars. The Decs. and R.A.'s given are for Jan. 1, 1889.

TABLE OF MAGNITUDE, DEC., AND R.A. OF THE PRINCIPAL STARS.

No. in Map.	Name.	Mag.	Dec.	An. Var.	R.A.	An. Var.
			° ′ ″	″	h. m. s.	s.
1	Alpherat, α Andromedæ	2.0	+ 28 28 39	+ 19.88	0 2 39	+ 3.09
2	Polaris, α Ursæ Minoris..	2.0	+ 88 42 59	+ 18.90	1 18 08	+ 23.15
3	γ Cassiopeiæ	2.0	+ 60 06 55	+ 19.56	0 50 01	+ 3.58
4	Algol, β Persei	2.7	+ 40 31 38	+ 14.12	3 0 57	+ 3.88
5	α Persei	2.0	+ 49 27 55	+ 13.10	3 16 24	+ 4.26
6	Aldebaran, α Tauri	1.0	+ 16 17 07	+ 7.52	4 29 33	+ 3.44
7	Capella, α Aurigæ	1.0	+ 45 53 03	+ 4.03	5 08 29	+ 4.42
8	α Arietis	2.0	+ 22 56 14	+ 17.7	2 0 55	+ 3.87
9	Rigel, β Orionis	1.0	− 8 19 50	+ 4.40	5 09 12	− 2.88
10	Betelgeuze, α Orionis	1.2	+ 7 23 8	+ 0.95	5 49 10	+ 3.25
11	Sirius, α Canis Majoris..	1.0	− 16 33 52	− 4.71	6 40 15	+ 2.64
12	Castor, α Geminorum..	1.7	+ 32 07 53	− 7.55	7 27 31	+ 3.84
13	Pollux, β Geminorum..	1.3	+ 28 17 37	− 8.41	7 38 31	+ 3.68
14	Procyon, α Canis Minoris	1.0	+ 5 30 32	− 8.99	7 33 29	+ 3.14
15	Regulus, α Leonis	1.3	+ 12 30 34	− 17.47	10 02 28	+ 3.20
16	α Ursæ Majoris	2.0	+ 62 21 0	− 19.36	10 56 52	+ 3.75
17	γ Ursæ Majoris	2.3	+ 54 18 42	− 20.03	11 47 59	+ 3.18
18	η Ursæ Majoris	2.0	+ 49 52 03	− 18.08	13 43 10	+ 2.37
19	Arcturus, α Bootis	1.0	+ 19 45 38	− 18.88	14 10 36	+ 2.73
20	Spica, α Virginis	1.0	− 10 34 54	− 18.90	13 19 21	+ 3.15
21	Antares, α Scorpii	1.3	− 26 11 06	− 8.30	16 22 36	+ 3.67
22	Vega, α Lyræ	1.0	+ 38 40 50	+ 3.17	18 33 11	+ 2.03
23	Altair, α Aquilæ	1.3	+ 8 34 32	+ 9.27	19 45 22	+ 2.93
24	α Cygni	1.7	+ 44 53 02	+ 12.72	20 37 39	+ 2.04
25	Fomalhaut, α P. Aust...	1.3	− 30 12 37	+ 18.99	22 51 81	+ 3.32
26	Markab, α Pegasi	2.0	+ 14 36 29	+ 19.30	22 59 14	+ 2.98

EXPLORATORY SURVEYING.

IN THE SOUTHERN HEMISPHERE WE ALSO HAVE—

Name.	Mag.	Dec.	An. Var.	R.A.	An. Var.
		° ′ ″	″	h. m. s.	s.
β Hydri	3.0	− 77 52 46	+ 20.28	0 19 54	+ 3.23
Achernar, α Eridani	1.0	− 57 48 03	+ 18.36	1 33 34	+ 2.23
Canopus, α Argus	1.0	− 52 38 07	− 1.87	6 21 29	+ 1.33
β Argus	1.5	− 69 15 36	− 14.80	9 11 59	+ 0.68
α Crucis	1.0	− 62 29 02	− 20.01	12 20 26	+ 3.29
β Centauri	1.0	− 59 50 14	− 17.59	13 55 59	+ 4.18
α Centauri	1.0	− 60 22 47	− 15.38	14 32 05	+ 4.05
α Trianguli Aust	2.0	− 68 49 21	− 7.16	16 36 55	+ 6.30
α Ophiuchi	2.0	+ 12 38 29	− 2.87	17 29 47	+ 2.78
α Gruis	2.0	− 47 29 53	+ 17.25	22 01 14	+ 3.81

In order better to recognize the positions of the stars at night, they may be pricked through on a sheet of paper, which, when turned backwards and held up towards the south, with the month at the lowest part, will correspond with the face of the sky at 10 P.M.

Part IV.

MISCELLANEOUS.

The following miscellaneous information may at times be found of service in the field to both the engineer and the explorer:

214. To find the Horse-power of Falling Water.

$$H.P. = 0.00189 \, QH,$$

where $Q =$ the number of cubic feet of water passing over the fall per minute, and $H =$ height of fall in feet.

Turbines can utilize about 75 p. c. of this H.P. Thus the *Effective* horse-power, i.e., available for useful work, = about .0014 QH.

215. To gauge a stream, roughly.

Take some body, which, when floating, will be almost entirely immersed, and throw it into the middle of the stream, in a part, if possible, unobstructed by reeds, etc., and free from slack-water, eddies, or counter-currents; and where the cross-section of the stream is fairly uniform. Observe the time T in seconds which the body takes to float a distance of 100 feet. Then if $A =$ the cross-section of the stream in square feet, and $Q =$ cubic feet of water that pass per minute,

$$Q = \frac{5000A}{T}.$$

This assumes that the middle surface velocity is to the mean velocity as 6 to 5, which is a fairly average ratio.

MISCELLANEOUS.

216. The Sustaining power of ordinary wooden piles in lbs. equals

$$\frac{FW}{8S},$$

where

$F =$ fall of hammer in inches,
$W =$ weight of hammer in lbs.,
$S =$ space driven by last blow in inches.

This formula is generally found to give results about as reliable as any general formula *can* give.

217. Supporting power of various materials.

Clay.................... 1.0 to 2.0 tons per sq. foot.
Sandy clay............. 2.0 to 4.0 " "
Sand................... 3.0 to 5.0 " "
Gravel................. 4.0 to 5.0 " "
Sandstone.............. 2.0 to 4.0 " "
Firm Rock.............   10.0 "   "

These are the pressures to which the above may usually be *safely* loaded.

218. Transverse strength of rectangular beams.
Let $L =$ length of beam in feet between points of support,
$b =$ breadth of beam in inches,
$d =$ depth of beam in inches,
$W =$ Load at centre of beam in lbs.,
$f =$ coefficient of modulus of rupture.
Then

$$W = \frac{bd^2f}{18L}; \quad d = \sqrt{\frac{18WL}{bf}}; \quad \text{and} \quad b = \frac{18WL}{d^2f}.$$

For the values of f see following table.

For example, if $b = 6''$, $d = 10''$, and $L = 20$ feet, if we take $f = 10,000$ lbs., by the above formula $W = 16,666$ lbs.; so that with a Factor of Safety of 6 we may *safely* load it at its centre, and consequently at any part of it, with a weight of 2778 lbs.

A beam will carry as a centre load only half the weight that it will bear distributed uniformly over it. So that, for instance, if we wish to know what total breadth we must give to a set of stringers, where $d = 16''$, in order safely to carry an ordinary train over a span of 15 feet, if we take $f = 10,000$ lbs., and the load per foot run as equivalent to 4000 lbs., we have as the equivalent value of W, 30,000 lbs. So that by the above formula $b =$ about 3 inches. Therefore, taking a factor of safety of 8, $b =$ about 24 inches; so that four $6'' \times 16''$ stringers may safely be used. The factor of safety usually adopted for wood varies from 5 to 10, according to the condition of the timber, the amount of impact caused by the load, and the possible amount of decay to which it will be subjected.

For spans, in railroad bridges, less than 10 feet, 5000 lbs. per foot run should *usually* be taken as the uniformly distributed load. In spans exceeding 15 feet 3500 lbs. is usually sufficient. These values take no account of the weight of the beams themselves.

VALUES OF f.

Material.	Lbs. per sq. in.	Material.	Lbs. per sq. in.
Ash	12,000 to 14,000	Red Pine	7100 to 9500
Birch	11,700	Spruce	9900 to 12,300
Blue Gum	18,000	Brit. Oak	12,000
Elm	6000 to 9700	Am. Red Oak	10,600

219. Natural Slopes of Earths.

Material.	Slope.	Material.	Slope.	Material.	Slope.
Gravel	40°	Vegetable Earth	28°	Ruble	45°
Dry Sand	38°	Compact Earth	50°	Clay (drained)	45°
Sand	22°	Shingle	39°	Clay (wet)	16°

220. Weight of Earths, Rocks, etc., per cubic yard.

Material.	Weight in lbs. per cu. yd.	Material.	Weight in lbs. per cu. yd.	Material.	Weight in lbs. per cu. yd.
Sand	3360	Clay	3470	Quarts	4590
Gravel	3360	Chalk	4030	Granite	4700
Mud	2800	Sandstone	4370	Trap	4700
Marl	2900	Shale	4480	Slate	4810

A cubic yard of water weighs about 1680 lbs.

221. Weight of Timber and Metals per cubic foot.

Material.	Weight in lbs. per cu. ft.	Material.	Weight in lbs. per cu. ft.	Material.	Weight in lbs. per cu. ft.
Elm, English	35	Pine, red	36	Iron, cast	450
Canadian Elm	45	" white	30	" wrought	482
Maple	42	Teak	50	Steel	490
English Oak	48	Spruce	30	Copper	550
American Oak	50	Larch	34	Lead	710

222. Mortar, Cement, etc. (common mixtures).

Mortar.—1 of lime to 2 or 3 of sharp river sand.

Coarse Mortar.—1 of lime to 4 of coarse gravelly sand.

Concrete.—1 of lime to 4 of gravel and 2 of sand.

Hydraulic Mortar.—1 of blue lias lime to $2\frac{1}{2}$ of burnt clay, ground together.

Beton.—1 of hydraulic mortar to $1\frac{1}{2}$ of angular stones.

Cement.—1 of sand to 1 of cement; or if great tenacity is required the sand may be omitted.

Portland Cement is composed of clayey mud and chalk ground together and afterwards calcined at a high temperature, and then ground to a fine powder.

NOTES.—For ordinary engineering work the following proportions make a good mortar:

1 measure of Lime;
3 to 5 measures of sand, according to the "hunger" of the sand,
1 measure of ashes, brick dust, or burnt clay.

For engineering work, if exposed to dampness, $\frac{1}{3}$ of the lime in the above should be replaced by hydraulic cement; whilst for work under water, 1 measure hydraulic cement to 2 measures of sand make a good mixture.

NOTES ON TIMBER.

223. Selection of standing trees. — "Scribner's Log Book."—"The principal circumstances which affect the quality of growing trees are *soil, climate*, and *aspect*.

"In a moist soil the wood is less firm, and decays sooner than in a dry, sandy soil; but in the latter the timber is seldom fine: the best is that which grows in a dark soil, mixed with stones and gravel. This remark does not apply to the poplar, willow, cypress, and other light woods which grow best in wet situations.

"Trees growing in the centre of a forest or on a plain are generally straighter and more free from limbs than those growing on the edge of the forest, in open ground, or on the sides of hills; but the former are at the same time less hard. The toughest part of a tree will always be found on the side next the north. The aspect most sheltered from prevalent winds is generally most favorable to the growth of timber. The vicinity of salt water is favorable to the strength and hardness of white oak.

"The selection of timber trees should be made before the fall of the leaf. A healthy tree is indicated by the top branches being vigorous, and well covered with leaves; the bark is clear, smooth, and of a uniform color. If the top has a regular, rounded form; if the bark is dull, scabby, and covered with white and red spots, caused by running water or sap,—the tree is unsound. The decay of the uppermost branches and the separation of the bark from the wood are infallible signs of the decline of the tree."

224. Defects of Timber Trees (especially of oak).—"*Sap*, the white wood next to the bark, which very soon rots, should never be used, except that of hickory. There are sometimes found rings of light-colored wood surrounded by good hard wood; this may be called the *second sap:* it should cause the rejection of the tree.

"*Brash-wood* is a defect generally consequent on the decline of the tree from age; the pores of the wood are open, the wood is reddish-colored, it breaks short without splinters, and the chips crumble to pieces.

"Wood which has *died before being felled* should in general be rejected; so should *knotty trees*, and those which are covered with tubercles, etc.

"*Twisted wood*, the grain of which ascends in a spiral form, is unfit for use in large scantling; but if the defect is not very decided, the wood may be used for naves, and for some light pieces.

"*Splits, checks, and cracks*, extending towards the centre, if deep and strongly marked, make the wood unfit for use, unless it is intended to be split.

"*Wind-shakes* are cracks separating the concentric layers of wood from each other; if the shake extends through the entire circle, it is a ruinous defect."

225. Felling Timber.—"The most suitable season for felling timber is that in which vegetation is at rest, which is the case in midwinter and in midsummer; recent opinions derived from facts incline to give preference to the latter season. The tree should be allowed to attain its full maturity before being felled; this period in oak timber is generally at the age of from 75 to 100 years, or upwards, according to circumstances. The age of hardwood is determined by the number of rings which may be counted in a section of the tree.

"The tree should be cut as near the ground as possible, the lower part being the best timber. The quality of the wood is in some degree indicated by the color, which should be nearly uniform in the heart wood, a little deeper toward the centre, and without transitions.

"Felled timber should be immediately stripped of its bark, and raised from the ground.

"As soon as practicable after the tree is felled the sap-wood should be taken off and the timber reduced, either by sawing or splitting, nearly to the dimensions required for use.

"The best method of preventing decay is the immediate removal of it to a dry situation, where it should be piled in such a manner as to secure a free circulation of air around it, but without exposure to the sun and wind. When thoroughly seasoned before cutting it up into small pieces, it is less liable to warp and twist in drying. When green, timber is not so *strong* as when thoroughly dry.

"Lumber containing much sap is not only weaker, but decays much sooner than that free from sap."

226. Seasoning and Preserving Timber.—"For the purpose of seasoning, timber should be piled under shelter, where it may be kept dry, but not exposed to a strong current of air; at the same time there should be a free circulation of air about the timber, with which view slats or blocks of wood should be placed between the pieces that lie over each other, near enough to prevent the timber from bending. The seasoning of timber requires from two to four years, according to its size.

"Gradual drying and seasoning in this manner is considered the most favorable to the durability and strength of timber.

"Timber of large dimensions is improved by *immersion in water* for some weeks. Oak timber loses about one fifth of its weight in seasoning, and about one third of its weight in becoming dry."

227. Decay of Timber.—There are three principal causes of decay of timber—dry-rot, wet-rot, and the "teredo navalis" and other worms.

Dry-rot does not usually occur where there is a free circulation of air, and if the timber is properly dried an *occasional* immersion in water should do no harm. Timber kept dry and well ventilated has been known to last for several hundred years without apparent deterioration. Dry-rot is caused by a species of wood fungus—*Merulius lachrymans*—which destroys the tensile and cohesive strength, gradually converting the timber into a fine powder.

Wet-rot.—This is the destructive agent at work more or less on all timber freely exposed to air and moisture. It is of two kinds:

A. *Chemical.*—In this case a slow combustion takes place, and by a gradual process of oxidation the wood slowly rots away.

B. *Mechanical.*—This is the more common form, and generally occurs near the water-line in timber subject to frequent immersion. It is the frequent alternate conditions of moisture and dryness that are most trying to timber, as is the case with metals. When timber is *constantly* under water, the action of the water dissolves a portion of its substance, which is made apparent by its becoming covered with a coating of slime, and this protects the interior. If, however, it is exposed to alternations of moisture and dryness, as is the case with piles in tidal waters, the dissolved parts being continually removed by evaporation and the action of the water, new surfaces are being frequently exposed for decomposition.

Piles driven in sea-water are frequently destroyed by the "teredo navalis," and also by another species of worm called the "limnoria." They both work from about the high-water mark to the surface of the mud.

228. To test Steel and Iron.—*Scientific American.*—Nitric acid will produce a black spot on steel; the darker the spot the harder the steel. Iron, on the contrary, remains bright if touched with nitric acid.

Good steel in its soft state has a curved fracture and a uniform gray lustre; in its hard state, a dull, silvery, uniform white. Cracks, threads, or sparkling particles denote bad quality.

Good steel will not bear a white heat without falling to pieces, and will crumble under the hammer at a *bright*-red heat, while at a *middling* heat it may be drawn out under the hammer to a fine point. Care should be taken that before attempting to draw it out to a point the fracture is not concave; and should it be so, the end should be filed to an obtuse point before operating. Steel should be drawn out to a fine point and plunged into cold water; the fractured point should scratch glass. To test its toughness, place a fragment on a block of cast-iron: if good, it may be driven by a blow of a hammer into the cast-iron; if poor, it will crush under the blow.

Tests of Iron.—A soft tough iron, if broken gradually, gives long silky fibres of leaden-gray hue, which twist together and cohere before breaking.

A medium even grain with fibres denotes good iron. Badly refined iron gives a short blackish fibre on fracture. A very fine grain denotes hard steely iron, likely to be cold-short and hard.

Coarse grain with bright crystallized fracture or discolored spots denotes cold-short, brittle iron, which works easily when heated and welds well. Cracks on the edge of a bar are indications of hot-short iron. Good iron is readily heated, is soft under the hammer, and throws out few sparks.

229. Strength of Rope.—The table on following page gives *some idea* of the strength of ordinary Manilla Rope.

It must be remembered that these values are for new ropes and that a few months' exposure to the weather will probably cause a decrease in the strength of 40 or 50 p. c. A factor of safety of 4 or 5 is generally employed to obtain their *safe working strength*.

Ropes made of good Italian hemp are considerably stronger than these.

TABLE OF MANILLA ROPE—3 STRANDS.

Size of Rope.		Breaking-strength in lbs.	Size of Rope.		Breaking-strength in lbs.
Diam. in inches.	Circum. in inches.		Diam. in inches.	Circum. in inches.	
$\tfrac{1}{4}$	0.71	375	$2\tfrac{1}{4}$	7.14	37,500
$\tfrac{1}{2}$	1.43	1,500	3	8.57	54,000
$\tfrac{3}{4}$	2.14	3,380	$3\tfrac{1}{4}$	10.0	73,600
1	2.86	6,000	4	11.4	96,000
$1\tfrac{1}{4}$	3.57	9,380	$4\tfrac{1}{4}$	12.1	121,000
$1\tfrac{1}{2}$	4.28	13,500	5	14.2	150,000
2	5.70	24,000	6	17.1	216,000

Wire Ropes.—The following table gives the strength of iron and cast-steel wire rope:

TABLE OF IRON AND CAST-STEEL WIRE ROPE.

Size of Rope.		Breaking-strength in lbs.		Size of Rope.		Breaking-strength in lbs.	
Diam. in In.	Circum. in In.	Iron.	C. Steel.	Diam. in In.	Circum. in In.	Iron.	C. Steel.
$\tfrac{1}{2}$	$1\tfrac{1}{2}$	6,960	15,000	$1\tfrac{1}{2}$	$4\tfrac{1}{2}$	78,000	154,000
$\tfrac{3}{4}$	$2\tfrac{1}{4}$	17,280	36,000	$1\tfrac{3}{4}$	$5\tfrac{1}{4}$	108,000	212,000
1	$3\tfrac{1}{4}$	32,000	66,000	2	6	130,000	250,000
$1\tfrac{1}{4}$	4	54,000	104,000	$2\tfrac{1}{4}$	$6\tfrac{3}{4}$	148,000	310,000

These ropes have 19 wires to the strand and hemp centres. One fifth of the above breaking-strength may be taken as the safe working strength.

For the strength of **Iron Rods** see Sec. 138.

230. Properties of the Circle.

Diameter $\times$ 3.14159 = circumference.
Diameter $\times$ .886226 = side of an equal square.
Diameter $\times$ .7071 = side of an inscribed square.
Diameter^2 $\times$ .7854 = area of circle.
Radius $\times$ 6.28318 = circumference.
Circumference $\times$ .31831 = diameter.
Circumference = 3.5449 $\sqrt{\text{area of circle.}}$
Diameter = 1.1283 $\sqrt{\text{area of circle.}}$
Length of arc = number of degrees $\times$ 0.017453 radius.
    Arc of 1° to rad. 1 = 0.01745329.
    Arc of 1′ to rad. 1 = 0.000290888.
    Arc of 1″ to rad. 1 = 0.000004848.
Degrees in arc whose length = radius = 57°.2957795.
    π = 3.1415926536;   Log π = 0.4971499.

MISCELLANEOUS.

231. PLANE TRIGONOMETRY.—In Fig. 96, if the angle $GAE = 90°$; then in the right-angled triangle ABC, if $AB = $ Radius $=$ unity,

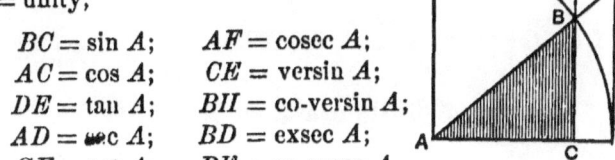

FIG. 96.

$BC = \sin A$; $AF = \operatorname{cosec} A$;
$AC = \cos A$; $CE = \operatorname{versin} A$;
$DE = \tan A$; $BH = \text{co-versin } A$;
$AD = \sec A$; $BD = \operatorname{exsec} A$;
$GF = \cot A$; $BF = \text{co-exsec } A$.

Therefore

$$\sin A = \frac{BC}{AB}; \quad \cos A = \frac{AC}{AB}; \quad \tan A = \frac{BC}{AC};$$

$$\operatorname{cosec} A = \frac{AB}{BC}; \quad \sec A = \frac{AB}{AC}; \quad \cot A = \frac{AC}{BC},$$

Thus,

$$\sin A = \frac{1}{\operatorname{cosec} A}; \quad \cos A = \frac{1}{\sec A}; \quad \tan A = \frac{1}{\cot A}.$$

An angle and its *Supplement* have the same *Sine* and *Cosecant;* but the *Tangents, Secants, Cosines* and *Cotangents*, though of equal length, are of contrary signs: so that in applying to obtuse angles trigonometrical formulæ which were originally intended for acute angles, the algebraic signs of the tangents, secants, cosines, and cotangents must be reversed.

The sine, secant, and tangent of an angle A are respectively equal to the cosine, cosecant, and cotangent of its complement (i.e., of $90° - A$).

$$AB^2 = AC^2 + BC^2; \quad B = 90° - A.$$

$$\text{Area of triangle} = \frac{AC \cdot BC}{2}.$$

Examples of Right-angled Triangles:
1. *Given* $A = 30°$, *and* $AC = 100$, find BC.

We see above that $\tan A = \dfrac{BC}{AC}$; therefore

$$BC = AC \tan A = 57.73.$$

2. *Find the sine of* 128°.

Since $\sin(180° - A) = \sin A$,

$$\sin 128° = \sin(180° - 52°) = \sin 52°,$$

which from the tables we find $= 0.788$.

Solution of Oblique-angled Triangles.

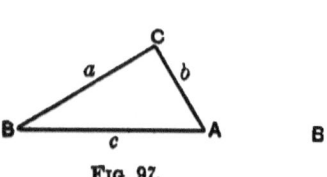
Fig. 97.

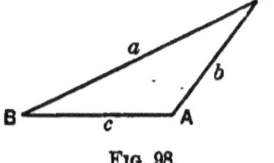
Fig. 98.

$$\frac{\sin A}{a} = \frac{\sin B}{b} = \frac{\sin C}{c}. \quad \ldots \ldots (1)$$

$$\tan\frac{A-B}{2} = \frac{a-b}{a+b}\tan\frac{A+B}{2}. \quad \ldots \ldots (2)$$

$$A = \frac{A+B}{2} + \frac{A-B}{2}. \quad \ldots \ldots (3)$$

$$B = \frac{A+B}{2} - \frac{A-B}{2}. \quad \ldots \ldots (4)$$

$$c = (a+b)\frac{\cos\dfrac{A+B}{2}}{\cos\dfrac{A-B}{2}}. \quad \ldots \ldots (5)$$

Let $\dfrac{a+b+c}{2} = s$; then

$$\operatorname{vers} A = \frac{2(s-b)(s-c)}{bc}. \quad \ldots \ldots (6)$$

$$\cos\frac{A}{2} = \sqrt{\frac{s(s-a)}{bc}}. \quad \ldots \ldots (7)$$

MISCELLANEOUS.

$$\text{Area of triangle} = \sqrt{s(s-a)(s-b)(s-c)}. \quad . \quad , \quad (8)$$

$$= \frac{ab}{2} \sin C. \quad . \quad . \quad . \quad . \quad . \quad . \quad (9)$$

$$= \frac{a^2 \sin B \sin C}{2 \sin A}. \quad . \quad . \quad . \quad . \quad (10)$$

$$A = 180° - (B + C). \quad . \quad . \quad . \quad . \quad (11)$$

The above **formulæ** are all that are required for the ordinary solution of plane triangles.

Remarks.—Though such a formula as No. 2 simply mentions A and B and their opposite sides, it holds equally well whether we substitute C for A, or C for B, provided that the sides are changed to correspond also. In Equations 2, 3, 4, and 5, A is intended to represent the *greater* angle of the two angles A and B.

Examples.—
1. *Given A, B, and b, find A.*
By Equation 1,
$$a = \frac{b \sin A}{\sin B}.$$

2. *Given B, c, and b, find C.*
By Equation 1,
$$\sin C = \frac{c \sin B}{b}.$$

3. *Given A, B, and c, find a.*
By Equation 11,
$$C = 180° - (A + B);$$
and by Eq. 1,
$$a = \frac{c \sin A}{\sin C}.$$

4. *Given B, a, and c, find A and b.*
By Eq. 2,
$$\tan \frac{A - C}{2} = \frac{a - c}{a + c} \tan \frac{A + C}{2};$$

from which we obtain the value of
$$\frac{A-C}{2};$$
and by Eq. 11,
$$\frac{A+C}{2} = 90° - \frac{B}{2};$$
therefore we can find A from Eq. 3.

Then by Eq. 5,
$$b = (a+c)\frac{\cos\frac{A+C}{2}}{\cos\frac{A-C}{2}}.$$

5. *Given a, b, and c, find B.*
By Eq. 6,
$$\text{vers } B = \frac{2(s-a)(s-c)}{ac};$$
or, we might equally well have used Eq. 7.

---

232. The following general equations are worth noting:

$$\sin A = \tan A \cos A = \sqrt{1-\cos^2 A} = 2\sin\frac{A}{2}\cos\frac{A}{2};$$

$$\cos A = \cot A \sin A = \sqrt{1-\sin^2 A} = 2\cos^2\frac{A}{2} - 1;$$

$$\tan A = \sin A \sec A = \frac{\text{vers } 2A}{\sin 2A} = \text{exsec } A \cot\frac{A}{2};$$

$$\cot A = \cos A \csc A = \frac{\sin 2A}{\text{vers } 2A} = \frac{\tan\frac{A}{2}}{\text{exsec } A};$$

$$\text{vers } A = 1 - \cos A = 2\sin^2\frac{A}{2} = \cos A \text{ exsec } A;$$

$$\text{exsec } A = \sec A - 1 = \tan A \tan\frac{A}{2} = \frac{\text{vers } A}{\cos A}.$$

233. Spherical Trigonometry.

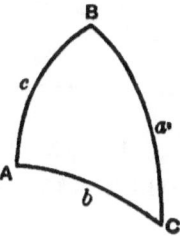

Fig. 99.

RIGHT-ANGLED TRIANGLES.—In Fig. 99 let $A = 90°$; then

$$\sin b = \sin a \sin B; \qquad \tan c = \tan a \cos B;$$

$$\cot C = \cos a \tan B; \qquad \tan c = \sin b \tan C;$$

$$\cos a = \cos b \cos c; \qquad \cos B = \cos b \sin C;$$

$$\tan a = \frac{\tan b}{\cos C}; \qquad \sin c = \frac{\tan b}{\tan B}; \qquad \sin a = \frac{\sin b}{\sin B};$$

$$\sin C = \frac{\cos B}{\cos b}; \qquad \cos c = \frac{\cos a}{\cos b}; \qquad \sin B = \frac{\sin b}{\sin a};$$

$$\cos C = \frac{\tan b}{\tan a}; \qquad \tan C = \frac{\tan c}{\sin b}; \qquad \tan B = \frac{\tan b}{\sin c};$$

$$\cos c = \frac{\cos C}{\sin B}; \qquad \cos b = \frac{\cos B}{\sin C}; \qquad \cos a = \frac{\cot C}{\tan B}$$

b and c are of the same species respectively as B and C.

Any side is greater than 90° if the other sides are of different species, and less than 90° if of the same species.

B or C is less than 90° if the containing sides are of the same species, and less than 90° if of different species.

Oblique-angled triangles.

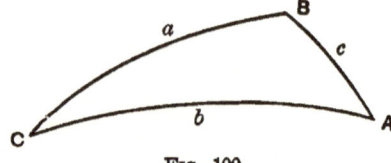

Fig. 100.

Let ABC in Fig. 100 represent any oblique-angled spherical triangle; then

$$\frac{\sin A}{\sin a} = \frac{\sin B}{\sin b} = \frac{\sin C}{\sin c}; \quad \ldots \quad (1)$$

$$\tan \frac{a+b}{2} = \tan \frac{c}{2} \frac{\cos \frac{A \sim B}{2}}{\cos \frac{A+B}{2}}; \quad \ldots \quad (2a)$$

$$\tan \frac{a \sim b}{2} = \tan \frac{c}{2} \frac{\sin \frac{A \sim B}{2}}{\sin \frac{A+B}{2}}; \quad \ldots \quad (2b)$$

$$\tan \frac{A+B}{2} = \cot \frac{C}{2} \frac{\cos \frac{a \sim b}{2}}{\cos \frac{a+b}{2}}; \quad \ldots \quad (3a)$$

$$\tan \frac{A \sim B}{2} = \cot \frac{C}{2} \frac{\sin \frac{a \sim b}{2}}{\sin \frac{a+b}{2}}; \quad \ldots \quad (3b)$$

$$\cos c = \cos a \cos b + \sin a \sin b \cos C; \quad \ldots \quad (4)$$

$$\sin \frac{A}{2} = \sqrt{\frac{\sin (s-b) \sin (s-c)}{\sin b \sin c}}; \quad \ldots \quad (5)$$

$$\sin \frac{a}{2} = \sqrt{\frac{\cos S \cos (S-A)}{\sin B \sin C}}; \quad \ldots \quad (6)$$

where $s = \dfrac{a+b+c}{2}$ and $S = \dfrac{A+B+C}{2}$.

APPENDIX.

NOTE A. (See Sec. 10.)

If we knew the average pressure in the cylinders we could find the propelling force of an engine at any speed, if not limited by adhesion, by the following rule :

Multiply together the square of the diameter of one piston in inches, the length of stroke in inches, and the mean pressure (above atmosphere) in lbs. per sq. in. The product divided by the diameter of a driver in inches gives the propelling force in lbs., ignoring "internal frictional resistances."

Theoretically, the mean effective cylinder-pressure in lbs. per sq. in. equals

$$\frac{P + 2.3P\,(\text{Log } S)}{S} - 15,$$

where $P =$ absolute boiler-pressure in lbs. per sq. in. and $S =$ Stroke $\div$ part of stroke before cut-off.

But owing to the contraction of the steam-ports, the initial cylinder-pressure always falls below the boiler-pressure. Similarly owing to the contraction of the exhaust-port, back-pressure always exists; and these are matters so purely of mechanical detail that no general rule can be given which would take them into consideration.

At 20 miles per hour, however, the effective initial cylinder pressure often equals only about 90 p. c. of the boiler-pressure, and at 50 m. p. h. about 60 p. c.

Thus if $P = 125$ lbs. per sq. in. and the stroke $= 24$ inches ; if steam is cut off at 6 inches, the theoretical mean cylinder-pressure $= 59$ lbs. per square inch, which at 50 m. p. h. will probably be reduced to about 36 lbs.: so that if the diameter of the piston $= 16$ inches, and of the driving-wheels 60 inches, the propelling force will equal 3680 lbs.; and if we deduct 10 p. c. from this for internal frictional resistances, the propelling force $= 3200$ lbs.

NOTE B. (See Sec. 19.)

In order to reduce the quantities used in Diagram II into the *same* units, say ton, mile, and hour, the ordinates of the curves must be multiplied by

$$\frac{(3600)^2}{2000 \times 5280} \times 32.2 = 40 \text{ (nearly)}$$

to reduce them to tons weight (2000 lbs.), in miles per hour units. Then, with the units selected, the equation of motion is

$$\frac{d}{dt}(OQ) = NQ - MQ.$$

But if x is the space passed over,

$$OQ = \frac{dx}{dt};$$

so that

$$\frac{d}{dt}(OQ) = OQ \frac{d}{dx}(OQ),$$

and therefore

$$\frac{OQ \cdot d(OQ)}{NQ - MQ} = dx,$$

the graphic process giving the integral. But with the scales used in Diagram II, instead of multiplying the ordinates as above, we can simply use as a scale 1 square inch = 1 mile, which practically comes to the same thing. If the horizontal scale were ten miles per hour to one inch, the scale then to be used would be 4 square inches = 1 mile; and this is often a more convenient scale to adopt.

NOTE C. (See Sec. 44.)

Messrs. W. and L. E. Gurley in their Manual give the following methods of adjusting the object-slide:

To Adjust the Object-slide of a Transit.—"Having set up and levelled the instrument, the line of collimation being also adjusted for objects from three hundred to five hundred feet distant, clamp the plates securely, and fix the vertical cross-wire upon an object as distant as may be distinctly seen; then, without disturbing the instrument, throw out the object-glass, so as to bring the vertical wire upon an object as near as the range of the telescope will allow. Having this clearly in mind, unclamp the limb, turn the instrument half-way around, reverse the eye-end of the telescope, clamp the limb, and with the tangent-screw bring the vertical wire again upon the near object; then draw in the object-glass slide until the distant object first sighted upon is brought into distinct vision. If the vertical wire strikes the same line as at first, the slide is correct for both near and remote objects; and, being itself straight, for all distances.

"But if there be an error, proceed as follows: First, with the thumb and forefinger twist off the thin brass tube that covers the screws. Next, with the screw-driver, turn the two screws on the opposite *sides* of the telescope, loosening one and tightening the other, so as apparently to increase the error, making, by estimation, one-half the correction required.

"Then go over the usual adjustment of the line of collimation, and having it completed, repeat the operation above described; first sighting upon the distant object, then finding a near one in line, and then reversing, making correction, etc., until the adjustment is complete."

To Adjust the Object-slide of a Y-Level.—"The maker selects an object as distant as may be distinctly observed, and upon it adjusts the line of collimation, making the centre of the wires to revolve without passing either above or below the point or line assumed.

"In this position, the slide will be drawn in nearly as far as the telescope-tube will allow.

"He then, with the pinion-head, moves out the slide until an object, distant about ten or fifteen feet, is brought clearly into view; again revolving the telescope in the Y's, he observes whether the wires will reverse upon this second object.

"Should this happen to be the case, he will assume that, as the line of collimation is in adjustment for these two distances, it will be so for all intermediate ones, since the bearings of the slide are supposed to be true, and their planes parallel with each other.

"If, however, as is most probable, either or both wires fail to reverse upon the second point, he must then, by estimation, remove half the error by the screws at right angles to the hair sought to be corrected, remembering, at the same time, that on account of the inverting property of the eye-piece he must move the slide in the direction which apparently increases the error. When both wires have thus been treated in succession, the line of collimation is adjusted on the near object, and the telescope again brought upon the most distant point; here the tube is again revolved, the reversion of the wires upon the object once more tested, and the correction, if necessary, made in precisely the same manner.

" He proceeds thus, until the wires will reverse upon both objects in succession; the line of collimation will then be in adjustment at these and all intermediate points, and by bringing the screw-heads, in the course of the operation, to a firm bearing upon the washers beneath them, the adjustable ring will be fastened so as for many years to need no further adjustment."

"*The centring of the eye-tube* is performed after the wires have been adjusted, and is effected by moving the ring, by means of the screws shown on the outside of the tube, until the intersection of the wires is brought into the centre of the field of view."

NOTE D. (See Sec. 57.)

The time at which any elongation will occur may be found by the formula

$$\cos h = \cot (\text{dec.}) \times \tan (\text{lat.}),$$

where $h =$ the hour-angle (see Sec. 182), h really being the supplement of the angle at P in the right-angled spherical triangle WZP (or EZP) in Fig. 10, the right angle being at W or E.

The angle h may be reduced to mean time as shown in Part III.

NOTE E. (See Sec. 57.)

To find the azimuth of two stars when in the same vertical plane (Polaris being one of them) proceed as follows:

Let $A =$ the difference in R.A. of the stars,
$d =$ the declination of Polaris,
and $D =$ the declination of the other star.

APPENDIX. 249

Find p and m from the formulæ

$$\tan m = \frac{\cos A}{\tan D}, \qquad p = \frac{\sin D}{\cos m};$$

then find a from the formula

$$\cos a = p \sin (d + m).$$

Then Z, the azimuth, is given by

$$\sin Z = \frac{\sin A \cos D \cos d}{\cos L \sin a},$$

where $L =$ the latitude of the place.

To find the interval of time which must elapse after the two stars are observed to be in the same vertical plane, before Polaris will be due north, find S from the equation

$$\sin S = \sin A \frac{\cos D}{\sin a}.$$

Then

$$\cot \frac{h}{2} = \frac{\cos \frac{L+d}{2}}{\sin \frac{L \sim d}{2}} \tan \frac{Z \sim S}{2},$$

where h is the hour-angle in sidereal time.

To find the interval in mean time, see Sec. 179.

The above steps may be easily traced by drawing the positions of the star, the pole, and the zenith.

It is not *necessary* to use Polaris; but if any other star is selected, d refers to the star whose declination is the greater.

NOTE F. (See Sec. 58.)

The true value of the convergence is given by the equation

$$\sin \frac{\text{convergence}}{2} = \sin \frac{\text{diff. of long.}}{2} \times \sin (\text{lat.}).$$

If the places are in different latitudes, as A and D in Fig. 12, we have the convergence = the difference in azimuth at

A and D, which we can find by solving the spherical triangle AND.

NOTE G. (See Sec. 189.)

The difference in altitude in seconds of arc, between the meridian altitude and the maximum altitude of a body, is equal to

$$\frac{d^2}{4a},$$

where

$$a = \frac{\cos \text{lat.} \cos \text{dec.} \times 1.964}{\sin (\text{lat.} - \text{dec.})},$$

and $d =$ the hourly change of declination in minutes of arc.

When the declination differs in sign from the latitude, it will be negative. If the body has its declination changing towards the north in the northern hemisphere or towards the south in the southern hemisphere, the meridian altitude precedes the maximum altitude, which will be the case between mid-winter and mid-summer; but if changing towards the south in the northern hemisphere, or towards the north in the southern, the maximum altitude occurs to the east of the meridian.

NOTE H. (See Sec. 24.)

Theoretically the train could just start and eventually attain the speed indicated, provided that the values of MN at all speeds lower than the given one are greater than the value of MN at the speed selected; but in order that the train may attain the speed required *in a reasonable time*, ample allowance must be made for inertia.

NOTE I. (See Sec. 45.)

Two very common causes of error in observing angles, which remain unaffected by the process of repetition, are (1) *station-twist*, due usually to some such cause as the action of the sun, which gives to the instrument a more or less steady motion in one direction, and (2) *general instability of the support*. The former may be eliminated by taking the mean of two sets of readings, one taken from left to right, and the other from right to left, and the latter by the application of a constant (c) obtained thus:

Let A be the reading of some required angle; then if B is the reading of the residuary angle,

$$c = \frac{360° - (A + B)}{2},$$

for the residuary angle *should* of course equal $360° - A$.

TABLES.

TABLE I.—RADII.

Deg.	Radius.	Deg.	Radius.	Deg.	Radius.	Deg.	Radius.	Deg.	Radius.
0° 0′	Infinite	1° 0′	5729.65	2° 0′	2864.93	3° 0′	1910.08	4° 0′	1432.69
1	343775.	1	5635.72	1	2841.26	1	1899.53	1	1426.74
2	171887.	2	5544.83	2	2817.97	2	1889.09	2	1420.85
3	114592.	3	5456.82	3	2795.06	3	1878.77	3	1415.01
4	85943.7	4	5371.56	4	2772.53	4	1868.56	4	1409.21
5	68754.9	5	5288.92	5	2750.35	5	1858.47	5	1403.46
6	57295.8	6	5208.79	6	2728.52	6	1848.48	6	1397.76
7	49110.7	7	5131.05	7	2707.04	7	1838.59	7	1392.10
8	42971.8	8	5055.59	8	2685.89	8	1828.82	8	1386.49
9	38197.2	9	4982.33	9	2665.08	9	1819.14	9	1380.92
10	34377.5	10	4911.15	10	2644.58	10	1809.57	10	1375.40
11	31252.3	11	4841.98	11	2624.39	11	1800.10	11	1369.92
12	28647.8	12	4774.74	12	2604.51	12	1790.73	12	1364.49
13	26444.2	13	4709.33	13	2584.93	13	1781.45	13	1359.10
14	24555.4	14	4645.69	14	2565.65	14	1772.27	14	1353.75
15	22918.3	15	4583.75	15	2546.64	15	1763.18	15	1348.45
16	21485.9	16	4523.44	16	2527.92	16	1754.19	16	1343.15
17	20222.1	17	4464.70	17	2509.47	17	1745.26	17	1337.65
18	19098.6	18	4407.46	18	2491.29	18	1736.48	18	1332.77
19	18093.4	19	4351.67	19	2473.37	19	1727.75	19	1327.63
20	17188.8	20	4297.28	20	2455.70	20	1719.12	20	1322.53
21	16370.2	21	4244.23	21	2438.29	21	1710.56	21	1317.46
22	15626.1	22	4192.47	22	2421.12	22	1702.10	22	1312.43
23	14946.7	23	4141.96	23	2404.19	23	1693.72	23	1307.45
24	14323.6	24	4092.66	24	2387.50	24	1685.42	24	1302.50
25	13751.0	25	4044.51	25	2371.04	25	1677.20	25	1297.58
26	13222.1	26	3997.49	26	2354.80	26	1669.06	26	1292.71
27	12732.4	27	3951.54	27	2338.78	27	1661.00	27	1287.87
28	12277.7	28	3906.54	28	2322.98	28	1653.01	28	1283.07
29	11854.3	29	3862.74	29	2307.39	29	1645.11	29	1278.30
30	11459.2	30	3819.83	30	2292.01	30	1637.28	30	1273.57
31	11089.6	31	3777.85	31	2276.84	31	1629.52	31	1268.87
32	10743.0	32	3736.79	32	2261.86	32	1621.84	32	1264.21
33	10417.5	33	3696.61	33	2247.08	33	1614.22	33	1259.58
34	10111.1	34	3657.29	34	2232.49	34	1606.68	34	1254.98
35	9822.18	35	3618.80	35	2218.09	35	1599.21	35	1250.42
36	9549.34	36	3581.10	36	2203.87	36	1591.81	36	1245.89
37	9291.29	37	3544.19	37	2189.84	37	1584.48	37	1241.40
38	9046.75	38	3508.02	38	2175.98	38	1577.21	38	1236.94
39	8814.78	39	3472.59	39	2162.30	39	1570.01	39	1232.51
40	8594.42	40	3437.87	40	2148.79	40	1562.88	40	1228.11
41	8384.80	41	3403.83	41	2135.44	41	1555.81	41	1223.74
42	8185.16	42	3370.46	42	2122.26	42	1548.80	42	1219.40
43	7994.81	43	3337.74	43	2109.24	43	1541.86	43	1215.30
44	7813.11	44	3305.65	44	2096.39	44	1534.98	44	1210.82
45	7639.49	45	3274.17	45	2083.68	45	1528.16	45	1206.57
46	7473.42	46	3243.29	46	2071.13	46	1521.40	46	1202.36
47	7314.41	47	3212.98	47	2058.73	47	1514.70	47	1198.17
48	7162.03	48	3183.23	48	2046.48	48	1508.06	48	1194.01
49	7015.87	49	3154.03	49	2034.37	49	1501.48	49	1189.88
50	6875.55	50	3125.36	50	2022.41	50	1494.95	50	1185.78
51	6740.74	51	3097.20	51	2010.59	51	1488.48	51	1181.71
52	6611.12	52	3069.55	52	1998.90	52	1482.07	52	1177.66
53	6486.38	53	3042.39	53	1987.35	53	1475.71	53	1173.65
54	6366.26	54	3015.71	54	1975.93	54	1469.41	54	1169.66
55	6250.51	55	2989.48	55	1964.64	55	1463.16	55	1165.70
56	6138.90	56	2963.71	56	1953.48	56	1456.96	56	1161.76
57	6031.20	57	2938.39	57	1942.44	57	1450.81	57	1157.85
58	5927.22	58	2913.49	58	1931.53	58	1444.72	58	1153.97
59	5826.76	59	2889.01	59	1920.75	59	1438.68	59	1150.11
60	5729.65	60	2864.93	60	1910.08	60	1432.69	60	1146.28

TABLE I.—RADII.

Deg.	Radius.	Deg.	Radius.	Deg.	Radius.	Deg.	Radius.	Deg.	Radius.
5° 0′	1146.28	6° 0′	955.366	7° 0′	819.020	8° 0′	716.779	9° 0′	637.275
1	1142.47	1	952.722	1	817.077	1	715.291	1	636.099
2	1138.69	2	950.093	2	815.144	2	713.810	2	634.928
3	1134.94	3	947.478	3	813.238	3	712.335	3	633.761
4	1131.21	4	944.877	4	811.303	4	710.865	4	632.599
5	1127.50	5	942.291	5	809.397	5	709.402	5	631.440
6	1123.82	6	939.719	6	807.499	6	707.945	6	630.286
7	1120.16	7	937.161	7	805.611	7	706.493	7	629.136
8	1116.52	8	934.616	8	803.731	8	705.048	8	627.991
9	1112.91	9	932.086	9	801.860	9	703.609	9	626.849
10	1109.33	10	929.569	10	799.997	10	702.175	10	625.712
11	1105.76	11	927.066	11	798.144	11	700.748	11	624.579
12	1102.22	12	924.576	12	796.299	12	699.326	12	623.450
13	1098.70	13	922.100	13	794.462	13	697.910	13	622.325
14	1095.20	14	919.637	14	792.634	14	696.499	14	621.203
15	1091.73	15	917.187	15	790.814	15	695.095	15	620.087
16	1088.28	16	914.750	16	789.003	16	693.696	16	618.974
17	1084.85	17	912.326	17	787.210	17	692.302	17	617.865
18	1081.44	18	909.915	18	785.405	18	690.914	18	616.760
19	1078.05	19	907.517	19	783.618	19	689.532	19	615.660
20	1074.68	20	905.131	20	781.840	20	688.156	20	614.563
21	1071.34	21	902.758	21	780.069	21	686.785	21	613.470
22	1068.01	22	900.397	22	778.307	22	685.419	22	612.380
23	1064.71	23	898.048	23	776.552	23	684.059	23	611.295
24	1061.43	24	895.712	24	774.806	24	682.704	24	610.214
25	1058.16	25	893.388	25	773.067	25	681.354	25	609.136
26	1054.92	26	891.076	26	771.336	26	680.010	26	608.062
27	1051.70	27	888.776	27	769.613	27	678.671	27	606.992
28	1048.48	28	886.488	28	767.897	28	677.338	28	605.926
29	1045.31	29	884.211	29	766.190	29	676.008	29	604.864
30	1042.14	30	881.946	30	764.489	30	674.686	30	603.805
31	1039.00	31	879.693	31	762.797	31	673.369	31	602.750
32	1035.87	32	877.451	32	761.112	32	672.056	32	601.698
33	1032.76	33	875.221	33	759.434	33	670.748	33	600.651
34	1029.67	34	873.002	34	757.764	34	669.446	34	599.607
35	1026.60	35	870.795	35	756.101	35	668.148	35	598.567
36	1023.55	36	868.598	36	754.445	36	666.856	36	597.530
37	1020.51	37	866.412	37	752.796	37	665.568	37	596.497
38	1017.49	38	864.238	38	751.155	38	664.286	38	595.467
39	1014.50	39	862.075	39	749.521	39	663.008	39	594.441
40	1011.51	40	859.922	30	747.894	40	661.736	40	593.419
41	1008.55	41	857.780	41	746.274	41	660.468	41	592.400
42	1005.60	42	855.648	42	744.661	42	659.205	42	591.384
43	1002.67	43	853.527	43	743.055	43	657.947	43	590.372
44	999.762	44	851.417	44	741.456	44	656.694	44	589.364
45	996.867	45	849.317	45	739.864	45	655.446	45	588.359
46	993.988	46	847.228	46	738.279	46	654.202	46	587.357
47	991.126	47	845.148	47	736.701	47	652.963	47	586.359
48	988.280	48	843.080	48	735.129	48	651.729	48	585.364
49	985.451	49	841.021	49	733.564	49	650.499	49	584.373
50	982.638	50	838.972	50	732.005	50	649.274	50	583.385
51	979.840	51	836.933	51	730.454	51	648.054	51	582.400
52	977.060	52	834.904	52	728.909	52	646.838	52	581.419
53	974.294	53	832.885	53	727.370	53	645.627	53	580.441
54	971.544	54	830.876	54	725.838	54	644.420	54	579.466
55	968.810	55	828.876	55	724.312	55	643.218	55	578.494
56	966.091	56	826.886	56	722.793	56	642.021	56	577.526
57	963.387	57	824.905	57	721.280	57	640.828	57	576.561
58	960.698	58	822.934	58	719.774	58	639.639	58	575.599
59	958.025	59	820.973	59	718.273	59	638.455	59	574.641
60	955.366	60	819.020	60	716.779	60	637.275	60	573.686

Deg.	Radius.	Deg.	Radius.	Deg.	Radius.	Deg.	Radius.	Deg.	Radius.
10° 0'	573.686	12° 0'	478.339	14° 0'	410.275	16° 0'	359.265	18° 0'	319.623
2	571.784	2	477.018	2	409.306	2	358.523	2	319.037
4	569.896	4	475.705	4	408.341	4	357.784	4	318.453
6	568.020	6	474.400	6	407.380	6	357.048	6	317.871
8	566.156	8	473.102	8	406.424	8	356.315	8	317.292
10	564.305	10	471.810	10	405.473	10	355.585	10	316.715
12	562.466	12	470.526	12	404.526	12	354.859	12	316.139
14	560.638	14	469.249	14	403.583	14	354.135	14	315.566
16	558.823	16	467.978	16	402.645	16	353.414	16	314.993
18	557.019	18	466.715	18	401.712	18	352.696	18	314.426
20	555.227	20	465.459	20	400.782	20	351.981	20	313.860
22	553.447	22	464.209	22	399.857	22	351.269	22	313.295
24	551.678	24	462.966	24	398.937	24	350.560	24	312.732
26	549.920	26	461.729	26	398.020	26	349.854	26	312.172
28	548.174	28	460.500	28	397.108	28	349.150	28	311.613
30	546.438	30	459.276	30	396.200	30	348.450	30	311.056
32	544.714	32	458.060	32	395.296	32	347.752	32	310.502
34	543.001	34	456.850	34	394.396	34	347.057	34	309.949
36	541.298	36	455.646	36	393.501	36	346.365	36	309.399
38	539.606	38	454.449	38	392.609	38	345.676	38	308.850
40	537.924	40	453.259	40	391.722	40	344.990	40	308.303
42	536.253	42	452.073	42	390.838	42	344.306	42	307.759
44	534.593	44	450.894	44	389.959	44	343.625	44	307.216
46	532.943	46	449.722	46	389.084	46	342.947	46	306.675
48	531.303	48	448.556	48	388.212	48	342.271	48	306.136
50	529.673	50	447.395	50	387.345	50	341.598	50	305.599
52	528.053	52	446.241	52	386.481	52	340.928	52	305.064
54	526.443	54	445.093	54	385.621	54	340.260	54	304.531
56	524.843	56	443.951	56	384.765	56	339.595	56	304.000
58	523.252	58	442.814	58	383.913	58	338.933	58	303.470
11° 0'	521.671	13° 0'	441.684	15° 0'	383.065	17° 0'	338.273	19° 0'	302.943
2	520.100	2	440.559	2	382.220	2	337.616	2	302.417
4	518.539	4	439.440	4	381.380	4	336.962	4	301.893
6	516.986	6	438.326	6	380.543	6	336.310	6	301.371
8	515.443	8	437.219	8	379.709	8	335.660	8	300.851
10	513.909	10	436.117	10	378.880	10	335.013	10	300.333
12	512.385	12	435.020	12	378.054	12	334.369	12	299.816
14	510.869	14	433.929	14	377.231	14	333.727	14	299.302
16	509.363	16	432.844	16	376.412	16	333.088	16	298.789
18	507.865	18	431.764	18	375.597	18	332.451	18	298.278
20	506.376	20	430.690	20	374.786	20	331.816	20	297.768
22	504.896	22	429.620	22	373.977	22	331.184	22	297.260
24	503.425	24	428.557	24	373.173	24	330.555	24	296.755
26	501.962	26	427.498	26	372.372	26	329.928	26	296.250
28	500.507	28	426.445	28	371.574	28	329.303	28	295.748
30	499.061	30	425.396	30	370.780	30	328.689	30	295.247
32	497.624	32	424.354	32	369.989	32	328.061	32	294.748
34	496.195	34	423.316	34	369.202	34	327.443	34	294.251
36	494.774	36	422.283	36	368.418	36	326.828	36	293.756
38	493.361	38	421.256	38	367.637	38	326.215	38	293.262
40	491.956	40	420.233	40	366.859	40	325.604	40	292.770
42	490.559	42	419.215	42	366.085	42	324.996	42	292.279
44	489.171	44	418.203	44	365.315	44	324.390	44	291.790
46	487.790	46	417.195	46	364.547	46	323.786	46	291.303
48	486.417	48	416.192	48	363.783	48	323.184	48	290.818
50	485.051	50	415.194	50	363.022	50	322.585	50	290.334
52	483.694	52	414.201	52	362.264	52	321.989	52	289.851
54	482.344	54	413.212	54	361.510	54	321.394	54	289.371
56	481.001	56	412.229	56	360.758	56	320.801	56	288.892
58	479.666	58	411.250	58	360.010	58	320.211	58	288.414
60	478.339	60	410.275	60	359.265	60	319.623	60	287.939

TABLE II.—TANGENTS AND EXTERNALS TO A 1° CURVE.

Angle. I.	Tangent. T.	External. E.	Angle. I.	Tangent. T.	External. E.	Angle. I.	Tangent. T.	External. E.
1°	50.00	.218	11	551.70	26.500	21°	1061.9	97.577
10′	58.34	.297	10′	560.11	27.313	10′	1070.6	99.155
20	66.67	.388	20	568.53	28.137	20	1079.2	100.75
30	75.01	.491	30	576.95	28.974	30	1087.8	102.35
40	83.34	.606	40	585.36	29.824	40	1096.4	103.97
50	91.68	.733	50	593.79	30.686	50	1105.1	105.60
2	100.01	.873	12	602.21	31.561	22	1113.7	107.24
10	108.35	1.024	10	610.64	32.447	10	1122.4	108.90
20	116.68	1.188	20	619.07	33.347	20	1131.0	110.57
30	125.02	1.364	30	627.50	34.259	30	1139.7	112.25
40	133.36	1.552	40	635.93	35.183	40	1148.4	113.95
50	141.70	1.752	50	644.37	36.120	50	1157.0	115.66
3	150.04	1.964	13	652.81	37.070	23	1165.7	117.38
10	158.38	2.188	10	661.25	38.031	10	1174.4	119.12
20	166.72	2.425	20	669.70	39.006	20	1183.1	120.87
30	175.06	2.674	30	678.15	39.993	30	1191.8	122.63
40	183.40	2.934	40	686.6	40.992	40	1200.5	124.41
50	191.74	3.207	50	695.06	42.004	50	1209.2	126.20
4	200.08	3.492	14	703.51	43.029	24	1217.9	128.00
10	208.43	3.790	10	711.97	44.066	10	1226.6	129.82
20	216.77	4.099	20	720.44	45.116	20	1235.3	131.65
30	225.12	4.421	30	728.90	46.178	30	1244.0	133.50
40	233.47	4.755	40	737.37	47.253	40	1252.8	135.35
50	241.81	5.100	50	745.85	48.341	50	1261.5	137.23
5	250.16	5.459	15	754.32	49.441	25	1270.2	139.11
10	258.51	5.829	10	762.80	50.554	10	1279.0	141.01
20	266.86	6.211	20	771.29	51.679	20	1287.7	142.93
30	275.21	6.606	30	779.77	52.818	30	1296.5	144.85
40	283.57	7.013	40	788.26	53.969	40	1305.3	146.79
50	291.92	7.432	50	796.75	55.132	50	1314.0	148.75
6	300.28	7.863	16	805.25	56.309	26	1322.8	150.71
10	308.64	8.307	10	813.75	57.498	10	1331.6	152.69
20	316.99	8.762	20	822.25	58.699	20	1340.4	154.69
30	325.35	9.230	30	830.76	59.914	30	1349.2	156.70
40	333.71	9.710	40	839.27	61.141	40	1358.0	158.72
50	342.08	10.202	50	847.78	62.381	50	1366.8	160.76
7	350.44	10.707	17	856.30	63.634	27	1375.6	162.81
10	358.81	11.224	10	864.82	64.900	10	1384.4	164.86
20	367.17	11.753	20	873.35	66.178	20	1393.2	166.95
30	375.54	12.294	30	881.88	67.470	30	1402.0	169.04
40	383.91	12.847	40	890.41	68.774	40	1410.9	171.15
50	392.28	13.413	50	898.95	70.091	50	1419.7	173.27
8	400.66	13.991	18	907.49	71.421	28	1428.6	175.41
10	409.03	14.582	10	916.03	72.764	10	1437.4	177.55
20	417.41	15.184	20	924.58	74.119	20	1446.3	179.72
30	425.79	15.799	30	933.13	75.488	30	1455.1	181.89
40	434.17	16.426	40	941.69	76.869	40	1464.0	184.08
50	442.55	17.065	50	950.25	78.264	50	1472.9	186.29
9	450.93	17.717	19	958.81	79.671	29	1481.8	188.51
10	459.32	18.381	10	967.38	81.092	10	1490.7	190.74
20	467.71	19.058	20	975.96	82.525	20	1499.6	192.99
30	476.10	19.746	30	984.53	83.972	30	1508.5	195.25
40	484.49	20.447	40	993.12	85.431	40	1517.4	197.53
50	492.88	21.161	50	1001.7	86.904	50	1526.3	199.82
10	501.28	21.887	20	1010.3	88.389	30	1535.3	202.12
10	509.68	22.624	10	1018.9	89.898	10	1544.2	204.44
20	518.08	23.375	20	1027.5	91.399	20	1553.1	206.77
30	526.48	24.138	30	1036.1	92.924	30	1562.1	209.12
40	534.89	24.913	40	1044.7	94.462	40	1571.0	211.48
50	543.29	25.700	50	1053.3	96.018	50	1580.0	213.86

Angle. I.	Tangent. T.	External. E.	Angle. I.	Tangent. T.	External. E.	Angle. I.	Tangent. T.	External. E.
31°	1589.0	216.25	41°	2142.2	387.38	51°	2732.9	618.39
10	1598.0	218.66	10'	2151.7	390.71	10'	2743.1	622.81
20	1606.9	221.08	20	2161.2	394.06	20	2753.4	627.24
30	1615.9	223.51	30	2170.8	397.43	30	2763.7	631.69
40	1624.9	225.96	40	2180.3	400.82	40	2773.9	636.17
50	1633.9	228.42	50	2189.9	404.22	50	2784.2	640.66
32	1643.0	230.90	42	2199.4	407.64	52	2794.5	645.17
10	1652.0	233.39	10	2209.0	411.07	10	2804.9	649.70
20	1661.0	235.90	20	2218.6	414.52	20	2815.2	654.25
30	1670.0	238.43	30	2228.1	417.99	30	2825.6	658.83
40	1679.1	240.96	40	2237.7	421.48	40	2835.9	663.42
50	1688.1	243.52	50	2247.3	424.98	50	2846.3	668.03
33	1697.2	246.08	43	2257.0	428.50	53	2856.7	672.66
10	1706.3	248.66	10	2266.6	432.04	10	2867.1	677.32
20	1715.3	251.26	20	2276.2	435.59	20	2877.5	681.99
30	1724.4	253.87	30	2285.9	439.16	30	2888.0	686.68
40	1733.5	256.50	40	2295.6	442.75	40	2898.4	691.40
50	1742.6	259.14	50	2305.2	446.35	50	2908.9	696.13
34	1751.7	261.80	44	2314.9	449.98	54	2919.4	700.89
10	1760.8	264.47	10	2324.6	453.62	10	2929.9	705.66
20	1770.0	267.16	20	2334.3	457.27	20	2940.4	710.46
30	1779.1	269.86	30	2344.1	460.95	30	2951.0	715.28
40	1788.2	272.58	40	2353.8	464.64	40	2961.5	720.11
50	1797.4	275.31	50	2363.5	468.35	50	2972.1	724.97
35	1806.6	278.05	45	2373.3	472.08	55	2982.7	729.85
10	1815.7	280.82	10	2383.1	475.82	10	2993.3	734.76
20	1824.9	283.60	20	2392.8	479.59	20	3003.9	739.68
30	1834.1	286.39	30	2402.6	483.37	30	3014.5	744.62
40	1843.3	289.20	40	2412.4	487.17	40	3025.2	749.59
50	1852.5	292.02	50	2422.3	490.98	50	3035.8	754.57
36	1861.7	294.86	46	2432.1	494.82	56	3046.5	759.58
10	1870.9	297.72	10	2441.9	498.67	10	3057.2	764.61
20	1880.1	300.59	20	2451.8	502.54	20	3067.9	769.66
30	1889.4	303.47	30	2461.7	506.42	30	3078.7	774.73
40	1898.6	306.37	40	2471.5	510.33	40	3089.4	779.83
50	1907.9	309.29	50	2481.4	514.25	50	3100.2	784.94
37	1917.1	312.22	47	2491.3	518.20	57	3110.9	790.08
10	1926.4	315.17	10	2501.2	522.16	10	3121.7	795.24
20	1935.7	318.13	20	2511.2	526.13	20	3132.6	800.42
30	1945.0	321.11	30	2521.1	530.13	30	3143.4	805.62
40	1954.3	324.11	40	2531.1	534.15	40	3154.2	810.85
50	1963.6	327.12	50	2541.0	538.18	50	3165.1	816.10
38	1972.9	330.15	48	2551.0	542.23	58	3176.0	821.37
10	1982.2	333.19	10	2561.0	546.30	10	3186.9	826.66
20	1991.5	336.25	20	2571.0	550.39	20	3197.8	831.98
30	2000.9	339.32	30	2581.0	554.50	30	3208.8	837.31
40	2010.2	342.41	40	2591.1	558.63	40	3219.7	842.67
50	2019.6	345.52	50	2601.1	562.77	50	3230.7	848.06
39	2029.0	348.64	49	2611.2	566.94	59	3241.7	853.46
10	2038.4	351.78	10	2621.2	571.12	10	3252.7	858.89
20	2047.8	354.94	20	2631.3	575.32	20	3263.7	864.34
30	2057.2	358.11	30	2641.4	579.54	30	3274.8	869.82
40	2066.6	361.29	40	2651.5	583.78	40	3285.8	875.32
50	2076.0	364.50	50	2661.6	588.04	50	3296.9	880.84
40	2085.4	367.72	50	2671.8	592.32	60	3308.0	886.38
10	2094.9	370.95	10	2681.9	596.62	10	3319.1	891.95
20	2104.3	374.20	20	2692.1	600.93	20	3330.3	897.54
30	2113.8	377.47	30	2702.3	605.27	30	3341.4	903.15
40	2123.3	380.76	40	2712.5	609.62	40	3352.6	908.79
50	2132.7	384.06	50	2722.7	614.00	50	3363.8	914.45

TABLE II.—TANGENTS AND EXTERNALS TO A 1° CURVE.

Angle. I.	Tangent. T.	External. E.	Angle. I.	Tangent. T.	External. E.	Angle. I.	Tangent. T.	External. E.
61°	3375.0	920.14	71°	4086.9	1308.2	81°	4898.6	1805.3
10'	3386.3	925.85	10'	4099.5	1315.6	10'	4908.0	1814.7
20	3397.5	931.58	20	4112.1	1322.9	20	4922.5	1824.1
30	3408.8	937.34	30	4124.8	1330.3	30	4937.0	1833.6
40	3420.1	943.12	40	4137.4	1337.7	40	4951.5	1843.1
50	3431.4	948.92	50	4150.1	1345.1	50	4966.1	1852.6
62	3442.7	954.75	72	4162.8	1352.6	82	4980.7	1862.2
10	3454.1	960.60	10	4175.6	1360.1	10	4995.4	1871.8
20	3465.4	966.48	20	4188.5	1367.6	20	5010.0	1881.5
30	3476.8	972.38	30	4201.2	1375.2	30	5024.8	1891.2
40	3488.3	978.31	40	4214.0	1382.8	40	5039.5	1900.9
50	3499.7	984.27	50	4226.8	1390.4	50	5054.3	1910.7
63	3511.1	990.24	73	4239.7	1398.0	83	5069.2	1920.5
10	3522.6	996.24	10	4252.6	1405.7	10	5084.0	1930.4
20	3534.1	1002.3	20	4265.6	1413.5	20	5099.0	1940.3
30	3545.6	1008.3	30	4278.5	1421.2	30	5113.9	1950.3
40	3557.2	1014.4	40	4291.5	1429.0	40	5128.9	1960.2
50	3568.7	1020.5	50	4304.6	1436.8	50	5143.9	1970.3
64	3580.3	1026.6	74	4317.6	1444.6	84	5159.0	1980.4
10	3591.9	1032.8	10	4330.7	1452.5	10	5174.1	1990.5
20	3603.5	1039.0	20	4343.8	1460.4	20	5189.3	2000.6
30	3615.1	1045.2	30	4356.9	1468.4	30	5204.2	2010.8
40	3626.8	1051.4	40	4370.1	1476.4	40	5219.7	2021.1
50	3638.5	1057.7	50	4383.3	1484.4	50	5234.9	2031.4
65	3650.2	1063.9	75	4396.5	1492.4	85	5250.3	2041.7
10	3661.9	1070.2	10	4409.8	1500.5	10	5265.6	2052.1
20	3673.7	1076.6	20	4423.1	1508.6	20	5281.0	2062.5
30	3685.4	1082.9	30	4436.4	1516.7	30	5296.4	2073.0
40	3697.2	1089.3	40	4449.7	1524.9	40	5311.9	2083.5
50	3709.0	1095.7	50	4463.1	1533.1	50	5327.4	2094.1
66	3720.9	1102.2	76	4476.5	1541.4	86	5343.0	2104.7
10	3732.7	1108.6	10	4489.9	1549.7	10	5358.6	2115.3
20	3744.6	1115.1	20	4503.4	1558.0	20	5374.2	2126.0
30	3756.5	1121.7	30	4516.9	1566.3	30	5389.9	2136.7
40	3768.5	1128.2	40	4530.4	1574.7	40	5405.6	2147.5
50	3780.4	1134.8	50	4544.0	1583.1	50	5421.4	2158.4
67	3792.4	1141.4	77	4557.6	1591.6	87	5437.2	2169.2
10	3804.4	1148.0	10	4571.2	1600.1	10	5453.1	2180.2
20	3816.4	1154.7	20	4584.8	1608.6	20	5469.0	2191.1
30	3828.4	1161.3	30	4598.5	1617.1	30	5484.9	2202.2
40	3840.5	1168.1	40	4612.2	1625.7	40	5500.9	2213.2
50	3852.6	1174.8	50	4626.0	1634.4	50	5517.0	2224.3
68	3864.7	1181.6	78	4639.8	1643.0	88	5533.1	2235.5
10	3876.8	1188.4	10	4653.6	1651.7	10	5549.2	2246.7
20	3889.0	1195.2	20	4667.4	1660.5	20	5565.4	2258.0
30	3901.2	1202.0	30	4681.3	1669.2	30	5581.6	2269.3
40	3913.4	1208.9	40	4695.2	1678.1	40	5597.8	2280.6
50	3925.6	1215.8	50	4709.2	1686.9	50	5614.2	2292.0
69	3937.9	1222.7	79	4723.2	1695.8	89	5630.5	2303.5
10	3950.2	1229.7	10	4737.2	1704.7	10	5646.9	2315.0
20	3962.5	1236.7	20	4751.2	1713.7	20	5663.4	2326.6
30	3974.8	1243.7	30	4765.3	1722.7	30	5679.9	2338.2
40	3987.2	1250.8	40	4779.4	1731.7	40	5696.4	2349.8
50	3999.5	1257.9	50	4793.6	1740.8	50	5713.0	2361.5
70	4011.9	1265.0	80	4807.7	1749.9	90	5729.7	2373.3
10	4024.4	1272.1	10	4822.0	1759.0	10	5746.3	2385.1
20	4036.8	1279.3	20	4836.2	1768.2	20	5763.1	2397.0
30	4049.3	1286.5	30	4850.5	1777.4	30	5779.9	2408.9
40	4061.8	1293.6	40	4864.8	1786.7	40	5796.7	2420.9
50	4074.4	1300.9	50	4879.2	1796.0	50	5813.6	2432.9

TABLE II.—TANGENTS AND EXTERNALS TO A 1° CURVE.

Angle. I.	Tangent. T.	External. E.	Angle. I.	Tangent. T.	External. E.	Angle. I.	Tangent. T.	External. E.
91°	5830.5	2444.9	97	6476.2	2917.3	103	7203.2	3474.4
10'	5847.5	2457.1	10	6495.2	2931.6	10	7224.7	3491.3
20	5864.6	2469.3	20	6514.3	2945.9	20	7246.3	3508.2
30	5881.7	2481.5	30	6533.4	2960.3	30	7268.0	3525.2
40	5898.8	2493.8	40	6552.6	2974.7	40	7289.8	3542.4
50	5916.0	2506.1	50	6571.9	2989.2	50	7311.7	3559.6
92	5933.2	2518.5	98	6591.2	3003.8	104	7333.6	3576.8
10	5950.5	2531.0	10	6610.6	3018.4	10	7355.6	3594.2
20	5967.9	2543.5	20	6630.1	3033.1	20	7377.8	3611.7
30	5985.3	2556.0	30	6649.6	3047.9	30	7399.9	3629.2
40	6002.7	2568.6	40	6669.2	3062.8	40	7422.2	3646.8
50	6020.2	2581.3	50	6688.8	3077.7	50	7444.6	3664.5
93	6037.8	2594.0	99	6708.6	3092.7	105	7467.0	3682.3
10	6055.4	2606.8	10	6728.4	3107.7	10	7489.6	3700.2
20	6073.1	2619.7	20	6748.2	3122.9	20	7512.2	3718.2
30	6090.8	2632.6	30	6768.1	3138.1	30	7534.9	3736.2
40	6108.6	2645.5	40	6788.1	3153.3	40	7557.7	3754.4
50	6126.4	2658.5	50	6808.2	3168.7	50	7580.5	3772.6
94	6144.3	2671.6	100	6828.3	3184.1	106	7603.5	3791.0
10	6162.2	2684.7	10	6848.5	3199.6	10	7626.6	3809.4
20	6180.2	2697.9	20	6868.8	3215.1	20	7649.7	3827.9
30	6198.3	2711.2	30	6889.2	3230.8	30	7672.9	3846.5
40	6216.4	2724.5	40	6909.6	3246.5	40	7696.3	3865.2
50	6234.6	2737.9	50	6930.1	3262.3	50	7719.7	3884.0
95	6252.8	2751.3	101°	6950.6	3278.1	107	7743.2	3902.9
10	6271.1	2764.8	10'	6971.3	3294.1	10	7766.8	3921.9
20	6289.4	2778.3	20	6992.0	3310.1	20	7790.5	3940.9
30	6307.9	2792.0	30	7012.7	3326.1	30	7814.3	3960.1
40	6326.3	2805.6	40	7033.6	3342.3	40	7838.1	3979.4
50	6344.8	2819.4	50	7054.5	3358.5	50	7862.1	3998.7
96	6363.4	2833.2	102	7075.5	3374.9	108	7886.2	4018.2
10	6382.1	2847.0	10	7096.6	3391.2	10	7910.4	4037.8
20	6400.8	2861.0	20	7117.8	3407.7	20	7934.6	4057.4
30	6419.5	2875.0	30	7139.0	3424.3	30	7959.0	4077.2
40	6438.4	2889.0	40	7160.3	3440.9	40	7983.5	4097.1
50	6457.3	2903.1	50	7181.7	3457.6	50	8008.0	4117.0

CORRECTIONS FOR TANGENTS AND EXTERNALS.

Ang I.	For Tangents, add						Ang I.	For Externals, add					
	5° Cur.	10° Cur.	15° Cur.	20° Cur.	25° Cur.	30° Cur.		5° Cur.	10° Cur.	15° Cur.	20° Cur.	25° Cur.	30° Cur.
10°	.03	.06	.09	.13	.16	.19	10°	.001	.003	.004	.006	.007	.008
20	.06	.13	.19	.26	.32	.39	20	.006	.011	.017	.022	.028	.034
30	.10	.19	.29	.39	.49	.59	30	.013	.025	.038	.051	.065	.078
40	.13	.26	.40	.53	.67	.80	40	.023	.046	.070	.093	.117	.141
50	.17	.34	.51	.68	.85	1 02	50	.037	.075	.116	.151	.189	.227
60	.21	.42	.63	.84	1.05	1.27	60	.056	.112	.168	.225	.283	.340
70	.25	.51	.76	1.02	1.28	1.54	70	.080	.159	.240	.321	.403	.485
80	.30	.61	.91	1.22	1.53	1.84	80	.110	.220	.332	.445	.558	.671
90	.36	.72	1.09	1.45	1.83	2.20	90	.149	.299	.450	.603	.756	.910
100	.43	.86	1.30	1.74	2 18	2.62	100	.200	.401	.604	.809	1.015	1.221
110	.51	1.03	1.56	2 08	2.61	3.14	110	.268	.536	.806	1.082	1.355	1.633
120	.62	1.25	1.93	2.52	3 16	3.81	120	.360	.721	1.086	1.456	1.825	2.197

(See Note on page 259.)

TABLE III.—TANGENTIAL OFFSETS 100 FT. ALONG THE CURVE.

Deg. of Curve.	0′	10′	20′	30′	40′	50′
0°	0.000	0.145	0.291	0.436	0.582	0.727
1°	0.873	1.018	1.164	1.309	1.454	1.600
2°	1.745	1.891	2.036	2.181	2.327	2.472
3°	2.618	2.763	2.908	3.054	3.199	3.345
4°	3.490	3.635	3.781	3.926	4.071	4.217
5°	4.362	4.507	4.653	4.798	4.943	5.088
6°	5.234	5.379	5.524	5.669	5.814	5.960
7°	6.105	6.250	6.395	6.540	6.685	6.831
8°	6.976	7.121	7.266	7.411	7.556	7.701
9°	7.846	7.991	8.136	8.281	8.426	8.571
10°	8.716	8.860	9.005	9.150	9.295	9.440
11°	9.585	9.729	9.874	10.019	10.164	10.308
12°	10.453	10.597	10.742	10.887	11.031	11.176
13°	11.320	11.465	11.609	11.754	11.898	12.043
14°	12.187	12.331	12.476	12.620	12.764	12.908
15°	13.053	13.197	13.341	13.485	13.629	13.773
16°	13.917	14.061	14.205	14.349	14.493	14.637
17°	14.781	14.925	15.069	15.212	15.356	15.500
18°	15.643	15.787	15.931	16.074	16.218	16.361
19°	16.505	16.648	16.792	16.935	17.078	17.222
20°	17.365	17.508	17.651	17.794	17.937	18.081
21°	18.224	18.367	18.509	18.652	18.795	18.938
22°	19.081	19.224	19.366	19.509	19.652	19.794
23°	19.937	20.079	20.222	20.364	20.507	20.649
24°	20.791	20.933	21.076	21.218	21.360	21.502

TABLE IV.—MID-ORDINATES TO A 100-FT. CHORD.

Deg. of Curve.	0	1	2	3	4	5	6	7	8	9
0°	0.000	0.218	0.436	0.655	0.873	1.091	1.309	1.528	1.746	1.965
10°	2.183	2.402	2.620	2.839	3.058	3.277	3.496	3.716	3.935	4.155
20°	4.374	4.594	4.814	5.035	5.255	5.476	5.697	5.918	6.139	6.360

Note.—As an example illustrating the use of Table II, suppose we require the value of T for a 5° curve, where $I = 40° 20'$. Then

$$T = \frac{2104.3}{5} + .13 = 420.99.$$

TABLE V.—LONG CHORDS.

Degree of Curve.	Actual Arc, One Station.	Long Chords.				
		2 Stations.	3 Stations.	4 Stations.	5 Stations.	6 Stations.
0° 10′	100.000	200.000	299.999	399.998	499.996	599.993
20	.000	199.999	299.997	399.992	499.983	599.970
30	.000	199.998	299.992	399.981	499.962	599.933
40	.001	199.997	299.986	399.966	499.932	599.882
50	.001	199.995	299.979	399.947	499.894	599.815
1	100.001	199.992	299.970	399.924	499.848	599.733
10	.002	199.990	299.959	399.896	499.793	599.637
20	.002	199.986	299.946	399.865	499.729	599.526
30	.003	199.983	299.932	399.829	499.657	599.401
40	.003	199.979	299.915	399.789	499.577	599.260
50	.004	199.974	299.898	399.744	499.488	599.105
2	100.005	199.970	299.878	399.695	499.391	598.934
10	.006	199.964	299.857	399.643	499.285	598.750
20	.007	199.959	299.834	399.586	499.171	598.550
30	.008	199.952	299.810	399.524	499.049	598.336
40	.009	199.946	299.783	399.459	498.918	598.106
50	.010	199.939	299.756	399.389	498.778	597.862
3	100.011	199.931	299.726	399.315	498.630	597.604
10	.013	199.924	299.695	399.237	498.474	597.331
20	.014	199.915	299.662	399.154	498.309	597.043
30	.015	199.907	299.627	399.068	498.136	596.740
40	.017	199.898	299.591	398.977	497.955	596.423
50	.019	199.888	299.553	398.882	497.765	596.091
4	100.020	199.878	299.513	398.782	497.566	595.744
10	.022	199.868	299.471	398.679	497.360	595.383
20	.024	199.857	299.428	398.571	497.145	595.007
30	.026	199.846	299.383	398.459	496.921	594.617
40	.028	199.834	299.337	398.343	496.689	594.212
50	.030	199.822	299.289	398.223	496.449	593.792
5	100.032	199.810	299.239	398.099	496.201	593.358
10	.034	199.797	299.187	397.970	495.944	592.909
20	.036	199.783	299.134	397.837	495.678	592.446
30	.038	199.770	299.079	397.700	495.405	591.968
40	.041	199.756	299.023	397.559	495.123	591.476
50	.043	199.741	298.964	397.413	494.832	590.970
6	100.046	199.726	298.904	397.264	494.534	590.449
10	.048	199.710	298.843	397.110	494.227	589.913
20	.051	199.695	298.779	396.952	493.912	589.364
30	.054	199.678	298.714	396.790	493.588	588.800
40	.056	199.662	298.648	396.623	493.257	588.221
50	.059	199.644	298.579	396.453	492.917	587.628
7	100.062	199.627	298.509	396.278	492.568	587.021
10	.065	199.609	298.438	396.099	492.212	586.400
20	.068	199.591	298.364	395.916	491.847	585.765
30	.071	199.572	298.289	395.729	491.474	585.115
40	.075	199.553	298.212	395.538	491.093	584.451
50	.078	199.533	298.134	395.342	490.704	583.773
8	100.081	199.513	298.054	395.142	490.306	583.081
10	.085	199.492	297.972	394.938	489.900	582.375
20	.088	199.471	297.888	394.731	489.486	581.654
30	.092	199.450	297.803	394.518	489.064	580.920
40	.095	199.428	297.716	394.302	488.634	580.172
50	.099	199.406	297.628	394.082	488.196	579.409
9	100.103	199.383	297.538	393.857	487.740	578.633
10	.107	199.360	297.446	393.629	487.294	577.843
20	.111	199.337	297.352	393.396	486.832	577.039
30	.115	199.313	297.257	393.159	486.361	576.222
40	.119	199.289	297.160	392.918	485.882	575.390
50	.123	199.264	297.062	392.673	485.395	574.545
10	100.127	199.239	296.962	392.424	484.900	573.686

TABLE V.—LONG CHORDS.

Degree of Curve.	Actual Arc, One Station.	Long Chords.				
		2 Stations.	3 Stations.	4 Stations.	5 Stations.	6 Stations.
10° 10′	100.131	199.213	296.860	392.171	484.397	572.813
20	.136	199.187	296.756	391.914	483.886	571.926
30	.140	199.161	296.651	391.652	483.367	571.027
40	.145	199.134	296.544	391.387	482.840	570.113
50	.149	199.107	296.436	391.117	482.305	569.186
11	100.154	199.079	296.325	390.843	481.762	568.245
10	.158	199.051	296.214	390.565	481.211	567.292
20	.163	199.023	296.100	390.284	480.653	566.324
30	.168	198.994	295.985	389.998	480.086	565.343
40	.173	198.964	295.868	389.708	479.511	564.349
50	.178	198.935	295.750	389.414	478.929	563.341
12	100.183	198.904	295.629	389.116	478.338	562.321
10	.188	198.874	295.508	388.814	477.740	561.287
20	.193	198.843	295.384	388.508	477.135	560.240
30	.199	198.811	295.259	388.197	476.521	559.180
40	.204	198.779	295.132	387.883	475.899	558.107
50	.209	198.747	295.004	387.565	475.270	557.020
13	100.215	198.714	294.874	387.243	474.633	555.921
10	.220	198.681	294.742	386.916	473.988	554.809
20	.226	198.648	294.609	386.586	473.336	553.684
30	.232	198.614	294.474	386.252	472.675	552.546
40	.237	198.579	294.337	385.914	472.007	551.395
50	.243	198.544	294.199	385.572	471.332	550.232
14	100.249	198.509	294.059	385.225	470.649	549.056
10	.255	198.474	293.918	384.875	469.958	547.867
20	.261	198.437	293.774	384.521	469.260	546.666
30	.267	198.401	293.629	384.163	468.554	545.452
40	.274	198.364	293.483	383.801	467.840	544.226
50	.280	198.327	293.335	383.435	467.119	542.987
15	100.286	198.289	293.185	383.065	466.390	541.736
10	.292	198.251	293.034	382.691	465.654	540.472
20	.299	198.212	292.881	382.313	464.911	539.196
30	.306	198.173	292.726	381.931	464.160	537.908
40	.312	198.134	292.570	381.546	463.401	536.608
50	.319	198.094	292.412	381.156	462.635	535.296
16	100.326	198.054	292.252	380.763	461.862	533.972
10	.333	198.013	292.091	380.365	461.081	532.635
20	.339	197.972	291.928	379.964	460.293	531.287
30	.346	197.930	291.764	379.559	459.498	529.927
40	.353	197.888	291.598	379.150	458.695	528.555
50	.361	197.846	291.430	378.737	457.886	527.171
17	100.368	197.803	291.261	378.320	457.069	525.776
10	.375	197.760	291.090	377.900	456.244	524.369
20	.382	197.716	290.918	377.475	455.413	522.950
30	.390	197.672	290.743	377.047	454.574	521.519
40	.397	197.628	290.568	376.615	453.728	520.078
50	.405	197.583	290.390	376.179	452.875	518.625
18	100.412	197.538	290.211	375.739	452.015	517.160
10	.420	197.492	290.031	375.295	451.147	515.685
20	.428	197.446	289.849	374.848	450.373	514.198
30	.436	197.399	289.665	374.397	449.392	512.699
40	.444	197.352	289.479	373.942	448.504	511.190
50	.452	197.305	289.293	373.483	447.608	509.670
19	100.460	197.256	289.104	373.021	446.706	508.139
10	.468	197.209	288.913	372.554	445.797	506.597
20	.476	197.160	288.722	372.084	444.881	505.043
30	.484	197.111	288.528	371.610	443.957	503.479
40	.493	197.062	288.333	371.133	443.028	501.905
50	.501	197.012	288.137	370.652	442.091	500.320
20	100.510	196.962	287.939	370.167	441.147	498.724

TABLE VI.—MID-ORDINATES TO LONG CHORDS.

Degree of Curve.	1 Station.	2 Stations.	3 Stations.	4 Stations.	5 Stations.	6 Stations.
0° 10′	.036	.145	.327	.582	.909	1.309
20	.073	.291	.654	1.164	1.818	2.618
30	.109	.436	.982	1.745	2.727	3.926
40	.145	.582	1.309	2.327	3.636	5.235
50	.182	.727	1.636	2.909	4.545	6.544
1	.218	.873	1.963	3.490	5.453	7.852
10	.255	1.018	2.291	4.072	6.362	9.160
20	.291	1.164	2.618	4.654	7.270	10.468
30	.327	1.309	2.945	5.235	8.179	11.775
40	.364	1.454	3.272	5.816	9.087	13.082
50	.400	1.600	3.599	6.398	9.994	14.389
2	.436	1.745	3.926	6.979	10.902	15.694
10	.473	1.891	4.253	7.560	11.809	17.000
20	.509	2.036	4.580	8.141	12.716	18.304
30	.545	2.181	4.907	8.722	13.623	19.608
40	.582	2.327	5.234	9.303	14.529	20.912
50	.618	2.472	5.561	9.883	15.435	22.214
3	.654	2.618	5.888	10.464	16.341	23.516
10	.691	2.763	6.215	11.044	17.246	24.817
20	.727	2.908	6.542	11.624	18.151	26.117
30	.763	3.054	6.868	12.204	19.055	27.416
40	.800	3.199	7.195	12.784	19.959	28.714
50	.836	3.345	7.522	13.363	20.863	30.012
4	.872	3.490	7.848	13.943	21.766	31.308
10	.909	3.635	8.175	14.522	22.668	32.603
20	.945	3.781	8.501	15.101	23.570	33.896
30	.982	3.926	8.828	15.680	24.471	35.189
40	1.018	4.071	9.154	16.258	25.372	36.480
50	1.054	4.217	9.480	16.837	26.272	37.770
5	1.091	4.362	9.807	17.415	27.171	39.059
10	1.127	4.507	10.133	17.992	28.070	40.346
20	1.164	4.653	10.459	18.570	28.968	41.631
30	1.200	4.798	10.785	19.147	29.866	42.916
40	1.237	4.943	11.111	19.724	30.762	44.198
50	1.273	5.088	11.436	20.301	31.658	45.479
6	1.309	5.234	11.762	20.877	32.553	46.759
10	1.346	5.379	12.088	21.453	33.448	48.037
20	1.382	5.524	12.413	22.029	34.341	49.313
30	1.418	5.669	12.739	22.604	35.234	50.587
40	1.455	5.814	13.064	23.179	36.126	51.860
50	1.491	5.960	13.389	23.754	37.017	53.130
7	1.528	6.105	13.715	24.328	37.907	54.399
10	1.564	6.250	14.040	24.902	38.796	55.666
20	1.600	6.395	14.365	25.476	39.684	56.931
30	1.637	6.540	14.689	26.049	40.571	58.193
40	1.673	6.685	15.014	26.622	41.458	59.454
50	1.710	6.831	15.339	27.195	42.343	60.712
8	1.746	6.976	15.663	27.767	43.227	61.969
10	1.782	7.121	15.988	28.338	44.110	63.223
20	1.819	7.266	16.312	28.910	44.992	64.475
30	1.855	7.411	16.636	29.481	45.873	65.724
40	1.892	7.556	16.960	30.051	46.753	66.972

TABLE VI.—MID-ORDINATES TO LONG CHORDS.

Degree of Curve.	1 Station.	2 Stations.	3 Stations.	4 Stations.	5 Stations.	6 Stations.
10° 10′	2.219	8.860	19.870	35.164	54.619	78.083
20	2.256	9.005	20.193	35.729	55.486	79.305
30	2.293	9.150	20.516	36.294	56.353	80.523
40	2.329	9.295	20.838	36.859	57.218	81.739
50	2.365	9.440	21.160	37.423	58.081	82.951
11	2.402	9.585	21.483	37.986	58.943	84.161
10	2.438	9.729	21.804	38.549	59.804	85.368
20	2.475	9.874	22.126	39.111	60.663	86.571
30	2.511	10.019	22.448	39.673	61.521	87.772
40	2.547	10.164	22.769	40.234	62.377	88.969
50	2.584	10.308	23.090	40.795	63.232	90.164
12	2.620	10.453	23.412	41.355	64.085	91.355
10	2.657	10.597	23.732	41.914	64.937	92.542
20	2.693	10.742	24.053	42.473	65.787	93.727
30	2.730	10.887	24.374	43.031	66.636	94.908
40	2.766	11.031	24.694	43.588	67.482	96.086
50	2.803	11.176	25.014	44.145	68.328	97.260
13	2.839	11.320	25.334	44.701	69.171	98.431
10	2.876	11.465	25.654	45.256	70.012	99.598
20	2.912	11.609	25.974	45.811	70.854	100.762
30	2.949	11.754	26.293	46.365	71.692	101.922
40	2.985	11.898	26.612	46.919	72.529	103.079
50	3.022	12.043	26.931	47.472	73.364	104.232
14	3.058	12.187	27.250	48.024	74.197	105.381
10	3.095	12.331	27.569	48.575	75.029	106.527
20	3.131	12.476	27.887	49.126	75.859	107.669
30	3.168	12.620	28.206	49.676	76.687	108.807
40	3.204	12.764	28.524	50.225	77.513	109.941
50	3.241	12.908	28.841	50.773	78.337	111.071
15	3.277	13.053	29.159	51.321	79.159	112.197
10	3.314	13.197	29.476	51.868	79.979	113.319
20	3.350	13.341	29.794	52.414	80.798	114.488
30	3.387	13.485	30.111	52.959	81.614	115.552
40	3.423	13.629	30.427	53.504	82.429	116.662
50	3.460	13.773	30.744	54.048	83.241	117.768
16	3.496	13.917	31.060	54.591	84.052	118.870
10	3.533	14.061	31.376	55.133	84.861	119.967
20	3.569	14.205	31.692	55.675	85.667	121.061
30	3.606	14.349	32.008	56.215	86.471	122.150
40	3.643	14.493	32.323	56.755	87.274	123.235
50	3.679	14.637	32.638	57.294	88.074	124.315
17	3.716	14.781	32.953	57.832	88.872	125.391
10	3.752	14.925	33.267	58.369	89.668	126.463
20	3.789	15.069	33.582	58.906	90.462	127.530
30	3.825	15.212	33.896	59.441	91.254	128.593
40	3.862	15.356	34.210	59.976	92.043	129.651
50	3.899	15.500	34.523	60.510	92.830	130.704
18	3.935	15.643	34.837	61.042	93.616	131.753
10	3.972	15.787	35.150	61.574	94.398	132.797
20	4.008	15.931	35.463	62.106	95.179	133.837
30	4.045	16.074	35.775	62.636	95.957	134.872
40	4.081	16.218	36.088	63.165	96.733	135.902

TABLE VII.—MINUTES IN DECIMALS OF A DEGREE.

′	0″	10″	15″	20″	30″	40″	45″	50″	′
0	.00000	.00278	.00417	.00556	.00833	.01111	.01250	.01389	0
1	.01667	.01944	.02083	.02222	.02500	.02778	.02917	.03055	1
2	.03333	.03611	.03750	.03889	.04167	.04444	.04583	.04722	2
3	.05000	.05278	.05417	.05556	.05833	.06111	.06250	.06389	3
4	.06667	.06944	.07083	.07222	.07500	.07778	.07917	.08056	4
5	.08333	.08611	.08750	.08889	.09167	.09444	.09583	.09722	5
6	.10000	.10278	.10417	.10556	.10833	.11111	.11250	.11389	6
7	.11667	.11944	.12083	.12222	.12500	.12778	.12917	.13056	7
8	.13333	.13611	.13750	.13889	.14167	.14444	.14583	.14722	8
9	.15000	.15278	.15417	.15556	.15833	.16111	.16250	.16389	9
10	.16667	.16944	.17083	.17222	.17500	.17778	.17917	.18056	10
11	.18333	.18611	.18750	.18889	.19167	.19444	.19583	.19722	11
12	.20000	.20278	.20417	.20556	.20833	.21111	.21250	.21389	12
13	.21667	.21944	.22083	.22222	.22500	.22778	.22917	.23056	13
14	.23333	.23611	.23750	.23889	.24167	.24444	.24583	.24722	14
15	.25000	.25278	.25417	.25556	.25833	.26111	.26250	.26389	15
16	.26667	.26944	.27083	.27222	.27500	.27778	.27917	.28056	16
17	.28333	.28611	.28750	.28889	.29167	.29444	.29583	.29722	17
18	.30000	.30278	.30417	.30556	.30833	.31111	.31250	.31389	18
19	.31667	.31944	.32083	.32222	.32500	.32778	.32917	.33056	19
20	.33333	.33611	.33750	.33889	.34167	.34444	.34583	.34722	20
21	.35000	.35278	.35417	.35556	.35833	.36111	.36250	.36389	21
22	.36667	.36944	.37083	.37222	.37500	.37778	.37917	.38056	22
23	.38333	.38611	.38750	.38889	.39167	.39444	.39583	.39722	23
24	.40000	.40278	.40417	.40556	.40833	.41111	.41250	.41389	24
25	.41667	.41944	.42083	.42222	.42500	.42778	.42917	.43056	25
26	.43333	.43611	.43750	.43889	.44167	.44444	.44583	.44722	26
27	.45000	.45278	.45417	.45556	.45833	.46111	.46250	.46389	27
28	.46667	.46944	.47083	.47222	.47500	.47778	.47917	.48056	28
29	.48333	.48611	.48750	.48889	.49167	.49444	.49583	.49722	29
30	.50000	.50278	.50417	.50556	.50833	.51111	.51250	.51389	30
31	.51667	.51944	.52083	.52222	.52500	.52778	.52917	.53056	31
32	.53333	.53611	.53750	.53889	.54167	.54444	.54583	.54722	32
33	.55000	.55278	.55417	.55556	.55833	.56111	.56250	.56389	33
34	.56667	.56944	.57083	.57222	.57500	.57778	.57917	.58056	34
35	.58333	.58611	.58750	.58889	.59167	.59444	.59583	.59722	35
36	.60000	.60278	.60417	.60556	.60833	.61111	.61250	.61389	36
37	.61667	.61944	.62083	.62222	.62500	.62778	.62917	.63056	37
38	.63333	.63611	.63750	.63889	.64167	.64444	.64583	.64722	38
39	.65000	.65278	.65417	.65556	.65833	.66111	.66250	.66389	39
40	.66667	.66944	.67083	.67222	.67500	.67778	.67917	.68056	40
41	.68333	.68611	.68750	.68889	.69167	.69444	.69583	.69722	41
42	.70000	.70278	.70417	.70556	.70833	.71111	.71250	.71389	42
43	.71667	.71944	.72083	.72222	.72500	.72778	.72917	.73056	43
44	.73333	.73611	.73750	.73889	.74167	.74444	.74583	.74722	44
45	.75000	.75278	.75417	.75556	.75833	.76111	.76250	.76389	45
46	.76667	.76944	.77083	.77222	.77500	.77778	.77917	.78056	46
47	.78333	.78611	.78750	.78889	.79167	.79444	.79583	.79722	47
48	.80000	.80278	.80417	.80556	.80833	.81111	.81250	.81389	48
49	.81667	.81944	.82083	.82222	.82500	.82778	.82917	.83056	49
50	.83333	.83611	.83750	.83889	.84167	.84444	.84583	.84722	50
51	.85000	.85278	.85417	.85556	.85833	.86111	.86250	.86389	51
52	.86667	.86944	.87083	.87222	.87500	.87778	.87917	.88056	52
53	.88333	.88611	.88750	.88889	.89167	.89444	.89583	.89722	53
54	.90000	.90278	.90417	.90556	.90833	.91111	.91250	.91389	54
55	.91667	.91944	.92083	.92222	.92500	.92778	.92917	.93056	55
56	.93333	.93611	.93750	.93889	.94167	.94444	.94583	.94722	56
57	.95000	.95278	.95417	.95556	.95833	.96111	.96250	.96389	57
58	.96667	.96944	.97083	.97222	.97500	.97778	.97917	.98056	58
59	.98333	.98611	.98750	.98889	.99167	.99444	.99583	.99722	59
′	0″	10″	15″	20″	30″	40″	45″	50″	′

TABLE VIII.—SQUARES, CUBES, SQUARE ROOTS, AND CUBE ROOTS.

No.	Squares.	Cubes.	Square Roots.	Cube Roots.	Reciprocals.
1	1	1	1.0000000	1.0000000	1.000000000
2	4	8	1.4142136	1.2599210	.500000000
3	9	27	1.7320508	1.4422496	.333333333
4	16	64	2.0000000	1.5874011	.250000000
5	25	125	2.2360680	1.7099759	.200000000
6	36	216	2.4494897	1.8171206	.166666667
7	49	343	2.6457513	1.9129312	.142857143
8	64	512	2.8284271	2.0000000	.125000000
9	81	729	3.0000000	2.0800837	.111111111
10	100	1000	3.1622777	2.1544347	.100000000
11	121	1331	3.3166248	2.2239801	.090909091
12	144	1728	3.4641016	2.2894286	.083333333
13	169	2197	3.6055513	2.3513347	.076923077
14	196	2744	3.7416574	2.4101422	.071428571
15	225	3375	3.8729833	2.4662121	.066666667
16	256	4096	4.0000000	2.5198421	.062500000
17	289	4913	4.1231056	2.5712816	.058823529
18	324	5832	4.2426407	2.6207414	.055555556
19	361	6859	4.3588989	2.6684016	.052631579
20	400	8000	4.4721360	2.7144177	.050000000
21	441	9261	4.5825757	2.7589243	.047619048
22	484	10648	4.6904158	2.8020393	.045454545
23	529	12167	4.7958315	2.8438670	.043478261
24	576	13824	4.8989795	2.8844991	.041666667
25	625	15625	5.0000000	2.9240177	.040000000
26	676	17576	5.0990195	2.9624960	.038461538
27	729	19683	5.1961524	3.0000000	.037037037
28	784	21952	5.2915026	3.0365889	.035714286
29	841	24389	5.3851648	3.0723168	.034482759
30	900	27000	5.4772256	3.1072325	.033333333
31	961	29791	5.5677644	3.1413806	.032258065
32	1024	32768	5.6568542	3.1748021	.031250000
33	1089	35937	5.7445626	3.2075343	.030303030
34	1156	39304	5.8309519	3.2396118	.029411765
35	1225	42875	5.9160798	3.2710663	.028571429
36	1296	46656	6.0000000	3.3019272	.027777778
37	1369	50653	6.0827625	3.3322218	.027027027
38	1444	54872	6.1644140	3.3619754	.026315789
39	1521	59319	6.2449980	3.3912114	.025641026
40	1600	64000	6.3245553	3.4199519	.025000000
41	1681	68921	6.4031242	3.4482172	.024390244
42	1764	74088	6.4807407	3.4760266	.023809524
43	1849	79507	6.5574385	3.5033981	.023255814
44	1936	85184	6.6332496	3.5303483	.022727273
45	2025	91125	6.7082039	3.5568933	.022222222
46	2116	97336	6.7823300	3.5830479	.021739130
47	2209	103823	6.8556546	3.6088261	.021276600
48	2304	110592	6.9282032	3.6342411	.020833333
49	2401	117649	7.0000000	3.6593057	.020408163
50	2500	125000	7.0710678	3.6840314	.020000000
51	2601	132651	7.1414284	3.7084298	.019607843
52	2704	140608	7.2111026	3.7325111	.019230769
53	2809	148877	7.2801099	3.7562858	.018867925
54	2916	157464	7.3484692	3.7797631	.018518519
55	3025	166375	7.4161985	3.8029525	.018181818
56	3136	175616	7.4833148	3.8258624	.017857143
57	3249	185193	7.5498344	3.8485011	.017543860
58	3364	195112	7.6157731	3.8708766	.017241379
59	3481	205379	7.6811457	3.8929965	.016949153
60	3600	216000	7.7459667	3.9148676	.016666667
61	3721	226981	7.8102497	3.9364972	.016393443
62	3844	238328	7.8740079	3.9578915	.016129032

TABLE VIII.—Continued.

No.	Squares.	Cubes.	Square Roots.	Cube Roots.	Reciprocals.
63	3969	250047	7.9372539	3.9790571	.015873016
64	4036	262144	8.0000000	4.0000000	.015625000
65	4225	274625	8.0622577	4.0207256	.015384615
66	4356	287496	8.1240384	4.0412401	.015151515
67	4489	300763	8.1853528	4.0615480	.014925373
68	4624	314432	8.2462113	4.0816551	.014705882
69	4761	328509	8.3066239	4.1015661	.014492754
70	4900	343000	8.3666003	4.1212853	.014285714
71	5041	357911	8.4261498	4.1408178	.014084507
72	5184	373248	8.4852814	4.1601676	.013888889
73	5329	389017	8.5440037	4.1793390	.013698630
74	5476	405224	8.6023253	4.1983364	.013513514
75	5625	421875	8.6602540	4.2171633	.013333333
76	5776	438976	8.7177979	4.2358236	.013157895
77	5929	456533	8.7749644	4.2543210	.012987013
78	6084	474552	8.8317609	4.2726586	.012820513
79	6241	493039	8.8881944	4.2908404	.012658228
80	6400	512000	8.9442719	4.3088695	.012500000
81	6561	531441	9.0000000	4.3267487	.012345679
82	6724	551368	9.0553851	4.3444815	.012195122
83	6889	571787	9.1104336	4.3620707	.012048193
84	7056	592704	9.1651514	4.3795191	.011904762
85	7225	614125	9.2195445	4.3968296	.011764706
86	7396	636056	9.2736185	4.4140049	.011627907
87	7569	658503	9.3273791	4.4310476	.011494253
88	7744	681472	9.3808315	4.4479602	.011363636
89	7921	704969	9.4339811	4.4647451	.011235955
90	8100	729000	9.4868330	4.4814047	.011111111
91	8281	753571	9.5393920	4.4979414	.010989011
92	8464	778688	9.5916630	4.5143574	.010869565
93	8649	804357	9.6436508	4.5306549	.010752688
94	8836	830584	9.6953597	4.5468359	.010638298
95	9025	857375	9.7467943	4.5629026	.010526316
96	9216	884736	9.7979590	4.5788570	.010416667
97	9409	912673	9.8488578	4.5947009	.010309278
98	9604	941192	9.8994949	4.6104363	.010204082
99	9801	970299	9.9498744	4.6260650	.010101010
100	10000	1000000	10.0000000	4.6415888	.010000000
101	10201	1030301	10.0498756	4.6570095	.009900990
102	10404	1061208	10.0995049	4.6723287	.009803922
103	10609	1092727	10.1488916	4.6875482	.009708738
104	10816	1124864	10.1980390	4.7026604	.009615385
105	11025	1157625	10.2469508	4.7176940	.009523810
106	11236	1191016	10.2956301	4.7326235	.009433962
107	11449	1225043	10.3440804	4.7474594	.009345794
108	11664	1259712	10.3923048	4.7622032	.009259259
109	11881	1295029	10.4403065	4.7768563	.009174312
110	12100	1331000	10.4880885	4.7914199	.009090909
111	12321	1367631	10.5356538	4.8058955	.009009009
112	12544	1404928	10.5830052	4.8202845	.008928571
113	12769	1442897	10.6301458	4.8345881	.008849558
114	12996	1481544	10.6770783	4.8488076	.008771930
115	13225	1520875	10.7238053	4.8629442	.008695652
116	13456	1560896	10.7703296	4.8769990	.008620690
117	13689	1601613	10.8166538	4.8909732	.008547009
118	13924	1643032	10.8627805	4.9048681	.008474576
119	14161	1685159	10.9087121	4.9186847	.008403361
120	14400	1728000	10.9544512	4.9324242	.008333333
121	14641	1771561	11.0000000	4.9460874	.008264463
122	14884	1815848	11.0453610	4.9596757	.008196721
123	15129	1860867	11.0905365	4.9731898	.008130081
124	15376	1906624	11.1355287	4.9866310	.008064516

No.	Squares.	Cubes.	Square Roots.	Cube Roots.	Reciprocals.
125	15625	1953125	11.1803399	5.0000000	.008000000
126	15876	2000376	11.2249722	5.0132979	.007936508
127	16129	2048383	11.2694277	5.0265257	.007874016
128	16384	2097152	11.3137085	5.0396842	.007812500
129	16641	2146689	11.3578167	5.0527743	.007751938
130	16900	2197000	11.4017543	5.0657970	.007692308
131	17161	2248091	11.4455231	5.0787531	.007633588
132	17424	2299968	11.4891253	5.0916434	.007575758
133	17689	2352637	11.5325626	5.1044687	.007518797
134	17956	2406104	11.5758369	5.1172299	.007462687
135	18225	2460375	11.6189500	5.1299278	.007407407
136	18496	2515456	11.6619038	5.1425632	.007352941
137	18769	2571353	11.7046999	5.1551367	.007299270
138	19044	2628072	11.7473401	5.1676493	.007246377
139	19321	2685619	11.7898261	5.1801015	.007194245
140	19600	2744000	11.8321596	5.1924941	.007142857
141	19881	2803221	11.8743421	5.2048279	.007092199
142	20164	2863288	11.9163753	5.2171034	.007042254
143	20449	2924207	11.9582607	5.2293215	.006993007
144	20736	2985984	12.0000000	5.2414828	.006944444
145	21025	3048625	12.0415946	5.2535879	.006896552
146	21316	3112136	12.0830460	5.2656374	.006849315
147	21609	3176523	12.1243557	5.2776321	.006802721
148	21904	3241792	12.1655251	5.2895725	.006756757
149	22201	3307949	12.2065556	5.3014592	.006711409
150	22500	3375000	12.2474487	5.3132928	.006666667
151	22801	3442951	12.2882057	5.3250740	.006622517
152	23104	3511808	12.3288280	5.3368033	.006578947
153	23409	3581577	12.3693169	5.3484812	.006535948
154	23716	3652264	12.4096736	5.3601084	.006493506
155	24025	3723875	12.4498996	5.3716854	.006451613
156	24336	3796416	12.4899960	5.3832126	.006410256
157	24649	3869893	12.5299641	5.3946907	.006369427
158	24964	3944312	12.5698051	5.4061202	.006329114
159	25281	4019679	12.6095202	5.4175015	.006289308
160	25600	4096000	12.6491106	5.4288352	.006250000
161	25921	4173281	12.6885775	5.4401218	.006211180
162	26244	4251528	12.7279221	5.4513618	.006172840
163	26569	4330747	12.7671453	5.4625556	.006134969
164	26896	4410944	12.8062485	5.4737037	.006097561
165	27225	4492125	12.8452326	5.4848066	.006060606
166	27556	4574296	12.8840987	5.4958647	.006024096
167	27889	4657463	12.9228480	5.5068784	.005988024
168	28224	4741632	12.9614814	5.5178484	.005952381
169	28561	4826809	13.0000000	5.5287748	.005917160
170	28900	4913000	13.0384048	5.5396583	.005882353
171	29241	5000211	13.0766968	5.5504991	.005847953
172	29584	5088448	13.1148770	5.5612978	.005813953
173	29929	5177717	13.1529464	5.5720546	.005780347
174	30276	5268024	13.1909060	5.5827702	.005747126
175	30625	5359375	13.2287566	5.5934447	.005714286
176	30976	5451776	13.2664992	5.6040787	.005681818
177	31329	5545233	13.3041347	5.6146724	.005649718
178	31684	5639752	13.3416641	5.6252263	.005617978
179	32041	5735339	13.3790882	5.6357408	.005586592
180	32400	5832000	13.4164079	5.6462162	.005555556
181	32761	5929741	13.4536240	5.6566528	.005524862
182	33124	6028568	13.4907376	5.6670511	.005494505
183	33489	6128487	13.5277493	5.6774114	.005464481
184	33856	6229504	13.5646600	5.6877340	.005434783
185	34225	6331625	13.6014705	5.6980192	.005405405
186	34596	6434856	13.6381817	5.7082675	.005376344

TABLE VIII.—*Continued.*

No.	Squares.	Cubes.	Square Roots.	Cube Roots.	Reciprocals.
187	34969	6539203	13.6747943	5.7184791	.005347594
188	35344	6644672	13.7113092	5.7286543	.005319149
189	35721	6751269	13.7477271	5.7387936	.005291005
190	36100	6859000	13.7840488	5.7488971	.005263158
191	36481	6967871	13.8202750	5.7589652	.005235602
192	36864	7077888	13.8564065	5.7689982	.005208333
193	37249	7189057	13.8924440	5.7789966	.005181347
194	37636	7301384	13.9283883	5.7889604	.005154639
195	38025	7414875	13.9642400	5.7988900	.005128205
196	38416	7529536	14.0000000	5.8087857	.005102041
197	38809	7645373	14.0356688	5.8186479	.005076142
198	39204	7762392	14.0712473	5.8284767	.005050505
199	39601	7880599	14.1067360	5.8382725	.005025126
200	40000	8000000	14.1421356	5.8480355	.005000000
201	40401	8120601	14.1774469	5.8577660	.004975124
202	40804	8242408	14.2126704	5.8674643	.004950495
203	41209	8365427	14.2478068	5.8771307	.004926108
204	41616	8489664	14.2828569	5.8867653	.004901961
205	42025	8615125	14.3178211	5.8963685	.004878049
206	42436	8741816	14.3527001	5.9059406	.004854369
207	42849	8869743	14.3874946	5.9154817	.004830918
208	43264	8998912	14.4222051	5.9249921	.004807692
209	43681	9129329	14.4568323	5.9344721	.004784689
210	44100	9261000	14.4913767	5.9439220	.004761905
211	44521	9393931	14.5258390	5.9533418	.004739336
212	44944	9528128	14.5602198	5.9627320	.004716981
213	45369	9663597	14.5945195	5.9720926	.004694836
214	45796	9800344	14.6287388	5.9814240	.004672897
215	46225	9938375	14.6628783	5.9907264	.004651163
216	46656	10077696	14.6969385	6.0000000	.004629630
217	47089	10218313	14.7309199	6.0092450	.004608295
218	47524	10360232	14.7648231	6.0184617	.004587156
219	47961	10503459	14.7986486	6.0276502	.004566210
220	48400	10648000	14.8323970	6.0368107	.004545455
221	48841	10793861	14.8660687	6.0459435	.004524887
222	49284	10941048	14.8996644	6.0550489	.004504505
223	49729	11089567	14.9331845	6.0641270	.004484305
224	50176	11239424	14.9666295	6.0731779	.004464286
225	50625	11390625	15.0000000	6.0822020	.004444444
226	51076	11543176	15.0332964	6.0911994	.004424779
227	51529	11697083	15.0665192	6.1001702	.004405286
228	51984	11852352	15.0996689	6.1091147	.004385965
229	52441	12008989	15.1327460	6.1180332	.004366812
230	52900	12167000	15.1657509	6.1269257	.004347826
231	53361	12326391	15.1986842	6.1357924	.004329004
232	53824	12487168	15.2315462	6.1446337	.004310345
233	54289	12649337	15.2643375	6.1534495	.004291845
234	54756	12812904	15.2970585	6.1622401	.004273504
235	55225	12977875	15.3297097	6.1710058	.004255319
236	55696	13144256	15.3622915	6.1797466	.004237288
237	56169	13312053	15.3948043	6.1884628	.004219409
238	56644	13481272	15.4272486	6.1971544	.004201681
239	57121	13651919	15.4596248	6.2058218	.004184100
240	57600	13824000	15.4919334	6.2144650	.004166667
241	58081	13997521	15.5241747	6.2230843	.004149378
242	58564	14172488	15.5563492	6.2316797	.004132231
243	59049	14348907	15.5884573	6.2402515	.004115226
244	59536	14526784	15.6204994	6.2487998	.004098361
245	60025	14706125	15.6524758	6.2573248	.004081633
246	60516	14886936	15.6843871	6.2658266	.004065041
247	61009	15069223	15.7162336	6.2743054	.004048583
248	61504	15252992	15.7480157	6.2827613	.004032258

TABLE VIII.—*Continued.*

No.	Squares.	Cubes.	Square Roots.	Cube Roots.	Reciprocals.
249	62001	15438249	15.7797338	6.2911946	.004016064
250	62500	15625000	15.8113883	6.2996053	.004000000
251	63001	15813251	15.8429795	6.3079935	.003984064
252	63504	16003008	15.8745079	6.3163596	.003968254
253	64009	16194277	15.9059737	6.3247035	.003952569
254	64516	16387064	15.9373775	6.3330256	.003937008
255	65025	16581375	15.9687194	6.3413257	.003921569
256	65536	16777216	16.0000000	6.3496042	.003906250
257	66049	16974593	16.0312195	6.3578611	.003891051
258	66564	17173512	16.0623784	6.3660968	.003875969
259	67081	17373979	16.0934769	6.3743111	.003861004
260	67600	17576000	16.1245155	6.3825043	.003846154
261	68121	17779581	16.1554944	6.3906765	.003831418
262	68644	17984728	16.1864141	6.3988279	.003816794
263	69169	18191447	16.2172747	6.4069585	.003802281
264	69696	18399744	16.2480768	6.4150687	.003787879
265	70225	18609625	16.2788206	6.4231583	.003773585
266	70756	18821096	16.3095064	6.4312276	.003759398
267	71289	19034163	16.3401346	6.4392767	.003745318
268	71824	19248832	16.3707055	6.4473057	.003731343
269	72361	19465109	16.4012195	6.4553148	.003717472
270	72900	19683000	16.4316767	6.4633041	.003703704
271	73441	19902511	16.4620776	6.4712736	.003690037
272	73984	20123648	16.4924225	6.4792236	.003676471
273	74529	20346417	16.5227116	6.4871541	.003663004
274	75076	20570824	16.5529454	6.4950653	.003649635
275	75625	20796875	16.5831240	6.5029572	.003636364
276	76176	21024576	16.6132477	6.5108300	.003623188
277	76729	21253933	16.6433170	6.5186839	.003610108
278	77284	21484952	16.6733320	6.5265189	.003597122
279	77841	21717639	16.7032931	6.5343351	.003584229
280	78400	21952000	16.7332005	6.5421326	.003571429
281	78961	22188041	16.7630546	6.5499116	.003558719
282	79524	22425768	16.7928556	6.5576722	.003546099
283	80089	22665187	16.8226038	6.5654144	.003533569
284	80656	22906304	16.8522995	6.5731385	.003521127
285	81225	23149125	16.8819430	6.5808443	.003508772
286	81796	23393656	16.9115345	6.5885323	.003496503
287	82369	23639903	16.9410743	6.5962023	.003484321
288	82944	23887872	16.9705627	6.6038545	.003472222
289	83521	24137569	17.0000000	6.6114890	.003460208
290	84100	24389000	17.0293864	6.6191060	.003448276
291	84681	24642171	17.0587221	6.6267054	.003436426
292	85264	24897088	17.0880075	6.6342874	.003424658
293	85849	25153757	17.1172428	6.6418522	.003412969
294	86436	25412184	17.1464282	6.6493998	.003401361
295	87025	25672375	17.1755640	6.6569302	.003389831
296	87616	25934336	17.2046505	6.6644437	.003378378
297	88209	26198073	17.2336879	6.6719403	.003367003
298	88804	26463592	17.2626765	6.6794200	.003355705
299	89401	26730899	17.2916165	6.6868831	.003344482
300	90000	27000000	17.3205081	6.6943295	.003333333
301	90601	27270901	17.3493516	6.7017593	.003322259
302	91204	27543608	17.3781472	6.7091729	.003311258
303	91809	27818127	17.4068952	6.7165700	.003300330
304	92416	28094464	17.4355958	6.7239508	.003289474
305	93025	28372625	17.4642492	6.7313155	.003278689
306	93636	28652616	17.4928557	6.7386641	.003267974
307	94249	28934443	17.5214155	6.7459967	.003257329
308	94864	29218112	17.5499288	6.7533134	.003246753
309	95481	29503629	17.5783958	6.7606143	.003236246
310	96100	29791000	17.6008169	6.7678095	.003225806

TABLE VIII.—*Continued.*

No.	Squares.	Cubes.	Square Roots.	Cube Roots.	Reciprocals.
311	96721	30080231	17.6351921	6.7751690	.003215434
312	97344	30371328	17.6635217	6.7824229	.003205128
313	97969	30664297	17.6918060	6.7896613	.003194888
314	98596	30959144	17.7200451	6.7968844	.003184713
315	99225	31255875	17.7482393	6.8040921	.003174603
316	99856	31554496	17.7763888	6.8112847	.003164557
317	100489	31855013	17.8044938	6.8184620	.003154574
318	101124	32157432	17.8325545	6.8256242	.003144654
319	101761	32461759	17.8605711	6.8327714	.003134796
320	102400	32768000	17.8885438	6.8399037	.003125000
321	103041	33076161	17.9164729	6.8470213	.003115265
322	103684	33386248	17.9443584	6.8541240	.003105590
323	104329	33698267	17.9722008	6.8612120	.003095975
324	104976	34012224	18.0000000	6.8682855	.003086420
325	105625	34328125	18.0277564	6.8753443	.003076923
326	106276	34645976	18.0554701	6.8823888	.003067485
327	106929	34965783	18.0831413	6.8894188	.003058104
328	107584	35287552	18.1107703	6.8964345	.003048780
329	108241	35611289	18.1383571	6.9034359	.003039514
330	108900	35937000	18.1659021	6.9104232	.003030303
331	109561	36264691	18.1934054	6.9173964	.003021148
332	110224	36594368	18.2208672	6.9243556	.003012048
333	110889	36926037	18.2482876	6.9313008	.003003003
334	111556	37259704	18.2756669	6.9382321	.002994012
335	112225	37595375	18.3030052	6.9451496	.002985075
336	112896	37933056	18.3303028	6.9520533	.002976190
337	113569	38272753	18.3575598	6.9589434	.002967359
338	114244	38614472	18.3847763	6.9658198	.002958580
339	114921	38958219	18.4119526	6.9726826	.002949853
340	115600	39304000	18.4390889	6.9795321	.002941176
341	116281	39651821	18.4661853	6.9863681	.002932551
342	116964	40001688	18.4932420	6.9931906	.002923977
343	117649	40353607	18.5202592	7.0000000	.002915452
344	118336	40707584	18.5472370	7.0067962	.002906977
345	119025	41063625	18.5741756	7.0135791	.002898551
346	119716	41421736	18.6010752	7.0203490	.002890173
347	120409	41781923	18.6279360	7.0271058	.002881844
348	121104	42144192	18.6547581	7.0338497	.002873563
349	121801	42508549	18.6815417	7.0405806	.002865330
350	122500	42875000	18.7082869	7.0472987	.002857143
351	123201	43243551	18.7349940	7.0540041	.002849003
352	123904	43614208	18.7616630	7.0606967	.002840909
353	124609	43986977	18.7882942	7.0673767	.002832861
354	125316	44361864	18.8148877	7.0740440	.002824859
355	126025	44738875	18.8414437	7.0806988	.002816901
356	126736	45118016	18.8679623	7.0873411	.002808989
357	127449	45499293	18.8944436	7.0939709	.002801120
358	128164	45882712	18.9208879	7.1005885	.002793296
359	128881	46268279	18.9472953	7.1071937	.002785515
360	129600	46656000	18.9736660	7.1137866	.002777778
361	130321	47045881	19.0000000	7.1203674	.002770083
362	131044	47437928	19.0262976	7.1269360	.002762431
363	131769	47832147	19.0525589	7.1334925	.002754821
364	132496	48228544	19.0787840	7.1400370	.002747253
365	133225	48627125	19.1049732	7.1465695	.002739726
366	133956	49027896	19.1311265	7.1530901	.002732240
367	134689	49430863	19.1572441	7.1595988	.002724796
368	135424	49836032	19.1833261	7.1660957	.002717391
369	136161	50243409	19.2093727	7.1725809	.002710027
370	136900	50653000	19.2353841	7.1790544	.002702703
371	137641	51064811	19.2613603	7.1855162	.002695418
372	138384	51478848	19.2873015	7.1919663	.002688172

TABLE VIII.—Continued.

No.	Squares.	Cubes.	Square Roots.	Cube Roots.	Reciprocals.
373	139129	51895117	19.3132079	7.1984050	.002680965
374	139876	52313624	19.3390796	7.2048322	.002673797
375	140625	52734375	19.3649167	7.2112479	.002666667
376	141376	53157376	19.3907194	7.2176322	.002659574
377	142129	53582633	19.4164878	7.2240450	.002652520
378	142884	54010152	19.4422221	7.2304268	.002645503
379	143641	54439939	19.4679223	7.2367972	.002638522
380	144400	54872000	19.4935887	7.2431565	.002631579
381	145161	55306341	19.5192213	7.2495045	.002624672
382	145924	55742968	19.5448203	7.2558415	.002617801
383	146689	56181887	19.5703858	7.2621675	.002610966
384	147456	56623104	19.5959179	7.2684824	.002604167
385	148225	57066625	19.6214169	7.2747864	.002597403
386	148996	57512456	19.6468827	7.2810794	.002590674
387	149769	57960603	19.6723156	7.2873617	.002583979
388	150544	58411072	19.6977156	7.2936330	.002577320
389	151321	58863869	19.7230829	7.2998936	.002570694
390	152100	59319000	19.7484177	7.3061436	.002564103
391	152881	59776471	19.7737199	7.3123828	.002557545
392	153664	60236288	19.7989899	7.3186114	.002551020
393	154449	60698457	19.8242276	7.3248295	.002544529
394	155236	61162984	19.8494332	7.3310369	.002538071
395	156025	61629875	19.8746069	7.3372339	.002531646
396	156816	62099136	19.8997487	7.3434205	.002525253
397	157609	62570773	19.9248588	7.3495966	.002518892
398	158404	63044792	19.9499373	7.3557624	.002512563
399	159201	63521199	19.9749844	7.3619178	.002506266
400	160000	64000000	20.0000000	7.3680630	.002500000
401	160801	64481201	20.0249844	7.3741979	.002493766
402	161604	64964808	20.0499377	7.3803227	.002487562
403	162409	65450827	20.0748599	7.3864373	.002481390
404	163216	65939264	20.0997512	7.3925418	.002475248
405	164025	66430125	20.1246118	7.3986363	.002469136
406	164836	66923416	20.1494417	7.4047206	.002463054
407	165649	67419143	20.1742410	7.4107950	.002457002
408	166464	67917312	20.1990099	7.4168595	.002450980
409	167281	68417929	20.2237484	7.4229142	.002444988
410	168100	68921000	20.2484567	7.4289589	.002439024
411	168921	69426531	20.2731349	7.4349938	.002433090
412	169744	69934528	20.2977831	7.4410189	.002427184
413	170569	70444997	20.3224014	7.4470342	.002421308
414	171396	70957944	20.3469899	7.4530399	.002415459
415	172225	71473375	20.3715488	7.4590359	.002409639
416	173056	71991296	20.3960781	7.4650223	.002403846
417	173889	72511713	20.4205779	7.4709991	.002398082
418	174724	73034632	20.4450483	7.4769664	.002392344
419	175561	73560059	20.4694895	7.4829242	.002386635
420	176400	74088000	20.4939015	7.4888724	.002380952
421	177241	74618461	20.5182845	7.4948113	.002375297
422	178084	75151448	20.5426386	7.5007406	.002369668
423	178929	75686967	20.5669638	7.5066607	.002364066
424	179776	76225024	20.5912603	7.5125715	.002358491
425	180625	76765625	20.6155281	7.5184730	.002352941
426	181476	77308776	20.6397674	7.5243652	.002347418
427	182329	77854483	20.6639783	7.5302482	.002341920
428	183184	78402752	20.6881609	7.5361221	.002336449
429	184041	78953589	20.7123152	7.5419867	.002331002
430	184900	79507000	20.7364414	7.5478423	.002325581
431	185761	80062991	20.7605395	7.5536888	.002320186
432	186624	80621568	20.7846097	7.5595263	.002314815
433	187489	81182737	20.8086520	7.5653548	.002309469
434	188356	81746504	20.8326667	7.5711743	.002304147

TABLE VIII.—*Continued.*

No.	Squares.	Cubes.	Square Roots.	Cube Roots.	Reciprocals.
435	189225	82312875	20.8566536	7.5769849	.002298851
436	190096	82881856	20.8806130	7.5827865	.002293578
437	190969	83453453	20.9045450	7.5885793	.002288330
438	191844	84027672	20.9284495	7.5943633	.002283105
439	192721	84604519	20.9523268	7.6001385	.002277904
440	193600	85184000	20.9761770	7.6059049	.002272727
441	194481	85766121	21.0000000	7.6116626	.002267574
442	195364	86350888	21.0237960	7.6174116	.002262443
443	196249	86938307	21.0475652	7.6231519	.002257336
444	197136	87528384	21.0713075	7.6288837	.002252252
445	198025	88121125	21.0950231	7.6346067	.002247191
446	198916	88716536	21.1187121	7.6403213	.002242152
447	199809	89314623	21.1423745	7.6460272	.002237136
448	200704	89915392	21.1660105	7.6517247	.002232143
449	201601	90518849	21.1896201	7.6574133	.002227171
450	202500	91125000	21.2132034	7.6630943	.002222222
451	203401	91733851	21.2367606	7.6687665	.002217295
452	204304	92345408	21.2602916	7.6744303	.002212389
453	205209	92959677	21.2837967	7.6800857	.002207506
454	206116	93576664	21.3072758	7.6857328	.002202643
455	207025	94196375	21.3307290	7.6913717	.002197802
456	207936	94818816	21.3541565	7.6970023	.002192982
457	208849	95443993	21.3775583	7.7026246	.002188184
458	209764	96071912	21.4009346	7.7082388	.002183406
459	210681	96702579	21.4242853	7.7138448	.002178649
460	211600	97336000	21.4476106	7.7194426	.002173913
461	212521	97972181	21.4709106	7.7250325	.002169197
462	213444	98611128	21.4941853	7.7306141	.002164502
463	214369	99252847	21.5174348	7.7361877	.002159827
464	215296	99897344	21.5406592	7.7417532	.002155172
465	216225	100544625	21.5638587	7.7473109	.002150538
466	217156	101194696	21.5870331	7.7528606	.002145923
467	218089	101847563	21.6101828	7.7584023	.002141328
468	219024	102503232	21.6333077	7.7639361	.002136752
469	219961	103161709	21.6564078	7.7694620	.002132196
470	220900	103823000	21.6794834	7.7749801	.002127660
471	221841	104487111	21.7025344	7.7804904	.002123142
472	222784	105154048	21.7255610	7.7859928	.002118644
473	223729	105823817	21.7485632	7.7914875	.002114165
474	224676	106496424	21.7715411	7.7969745	.002109705
475	225625	107171875	21.7944947	7.8024538	.002105263
476	226576	107850176	21.8174242	7.8079254	.002100840
477	227529	108531333	21.8403297	7.8133892	.002096436
478	228484	109215352	21.8632111	7.8188456	.002092050
479	229441	109902239	21.8860686	7.8242942	.002087683
480	230400	110592000	21.9089023	7.8297353	.002083333
481	231361	111284641	21.9317122	7.8351688	.002079002
482	232324	111980168	21.9544984	7.8405949	.002074689
483	233289	112678587	21.9772610	7.8460134	.002070393
484	234256	113379904	22.0000000	7.8514244	.002066116
485	235225	114084125	22.0227155	7.8568281	.002061856
486	236196	114791256	22.0454077	7.8622242	.002057613
487	237169	115501303	22.0680765	7.8676130	.002053388
488	238144	116214272	22.0907220	7.8729944	.002049180
489	239121	116930169	22.1133444	7.8783684	.002044990
490	240100	117649000	22.1359436	7.8837352	.002040816
491	241081	118370771	22.1585198	7.8890946	.002036660
492	242064	119095488	22.1810730	7.8944468	.002032520
493	243049	119823157	22.2036033	7.8997917	.002028398
494	244036	120553784	22.2261108	7.9051294	.002024291
495	245025	121287375	22.2485955	7.9104599	.002020202
496	246016	122023936	22.2710575	7.9157832	.002016129

TABLE VIII.—*Continued.*

No.	Squares.	Cubes.	Square Roots.	Cube Roots.	Reciprocals.
497	247009	122763473	22.2934968	7.9210094	.002012072
498	248004	123505992	22.3159136	7.9264085	.002008032
499	249001	124251499	22.3383079	7.9317104	.002004008
500	250000	125000000	22.3606798	7.9370053	.002000000
501	251001	125751501	22.3830293	7.9422931	.001996008
502	252004	126506008	22.4053565	7.9475739	.001992032
503	253009	127263527	22.4276615	7.9528477	.001988072
504	254016	128024064	22.4499443	7.9581144	.001984127
505	255025	128787625	22.4722051	7.9633743	.001980198
506	256036	129554216	22.4944438	7.9686271	.001976285
507	257049	130323843	22.5166605	7.9738731	.001972387
508	258064	131096512	22.5388553	7.9791122	.001968504
509	259081	131872229	22.5610283	7.9843444	.001964637
510	260100	132651000	22.5831796	7.9895697	.001960784
511	261121	133432831	22.6053091	7.9947883	.001956947
512	262144	134217728	22.6274170	8.0000000	.001953125
513	263169	135005697	22.6495033	8.0052049	.001949318
514	264196	135796744	22.6715681	8.0104032	.001945525
515	265225	136590875	22.6936114	8.0155946	.001941748
516	266256	137388096	22.7156334	8.0207794	.001937984
517	267289	138188413	22.7376340	8.0259574	.001934236
518	268324	138991832	22.7596134	8.0311287	.001930502
519	269361	139798359	22.7815715	8.0362935	.001926782
520	270400	140608000	22.8035085	8.0414515	.001923077
521	271441	141420761	22.8254244	8.0466030	.001919386
522	272484	142236648	22.8473193	8.0517479	.001915709
523	273529	143055667	22.8691933	8.0568862	.001912046
524	274576	143877824	22.8910463	8.0620180	.001908397
525	275625	144703125	22.9128785	8.0671432	.001904762
526	276676	145531576	22.9346899	8.0722620	.001901141
527	277729	146363183	22.9564806	8.0773743	.001897533
528	278784	147197952	22.9782506	8.0824800	.001893939
529	279841	148035889	23.0000000	8.0875794	.001890359
530	280900	148877000	23.0217289	8.0926723	.001886792
531	281961	149721291	23.0434372	8.0977589	.001883239
532	283024	150568768	23.0651252	8.1028390	.001879699
533	284089	151419437	23.0867928	8.1079128	.001876173
534	285156	152273304	23.1084400	8.1129803	.001872659
535	286225	153130375	23.1300670	8.1180414	.001869159
536	287296	153990656	23.1516738	8.1230962	.001865672
537	288369	154854153	23.1732605	8.1281447	.001862197
538	289444	155720872	23.1948270	8.1331870	.001858736
539	290521	156590819	23.2163735	8.1382230	.001855288
540	291600	157464000	23.2379001	8.1432529	.001851852
541	292681	158340421	23.2594067	8.1482765	.001848429
542	293764	159220088	23.2808935	8.1532939	.001845018
543	294849	160103007	23.3023604	8.1583051	.001841621
544	295936	160989184	23.3238076	8.1633102	.001838235
545	297025	161878625	23.3452351	8.1683092	.001834862
546	298116	162771336	23.3666429	8.1733020	.001831502
547	299209	163667323	23.3880311	8.1782888	.001828154
548	300304	164566592	23.4093998	8.1832695	.001824818
549	301401	165469149	23.4307490	8.1882441	.001821494
550	302500	166375000	23.4520788	8.1932127	.001818182
551	303601	167284151	23.4733892	8.1981752	.001814882
552	304704	168196608	23.4946802	8.2031319	.001811594
553	305809	169112377	23.5159520	8.2080825	.001808318
554	306916	170031464	23.5372046	8.2130271	.001805054
555	308025	170953875	23.5584380	8.2179657	.001801802
556	309136	171879616	23.5796522	8.2228985	.001798561
557	310249	172808693	23.6008474	8.2278254	.001795332
558	311364	173741112	23.6220236	8.2327463	.001792115

TABLE VIII.—Continued.

No.	Squares.	Cubes.	Square Roots.	Cube Roots.	Reciprocals.
559	312481	174676879	23.6431808	8.2376614	.001788909
560	313600	175616000	23.6643191	8.2425706	.001785714
561	314721	176558481	23.6854386	8.2474740	.001782531
562	315844	177504328	23.7065392	8.2523715	.001779359
563	316969	178453547	23.7276210	8.2572633	.001776199
564	318096	179406144	23.7486842	8.2621492	.001773050
565	319225	180362125	23.7697286	8.2670294	.001769912
566	320356	181321496	23.7907545	8.2719039	.001766784
567	321489	182284263	23.8117618	8.2767726	.001763668
568	322624	183250432	23.8327506	8.2816355	.001760563
569	323761	184220000	23.8537209	8.2864928	.001757469
570	324900	185193000	23.8746728	8.2913444	.001754386
571	326041	186169411	23.8956063	8.2961903	.001751313
572	327184	187149248	23.9165215	8.3010304	.001748252
573	328329	188132517	23.9374184	8.3058651	.001745201
574	329476	189119224	23.9582971	8.3106941	.001742160
575	330625	190109375	23.9791576	8.3155175	.001739130
576	331776	191102976	24.0000000	8.3203353	.001736111
577	332929	192100033	24.0208243	8.3251475	.001733102
578	334084	193100552	24.0416306	8.3299542	.001730104
579	335241	194104529	24.0624188	8.3347553	.001727116
580	336400	195112000	24.0831891	8.3395509	.001724138
581	337561	196122941	24.1039416	8.3443410	.001721170
582	338724	197137368	24.1246762	8.3491256	.001718213
583	339889	198155287	24.1453929	8.3539047	.001715266
584	341056	199176704	24.1660919	8.3586784	.001712329
585	342225	200201625	24.1867732	8.3634466	.001709402
586	343396	201230056	24.2074369	8.3682095	.001706485
587	344569	202262003	24.2280829	8.3729668	.001703578
588	345744	203297472	24.2487113	8.3777188	.001700680
589	346921	204336469	24.2693222	8.3824653	.001697793
590	348100	205379000	24.2899156	8.3872065	.001694915
591	349281	206425071	24.3104916	8.3919423	.001692047
592	350464	207474688	24.3310501	8.3966729	.001689189
593	351649	208527857	24.3515913	8.4013981	.001686341
594	352836	209584584	24.3721152	8.4061180	.001683502
595	354025	210644875	24.3926218	8.4108326	.001680672
596	355216	211708736	24.4131112	8.4155419	.001677852
597	356409	212776173	24.4335834	8.4202460	.001675042
598	357604	213847192	24.4540385	8.4249448	.001672241
599	358801	214921799	24.4744765	8.4296383	.001669449
600	360000	216000000	24.4948974	8.4343267	.001666667
601	361201	217081801	24.5153013	8.4390098	.001663894
602	362404	218167208	24.5356883	8.4436877	.001661130
603	363609	219256227	24.5560583	8.4483605	.001658375
604	364816	220348864	24.5764115	8.4530281	.001655629
605	366025	221445125	24.5967478	8.4576906	.001652893
606	367236	222545016	24.6170673	8.4623479	.001650165
607	368449	223648543	24.6373700	8.4670001	.001647446
608	369664	224755712	24.6576560	8.4716471	.001644737
609	370881	225866529	24.6779254	8.4762892	.001642036
610	372100	226981000	24.6981781	8.4809261	.001639344
611	373321	228099131	24.7184142	8.4855579	.001636661
612	374544	229220928	24.7386338	8.4901848	.001633987
613	375769	230346397	24.7588368	8.4948065	.001631321
614	376996	231475544	24.7790234	8.4994233	.001628664
615	378225	232608375	24.7991935	8.5040350	.001626016
616	379456	233744896	24.8193473	8.5086417	.001623377
617	380689	234885113	24.8394847	8.5132435	.001620746
618	381924	236029032	24.8596058	8.5178403	.001618123
619	383161	237176659	24.8797106	8.5224321	.001615509
620	384400	238328000	24.8997992	8.5270189	.001612903

TABLE VIII.—Continued.

No.	Squares.	Cubes.	Square Roots.	Cube Roots.	Reciprocals.
621	385641	239483061	24.9198716	8.5316009	.001610306
622	386884	240641848	24.9399278	8.5361780	.001607717
623	388129	241804367	24.9599679	8.5407501	.001605136
624	389376	242970624	24.9799920	8.5453173	.001602564
625	390625	244140625	25.0000000	8.5498797	.001600000
626	391876	245314376	25.0199920	8.5544372	.001597444
627	393129	246491883	25.0399681	8.5589899	.001594896
628	394384	247673152	25.0599282	8.5635377	.001592357
629	395641	248858189	25.0798724	8.5680807	.001589825
630	396900	250047000	25.0998008	8.5726189	.001587302
631	398161	251239591	25.1197134	8.5771523	.001584786
632	399424	252435968	25.1396102	8.5816809	.001582278
633	400689	253636137	25.1594913	8.5862047	.001579779
634	401956	254840104	25.1793566	8.5907238	.001577287
635	403225	256047875	25.1992063	8.5952380	.001574803
636	404496	257259456	25.2190404	8.5997476	.001572327
637	405769	258474853	25.2388589	8.6042525	.001569859
638	407044	259694072	25.2586619	8.6087526	.001567398
639	408321	260917119	25.2784493	8.6132480	.001564945
640	409600	262144000	25.2982213	8.6177388	.001562500
641	410881	263374721	25.3179778	8.6222248	.001560062
642	412164	264609288	25.3377189	8.6267063	.001557632
643	413449	265847707	25.3574447	8.6311830	.001555210
644	414736	267089984	25.3771551	8.6356551	.001552795
645	416025	268336125	25.3968502	8.6401226	.001550388
646	417316	269586136	25.4165301	8.6445855	.001547988
647	418609	270840023	25.4361947	8.6490437	.001545595
648	419904	272097792	25.4558441	8.6534974	.001543210
649	421201	273359449	25.4754784	8.6579465	.001540832
650	422500	274625000	25.4950976	8.6623911	.001538462
651	423801	275894451	25.5147016	8.6668310	.001536098
652	425104	277167808	25.5342907	8.6712665	.001533742
653	426409	278445077	25.5538647	8.6756974	.001531394
654	427716	279726264	25.5734237	8.6801237	.001529052
655	429025	281011375	25.5929678	8.6845456	.001526718
656	430336	282300416	25.6124969	8.6889630	.001524390
657	431649	283593393	25.6320112	8.6933759	.001522070
658	432964	284890312	25.6515107	8.6977843	.001519757
659	434281	286191179	25.6709953	8.7021882	.001517451
660	435600	287496000	25.6904652	8.7065877	.001515152
661	436921	288804781	25.7099203	8.7109827	.001512859
662	438244	290117528	25.7293607	8.7153734	.001510574
663	439569	291434247	25.7487864	8.7197596	.001508296
664	440896	292754944	25.7681975	8.7241414	.001506024
665	442225	294079625	25.7875939	8.7285187	.001503759
666	443556	295408296	25.8069758	8.7328918	.001501502
667	444889	296740963	25.8263431	8.7372604	.001499250
668	446224	298077632	25.8456960	8.7416246	.001497006
669	447561	299418309	25.8650343	8.7459846	.001494768
670	448900	300763000	25.8843582	8.7503401	.001492537
671	450241	302111711	25.9036677	8.7546913	.001490313
672	451584	303464448	25.9229628	8.7590383	.001488095
673	452929	304821217	25.9422435	8.7633809	.001485884
674	454276	306183024	25.9615100	8.7677192	.001483680
675	455625	307546875	25.9807621	8.7720532	.001481481
676	456976	308915776	26.0000000	8.7763830	.001479290
677	458329	310288733	26.0192237	8.7807084	.001477105
678	459684	311665752	26.0384331	8.7850296	.001474926
679	461041	313046839	26.0576284	8.7893466	.001472754
680	462400	314432000	26.0768096	8.7936593	.001470588
681	463761	315821241	26.0959767	8.7979679	.001468429
682	465124	317214568	26.1151297	8.8022721	.001466276

TABLE VIII.—*Continued.*

No.	Squares.	Cubes.	Square Roots.	Cube Roots.	Reciprocals.
683	466489	318611987	26.1342687	8.8065722	.001464129
684	467856	320013504	26.1533937	8.8108081	.001461988
685	469225	321419125	26.1725047	8.8151598	.001459854
686	470596	322828856	26.1916017	8.8194474	.001457726
687	471969	324242703	26.2106848	8.8237307	.001455604
688	473344	325660672	26.2297541	8.8280099	.001453488
689	474721	327082769	26.2488095	8.8322850	.001451379
690	476100	328509000	26.2678511	8.8365559	.001449275
691	477481	329939371	26.2868789	8.8408227	.001447178
692	478864	331373888	26.3058929	8.8450854	.001445087
693	480249	332812557	26.3248932	8.8493440	.001443001
694	481636	334255384	26.3438797	8.8535985	.001440922
695	483025	335702375	26.3628527	8.8578489	.001438849
696	484416	337153536	26.3818119	8.8620952	.001436782
697	485809	338608873	26.4007576	8.8663375	.001434720
698	487204	340068392	26.4196896	8.8705757	.001432665
699	488601	341532099	26.4386081	8.8748099	.001430615
700	490000	343000000	26.4575131	8.8790400	.001428571
701	491401	344472101	26.4764046	8.8832661	.001426534
702	492804	345948408	26.4952826	8.8874882	.001424501
703	494209	347428927	26.5141472	8.8917063	.001422475
704	495616	348913664	26.5329983	8.8959204	.001420455
705	497025	350402625	26.5518361	8.9001304	.001418440
706	498436	351895816	26.5706605	8.9043366	.001416431
707	499849	353393243	26.5894716	8.9085387	.001414427
708	501264	354894912	26.6082694	8.9127369	.001412430
709	502681	356400829	26.6270539	8.9169311	.001410437
710	504100	357911000	26.6458252	8.9211214	.001408451
711	505521	359425431	26.6645833	8.9253078	.001406470
712	506944	360944128	26.6833281	8.9294902	.001404494
713	508369	362467097	26.7020598	8.9336687	.001402525
714	509796	363994344	26.7207784	8.9378433	.001400560
715	511225	365525875	26.7394839	8.9420140	.001398601
716	512656	367061696	26.7581763	8.9461809	.001396648
717	514089	368601813	26.7768557	8.9503438	.001394700
718	515524	370146232	26.7955220	8.9545029	.001392758
719	516961	371694959	26.8141754	8.9586581	.001390821
720	518400	373248000	26.8328157	8.9628095	.001388889
721	519841	374805361	26.8514432	8.9669570	.001386963
722	521284	376367048	26.8700577	8.9711007	.001385042
723	522729	377933067	26.8886593	8.9752406	.001383126
724	524176	379503424	26.9072481	8.9793766	.001381215
725	525625	381078125	26.9258240	8.9835089	.001379310
726	527076	382657176	26.9443872	8.9876373	.001377410
727	528529	384240583	26.9629375	8.9917620	.001375516
728	529984	385828352	26.9814751	8.9958829	.001373626
729	531441	387420489	27.0000000	9.0000000	.001371742
730	532900	389017000	27.0185122	9.0041134	.001369863
731	534361	390617891	27.0370117	9.0082229	.001367989
732	535824	392223168	27.0554985	9.0123288	.001366120
733	537289	393832837	27.0739727	9.0164309	.001364256
734	538756	395446904	27.0924344	9.0205293	.001362398
735	540225	397065375	27.1108834	9.0246239	.001360544
736	541696	398688256	27.1293199	9.0287149	.001358696
737	543169	400315553	27.1477439	9.0328021	.001356852
738	544644	401947272	27.1661554	9.0368857	.001355014
739	546121	403583419	27.1845544	9.0409655	.001353180
740	547600	405224000	27.2029410	9.0450419	.001351351
741	549081	406869021	27.2213152	9.0491142	.001349528
742	550564	408518488	27.2396769	9.0531831	.001347709
743	552049	410172407	27.2580263	9.0572482	.001345895
744	553536	411830784	27.2763634	9.0613098	.001344086

TABLE VIII.—*Continued.*

No.	Squares.	Cubes.	Square Roots.	Cube Roots.	Reciprocals.
745	555025	413493625	27.2946881	9.0653677	.001342282
746	556516	415160936	27.3130006	9.0694220	.001340483
747	558009	416832723	27.3313007	9.0734726	.001338688
748	559504	418508992	27.3495887	9.0775197	.001336898
749	561001	420189749	27.3678644	9.0815631	.001335113
750	562500	421875000	27.3861279	9.0856030	.001333333
751	564001	423564751	27.4043792	9.0896392	.001331558
752	565504	425259008	27.4226184	9.0936719	.001329787
753	567009	426957777	27.4408455	9.0977010	.001328021
754	568516	428661064	27.4590604	9.1017265	.001326260
755	570025	430368875	27.4772632	9.1057485	.001324503
756	571536	432081216	27.4954542	9.1097669	.001322751
757	573049	433798093	27.5136330	9.1137818	.001321004
758	574564	435519512	27.5317998	9.1177931	.001319261
759	576081	437245479	27.5499546	9.1218010	.001317523
760	577600	438976000	27.5680975	9.1258053	.001315789
761	579121	440711081	27.5862284	9.1298061	.001314060
762	580644	442450728	27.6043475	9.1338034	.001312336
763	582169	444194947	27.6224546	9.1377971	.001310616
764	583696	445943744	27.6405499	9.1417874	.001308901
765	585225	447697125	27.6586334	9.1457742	.001307190
766	586756	449455096	27.6767050	9.1497576	.001305483
767	588289	451217663	27.6947648	9.1537375	.001303781
768	589824	452984832	27.7128129	9.1577139	.001302083
769	591361	454756609	27.7308492	9.1616869	.001300390
770	592900	456533000	27.7488739	9.1656565	.001298701
771	594441	458314011	27.7668868	9.1696225	.001297017
772	595984	460099648	27.7848880	9.1735852	.001295337
773	597529	461889917	27.8028775	9.1775445	.001293661
774	599076	463684824	27.8208555	9.1815003	.001291990
775	600625	465484375	27.8388218	9.1854527	.001290323
776	602176	467288576	27.8567766	9.1894018	.001288660
777	603729	469097433	27.8747197	9.1933474	.001287001
778	605284	470910952	27.8926514	9.1972897	.001285347
779	606841	472729139	27.9105715	9.2012286	.001283697
780	608400	474552000	27.9284801	9.2051641	.001282051
781	609961	476379541	27.9463772	9.2090962	.001280410
782	611524	478211768	27.9642629	9.2130250	.001278772
783	613089	480048687	27.9821372	9.2169505	.001277139
784	614656	481890304	28.0000000	9.2208726	.001275510
785	616225	483736625	28.0178515	9.2247914	.001273885
786	617796	485587656	28.0356915	9.2287068	.001272265
787	619369	487443403	28.0535203	9.2326189	.001270648
788	620944	489303872	28.0713377	9.2365277	.001269036
789	622521	491169069	28.0891438	9.2404333	.001267427
790	624100	493039000	28.1069386	9.2443355	.001265823
791	625681	494913671	28.1247222	9.2482344	.001264223
792	627264	496793088	28.1424946	9.2521300	.001262626
793	628849	498677257	28.1602557	9.2560224	.001261034
794	630436	500566184	28.1780056	9.2599114	.001259446
795	632025	502459875	28.1957444	9.2637973	.001257862
796	633616	504358336	28.2134720	9.2676798	.001256281
797	635209	506261573	28.2311884	9.2715592	.001254705
798	636804	508169592	28.2488938	9.2754352	.001253133
799	638401	510082399	28.2665881	9.2793081	.001251564
800	640000	512000000	28.2842712	9.2831777	.001250000
801	641601	513922401	28.3019434	9.2870440	.001248439
802	643204	515849608	28.3196045	9.2909072	.001246883
803	644809	517781627	28.3372546	9.2947671	.001245330
804	646416	519718464	28.3548938	9.2986239	.001243781
805	648025	521660125	28.3725219	9.3024775	.001242236
806	649636	523606616	28.3901391	9.3063278	.001240695

TABLE VIII.—*Continued*.

No.	Squares.	Cubes.	Square Roots.	Cube Roots.	Reciprocals.
807	651249	525557943	28.4077454	9.3101750	.001239157
808	652864	527514112	28.4253408	9.3140190	.001237624
809	654481	529475129	28.4429253	9.3178599	.001236094
810	656100	531441000	28.4604989	9.3216975	.001234568
811	657721	533411731	28.4780617	9.3255320	.001233046
812	659344	535387328	28.4956137	9.3293634	.001231527
813	660969	537367797	28.5131549	9.3331916	.001230012
814	662596	539353144	28.5306852	9.3370167	.001228501
815	664225	541343375	28.5482048	9.3408386	.001226994
816	665856	543338496	28.5657137	9.3446575	.001225490
817	667489	545338513	28.5832119	9.3484731	.001223990
818	669124	547343432	28.6006993	9.3522857	.001222494
819	670761	549353259	28.6181760	9.3560952	.001221001
820	672400	551368000	28.6356421	9.3599016	.001219512
821	674041	553387661	28.6530976	9.3637049	.001218027
822	675684	555412248	28.6705424	9.3675051	.001216545
823	677329	557441767	28.6879766	9.3713022	.001215067
824	678976	559476224	28.7054002	9.3750963	.001213592
825	680625	561515625	28.7228132	9.3788873	.001212121
826	682276	563559976	28.7402157	9.3826752	.001210654
827	683929	565609283	28.7576077	9.3864600	.001209190
828	685584	567663552	28.7749891	9.3902419	.001207729
829	687241	569722789	28.7923601	9.3940206	.001206273
830	688900	571787000	28.8007206	9.3977964	.001204819
831	690561	573856191	28.8270706	9.4015691	.001203369
832	692224	575930368	28.8444102	9.4053387	.001201923
833	693889	578009537	28.8617394	9.4091054	.001200480
834	695556	580093704	28.8790582	9.4128690	.001199041
835	697225	582182875	28.8963666	9.4166297	.001197605
836	698896	584277056	28.9136646	9.4203873	.001196172
837	700569	586376253	28.9309523	9.4241420	.001194743
838	702244	588480472	28.9482297	9.4278936	.001193317
839	703921	590589719	28.9654967	9.4316423	.001191895
840	705600	592704000	28.9827535	9.4353880	.001190476
841	707281	594823321	29.0000000	9.4391307	.001189061
842	708964	596947688	29.0172363	9.4428704	.001187648
843	710649	599077107	29.0344623	9.4466072	.001186240
844	712336	601211584	29.0516781	9.4503410	.001184834
845	714025	603351125	29.0688837	9.4540719	.001183432
846	715716	605495736	29.0860791	9.4577999	.001182033
847	717409	607645423	29.1032644	9.4615249	.001180638
848	719104	609800192	29.1204396	9.4652470	.001179245
849	720801	611960049	29.1376046	9.4689661	.001177856
850	722500	614125000	29.1547595	9.4726824	.001176471
851	724201	616295051	29.1719043	9.4763957	.001175088
852	725904	618470208	29.1890390	9.4801061	.001173709
853	727609	620650477	29.2061637	9.4838136	.001172333
854	729316	622835864	29.2232784	9.4875182	.001170960
855	731025	625026375	29.2403830	9.4912200	.001169591
856	732736	627222016	29.2574777	9.4949188	.001168224
857	734449	629422793	29.2745623	9.4986147	.001166861
858	736164	631628712	29.2916370	9.5023078	.001165501
859	737881	633839779	29.3087018	9.5059980	.001164144
860	739600	636056000	29.3257566	9.5096854	.001162791
861	741321	638277381	29.3428015	9.5133699	.001161440
862	743044	640503928	29.3598365	9.5170515	.001160093
863	744769	642735647	29.3768616	9.5207303	.001158749
864	746496	644972544	29.3938769	9.5244063	.001157407
865	748225	647214625	29.4108823	9.5280794	.001156069
866	749956	649461896	29.4278779	9.5317497	.001154734
867	751689	651714363	29.4448637	9.5354172	.001153403
868	753424	653972032	29.4618397	9.5390818	.001152074

TABLE VIII.—Continued.

No.	Squares.	Cubes.	Square Roots.	Cube Roots.	Reciprocals.
869	755161	656234909	29.4788059	9.5427437	.001150748
870	756900	658503000	29.4957624	9.5464027	.001149425
871	758641	660776311	29.5127091	9.5500580	.001148100
872	760384	663054848	29.5296461	9.5537123	.001146789
873	762129	665338617	29.5465734	9.5573630	.001145475
874	763876	667627624	29.5634910	9.5610108	.001144165
875	765625	669921875	29.5803989	9.5646550	.001142857
876	767376	672221376	29.5972972	9.5682982	.001141553
877	769129	674526133	29.6141858	9.5719377	.001140251
878	770884	676836152	29.6310648	9.5755745	.001138952
879	772641	679151439	29.6479342	9.5792085	.001137656
880	774400	681472000	29.6647939	9.5828397	.001136364
881	776161	683797841	29.6816442	9.5864682	.001135074
882	777924	686128968	29.6984848	9.5900939	.001133787
883	779689	688465387	29.7153159	9.5937169	.001132503
884	781456	690807104	29.7321375	9.5973373	.001131222
885	783225	693154125	29.7489496	9.6009548	.001129944
886	784996	695506456	29.7657521	9.6045696	.001128668
887	786769	697864103	29.7825452	9.6081817	.001127396
888	788544	700227072	29.7993289	9.6117911	.001126126
889	790321	702595369	29.8161030	9.6153977	.001124859
890	792100	704969000	29.8328678	9.6190017	.001123596
891	793881	707347971	29.8496231	9.6226030	.001122334
892	795664	709732288	29.8663690	9.6262016	.001121076
893	797449	712121957	29.8831056	9.6297975	.001119821
894	799236	714516984	29.8998328	9.6333907	.001118568
895	801025	716917375	29.9165506	9.6369812	.001117318
896	802816	719323136	29.9332591	9.6405690	.001116071
897	804609	721734273	29.9499583	9.6441542	.001114827
898	806404	724150792	29.9666481	9.6477367	.001113586
899	808201	726572699	29.9833287	9.6513166	.001112347
900	810000	729000000	30.0000000	9.6548938	.001111111
901	811801	731432701	30.0166620	9.6584684	.001109878
902	813604	733870808	30.0333148	9.6620403	.001108647
903	815409	736314327	30.0499584	9.6656096	.001107420
904	817216	738763264	30.0665928	9.6691762	.001106195
905	819025	741217625	30.0832179	9.6727403	.001104972
906	820836	743677416	30.0998339	9.6763017	.001103753
907	822649	746142643	30.1164407	9.6798604	.001102536
908	824464	748613312	30.1330383	9.6834166	.001101322
909	826281	751089429	30.1496269	9.6869701	.001100110
910	828100	753571000	30.1662063	9.6905211	.001098901
911	829921	756058031	30.1827765	9.6940694	.001097695
912	831744	758550528	30.1993377	9.6976151	.001096491
913	833569	761048497	30.2158899	9.7011583	.001095290
914	835396	763551944	30.2324329	9.7046989	.001094092
915	837225	766060875	30.2489669	9.7082369	.001092896
916	839056	768575296	30.2654919	9.7117723	.001091703
917	840889	771095213	30.2820079	9.7153051	.001090513
918	842724	773620632	30.2985148	9.7188354	.001089325
919	844561	776151559	30.3150128	9.7223631	.001088139
920	846400	778688000	30.3315018	9.7258883	.001086957
921	848241	781229961	30.3479818	9.7294109	.001085776
922	850084	783777448	30.3644529	9.7329309	.001084599
923	851929	786330467	30.3809151	9.7364484	.001083423
924	853776	788889024	30.3973683	9.7399634	.001082251
925	855625	791453125	30.4138127	9.7434758	.001081081
926	857476	794022776	30.4302481	9.7469857	.001079914
927	859329	796597983	30.4466747	9.7504930	.001078749
928	861184	799178752	30.4630924	9.7539979	.001077586
929	863041	801765089	30.4795013	9.7575002	.001076426
930	864900	804357000	30.4959014	9.7610001	.001075269

TABLE VIII.—Continued.

No.	Squares.	Cubes.	Square Roots.	Cube Roots.	Reciprocals.
931	866761	806954491	30.5122026	9.7644974	.001074114
932	868624	809557568	30.5286750	9.7679922	.001072961
933	870489	812166237	30.5450487	9.7714845	.001071811
934	872356	814780504	30.5614136	9.7749743	.001070664
935	874225	817400375	30.5777697	9.7784616	.001069519
936	876096	820025856	30.5941171	9.7819466	.001068376
937	877969	822656953	30.6104557	9.7854288	.001067236
938	879844	825293672	30.6267857	9.7889087	.001066098
939	881721	827936019	30.6431069	9.7923861	.001064963
940	883600	830584000	30.6594194	9.7958611	.001063830
941	885481	833237621	30.6757233	9.7993336	.001062699
942	887364	835896888	30.6920185	9.8028036	.001061571
943	889249	838561807	30.7083051	9.8062711	.001060445
944	891136	841232384	30.7245830	9.8097362	.001059322
945	893025	843908625	30.7408523	9.8131989	.001058201
946	894916	846590536	30.7571130	9.8166591	.001057082
947	896809	849278123	30.7733651	9.8201169	.001055966
948	898704	851971392	30.7896086	9.8235723	.001054852
949	900601	854670349	30.8058436	9.8270252	.001053741
950	902500	857375000	30.8220700	9.8304757	.001052632
951	904401	860085351	30.8382879	9.8339238	.001051525
952	906304	862801408	30.8544972	9.8373695	.001050420
953	908209	865523177	30.8706981	9.8408127	.001049318
954	910116	868250664	30.8868904	9.8442536	.001048218
955	912025	870983875	30.9030743	9.8476920	.001047120
956	913936	873722816	30.9192497	9.8511280	.001046025
957	915849	876467493	30.9354166	9.8545617	.001044932
958	917764	879217912	30.9515751	9.8579929	.001043841
959	919681	881974079	30.9677251	9.8614218	.001042753
960	921600	884736000	30.9838668	9.8648483	.001041667
961	923521	887503681	31.0000000	9.8682724	.001040583
962	925444	890277128	31.0161248	9.8716941	.001039501
963	927369	893056347	31.0322413	9.8751135	.001038422
964	929296	895841344	31.0483494	9.8785305	.001037344
965	931225	898632125	31.0644491	9.8819451	.001036269
966	933156	901428696	31.0805405	9.8853574	.001035197
967	935089	904231063	31.0966236	9.8887673	.001034126
968	937024	907039232	31.1126984	9.8921749	.001033058
969	938961	909853209	31.1287648	9.8955801	.001031992
970	940900	912673000	31.1448230	9.8989830	.001030928
971	942841	915498611	31.1608729	9.9023835	.001029866
972	944784	918330048	31.1769145	9.9057817	.001028807
973	946729	921167317	31.1929477	9.9091776	.001027749
974	948676	924010424	31.2089731	9.9125712	.001026694
975	950625	926859375	31.2249900	9.9159624	.001025641
976	952576	929714176	31.2409987	9.9193513	.001024590
977	954529	932574833	31.2569992	9.9227379	.001023541
978	956484	935441352	31.2729915	9.9261222	.001022495
979	958441	938313739	31.2889757	9.9295042	.001021450
980	960400	941192000	31.3049517	9.9328839	.001020408
981	962361	944076141	31.3209195	9.9362613	.001019368
982	964324	946966168	31.3368792	9.9396363	.001018330
983	966289	949862087	31.3528308	9.9430092	.001017294
984	968256	952763904	31.3687743	9.9463797	.001016260
985	970225	955671625	31.3847097	9.9497479	.001015228
986	972196	958585256	31.4006369	9.9531138	.001014199
987	974169	961504803	31.4165561	9.9564775	.001013171
988	976144	964430272	31.4324673	9.9598389	.001012146
989	978121	967361669	31.4483704	9.9631981	.001011122
990	980100	970299000	31.4642654	9.9665549	.001010101
991	982081	973242271	31.4801525	9.9699095	.001009082
992	984064	976191488	31.4960315	9.9732619	.001008065

TABLE VIII.—Continued.

No.	Squares.	Cubes.	Square Roots.	Cube Roots.	Reciprocals.
993	986049	979146657	31.5119025	9.9766120	.001007049
994	988036	982107784	31.5277655	9.9799599	.001006036
995	990025	985074875	31.5436206	9.9833055	.001005025
996	992016	988047936	31.5594677	9.9866488	.001004016
997	994009	991026973	31.5753068	9.9899900	.001003009
998	996004	994011992	31.5911380	9.9933289	.001002004
999	998001	997002999	31.6069613	9.9966656	.001001001
1000	1000000	1000000000	31.6227766	10.0000000	.001000000
1001	1002001	1003003001	31.6385840	10.0033322	.0009990010
1002	1004004	1006012008	31.6543836	10.0066622	.0009980040
1003	1006009	1009027027	31.6701752	10.0099899	.0009970090
1004	1008016	1012048064	31.6859590	10.0133155	.0009960159
1005	1010025	1015075125	31.7017349	10.0166389	.0009950249
1006	1012036	1018108216	31.7175030	10.0199601	.0009940358
1007	1014049	1021147343	31.7332633	10.0232791	.0009930487
1008	1016064	1024192512	31.7490157	10.0265958	.0009920635
1009	1018081	1027243729	31.7647603	10.0299104	.0009910803
1010	1020100	1030301000	31.7804972	10.0332228	.0009900990
1011	1022121	1033364331	31.7962262	10.0365330	.0009891197
1012	1024144	1036433728	31.8119474	10.0398410	.0009881423
1013	1026169	1039509197	31.8276609	10.0431469	.0009871668
1014	1028196	1042590744	31.8433666	10.0464506	.0009861933
1015	1030225	1045678375	31.8590646	10.0497521	.0009852217
1016	1032256	1048772096	31.8747549	10.0530514	.0009842520
1017	1034289	1051871913	31.8904374	10.0563485	.0009832842
1018	1036324	1054977832	31.9061123	10.0596435	.0009823183
1019	1038361	1058089859	31.9217794	10.0629364	.0009813543
1020	1040400	1061208000	31.9374388	10.0662271	.0009803922
1021	1042441	1064332261	31.9530906	10.0695156	.0009794319
1022	1044484	1067462648	31.9687347	10.0728020	.0009784736
1023	1046529	1070599167	31.9843712	10.0760863	.0009775171
1024	1048576	1073741824	32.0000000	10.0793684	.0009765625
1025	1050625	1076890625	32.0156212	10.0826484	.0009756098
1026	1052676	1080045576	32.0312348	10.0859262	.0009746589
1027	1054729	1083206683	32.0468407	10.0892019	.0009737098
1028	1056784	1086373952	32.0624391	10.0924755	.0009727626
1029	1058841	1089547389	32.0780298	10.0957469	.0009718173
1030	1060900	1092727000	32.0936131	10.0990163	.0009708738
1031	1062961	1095912791	32.1091887	10.1022835	.0009699321
1032	1065024	1099104768	32.1247568	10.1055487	.00 9689022
1033	1067089	1102302937	32.1403173	10.1088117	.0009680542
1034	1069156	1105507304	32.1558704	10.1120726	.0009671180
1035	1071225	1108717875	32.1714159	10.1153314	.0009661836
1036	1073296	1111934656	32.1869539	10.1185882	.0009652510
1037	1075369	1115157653	32.2024844	10.1218428	.0009643202
1038	1077444	1118386872	32.2180074	10.1250953	.0009633911
1039	1079521	1121622319	32.2335229	10.1283457	.0009624639
1040	1081600	1124864000	32.2490310	10.1315941	.0009615385
1041	1083681	1128111921	32.2645316	10.1348403	.0009606148
1042	1085764	1131366088	32.2800248	10.1380845	.0009596929
1043	1087849	1134626507	32.2955105	10.1413266	.0009587738
1044	1089936	1137893184	32.3109888	10.1445667	.0009578544
1045	1092025	1141166125	32.3264598	10.1478047	.0009569378
1046	1094116	1144445336	32.3419233	10.1510406	.0009560229
1047	1096209	1147730823	32.3573794	10.1542744	.0009551098
1048	1098304	1151022592	32.3728281	10.1575062	.0009541985
1049	1100401	1154320649	32.3882695	10.1607359	.0009532888
1050	1102500	1157625000	32.4037035	10.1639636	.0009523810
1051	1104601	1160935651	32.4191301	10.1671893	.0009514748
1052	1106704	1164252608	32.4345495	10.1704129	.0009505703
1053	1108809	1167575877	32.4499615	10.1736344	.0009496676
1054	1110916	1170905464	32.4653662	10.1768539	.0009487666

TABLE IX.—LOGARITHM OF NUMBERS FROM 0 TO 1000.

No.	0	1	2	3	4	5	6	7	8	9
0	0	00000	30103	47712	60206	69897	77815	84510	90309	95424
10	00000	00432	00860	01284	01703	02119	02530	02928	03342	03743
11	04139	04532	04922	05307	05690	06070	06446	06819	07188	07555
12	07918	08279	08637	08990	09342	09691	10037	10380	10721	11059
13	11394	11727	12057	12385	12710	13033	13354	13672	13988	14301
14	14613	14922	15229	15533	15836	16137	16435	16732	17026	17319
15	17609	17898	18184	18469	18752	19033	19312	19590	19866	20140
16	20412	20683	20952	21219	21484	21748	22011	22272	22531	22789
17	23045	23300	23553	23805	24055	24304	24551	24797	25042	25285
18	25527	25768	26007	26245	26482	26717	26951	27184	27416	27646
19	27875	28103	28330	28556	28780	29003	29226	29447	29667	29885
20	30103	30320	30535	30749	30963	31175	31386	31597	31806	32015
21	32222	32428	32633	32838	33041	33244	33445	33646	33846	34044
22	34242	34439	34635	34830	35025	35218	35411	35603	35793	35984
23	36173	36361	36549	36736	36922	37107	37291	37475	37658	37840
24	38021	38202	38382	38561	38739	38916	39094	39270	39445	39619
25	39794	39967	40140	40312	40483	40654	40824	40993	41162	41330
26	41497	41664	41830	41996	42160	42325	42488	42651	42813	42975
27	43136	43297	43457	43616	43775	43933	44091	44248	44404	44560
28	44716	44871	45025	45179	45332	45484	45637	45788	45939	46090
29	46240	46389	46538	46687	46835	46982	47129	47276	47422	47567
30	47712	47857	48001	48144	48287	48430	48572	48714	48855	48996
31	49136	49276	49415	49554	49693	49831	49969	50106	50243	50379
32	50515	50651	50786	50920	51055	51189	51322	51455	51587	51720
33	51851	51983	52114	52244	52375	52504	52634	52763	52892	53020
34	53148	53275	53403	53529	53656	53782	53908	54033	54158	54283
35	54407	54531	54654	54777	54900	55022	55145	55267	55388	55509
36	55630	55751	55871	55991	56110	56229	56348	56467	56585	56703
37	56820	56937	57054	57171	57287	57403	57519	57634	57749	57863
38	57978	58093	58206	58320	58433	58546	58659	58771	58883	58995
39	59106	59218	59328	59439	59550	59660	59770	59879	59989	60097
40	60206	60314	60423	60531	60638	60745	60853	60959	61066	61172
41	61278	61384	61490	61595	61700	61805	61909	62014	62118	62221
42	62325	62428	62531	62634	62737	62839	62941	63043	63144	63246
43	63347	63448	63548	63649	63749	63849	63949	64048	64147	64246
44	64345	64444	64542	64640	64738	64836	64933	65031	65128	65225
45	65321	65418	65514	65609	65706	65801	65896	65992	66087	66181
46	66276	66370	66404	66558	66652	66745	66839	66932	67025	67117
47	67210	67302	67394	67486	67578	67669	67761	67852	67943	68034
48	68124	68215	68305	68395	68485	68574	68664	68753	68842	68931
49	69020	69108	69197	69285	69373	69461	69548	69636	69723	69810
50	69897	69984	70070	70157	70243	70329	70415	70501	70586	70672

TABLE IX—*Continued.*—LOGARITHM OF NUMBERS FROM 0 TO 1000.

No.	0	1	2	3	4	5	6	7	8	9
51	70757	70842	70927	71012	71096	71181	71265	71349	71433	71517
52	71600	71684	71767	71850	71933	72016	72099	72181	72263	72346
53	72428	72509	72591	72673	72754	72835	72916	72997	73078	73159
54	73239	73320	73390	73480	73560	73639	73719	73799	73878	73957
55	74036	74115	74194	74273	74351	74429	74507	74586	74663	74741
56	74819	74896	74974	75051	75128	75205	75282	75358	75435	75511
57	75587	75664	75740	75815	75891	75967	76042	76118	76193	76268
58	76343	76418	76492	76567	76641	76716	76790	76864	76938	77012
59	77085	77159	77232	77305	77379	77452	77525	77597	77610	77743
60	77815	77887	77960	78032	78104	78176	78247	78319	78390	78462
61	78533	78604	78675	78746	78817	78888	78958	79029	79099	79169
62	79239	79309	79379	79449	79518	79588	79657	79727	79796	79865
63	79934	80003	80072	80140	80209	80277	80346	80414	80482	80550
64	80618	80686	80754	80821	80889	80956	81023	81090	81158	81224
65	81291	81358	81425	81491	81558	81624	81690	81757	81823	81889
66	81954	82020	82086	82151	82217	82282	82347	82413	82478	82543
67	82607	82672	82737	82802	82866	82930	82995	83059	83123	83187
68	83251	83315	83378	83442	83506	83569	83632	83696	83759	83822
69	83885	83948	84011	84073	84136	84198	84261	84323	84386	84448
70	84510	84572	84634	84696	84757	84819	84880	84942	85003	85065
71	85126	85187	85248	85309	85370	85431	85491	85552	85612	85673
72	85733	85794	85854	85914	85974	86034	86094	86153	86213	86273
73	86332	86392	86451	86510	86570	86629	86688	86747	86806	86864
74	86923	86982	87040	87099	87157	87216	87274	87332	87390	87448
75	87506	87564	87622	87680	87737	87795	87852	87910	87967	88024
76	88081	88138	88196	88252	88309	88366	88423	88480	88536	88593
77	88649	88705	88762	88818	88874	88930	88986	89042	89098	89154
78	89209	89265	89321	89376	89432	89487	89542	89597	89653	89708
79	89763	89818	89873	89927	89982	90037	90091	90146	90200	90255
80	90309	90363	90417	90472	90526	90580	90634	90687	90741	90795
81	90848	90902	90956	91009	91062	91116	91169	91222	91275	91328
82	91381	91434	91487	91540	91593	91645	91698	91751	91803	91855
83	91908	91960	92012	92065	92117	92169	92221	92273	92324	92376
84	92428	92480	92531	92583	92634	92686	92737	92789	92840	92891
85	92942	92993	93044	93095	93146	93197	93247	93298	93349	93399
86	93450	93500	93551	93601	93651	93702	93752	93802	93852	93902
87	93952	94002	94052	94101	94151	94201	94250	94300	94349	94398
88	94448	94498	94547	94596	94645	94694	94743	94792	94841	94890
89	94939	94988	95036	95085	95134	95182	95231	95279	95328	95376
90	95424	95472	95521	95569	95617	95665	95713	95761	95800	95856
91	95904	95952	95999	96047	96095	96142	96190	96237	96284	96332
92	96379	96426	96473	96520	96567	96614	96661	96708	96755	96802
93	96848	96895	96942	96988	97035	97081	97128	97174	97220	97267
94	97313	97359	97405	97451	97497	97543	97589	97635	97681	97727
95	97772	97818	97864	97909	97955	98000	98046	98091	98137	98182
96	98227	98272	98318	98363	98408	98453	98498	98543	98588	98632
97	98677	98722	98767	98811	98856	98900	98945	98989	99034	99078
98	99123	99167	99211	99255	99300	99344	99388	99432	99476	99520
99	99564	99607	99651	99695	99739	99782	99826	99870	99913	99957

NOTE TO TABLES OF TRIGONOMETRIC FUNCTIONS.

In the following Tables the values of Sines, Cosines, Tangents, Cotangents, Versines, and Exsecants are carried only to 5 places of decimals; the Table of Secants and Cosecants, however, is given to 7 places of decimals, and from it more accurate determinations of the Sines, etc., may be obtained, if for any special purpose they be required. For, by Secs. 231 and 232,

$$\sin A = \frac{1}{\operatorname{cosec} A}; \quad \cos A = \frac{1}{\sec A}; \quad \tan A = \frac{\sec A}{\operatorname{cosec} A};$$

$$\operatorname{vers} A = 1 - \frac{1}{\sec A}; \quad \operatorname{exsec} A = \sec A - 1; \quad \cot A = \frac{\operatorname{cosec} A}{\sec A}.$$

TABLE X.—SINES AND COSINES.

′	0° Sine	0° Cosin	1° Sine	1° Cosin	2° Sine	2° Cosin	3° Sine	3° Cosin	4° Sine	4° Cosin	
0	.00000	One.	.01745	.99985	.03490	.99939	.05234	.99863	.06976	.99756	60
1	.00029	One.	.01774	.99984	.03519	.99938	.05263	.99861	.07005	.99754	59
2	.00058	One.	.01803	.99984	.03548	.99937	.05292	.99860	.07034	.99752	58
3	.00087	One.	.01832	.99983	.03577	.99936	.05321	.99858	.07063	.99750	57
4	.00116	One.	.01862	.99983	.03606	.99935	.05350	.99857	.07092	.99748	56
5	.00145	One.	.01891	.99982	.03635	.99934	.05379	.99855	.07121	.99746	55
6	.00175	One.	.01920	.99982	.03664	.99933	.05408	.99854	.07150	.99744	54
7	.00204	One.	.01949	.99981	.03693	.99932	.05437	.99852	.07179	.99742	53
8	.00233	One.	.01978	.99980	.03723	.99931	.05466	.99851	.07208	.99740	52
9	.00262	One.	.02007	.99980	.03752	.99930	.05495	.99849	.07237	.99738	51
10	.00291	One.	.02036	.99979	.03781	.99929	.05524	.99847	.07266	.99736	50
11	.00320	.99999	.02065	.99979	.03810	.99927	.05553	.99846	.07295	.99734	49
12	.00349	.99999	.02094	.99978	.03839	.99926	.05582	.99844	.07324	.99731	48
13	.00378	.99999	.02123	.99977	.03868	.99925	.05611	.99842	.07353	.99729	47
14	.00407	.99999	.02152	.99977	.03897	.99924	.05640	.99841	.07382	.99727	46
15	.00436	.99999	.02181	.99976	.03926	.99923	.05669	.99839	.07411	.99725	45
16	.00465	.99999	.02211	.99976	.03955	.99922	.05698	.99838	.07440	.99723	44
17	.00495	.99999	.02240	.99975	.03984	.99921	.05727	.99836	.07469	.99721	43
18	.00524	.99999	.02269	.99974	.04013	.99919	.05756	.99834	.07498	.99719	42
19	.00553	.99998	.02298	.99974	.04042	.99918	.05785	.99833	.07527	.99716	41
20	.00582	.99998	.02327	.99973	.04071	.99917	.05814	.99831	.07556	.99714	40
21	.00611	.99998	.02356	.99972	.04100	.99916	.05844	.99829	.07585	.99712	39
22	.00640	.99998	.02385	.99972	.04129	.99915	.05873	.99827	.07614	.99710	38
23	.00669	.99998	.02414	.99971	.04159	.99913	.05902	.99826	.07643	.99708	37
24	.00698	.99998	.02443	.99970	.04188	.99912	.05931	.99824	.07672	.99705	36
25	.00727	.99997	.02472	.99969	.04217	.99911	.05960	.99822	.07701	.99703	35
26	.00756	.99997	.02501	.99969	.04246	.99910	.05989	.99821	.07730	.99701	34
27	.00785	.99997	.02530	.99968	.04275	.99909	.06018	.99819	.07759	.99699	33
28	.00814	.99997	.02560	.99967	.04304	.99907	.06047	.99817	.07788	.99696	32
29	.00844	.99996	.02589	.99966	.04333	.99906	.06076	.99815	.07817	.99694	31
30	.00873	.99996	.02618	.99966	.04362	.99905	.06105	.99813	.07846	.99692	30
31	.00902	.99996	.02647	.99965	.04391	.99904	.06134	.99812	.07875	.99689	29
32	.00931	.99996	.02676	.99964	.04420	.99902	.06163	.99810	.07904	.99687	28
33	.00960	.99995	.02705	.99963	.04449	.99901	.06192	.99808	.07933	.99685	27
34	.00989	.99995	.02734	.99963	.04478	.99900	.06221	.99806	.07962	.99683	26
35	.01018	.99995	.02763	.99962	.04507	.99898	.06250	.99804	.07991	.99680	25
36	.01047	.99995	.02792	.99961	.04536	.99897	.06279	.99803	.08020	.99678	24
37	.01076	.99994	.02821	.99960	.04565	.99896	.06308	.99801	.08049	.99676	23
38	.01105	.99994	.02850	.99959	.04594	.99894	.06337	.99799	.08078	.99673	22
39	.01134	.99994	.02879	.99959	.04623	.99893	.06366	.99797	.08107	.99671	21
40	.01164	.99993	.02908	.99958	.04653	.99892	.06395	.99795	.08136	.99668	20
41	.01193	.99993	.02938	.99957	.04682	.99890	.06424	.99793	.08165	.99666	19
42	.01222	.99993	.02967	.99956	.04711	.99889	.06453	.99792	.08194	.99664	18
43	.01251	.99992	.02996	.99955	.04740	.99888	.06482	.99790	.08223	.99661	17
44	.01280	.99992	.03025	.99954	.04769	.99886	.06511	.99788	.08252	.99659	16
45	.01309	.99991	.03054	.99953	.04798	.99885	.06540	.99786	.08281	.99657	15
46	.01338	.99991	.03083	.99952	.04827	.99883	.06569	.99784	.08310	.99654	14
47	.01367	.99991	.03112	.99952	.04856	.99882	.06598	.99782	.08339	.99652	13
48	.01396	.99990	.03141	.99951	.04885	.99881	.06627	.99780	.08368	.99649	12
49	.01425	.99990	.03170	.99950	.04914	.99879	.06656	.99778	.08397	.99647	11
50	.01454	.99989	.03199	.99949	.04943	.99878	.06685	.99776	.08426	.99644	10
51	.01483	.99989	.03228	.99948	.04972	.99876	.06714	.99774	.08455	.99642	9
52	.01513	.99989	.03257	.99947	.05001	.99875	.06743	.99772	.08484	.99639	8
53	.01542	.99988	.03286	.99946	.05030	.99873	.06773	.99770	.08513	.99637	7
54	.01571	.99988	.03316	.99945	.05059	.99872	.06802	.99768	.08542	.99635	6
55	.01600	.99987	.03345	.99944	.05088	.99870	.06831	.99766	.08571	.99632	5
56	.01629	.99987	.03374	.99943	.05117	.99869	.06860	.99764	.08600	.99630	4
57	.01658	.99986	.03403	.99942	.05146	.99867	.06889	.99762	.08629	.99627	3
58	.01687	.99986	.03432	.99941	.05175	.99866	.06918	.99760	.08658	.99625	2
59	.01716	.99985	.03461	.99940	.05205	.99864	.06947	.99758	.08687	.99622	1
60	.01745	.99985	.03490	.99939	.05234	.99863	.06976	.99756	.08716	.99619	0
′	Cosin	Sine	Cosin	Sine	Cosin	Sine	Cosin	Sine	Cosin	Sine	′
	89°		88°		87°		86°		85°		

285

TABLE X.—SINES AND COSINES.

′	5° Sine	5° Cosin	6° Sine	6° Cosin	7° Sine	7° Cosin	8° Sine	8° Cosin	9° Sine	9° Cosin	′
0	.08716	.99619	.10453	.99452	.12187	.99255	.13917	.99027	.15643	.98769	60
1	.08745	.99617	.10482	.99449	.12216	.99251	.13946	.99023	.15672	.98764	59
2	.08774	.99614	.10511	.99446	.12245	.99248	.13975	.99019	.15701	.98760	58
3	.08803	.99612	.10540	.99443	.12274	.99244	.14004	.99015	.15730	.98755	57
4	.08831	.99609	.10569	.99440	.12302	.99240	.14033	.99011	.15758	.98751	56
5	.08860	.99607	.10597	.99437	.12331	.99237	.14061	.99006	.15787	.98746	55
6	.08889	.99604	.10626	.99434	.12360	.99233	.14090	.99002	.15816	.98741	54
7	.08918	.99602	.10655	.99431	.12389	.99230	.14119	.98998	.15845	.98737	53
8	.08947	.99599	.10684	.99428	.12418	.99226	.14148	.98994	.15873	.98732	52
9	.08976	.99596	.10713	.99424	.12447	.99222	.14177	.98990	.15902	.98728	51
10	.09005	.99594	.10742	.99421	.12476	.99219	.14205	.98986	.15931	.98723	50
11	.09034	.99591	.10771	.99418	.12504	.99215	.14234	.98982	.15959	.98718	49
12	.09063	.99588	.10800	.99415	.12533	.99211	.14263	.98978	.15988	.98714	48
13	.09092	.99586	.10829	.99412	.12562	.99208	.14292	.98973	.16017	.98709	47
14	.09121	.99583	.10858	.99409	.12591	.99204	.14320	.98969	.16046	.98704	46
15	.09150	.99580	.10887	.99406	.12620	.99200	.14349	.98965	.16074	.98700	45
16	.09179	.99578	.10916	.99402	.12649	.99197	.14378	.98961	.16103	.98695	44
17	.09208	.99575	.10945	.99399	.12678	.99193	.14407	.98957	.16132	.98690	43
18	.09237	.99572	.10973	.99396	.12706	.99189	.14436	.98953	.16160	.98686	42
19	.09266	.99570	.11002	.99393	.12735	.99186	.14464	.98948	.16189	.98681	41
20	.09295	.99567	.11031	.99390	.12764	.99182	.14493	.98944	.16218	.98676	40
21	.09324	.99564	.11060	.99386	.12793	.99178	.14522	.98940	.16246	.98671	39
22	.09353	.99562	.11089	.99383	.12822	.99175	.14551	.98936	.16275	.98667	38
23	.09382	.99559	.11118	.99380	.12851	.99171	.14580	.98931	.16304	.98662	37
24	.09411	.99556	.11147	.99377	.12880	.99167	.14608	.98927	.16333	.98657	36
25	.09440	.99553	.11176	.99374	.12908	.99163	.14637	.98923	.16361	.98652	35
26	.09469	.99551	.11205	.99370	.12937	.99160	.14666	.98919	.16390	.98648	34
27	.09498	.99548	.11234	.99367	.12966	.99156	.14695	.98914	.16419	.98643	33
28	.09527	.99545	.11263	.99364	.12995	.99152	.14723	.98910	.16447	.98638	32
29	.09556	.99542	.11291	.99360	.13024	.99148	.14752	.98906	.16476	.98633	31
30	.09585	.99540	.11320	.99357	.13053	.99144	.14781	.98902	.16505	.98629	30
31	.09614	.99537	.11349	.99354	.13081	.99141	.14810	.98897	.16533	.98624	29
32	.09642	.99534	.11378	.99351	.13110	.99137	.14838	.98893	.16562	.98619	28
33	.09671	.99531	.11407	.99347	.13139	.99133	.14867	.98889	.16591	.98614	27
34	.09700	.99528	.11436	.99344	.13168	.99129	.14896	.98884	.16620	.98609	26
35	.09729	.99526	.11465	.99341	.13197	.99125	.14925	.98880	.16648	.98604	25
36	.09758	.99523	.11494	.99337	.13226	.99122	.14954	.98876	.16677	.98600	24
37	.09787	.99520	.11523	.99334	.13254	.99118	.14982	.98871	.16706	.98595	23
38	.09816	.99517	.11552	.99331	.13283	.99114	.15011	.98867	.16734	.98590	22
39	.09845	.99514	.11580	.99327	.13312	.99110	.15040	.98863	.16763	.98585	21
40	.09874	.99511	.11609	.99324	.13341	.99106	.15069	.98858	.16792	.98580	20
41	.09903	.99508	.11638	.99320	.13370	.99102	.15097	.98854	.16820	.98575	19
42	.09932	.99506	.11667	.99317	.13399	.99098	.15126	.98849	.16849	.98570	18
43	.09961	.99503	.11696	.99314	.13427	.99094	.15155	.98845	.16878	.98565	17
44	.09990	.99500	.11725	.99310	.13456	.99091	.15184	.98841	.16906	.98561	16
45	.10019	.99497	.11754	.99307	.13485	.99087	.15212	.98836	.16935	.98556	15
46	.10048	.99494	.11783	.99303	.13514	.99083	.15241	.98832	.16964	.98551	14
47	.10077	.99491	.11812	.99300	.13543	.99079	.15270	.98827	.16992	.98546	13
48	.10106	.99488	.11840	.99297	.13572	.99075	.15299	.98823	.17021	.98541	12
49	.10135	.99485	.11869	.99293	.13600	.99071	.15327	.98818	.17050	.98536	11
50	.10164	.99482	.11898	.99290	.13629	.99067	.15356	.98814	.17078	.98531	10
51	.10192	.99479	.11927	.99286	.13658	.99063	.15385	.98809	.17107	.98526	9
52	.10221	.99476	.11956	.99283	.13687	.99059	.15414	.98805	.17136	.98521	8
53	.10250	.99473	.11985	.99279	.13716	.99055	.15442	.98800	.17164	.98516	7
54	.10279	.99470	.12014	.99276	.13744	.99051	.15471	.98796	.17193	.98511	6
55	.10308	.99467	.12043	.99272	.13773	.99047	.15500	.98791	.17222	.98506	5
56	.10337	.99464	.12071	.99269	.13802	.99043	.15529	.98787	.17250	.98501	4
57	.10366	.99461	.12100	.99265	.13831	.99039	.15557	.98782	.17279	.98496	3
58	.10395	.99458	.12129	.99262	.13860	.99035	.15586	.98778	.17308	.98491	2
59	.10424	.99455	.12158	.99258	.13889	.99031	.15615	.98773	.17336	.98486	1
60	.10453	.99452	.12187	.99255	.13917	.99027	.15643	.98769	.17365	.98481	0
′	Cosin	Sine	Cosin	Sine	Cosin	Sine	Cosin	Sine	Cosin	Sine	′
	84°		83°		82°		81°		80°		

TABLE X.—SINES AND COSINES.

′	10°		11°		12°		13°		14°		′
	Sine	Cosin	Sine	Cosin	Sine	Cosin	Sine	Cosin	Sine	Cosin	
0	.17365	.98481	.19081	.98163	.20791	.97815	.22495	.97437	.24192	.97030	60
1	.17393	.98476	.19109	.98157	.20820	.97809	.22523	.97430	.24220	.97023	59
2	.17422	.98471	.19138	.98152	.20848	.97803	.22552	.97424	.24249	.97015	58
3	.17451	.98466	.19167	.98146	.20877	.97797	.22580	.97417	.24277	.97008	57
4	.17479	.98461	.19195	.98140	.20905	.97791	.22608	.97411	.24305	.97001	56
5	.17508	.98455	.19224	.98135	.20933	.97784	.22637	.97404	.24333	.96994	55
6	.17537	.98450	.19252	.98129	.20962	.97778	.22665	.97398	.24362	.96987	54
7	.17565	.98445	.19281	.98124	.20990	.97772	.22693	.97391	.24390	.96980	53
8	.17594	.98440	.19309	.98118	.21019	.97766	.22722	.97384	.24418	.96973	52
9	.17623	.98435	.19338	.98112	.21047	.97760	.22750	.97378	.24446	.96966	51
10	.17651	.98430	.19366	.98107	.21076	.97754	.22778	.97371	.24474	.96959	50
11	.17680	.98425	.19395	.98101	.21104	.97748	.22807	.97365	.24503	.96952	49
12	.17708	.98420	.19423	.98096	.21132	.97742	.22835	.97358	.24531	.96945	48
13	.17737	.98414	.19452	.98090	.21161	.97735	.22863	.97351	.24559	.96937	47
14	.17766	.98409	.19481	.98084	.21189	.97729	.22892	.97345	.24587	.96930	46
15	.17794	.98404	.19509	.98079	.21218	.97723	.22920	.97338	.24615	.96923	45
16	.17823	.98399	.19538	.98073	.21246	.97717	.22948	.97331	.24644	.96916	44
17	.17852	.98394	.19566	.98067	.21275	.97711	.22977	.97325	.24672	.96909	43
18	.17880	.98389	.19595	.98061	.21303	.97705	.23005	.97318	.24700	.96902	42
19	.17909	.98383	.19623	.98056	.21331	.97698	.23033	.97311	.24728	.96894	41
20	.17937	.98378	.19652	.98050	.21360	.97692	.23062	.97304	.24756	.96887	40
21	.17966	.98373	.19680	.98044	.21388	.97686	.23090	.97298	.24784	.96880	39
22	.17995	.98368	.19709	.98039	.21417	.97680	.23118	.97291	.24813	.96873	38
23	.18023	.98362	.19737	.98033	.21445	.97673	.23146	.97284	.24841	.96866	37
24	.18052	.98357	.19766	.98027	.21474	.97667	.23175	.97278	.24869	.96858	36
25	.18081	.98352	.19794	.98021	.21502	.97661	.23203	.97271	.24897	.96851	35
26	.18109	.98347	.19823	.98016	.21530	.97655	.23231	.97264	.24925	.96844	34
27	.18138	.98341	.19851	.98010	.21559	.97648	.23260	.97257	.24954	.96837	33
28	.18166	.98336	.19880	.98004	.21587	.97642	.23288	.97251	.24982	.96829	32
29	.18195	.98331	.19908	.97998	.21616	.97636	.23316	.97244	.25010	.96822	31
30	.18224	.98325	.19937	.97992	.21644	.97630	.23345	.97237	.25038	.96815	30
31	.18252	.98320	.19965	.97987	.21672	.97623	.23373	.97230	.25066	.96807	29
32	.18281	.98315	.19994	.97981	.21701	.97617	.23401	.97223	.25094	.96800	28
33	.18309	.98310	.20022	.97975	.21729	.97611	.23429	.97217	.25122	.96793	27
34	.18338	.98304	.20051	.97969	.21758	.97604	.23458	.97210	.25151	.96786	26
35	.18367	.98299	.20079	.97963	.21786	.97598	.23486	.97203	.25179	.96778	25
36	.18395	.98294	.20108	.97958	.21814	.97592	.23514	.97196	.25207	.96771	24
37	.18424	.98288	.20136	.97952	.21843	.97585	.23542	.97189	.25235	.96764	23
38	.18452	.98283	.20165	.97946	.21871	.97579	.23571	.97182	.25263	.96756	22
39	.18481	.98277	.20193	.97940	.21899	.97573	.23599	.97176	.25291	.96749	21
40	.18509	.98272	.20222	.97934	.21928	.97566	.23627	.97169	.25320	.96742	20
41	.18538	.98267	.20250	.97928	.21956	.97560	.23656	.97162	.25348	.96734	19
42	.18567	.98261	.20279	.97922	.21985	.97553	.23684	.97155	.25376	.96727	18
43	.18595	.98256	.20307	.97916	.22013	.97547	.23712	.97148	.25404	.96719	17
44	.18624	.98250	.20336	.97910	.22041	.97541	.23740	.97141	.25432	.96712	16
45	.18652	.98245	.20364	.97905	.22070	.97534	.23769	.97134	.25460	.96705	15
46	.18681	.98240	.20393	.97899	.22098	.97528	.23797	.97127	.25488	.96697	14
47	.18710	.98234	.20421	.97893	.22126	.97521	.23825	.97120	.25516	.96690	13
48	.18738	.98229	.20450	.97887	.22155	.97515	.23853	.97113	.25545	.96682	12
49	.18767	.98223	.20478	.97881	.22183	.97508	.23882	.97106	.25573	.96675	11
50	.18795	.98218	.20507	.97875	.22212	.97502	.23910	.97100	.25601	.96667	10
51	.18824	.98212	.20535	.97869	.22240	.97496	.23938	.97093	.25629	.96660	9
52	.18852	.98207	.20563	.97863	.22268	.97489	.23966	.97086	.25657	.96653	8
53	.18881	.98201	.20592	.97857	.22297	.97483	.23995	.97079	.25685	.96645	7
54	.18910	.98196	.20620	.97851	.22325	.97476	.24023	.97072	.25713	.96638	6
55	.18938	.98190	.20649	.97845	.22353	.97470	.24051	.97065	.25741	.96630	5
56	.18967	.98185	.20677	.97839	.22382	.97463	.24079	.97058	.25769	.96623	4
57	.18995	.98179	.20706	.97833	.22410	.97457	.24108	.97051	.25798	.96615	3
58	.19024	.98174	.20734	.97827	.22438	.97450	.24136	.97044	.25826	.96608	2
59	.19052	.98168	.20763	.97821	.22467	.97444	.24164	.97037	.25854	.96600	1
60	.19081	.98163	.20791	.97815	.22495	.97437	.24192	.97030	.25882	.96593	0
	Cosin	Sine	Cosin	Sine	Cosin	Sine	Cosin	Sine	Cosin	Sine	′
′	79°		78°		77°		76°		75°		

TABLE X.—SINES AND COSINES.

′	15°		16°		17°		18°		19°		′
	Sine	Cosin	Sine	Cosin	Sine	Cosin	Sine	Cosin	Sine	Cosin	
0	.25882	.96593	.27564	.96126	.29237	.95630	.30902	.95106	.32557	.94552	60
1	.25910	.96585	.27592	.96118	.29265	.95622	.30929	.95097	.32584	.94542	59
2	.25938	.96578	.27620	.96110	.29293	.95613	.30957	.95088	.32612	.94533	58
3	.25966	.96570	.27648	.96102	.29321	.95605	.30985	.95079	.32639	.94523	57
4	.25994	.96562	.27676	.96094	.29348	.95596	.31012	.95070	.32667	.94514	56
5	.26022	.96555	.27704	.96086	.29376	.95588	.31040	.95061	.32694	.94504	55
6	.26050	.96547	.27731	.96078	.29404	.95579	.31068	.95052	.32722	.94495	54
7	.26079	.96540	.27759	.96070	.29432	.95571	.31095	.95043	.32749	.94485	53
8	.26107	.96532	.27787	.96062	.29460	.95562	.31123	.95033	.32777	.94476	52
9	.26135	.96524	.27815	.96054	.29487	.95554	.31151	.95024	.32804	.94466	51
10	.26163	.96517	.27843	.96046	.29515	.95545	.31178	.95015	.32832	.94457	50
11	.26191	.96509	.27871	.96037	.29543	.95536	.31206	.95006	.32859	.94447	49
12	.26219	.96502	.27899	.96029	.29571	.95528	.31233	.94997	.32887	.94438	48
13	.26247	.96494	.27927	.96021	.29599	.95519	.31261	.94988	.32914	.94428	47
14	.26275	.96486	.27955	.96013	.29626	.95511	.31289	.94979	.32942	.94418	46
15	.26303	.96479	.27983	.96005	.29654	.95502	.31316	.94970	.32969	.94409	45
16	.26331	.96471	.28011	.95997	.29682	.95493	.31344	.94961	.32997	.94399	44
17	.26359	.96463	.28039	.95989	.29710	.95485	.31372	.94952	.33024	.94390	43
18	.26387	.96456	.28067	.95981	.29737	.95476	.31399	.94943	.33051	.94380	42
19	.26415	.96448	.28095	.95972	.29765	.95467	.31427	.94933	.33079	.94370	41
20	.26443	.96440	.28123	.95964	.29793	.95459	.31454	.94924	.33106	.94361	40
21	.26471	.96433	.28150	.95956	.29821	.95450	.31482	.94915	.33134	.94351	39
22	.26500	.96425	.28178	.95948	.29849	.95441	.31510	.94906	.33161	.94342	38
23	.26528	.96417	.28206	.95940	.29876	.95433	.31537	.94897	.33189	.94332	37
24	.26556	.96410	.28234	.95931	.29904	.95424	.31565	.94888	.33216	.94322	36
25	.26584	.96402	.28262	.95923	.29932	.95415	.31593	.94878	.33244	.94313	35
26	.26612	.96394	.28290	.95915	.29960	.95407	.31620	.94869	.33271	.94303	34
27	.26640	.96386	.28318	.95907	.29987	.95398	.31648	.94860	.33298	.94293	33
28	.26668	.96379	.28346	.95898	.30015	.95389	.31675	.94851	.33326	.94284	32
29	.26696	.96371	.28374	.95890	.30043	.95380	.31703	.94842	.33353	.94274	31
30	.26724	.96363	.28402	.95882	.30071	.95372	.31730	.94832	.33381	.94264	30
31	.26752	.96355	.28429	.95874	.30098	.95363	.31758	.94823	.33408	.94254	29
32	.26780	.96347	.28457	.95865	.30126	.95354	.31786	.94814	.33436	.94245	28
33	.26808	.96340	.28485	.95857	.30154	.95345	.31813	.94805	.33463	.94235	27
34	.26836	.96332	.28513	.95849	.30182	.95337	.31841	.94795	.33490	.94225	26
35	.26864	.96324	.28541	.95841	.30209	.95328	.31868	.94786	.33518	.94215	25
36	.26892	.96316	.28569	.95832	.30237	.95319	.31896	.94777	.33545	.94206	24
37	.26920	.96308	.28597	.95824	.30265	.95310	.31923	.94768	.33573	.94196	23
38	.26948	.96301	.28625	.95816	.30292	.95301	.31951	.94758	.33600	.94186	22
39	.26976	.96293	.28652	.95807	.30320	.95293	.31979	.94749	.33627	.94176	21
40	.27004	.96285	.28680	.95799	.30348	.95284	.32006	.94740	.33655	.94167	20
41	.27032	.96277	.28708	.95791	.30376	.95275	.32034	.94730	.33682	.94157	19
42	.27060	.96269	.28736	.95782	.30403	.95266	.32061	.94721	.33710	.94147	18
43	.27088	.96261	.28764	.95774	.30431	.95257	.32089	.94712	.33737	.94137	17
44	.27116	.96253	.28792	.95766	.30459	.95248	.32116	.94702	.33764	.94127	16
45	.27144	.96246	.28820	.95757	.30486	.95240	.32144	.94693	.33792	.94118	15
46	.27172	.96238	.28847	.95749	.30514	.95231	.32171	.94684	.33819	.94108	14
47	.27200	.96230	.28875	.95740	.30542	.95222	.32199	.94674	.33846	.94098	13
48	.27228	.96222	.28903	.95732	.30570	.95213	.32227	.94665	.33874	.94088	12
49	.27256	.96214	.28931	.95724	.30597	.95204	.32254	.94656	.33901	.94078	11
50	.27284	.96206	.28959	.95715	.30625	.95195	.32282	.94646	.33929	.94068	10
51	.27312	.96198	.28987	.95707	.30653	.95186	.32309	.94637	.33956	.94058	9
52	.27340	.96190	.29015	.95698	.30680	.95177	.32337	.94627	.33983	.94049	8
53	.27368	.96182	.29042	.95690	.30708	.95168	.32364	.94618	.34011	.94039	7
54	.27396	.96174	.29070	.95681	.30736	.95159	.32392	.94609	.34038	.94029	6
55	.27424	.96166	.29098	.95673	.30763	.95150	.32419	.94599	.34065	.94019	5
56	.27452	.96158	.29126	.95664	.30791	.95142	.32447	.94590	.34093	.94009	4
57	.27480	.96150	.29154	.95656	.30819	.95133	.32474	.94580	.34120	.93999	3
58	.27508	.96142	.29182	.95647	.30846	.95124	.32502	.94571	.34147	.93989	2
59	.27536	.96134	.29209	.95639	.30874	.95115	.32529	.94561	.34175	.93979	1
60	.27564	.96126	.29237	.95630	.30902	.95106	.32557	.94552	.34202	.93969	0
	Cosin	Sine	Cosin	Sine	Cosin	Sine	Cosin	Sine	Cosin	Sine	′
	74°		73°		72°		71°		70°		

288

TABLE X.—SINES AND COSINES.

′	20° Sine	20° Cosin	21° Sine	21° Cosin	22° Sine	22° Cosin	23° Sine	23° Cosin	24° Sine	24° Cosin	′
0	.34202	.93969	.35837	.93358	.37461	.92718	.39073	.92050	.40674	.91355	60
1	.34229	.93959	.35864	.93348	.37488	.92707	.39100	.92039	.40700	.91343	59
2	.34257	.93949	.35891	.93337	.37515	.92697	.39127	.92028	.40727	.91331	58
3	.34284	.93939	.35918	.93327	.37542	.92686	.39153	.92016	.40753	.91319	57
4	.34311	.93929	.35945	.93316	.37569	.92675	.39180	.92005	.40780	.91307	56
5	.34339	.93919	.35973	.93306	.37595	.92664	.39207	.91994	.40806	.91295	55
6	.34366	.93909	.36000	.93295	.37622	.92653	.39234	.91982	.40833	.91283	54
7	.34393	.93899	.36027	.93285	.37649	.92642	.39260	.91971	.40860	.91272	53
8	.34421	.93889	.36054	.93274	.37676	.92631	.39287	.91959	.40886	.91260	52
9	.34448	.93879	.36081	.93264	.37703	.92620	.39314	.91948	.40913	.91248	51
10	.34475	.93869	.36108	.93253	.37730	.92609	.39341	.91936	.40939	.91236	50
11	.34503	.93859	.36135	.93243	.37757	.92598	.39367	.91925	.40966	.91224	49
12	.34530	.93849	.36162	.93232	.37784	.92587	.39394	.91914	.40992	.91212	48
13	.34557	.93839	.36190	.93222	.37811	.92576	.39421	.91902	.41019	.91200	47
14	.34584	.93829	.36217	.93211	.37838	.92565	.39448	.91891	.41045	.91188	46
15	.34612	.93819	.36244	.93201	.37865	.92554	.39474	.91879	.41072	.91176	45
16	.34639	.93809	.36271	.93190	.37892	.92543	.39501	.91868	.41098	.91164	44
17	.34666	.93799	.36298	.93180	.37919	.92532	.39528	.91856	.41125	.91152	43
18	.34694	.93789	.36325	.93169	.37946	.92521	.39555	.91845	.41151	.91140	42
19	.34721	.93779	.36352	.93159	.37973	.92510	.39581	.91833	.41178	.91128	41
20	.34748	.93769	.36379	.93148	.37999	.92499	.39608	.91822	.41204	.91116	40
21	.34775	.93759	.36406	.93137	.38026	.92488	.39635	.91810	.41231	.91104	39
22	.34803	.93748	.36434	.93127	.38053	.92477	.39661	.91799	.41257	.91092	38
23	.34830	.93738	.36461	.93116	.38080	.92466	.39688	.91787	.41284	.91080	37
24	.34857	.93728	.36488	.93106	.38107	.92455	.39715	.91775	.41310	.91068	36
25	.34884	.93718	.36515	.93095	.38134	.92444	.39741	.91764	.41337	.91056	35
26	.34912	.93708	.36542	.93084	.38161	.92432	.39768	.91752	.41363	.91044	34
27	.34939	.93698	.36569	.93074	.38188	.92421	.39795	.91741	.41390	.91032	33
28	.34966	.93688	.36596	.93063	.38215	.92410	.39822	.91729	.41416	.91020	32
29	.34993	.93677	.36623	.93052	.38241	.92399	.39848	.91718	.41443	.91008	31
30	.35021	.93667	.36650	.93042	.38268	.92388	.39875	.91706	.41469	.90996	30
31	.35048	.93657	.36677	.93031	.38295	.92377	.39902	.91694	.41496	.90984	29
32	.35075	.93647	.36704	.93020	.38322	.92366	.39928	.91683	.41522	.90972	28
33	.35102	.93637	.36731	.93010	.38349	.92355	.39955	.91671	.41549	.90960	27
34	.35130	.93626	.36758	.92999	.38376	.92343	.39982	.91660	.41575	.90948	26
35	.35157	.93616	.36785	.92988	.38403	.92332	.40008	.91648	.41602	.90936	25
36	.35184	.93606	.36812	.92978	.38430	.92321	.40035	.91636	.41628	.90924	24
37	.35211	.93596	.36839	.92967	.38456	.92310	.40062	.91625	.41655	.90911	23
38	.35239	.93585	.36867	.92956	.38483	.92299	.40088	.91613	.41681	.90899	22
39	.35266	.93575	.36894	.92945	.38510	.92287	.40115	.91601	.41707	.90887	21
40	.35293	.93565	.36921	.92935	.38537	.92276	.40141	.91590	.41734	.90875	20
41	.35320	.93555	.36948	.92924	.38564	.92265	.40168	.91578	.41760	.90863	19
42	.35347	.93544	.36975	.92913	.38591	.92254	.40195	.91566	.41787	.90851	18
43	.35375	.93534	.37002	.92902	.38617	.92243	.40221	.91555	.41813	.90839	17
44	.35402	.93524	.37029	.92892	.38644	.92231	.40248	.91543	.41840	.90826	16
45	.35429	.93514	.37056	.92881	.38671	.92220	.40275	.91531	.41866	.90814	15
46	.35456	.93503	.37083	.92870	.38698	.92209	.40301	.91519	.41892	.90802	14
47	.35484	.93493	.37110	.92859	.38725	.92198	.40328	.91508	.41919	.90790	13
48	.35511	.93483	.37137	.92849	.38752	.92186	.40355	.91496	.41945	.90778	12
49	.35538	.93472	.37164	.92838	.38778	.92175	.40381	.91484	.41972	.90766	11
50	.35565	.93462	.37191	.92827	.38805	.92164	.40408	.91472	.41998	.90753	10
51	.35592	.93452	.37218	.92816	.38832	.92152	.40434	.91461	.42024	.90741	9
52	.35619	.93441	.37245	.92805	.38859	.92141	.40461	.91449	.42051	.90729	8
53	.35647	.93431	.37272	.92794	.38886	.92130	.40488	.91437	.42077	.90717	7
54	.35674	.93420	.37299	.92784	.38912	.92119	.40514	.91425	.42104	.90704	6
55	.35701	.93410	.37326	.92773	.38939	.92107	.40541	.91414	.42130	.90692	5
56	.35728	.93400	.37353	.92762	.38966	.92096	.40567	.91402	.42156	.90680	4
57	.35755	.93389	.37380	.92751	.38993	.92085	.40594	.91390	.42183	.90668	3
58	.35782	.93379	.37407	.92740	.39020	.92073	.40621	.91378	.42209	.90655	2
59	.35810	.93368	.37434	.92729	.39046	.92062	.40647	.91366	.42235	.90643	1
60	.35837	.93358	.37461	.92718	.39073	.92050	.40674	.91355	.42262	.90631	0
′	Cosin	Sine	Cosin	Sine	Cosin	Sine	Cosin	Sine	Cosin	Sine	′
	69°		68°		67°		66°		65°		

TABLE X.—SINES AND COSINES.

′	25° Sine	Cosin	26° Sine	Cosin	27° Sine	Cosin	28° Sine	Cosin	29° Sine	Cosin	′
0	.42262	.90631	.43837	.89879	.45399	.89101	.46947	.88295	.48481	.87462	60
1	.42288	.90618	.43863	.89867	.45425	.89087	.46973	.88281	.48506	.87448	59
2	.42315	.90606	.43889	.89854	.45451	.89074	.46999	.88267	.48532	.87434	58
3	.42341	.90594	.43916	.89841	.45477	.89061	.47024	.88254	.48557	.87420	57
4	.42367	.90582	.43942	.89828	.45503	.89048	.47050	.88240	.48583	.87406	56
5	.42394	.90569	.43968	.89816	.45529	.89035	.47076	.88226	.48608	.87391	55
6	.42420	.90557	.43994	.89803	.45554	.89021	.47101	.88213	.48634	.87377	54
7	.42446	.90545	.44020	.89790	.45580	.89008	.47127	.88199	.48659	.87363	53
8	.42473	.90532	.44046	.89777	.45606	.88995	.47153	.88185	.48684	.87349	52
9	.42499	.90520	.44072	.89764	.45632	.88981	.47178	.88172	.48710	.87335	51
10	.42525	.90507	.44098	.89752	.45658	.88968	.47204	.88158	.48735	.87321	50
11	.42552	.90495	.44124	.89739	.45684	.88955	.47229	.88144	.48761	.87306	49
12	.42578	.90483	.44151	.89726	.45710	.88942	.47255	.88130	.48786	.87292	48
13	.42604	.90470	.44177	.89713	.45736	.88928	.47281	.88117	.48811	.87278	47
14	.42631	.90458	.44203	.89700	.45762	.88915	.47306	.88103	.48837	.87264	46
15	.42657	.90446	.44229	.89687	.45787	.88902	.47332	.88089	.48862	.87250	45
16	.42683	.90433	.44255	.89674	.45813	.88888	.47358	.88075	.48888	.87235	44
17	.42709	.90421	.44281	.89662	.45839	.88875	.47383	.88062	.48913	.87221	43
18	.42736	.90408	.44307	.89649	.45865	.88862	.47409	.88048	.48938	.87207	42
19	.42762	.90396	.44333	.89636	.45891	.88848	.47434	.88034	.48964	.87193	41
20	.42788	.90383	.44359	.89623	.45917	.88835	.47460	.88020	.48989	.87178	40
21	.42815	.90371	.44385	.89610	.45942	.88822	.47486	.88006	.49014	.87164	39
22	.42841	.90358	.44411	.89597	.45968	.88808	.47511	.87993	.49040	.87150	38
23	.42867	.90346	.44437	.89584	.45994	.88795	.47537	.87979	.49065	.87136	37
24	.42894	.90334	.44464	.89571	.46020	.88782	.47562	.87965	.49090	.87121	36
25	.42920	.90321	.44490	.89558	.46046	.88768	.47588	.87951	.49116	.87107	35
26	.42946	.90309	.44516	.89545	.46072	.88755	.47614	.87937	.49141	.87093	34
27	.42972	.90296	.44542	.89532	.46097	.88741	.47639	.87923	.49166	.87079	33
28	.42999	.90284	.44568	.89519	.46123	.88728	.47665	.87909	.49192	.87064	32
29	.43025	.90271	.44594	.89506	.46149	.88715	.47690	.87896	.49217	.87050	31
30	.43051	.90259	.44620	.89493	.46175	.88701	.47716	.87882	.49242	.87036	30
31	.43077	.90246	.44646	.89480	.46201	.88688	.47741	.87868	.49268	.87021	29
32	.43104	.90233	.44672	.89467	.46226	.88674	.47767	.87854	.49293	.87007	28
33	.43130	.90221	.44698	.89454	.46252	.88661	.47793	.87840	.49318	.86993	27
34	.43156	.90208	.44724	.89441	.46278	.88647	.47818	.87826	.49344	.86978	26
35	.43182	.90196	.44750	.89428	.46304	.88634	.47844	.87812	.49369	.86964	25
36	.43209	.90183	.44776	.89415	.46330	.88620	.47869	.87798	.49394	.86949	24
37	.43235	.90171	.44802	.89402	.46355	.88607	.47895	.87784	.49419	.86935	23
38	.43261	.90158	.44828	.89389	.46381	.88593	.47920	.87770	.49445	.86921	22
39	.43287	.90146	.44854	.89376	.46407	.88580	.47946	.87756	.49470	.86906	21
40	.43313	.90133	.44880	.89363	.46433	.88566	.47971	.87743	.49495	.86892	20
41	.43340	.90120	.44906	.89350	.46458	.88553	.47997	.87729	.49521	.86878	19
42	.43366	.90108	.44932	.89337	.46484	.88539	.48022	.87715	.49546	.86863	18
43	.43392	.90095	.44958	.89324	.46510	.88526	.48048	.87701	.49571	.86849	17
44	.43418	.90082	.44984	.89311	.46536	.88512	.48073	.87687	.49596	.86834	16
45	.43445	.90070	.45010	.89298	.46561	.88499	.48099	.87673	.49622	.86820	15
46	.43471	.90057	.45036	.89285	.46587	.88485	.48124	.87659	.49647	.86805	14
47	.43497	.90045	.45062	.89272	.46613	.88472	.48150	.87645	.49672	.86791	13
48	.43523	.90032	.45088	.89259	.46639	.88458	.48175	.87631	.49697	.86777	12
49	.43549	.90019	.45114	.89245	.46664	.88445	.48201	.87617	.49723	.86762	11
50	.43575	.90007	.45140	.89232	.46690	.88431	.48226	.87603	.49748	.86748	10
51	.43602	.89994	.45166	.89219	.46716	.88417	.48252	.87589	.49773	.86733	9
52	.43628	.89981	.45192	.89206	.46742	.88404	.48277	.87575	.49798	.86719	8
53	.43654	.89968	.45218	.89193	.46767	.88390	.48303	.87561	.49824	.86704	7
54	.43680	.89956	.45243	.89180	.46793	.88377	.48328	.87546	.49849	.86690	6
55	.43706	.89943	.45269	.89167	.46819	.88363	.48354	.87532	.49874	.86675	5
56	.43733	.89930	.45295	.89153	.46844	.88349	.48379	.87518	.49899	.86661	4
57	.43759	.89918	.45321	.89140	.46870	.88336	.48405	.87504	.49924	.86646	3
58	.43785	.89905	.45347	.89127	.46896	.88322	.48430	.87490	.49950	.86632	2
59	.43811	.89892	.45373	.89114	.46921	.88308	.48456	.87476	.49975	.86617	1
60	.43837	.89879	.45399	.89101	.46947	.88295	.48481	.87462	.50000	.86603	0
′	Cosin	Sine	Cosin	Sine	Cosin	Sine	Cosin	Sine	Cosin	Sine	′
	64°		63°		62°		61°		60°		

TABLE X.—SINES AND COSINES.

′	30°		31°		32°		33°		34°		′
	Sine	Cosin	Sine	Cosin	Sine	Cosin	Sine	Cosin	Sine	Cosin	
0	.50000	.86603	.51504	.85717	.52992	.84805	.54464	.83867	.55919	.82904	60
1	.50025	.86588	.51529	.85702	.53017	.84789	.54488	.83851	.55943	.82887	59
2	.50050	.86573	.51554	.85687	.53041	.84774	.54513	.83835	.55968	.82871	58
3	.50076	.86559	.51579	.85672	.53066	.84759	.54537	.83819	.55992	.82855	57
4	.50101	.86544	.51604	.85657	.53091	.84743	.54561	.83804	.56016	.82839	56
5	.50126	.86530	.51628	.85642	.53115	.84728	.54586	.83788	.56040	.82822	55
6	.50151	.86515	.51653	.85627	.53140	.84712	.54610	.83772	.56064	.82806	54
7	.50176	.86501	.51678	.85612	.53164	.84697	.54635	.83756	.56088	.82790	53
8	.50201	.86486	.51703	.85597	.53189	.84681	.54659	.83740	.56112	.82773	52
9	.50227	.86471	.51728	.85582	.53214	.84666	.54683	.83724	.56136	.82757	51
10	.50252	.86457	.51753	.85567	.53238	.84650	.54708	.83708	.56160	.82741	50
11	.50277	.86442	.51778	.85551	.53263	.84635	.54732	.83692	.56184	.82724	49
12	.50302	.86427	.51803	.85536	.53288	.84619	.54756	.83676	.56208	.82708	48
13	.50327	.86413	.51828	.85521	.53312	.84604	.54781	.83660	.56232	.82692	47
14	.50352	.86398	.51852	.85506	.53337	.84588	.54805	.83645	.56256	.82675	46
15	.50377	.86384	.51877	.85491	.53361	.84573	.54829	.83629	.56280	.82659	45
16	.50403	.86369	.51902	.85476	.53386	.84557	.54854	.83613	.56305	.82643	44
17	.50428	.86354	.51927	.85461	.53411	.84542	.54878	.83597	.56329	.82626	43
18	.50453	.86340	.51952	.85446	.53435	.84526	.54902	.83581	.56353	.82610	42
19	.50478	.86325	.51977	.85431	.53460	.84511	.54927	.83565	.56377	.82593	41
20	.50503	.86310	.52002	.85416	.53484	.84495	.54951	.83549	.56401	.82577	40
21	.50528	.86295	.52026	.85401	.53509	.84480	.54975	.83533	.56425	.82561	39
22	.50553	.86281	.52051	.85385	.53534	.84464	.54999	.83517	.56449	.82544	38
23	.50578	.86266	.52076	.85370	.53558	.84448	.55024	.83501	.56473	.82528	37
24	.50603	.86251	.52101	.85355	.53583	.84433	.55048	.83485	.56497	.82511	36
25	.50628	.86237	.52126	.85340	.53607	.84417	.55072	.83469	.56521	.82495	35
26	.50654	.86222	.52151	.85325	.53632	.84402	.55097	.83453	.56545	.82478	34
27	.50679	.86207	.52175	.85310	.53656	.84386	.55121	.83437	.56569	.82462	33
28	.50704	.86192	.52200	.85294	.53681	.84370	.55145	.83421	.56593	.82446	32
29	.50729	.86178	.52225	.85279	.53705	.84355	.55169	.83405	.56617	.82429	31
30	.50754	.86163	.52250	.85264	.53730	.84339	.55194	.83389	.56641	.82413	30
31	.50779	.86148	.52275	.85249	.53754	.84324	.55218	.83373	.56665	.82396	29
32	.50804	.86133	.52299	.85234	.53779	.84308	.55242	.83356	.56689	.82380	28
33	.50829	.86119	.52324	.85218	.53804	.84292	.55266	.83340	.56713	.82363	27
34	.50854	.86104	.52349	.85203	.53828	.84277	.55291	.83324	.56736	.82347	26
35	.50879	.86089	.52374	.85188	.53853	.84261	.55315	.83308	.56760	.82330	25
36	.50904	.86074	.52399	.85173	.53877	.84245	.55339	.83292	.56784	.82314	24
37	.50929	.86059	.52423	.85157	.53902	.84230	.55363	.83276	.56808	.82297	23
38	.50954	.86045	.52448	.85142	.53926	.84214	.55388	.83260	.56832	.82281	22
39	.50979	.86030	.52473	.85127	.53951	.84198	.55412	.83244	.56856	.82264	21
40	.51004	.86015	.52498	.85112	.53975	.84182	.55436	.83228	.56880	.82248	20
41	.51029	.86000	.52522	.85096	.54000	.84167	.55460	.83212	.56904	.82231	19
42	.51054	.85985	.52547	.85081	.54024	.84151	.55484	.83195	.56928	.82214	18
43	.51079	.85970	.52572	.85066	.54049	.84135	.55509	.83179	.56952	.82198	17
44	.51104	.85956	.52597	.85051	.54073	.84120	.55533	.83163	.56976	.82181	16
45	.51129	.85941	.52621	.85035	.54097	.84104	.55557	.83147	.57000	.82165	15
46	.51154	.85926	.52646	.85020	.54122	.84088	.55581	.83131	.57024	.82148	14
47	.51179	.85911	.52671	.85005	.54146	.84072	.55605	.83115	.57047	.82132	13
48	.51204	.85896	.52696	.84989	.54171	.84057	.55630	.83098	.57071	.82115	12
49	.51229	.85881	.52720	.84974	.54195	.84041	.55654	.83082	.57095	.82098	11
50	.51254	.85866	.52745	.84959	.54220	.84025	.55678	.83066	.57119	.82082	10
51	.51279	.85851	.52770	.84943	.54244	.84009	.55702	.83050	.57143	.82065	9
52	.51304	.85836	.52794	.84928	.54269	.83994	.55726	.83034	.57167	.82048	8
53	.51329	.85821	.52819	.84913	.54293	.83978	.55750	.83017	.57191	.82032	7
54	.51354	.85806	.52844	.84897	.54317	.83962	.55775	.83001	.57215	.82015	6
55	.51379	.85792	.52869	.84882	.54342	.83946	.55799	.82985	.57238	.81999	5
56	.51404	.85777	.52893	.84866	.54366	.83930	.55823	.82969	.57262	.81982	4
57	.51429	.85762	.52918	.84851	.54391	.83915	.55847	.82953	.57286	.81965	3
58	.51454	.85747	.52943	.84836	.54415	.83899	.55871	.82936	.57310	.81949	2
59	.51479	.85732	.52967	.84820	.54440	.83883	.55895	.82920	.57334	.81932	1
60	.51504	.85717	.52992	.84805	.54464	.83867	.55919	.82904	.57358	.81915	0
	Cosin	Sine	Cosin	Sine	Cosin	Sine	Cosin	Sine	Cosin	Sine	
′	59°		58°		57°		56°		55°		′

TABLE X.—SINES AND COSINES.

,	35° Sine	35° Cosin	36° Sine	36° Cosin	37° Sine	37° Cosin	38° Sine	38° Cosin	39° Sine	39° Cosin	,
0	.57358	.81915	.58779	.80902	.60182	.79864	.61566	.78801	.62932	.77715	60
1	.57381	.81899	.58802	.80885	.60205	.79846	.61589	.78783	.62955	.77696	59
2	.57405	.81882	.58826	.80867	.60228	.79829	.61612	.78765	.62977	.77678	58
3	.57429	.81865	.58849	.80850	.60251	.79811	.61635	.78747	.63000	.77660	57
4	.57453	.81848	.58873	.80833	.60274	.79793	.61658	.78729	.63022	.77641	56
5	.57477	.81832	.58896	.80816	.60298	.79776	.61681	.78711	.63045	.77623	55
6	.57501	.81815	.58920	.80799	.60321	.79758	.61704	.78694	.63068	.77605	54
7	.57524	.81798	.58943	.80782	.60344	.79741	.61726	.78676	.63090	.77586	53
8	.57548	.81782	.58967	.80765	.60367	.79723	.61749	.78658	.63113	.77568	52
9	.57572	.81765	.58990	.80748	.60390	.79706	.61772	.78640	.63135	.77550	51
10	.57596	.81748	.59014	.80730	.60414	.79688	.61795	.78622	.63158	.77531	50
11	.57619	.81731	.59037	.80713	.60437	.79671	.61818	.78604	.63180	.77513	49
12	.57643	.81714	.59061	.80696	.60460	.79653	.61841	.78586	.63203	.77494	48
13	.57667	.81698	.59084	.80679	.60483	.79635	.61864	.78568	.63225	.77476	47
14	.57691	.81681	.59108	.80662	.60506	.79618	.61887	.78550	.63248	.77458	46
15	.57715	.81664	.59131	.80644	.60529	.79600	.61909	.78532	.63271	.77439	45
16	.57738	.81647	.59154	.80627	.60553	.79583	.61932	.78514	.63293	.77421	44
17	.57762	.81631	.59178	.80610	.60576	.79565	.61955	.78496	.63316	.77402	43
18	.57786	.81614	.59201	.80593	.60599	.79547	.61978	.78478	.63338	.77384	42
19	.57810	.81597	.59225	.80576	.60622	.79530	.62001	.78460	.63361	.77366	41
20	.57833	.81580	.59248	.80558	.60645	.79512	.62024	.78442	.63383	.77347	40
21	.57857	.81563	.59272	.80541	.60668	.79494	.62046	.78424	.63406	.77329	39
22	.57881	.81546	.59295	.80524	.60691	.79477	.62069	.78405	.63428	.77310	38
23	.57904	.81530	.59318	.80507	.60714	.79459	.62092	.78387	.63451	.77292	37
24	.57928	.81513	.59342	.80489	.60738	.79441	.62115	.78369	.63473	.77273	36
25	.57952	.81496	.59365	.80472	.60761	.79424	.62138	.78351	.63496	.77255	35
26	.57976	.81479	.59389	.80455	.60784	.79406	.62160	.78333	.63518	.77236	34
27	.57999	.81462	.59412	.80438	.60807	.79388	.62183	.78315	.63540	.77218	33
28	.58023	.81445	.59436	.80420	.60830	.79371	.62206	.78297	.63563	.77199	32
29	.58047	.81428	.59459	.80403	.60853	.79353	.62229	.78279	.63585	.77181	31
30	.58070	.81412	.59482	.80386	.60876	.79335	.62251	.78261	.63608	.77162	30
31	.58094	.81395	.59506	.80368	.60899	.79318	.62274	.78243	.63630	.77144	29
32	.58118	.81378	.59529	.80351	.60922	.79300	.62297	.78225	.63653	.77125	28
33	.58141	.81361	.59552	.80334	.60945	.79282	.62320	.78206	.63675	.77107	27
34	.58165	.81344	.59576	.80316	.60968	.79264	.62342	.78188	.63698	.77088	26
35	.58189	.81327	.59599	.80299	.60991	.79247	.62365	.78170	.63720	.77070	25
36	.58212	.81310	.59622	.80282	.61015	.79229	.62388	.78152	.63742	.77051	24
37	.58236	.81293	.59646	.80264	.61038	.79211	.62411	.78134	.63765	.77033	23
38	.58260	.81276	.59669	.80247	.61061	.79193	.62433	.78116	.63787	.77014	22
39	.58283	.81259	.59693	.80230	.61084	.79176	.62456	.78098	.63810	.76996	21
40	.58307	.81242	.59716	.80212	.61107	.79158	.62479	.78079	.63832	.76977	20
41	.58330	.81225	.59739	.80195	.61130	.79140	.62502	.78061	.63854	.76959	19
42	.58354	.81208	.59763	.80178	.61153	.79122	.62524	.78043	.63877	.76940	18
43	.58378	.81191	.59786	.80160	.61176	.79105	.62547	.78025	.63899	.76921	17
44	.58401	.81174	.59809	.80143	.61199	.79087	.62570	.78007	.63922	.76903	16
45	.58425	.81157	.59832	.80125	.61222	.79069	.62592	.77988	.63944	.76884	15
46	.58449	.81140	.59856	.80108	.61245	.79051	.62615	.77970	.63966	.76866	14
47	.58472	.81123	.59879	.80091	.61268	.79033	.62638	.77952	.63989	.76847	13
48	.58496	.81106	.59902	.80073	.61291	.79016	.62660	.77934	.64011	.76828	12
49	.58519	.81089	.59926	.80056	.61314	.78998	.62683	.77916	.64033	.76810	11
50	.58543	.81072	.59949	.80038	.61337	.78980	.62706	.77897	.64056	.76791	10
51	.58567	.81055	.59972	.80021	.61360	.78962	.62728	.77879	.64078	.76772	9
52	.58590	.81038	.59995	.80003	.61383	.78944	.62751	.77861	.64100	.76754	8
53	.58614	.81021	.60019	.79986	.61406	.78926	.62774	.77843	.64123	.76735	7
54	.58637	.81004	.60042	.79968	.61429	.78908	.62796	.77824	.64145	.76717	6
55	.58661	.80987	.60065	.79951	.61451	.78891	.62819	.77806	.64167	.76698	5
56	.58684	.80970	.60089	.79934	.61474	.78873	.62842	.77788	.64190	.76679	4
57	.58708	.80953	.60112	.79916	.61497	.78855	.62864	.77769	.64212	.76661	3
58	.58731	.80936	.60135	.79899	.61520	.78837	.62887	.77751	.64234	.76642	2
59	.58755	.80919	.60158	.79881	.61543	.78819	.62909	.77733	.64256	.76623	1
60	.58779	.80902	.60182	.79864	.61566	.78801	.62932	.77715	.64279	.76604	0
,	Cosin	Sine	Cosin	Sine	Cosin	Sine	Cosin	Sine	Cosin	Sine	,
	54°		53°		52°		51°		50°		

TABLE X.—SINES AND COSINES.

′	40°		41°		42°		43°		44°		′
	Sine	Cosin	Sine	Cosin	Sine	Cosin	Sine	Cosin	Sine	Cosin	
0	.64279	.76604	.65606	.75471	.66913	.74314	.68200	.73135	.69466	.71934	60
1	.64301	.76586	.65628	.75452	.66935	.74295	.68221	.73116	.69487	.71914	59
2	.64323	.76567	.65650	.75433	.66956	.74276	.68242	.73096	.69508	.71894	58
3	.64346	.76548	.65672	.75414	.66978	.74256	.68264	.73076	.69529	.71873	57
4	.64368	.76530	.65694	.75395	.66999	.74237	.68285	.73056	.69549	.71853	56
5	.64390	.76511	.65716	.75375	.67021	.74217	.68306	.73036	.69570	.71833	55
6	.64412	.76492	.65738	.75356	.67043	.74198	.68327	.73016	.69591	.71813	54
7	.64435	.76473	.65759	.75337	.67064	.74178	.68349	.72996	.69612	.71792	53
8	.64457	.76455	.65781	.75318	.67086	.74159	.68370	.72976	.69633	.71772	52
9	.64479	.76436	.65803	.75299	.67107	.74139	.68391	.72957	.69654	.71752	51
10	.64501	.76417	.65825	.75280	.67129	.74120	.68412	.72937	.69675	.71732	50
11	.64524	.76398	.65847	.75261	.67151	.74100	.68434	.72917	.69696	.71711	49
12	.64546	.76380	.65869	.75241	.67172	.74080	.68455	.72897	.69717	.71691	48
13	.64568	.76361	.65891	.75222	.67194	.74061	.68476	.72877	.69737	.71671	47
14	.64590	.76342	.65913	.75203	.67215	.74041	.68497	.72857	.69758	.71650	46
15	.64612	.76323	.65935	.75184	.67237	.74022	.68518	.72837	.69779	.71630	45
16	.64635	.76304	.65956	.75165	.67258	.74002	.68539	.72817	.69800	.71610	44
17	.64657	.76286	.65978	.75146	.67280	.73983	.68561	.72797	.69821	.71590	43
18	.64679	.76267	.66000	.75126	.67301	.73963	.68582	.72777	.69842	.71569	42
19	.64701	.76248	.66022	.75107	.67323	.73944	.68603	.72757	.69862	.71549	41
20	.64723	.76229	.66044	.75088	.67344	.73924	.68624	.72737	.69883	.71529	40
21	.64746	.76210	.66066	.75069	.67366	.73904	.68645	.72717	.69904	.71508	39
22	.64768	.76192	.66088	.75050	.67387	.73885	.68666	.72697	.69925	.71488	38
23	.64790	.76173	.66109	.75030	.67409	.73865	.68688	.72677	.69946	.71468	37
24	.64812	.76154	.66131	.75011	.67430	.73846	.68709	.72657	.69966	.71447	36
25	.64834	.76135	.66153	.74992	.67452	.73826	.68730	.72637	.69987	.71427	35
26	.64856	.76116	.66175	.74973	.67473	.73806	.68751	.72617	.70008	.71407	34
27	.64878	.76097	.66197	.74953	.67495	.73787	.68772	.72597	.70029	.71386	33
28	.64901	.76078	.66218	.74934	.67516	.73767	.68793	.72577	.70049	.71366	32
29	.64923	.76059	.66240	.74915	.67538	.73747	.68814	.72557	.70070	.71345	31
30	.64945	.76041	.66262	.74896	.67559	.73728	.68835	.72537	.70091	.71325	30
31	.64967	.76022	.66284	.74876	.67580	.73708	.68857	.72517	.70112	.71305	29
32	.64989	.76003	.66306	.74857	.67602	.73688	.68878	.72497	.70132	.71284	28
33	.65011	.75984	.66327	.74838	.67623	.73669	.68899	.72477	.70153	.71264	27
34	.65033	.75965	.66349	.74818	.67645	.73649	.68920	.72457	.70174	.71243	26
35	.65055	.75946	.66371	.74799	.67666	.73629	.68941	.72437	.70195	.71223	25
36	.65077	.75927	.66393	.74780	.67688	.73610	.68962	.72417	.70215	.71203	24
37	.65100	.75908	.66414	.74760	.67709	.73590	.68983	.72397	.70236	.71182	23
38	.65122	.75889	.66436	.74741	.67730	.73570	.69004	.72377	.70257	.71162	22
39	.65144	.75870	.66458	.74722	.67752	.73551	.69025	.72357	.70277	.71141	21
40	.65166	.75851	.66480	.74703	.67773	.73531	.69046	.72337	.70298	.71121	20
41	.65188	.75832	.66501	.74683	.67795	.73511	.69067	.72317	.70319	.71100	19
42	.65210	.75813	.66523	.74664	.67816	.73491	.69088	.72297	.70339	.71080	18
43	.65232	.75794	.66545	.74644	.67837	.73472	.69109	.72277	.70360	.71059	17
44	.65254	.75775	.66566	.74625	.67859	.73452	.69130	.72257	.70381	.71039	16
45	.65276	.75756	.66588	.74606	.67880	.73432	.69151	.72236	.70401	.71019	15
46	.65298	.75738	.66610	.74586	.67901	.73413	.69172	.72216	.70422	.70998	14
47	.65320	.75719	.66632	.74567	.67923	.73393	.69193	.72196	.70443	.70978	13
48	.65342	.75700	.66653	.74548	.67944	.73373	.69214	.72176	.70463	.70957	12
49	.65364	.75680	.66675	.74528	.67965	.73353	.69235	.72156	.70484	.70937	11
50	.65386	.75661	.66697	.74509	.67987	.73333	.69256	.72136	.70505	.70916	10
51	.65408	.75642	.66718	.74489	.68008	.73314	.69277	.72116	.70525	.70896	9
52	.65430	.75623	.66740	.74470	.68029	.73294	.69298	.72095	.70546	.70875	8
53	.65452	.75604	.66762	.74451	.68051	.73274	.69319	.72075	.70567	.70855	7
54	.65474	.75585	.66783	.74431	.68072	.73254	.69340	.72055	.70587	.70834	6
55	.65496	.75566	.66805	.74412	.68093	.73234	.69361	.72035	.70608	.70813	5
56	.65518	.75547	.66827	.74392	.68115	.73215	.69382	.72015	.70628	.70793	4
57	.65540	.75528	.66848	.74373	.68136	.73195	.69403	.71995	.70649	.70772	3
58	.65562	.75509	.66870	.74353	.68157	.73175	.69424	.71974	.70670	.70752	2
59	.65584	.75490	.66891	.74334	.68179	.73155	.69445	.71954	.70690	.70731	1
60	.65606	.75471	.66913	.74314	.68200	.73135	.69466	.71934	.70711	.70711	0
′	Cosin	Sine	Cosin	Sine	Cosin	Sine	Cosin	Sine	Cosin	Sine	′
	49°		48°		47°		46°		45°		

TABLE XI.—SECANTS AND COSECANTS.

Secants.

′	0°	1°	2°	3°	4°	5°	′
0	1·0000000	1·0001523	1·0006095	1·0013723	1·0024419	1·0038198	60
1	1·0000000	1·0001574	1·0006198	1·0013877	1·0024623	1·0038454	59
2	1·0000002	1·0001627	1·0006300	1·0014030	1·0024829	1·0038711	58
3	1·0000004	1·0001679	1·0006404	1·0014185	1·0025035	1·0038969	57
4	1·0000007	1·0001733	1·0006509	1·0014341	1·0025241	1·0039227	56
5	1·0000011	1·0001788	1·0006614	1·0014497	1·0025449	1·0039486	55
6	1·0000015	1·0001843	1·0006721	1·0014655	1·0025658	1·0039747	54
7	1·0000021	1·0001900	1·0006828	1·0014813	1·0025867	1·0040008	53
8	1·0000027	1·0001957	1·0006936	1·0014972	1·0026078	1·0040270	52
9	1·0000034	1·0002015	1·0007045	1·0015132	1·0026289	1·0040533	51
10	1·0000042	1·0002073	1·0007154	1·0015293	1·0026501	1·0040796	50
11	1·0000051	1·0002133	1·0007265	1·0015454	1·0026714	1·0041061	49
12	1·0000061	1·0002194	1·0007376	1·0015617	1·0026928	1·0041326	48
13	1·0000072	1·0002255	1·0007489	1·0015780	1·0027142	1·0041592	47
14	1·0000083	1·0002317	1·0007602	1·0015944	1·0027358	1·0041859	46
15	1·0000095	1·0002380	1·0007716	1·0016109	1·0027574	1·0042127	45
16	1·0000108	1·0002444	1·0007830	1·0016275	1·0027791	1·0042396	44
17	1·0000122	1·0002509	1·0007946	1·0016442	1·0028009	1·0042666	43
18	1·0000137	1·0002575	1·0008063	1·0016609	1·0028228	1·0042937	42
19	1·0000153	1·0002641	1·0008180	1·0016778	1·0028448	1·0043208	41
20	1·0000169	1·0002708	1·0008298	1·0016947	1·0028669	1·0043480	40
21	1·0000187	1·0002776	1·0008417	1·0017117	1·0028890	1·0043753	39
22	1·0000205	1·0002845	1·0008537	1·0017288	1·0029112	1·0044028	38
23	1·0000224	1·0002915	1·0008658	1·0017460	1·0029336	1·0044302	37
24	1·0000244	1·0002986	1·0008779	1·0017633	1·0029560	1·0044578	36
25	1·0000264	1·0003058	1·0008902	1·0017806	1·0029785	1·0044855	35
26	1·0000286	1·0003130	1·0009025	1·0017981	1·0030010	1·0045132	34
27	1·0000308	1·0003203	1·0009149	1·0018156	1·0030237	1·0045411	33
28	1·0000332	1·0003277	1·0009274	1·0018332	1·0030464	1·0045690	32
29	1·0000356	1·0003352	1·0009400	1·0018509	1·0030693	1·0045970	31
30	1·0000381	1·0003428	1·0009527	1·0018687	1·0030922	1·0046251	30
31	1·0000407	1·0003505	1·0009654	1·0018866	1·0031152	1·0046533	29
32	1·0000433	1·0003582	1·0009783	1·0019045	1·0031383	1·0046815	28
33	1·0000461	1·0003660	1·0009912	1·0019225	1·0031615	1·0047099	27
34	1·0000489	1·0003739	1·0010042	1·0019407	1·0031847	1·0047383	26
35	1·0000518	1·0003820	1·0010173	1·0019589	1·0032081	1·0047669	25
36	1·0000548	1·0003900	1·0010305	1·0019772	1·0032315	1·0047955	24
37	1·0000579	1·0003982	1·0010438	1·0019956	1·0032551	1·0048242	23
38	1·0000611	1·0004065	1·0010571	1·0020140	1·0032787	1·0048530	22
39	1·0000644	1·0004148	1·0010705	1·0020326	1·0033024	1·0048819	21
40	1·0000677	1·0004232	1·0010841	1·0020512	1·0033261	1·0049108	20
41	1·0000711	1·0004317	1·0010977	1·0020699	1·0033500	1·0049399	19
42	1·0000746	1·0004403	1·0011114	1·0020887	1·0033740	1·0049690	18
43	1·0000782	1·0004490	1·0011251	1·0021076	1·0033980	1·0049982	17
44	1·0000819	1·0004578	1·0011390	1·0021266	1·0034221	1·0050275	16
45	1·0000857	1·0004666	1·0011529	1·0021457	1·0034463	1·0050569	15
46	1·0000895	1·0004756	1·0011670	1·0021648	1·0034706	1·0050864	14
47	1·0000935	1·0004846	1·0011811	1·0021841	1·0034950	1·0051160	13
48	1·0000975	1·0004937	1·0011953	1·0022034	1·0035195	1·0051456	12
49	1·0001016	1·0005029	1·0012096	1·0022228	1·0035440	1·0051754	11
50	1·0001058	1·0005121	1·0012239	1·0022423	1·0035687	1·0052052	10
51	1·0001101	1·0005215	1·0012384	1·0022619	1·0035934	1·0052351	9
52	1·0001144	1·0005309	1·0012529	1·0022815	1·0036182	1·0052651	8
53	1·0001189	1·0005405	1·0012676	1·0023013	1·0036431	1·0052952	7
54	1·0001234	1·0005501	1·0012823	1·0023211	1·0036681	1·0053254	6
55	1·0001280	1·0005598	1·0012971	1·0023410	1·0036932	1·0053557	5
56	1·0001327	1·0005696	1·0013120	1·0023610	1·0037183	1·0053860	4
57	1·0001375	1·0005794	1·0013269	1·0023811	1·0037436	1·0054164	3
58	1·0001423	1·0005894	1·0013420	1·0024013	1·0037689	1·0054470	2
59	1·0001473	1·0005994	1·0013571	1·0024216	1·0037943	1·0054776	1
60	1·0001523	1·0006095	1·0013723	1·0024419	1·0038198	1·0055083	0
′	89°	88°	87°	86°	85°	84°	′

Cosecants.

TABLE XI.—SECANTS AND COSECANTS.

Secants.

′	6°	7°	8°	9°	10°	11°	′
0		1·0075098	1·0098276	1·0124651	1·0154266	1·0187167	60
1	1·0055083	1·0075459	1·0098689	1·0125118	1·0154787	1·0187743	59
2	1·0055391	1·0075820	1·0099103	1·0125586	1·0155310	1·0188321	58
3	1·0055699	1·0076182	1·0099518	1·0126055	1·0155833	1·0188899	57
4	1·0056009	1·0076545	1·0099934	1·0126524	1·0156357	1·0189478	56
5	1·0056319	1·0076908	1·0100351	1·0126995	1·0156882	1·0190059	55
6	1·0056631	1·0077273	1·0100769	1·0127466	1·0157408	1·0190640	54
7	1·0056943	1·0077639	1·0101187	1·0127939	1·0157934	1·0191222	53
8	1·0057256	1·0078005	1·0101607	1·0128413	1·0158463	1·0191805	52
9	1·0057570	1·0078372	1·0102027	1·0128886	1·0158991	1·0192389	51
10	1·0057885	1·0078741	1·0102449	1·0129361	1·0159520	1·0192973	50
11	1·0058200	1·0079110	1·0102871	1·0129837	1·0160050	1·0193559	49
12	1·0058517	1·0079480	1·0103294	1·0130314	1·0160582	1·0194146	48
13	1·0058834	1·0079851	1·0103718	1·0130791	1·0161114	1·0194734	47
14	1·0059153	1·0080222	1·0104143	1·0131270	1·0161647	1·0195322	46
15	1·0059472	1·0080595	1·0104568	1·0131750	1·0162181	1·0195912	45
16	1·0059792	1·0080968	1·0104995	1·0132230	1·0162716	1·0196502	44
17	1·0060113	1·0081343	1·0105423	1·0132711	1·0163252	1·0197093	43
18	1·0060435	1·0081718	1·0105851	1·0133194	1·0163789	1·0197685	42
19	1·0060757	1·0082094	1·0106280	1·0133677	1·0164327	1·0198279	41
20	1·0061081	1·0082471	1·0106710	1·0134161	1·0164865	1·0198873	40
21	1·0061405	1·0082849	1·0107141	1·0134646	1·0165405	1·0199468	39
22	1·0061731	1·0083228	1·0107573	1·0135132	1·0165946	1·0200064	38
23	1·0062057	1·0083607	1·0108006	1·0135618	1·0166487	1·0200661	37
24	1·0062384	1·0083988	1·0108440	1·0136106	1·0167029	1·0201259	36
25	1·0062712	1·0084369	1·0108875	1·0136595	1·0167573	1·0201858	35
26	1·0063040	1·0084752	1·0109310	1·0137084	1·0168117	1·0202457	34
27	1·0063370	1·0085135	1·0109747	1·0137574	1·0168662	1·0203058	33
28	1·0063701	1·0085519	1·0110184	1·0138066	1·0169208	1·0203660	32
29	1·0064032	1·0085904	1·0110622	1·0138558	1·0169755	1·0204262	31
30	1·0064364	1·0086290	1·0111061	1·0139051	1·0170303	1·0204866	30
31	1·0064697	1·0086676	1·0111501	1·0139545	1·0170851	1·0205470	29
32	1·0065031	1·0087064	1·0111942	1·0140040	1·0171401	1·0206075	28
33	1·0065366	1·0087452	1·0112384	1·0140536	1·0171952	1·0206682	27
34	1·0065702	1·0087842	1·0112827	1·0141033	1·0172503	1·0207289	26
35	1·0066039	1·0088232	1·0113270	1·0141530	1·0173056	1·0207897	25
36	1·0066376	1·0088623	1·0113715	1·0142029	1·0173609	1·0208506	24
37	1·0066714	1·0089015	1·0114160	1·0142528	1·0174163	1·0209116	23
38	1·0067054	1·0089408	1·0114606	1·0143028	1·0174719	1·0209727	22
39	1·0067394	1·0089802	1·0115054	1·0143530	1·0175275	1·0210339	21
40	1·0067735	1·0090196	1·0115502	1·0144032	1·0175832	1·0210952	20
41	1·0068077	1·0090592	1·0115951	1·0144535	1·0176390	1·0211566	19
42	1·0068419	1·0090988	1·0116400	1·0145039	1·0176949	1·0212180	18
43	1·0068763	1·0091386	1·0116851	1·0145544	1·0177509	1·0212796	17
44	1·0069108	1·0091784	1·0117303	1·0146050	1·0178069	1·0213413	16
45	1·0069453	1·0092183	1·0117755	1·0146556	1·0178631	1·0214030	15
46	1·0069799	1·0092583	1·0118209	1·0147064	1·0179194	1·0214649	14
47	1·0070146	1·0092984	1·0118663	1·0147573	1·0179757	1·0215268	13
48	1·0070494	1·0093386	1·0119118	1·0148082	1·0180321	1·0215888	12
49	1·0070843	1·0093788	1·0119575	1·0148592	1·0180887	1·0216510	11
50	1·0071193	1·0094192	1·0120032	1·0149103	1·0181453	1·0217132	10
51	1·0071544	1·0094596	1·0120489	1·0149616	1·0182020	1·0217755	9
52	1·0071895	1·0095001	1·0120948	1·0150129	1·0182588	1·0218379	8
53	1·0072248	1·0095408	1·0121408	1·0150643	1·0183158	1·0219004	7
54	1·0072601	1·0095815	1·0121869	1·0151158	1·0183728	1·0219630	6
55	1·0072955	1·0096223	1·0122330	1·0151673	1·0184298	1·0220257	5
56	1·0073310	1·0096631	1·0122793	1·0152190	1·0184870	1·0220885	4
57	1·0073666	1·0097041	1·0123256	1·0152708	1·0185443	1·0221514	3
58	1·0074023	1·0097452	1·0123720	1·0153226	1·0186017	1·0222144	2
59	1·0074380	1·0097863	1·0124185	1·0153746	1·0186591	1·0222774	1
60	1·0074739	1·0098276	1·0124651	1·0154266	1·0187167	1·0223406	0
	1·0075098						
′	83°	82°	81°	80°	79°	78°	′

Cosecants.

TABLE XI.—SECANTS AND COSECANTS.

Secants.

′	12°	13°	14°	15°	16°	17°	′
0	1·0223406	1·0263041	1·0306136	1·0352762	1·0402994	1·0456918	60
1	1·0224039	1·0263731	1·0306884	1·0353569	1·0403863	1·0457848	59
2	1·0224672	1·0264421	1·0307633	1·0354378	1·0404732	1·0458780	58
3	1·0225307	1·0265113	1·0308383	1·0355187	1·0405602	1·0459712	57
4	1·0225942	1·0265806	1·0309134	1·0355998	1·0406473	1·0460646	56
5	1·0226578	1·0266499	1·0309886	1·0356809	1·0407346	1·0461581	55
6	1·0227216	1·0267194	1·0310639	1·0357621	1·0408219	1·0462516	54
7	1·0227854	1·0267889	1·0311393	1·0358435	1·0409094	1·0463453	53
8	1·0228493	1·0268586	1·0312147	1·0359249	1·0409969	1·0464391	52
9	1·0229133	1·0269283	1·0312903	1·0360065	1·0410845	1·0465330	51
10	1·0229774	1·0269982	1·0313660	1·0360881	1·0411723	1·0466270	50
11	1·0230416	1·0270681	1·0314418	1·0361699	1·0412601	1·0467211	49
12	1·0231059	1·0271381	1·0315177	1·0362517	1·0413481	1·0468153	48
13	1·0231703	1·0272082	1·0315936	1·0363337	1·0414362	1·0469096	47
14	1·0232348	1·0272785	1·0316697	1·0364157	1·0415243	1·0470040	46
15	1·0232994	1·0273488	1·0317459	1·0364979	1·0416126	1·0470986	45
16	1·0233641	1·0274192	1·0318222	1·0365801	1·0417009	1·0471932	44
17	1·0234288	1·0274897	1·0318985	1·0366625	1·0417894	1·0472879	43
18	1·0234937	1·0275603	1·0319750	1·0367449	1·0418780	1·0473828	42
19	1·0235587	1·0276310	1·0320516	1·0368275	1·0419667	1·0474777	41
20	1·0236237	1·0277018	1·0321282	1·0369101	1·0420554	1·0475728	40
21	1·0236889	1·0277727	1·0322050	1·0369929	1·0421443	1·0476679	39
22	1·0237541	1·0278437	1·0322818	1·0370757	1·0422333	1·0477632	38
23	1·0238195	1·0279148	1·0323588	1·0371587	1·0423224	1·0478586	37
24	1·0238849	1·0279860	1·0324359	1·0372417	1·0424116	1·0479540	36
25	1·0239504	1·0280573	1·0325130	1·0373249	1·0425009	1·0480496	35
26	1·0240161	1·0281287	1·0325903	1·0374082	1·0425903	1·0481453	34
27	1·0240818	1·0282002	1·0326676	1·0374915	1·0426799	1·0482411	33
28	1·0241476	1·0282717	1·0327451	1·0375750	1·0427694	1·0483370	32
29	1·0242135	1·0283434	1·0328227	1·0376585	1·0428591	1·0484330	31
30	1·0242795	1·0284152	1·0329003	1·0377422	1·0429489	1·0485291	30
31	1·0243456	1·0284871	1·0329781	1·0378260	1·0430388	1·0486253	29
32	1·0244118	1·0285590	1·0330559	1·0379098	1·0431289	1·0487217	28
33	1·0244781	1·0286311	1·0331339	1·0379938	1·0432190	1·0488181	27
34	1·0245445	1·0287033	1·0332119	1·0380779	1·0433092	1·0489146	26
35	1·0246110	1·0287755	1·0332901	1·0381621	1·0433995	1·0490113	25
36	1·0246776	1·0288479	1·0333683	1·0382463	1·0434900	1·0491080	24
37	1·0247442	1·0289203	1·0334467	1·0383307	1·0435805	1·0492049	23
38	1·0248110	1·0289929	1·0335251	1·0384152	1·0436712	1·0493019	22
39	1·0248779	1·0290655	1·0336037	1·0384998	1·0437619	1·0493989	21
40	1·0249448	1·0291383	1·0336823	1·0385844	1·0438528	1·0494961	20
41	1·0250119	1·0292111	1·0337611	1·0386692	1·0439437	1·0495934	19
42	1·0250790	1·0292840	1·0338399	1·0387541	1·0440348	1·0496908	18
43	1·0251463	1·0293571	1·0339188	1·0388391	1·0441259	1·0497883	17
44	1·0252136	1·0294302	1·0339979	1·0389242	1·0442172	1·0498859	16
45	1·0252811	1·0295034	1·0340770	1·0390094	1·0443086	1·0499836	15
46	1·0253486	1·0295768	1·0341563	1·0390947	1·0444001	1·0500815	14
47	1·0254162	1·0296502	1·0342356	1·0391800	1·0444917	1·0501794	13
48	1·0254839	1·0297237	1·0343151	1·0392655	1·0445833	1·0502774	12
49	1·0255518	1·0297973	1·0343946	1·0393511	1·0446751	1·0503756	11
50	1·0256197	1·0298711	1·0344743	1·0394368	1·0447670	1·0504738	10
51	1·0256877	1·0299449	1·0345540	1·0395226	1·0448590	1·0505722	9
52	1·0257558	1·0300188	1·0346338	1·0396085	1·0449511	1·0506706	8
53	1·0258240	1·0300928	1·0347138	1·0396945	1·0450433	1·0507692	7
54	1·0258923	1·0301669	1·0347938	1·0397806	1·0451357	1·0508679	6
55	1·0259607	1·0302411	1·0348740	1·0398669	1·0452281	1·0509667	5
56	1·0260292	1·0303154	1·0349542	1·0399532	1·0453206	1·0510656	4
57	1·0260978	1·0303898	1·0350346	1·0400396	1·0454133	1·0511646	3
58	1·0261665	1·0304643	1·0351150	1·0401261	1·0455060	1·0512637	2
59	1·0262353	1·0305389	1·0351955	1·0402127	1·0455989	1·0513629	1
60	1·0263041	1·0306136	1·0352762	1·0402994	1·0456918	1·0514622	0
′	77°	76°	75°	74°	73°	72°	′

Cosecants.

TABLE XI.—SECANTS AND COSECANTS.

SECANTS.

′	18°	19°	20°	21°	22°	23°	′
0	1·0514622	1·0576207	1·0641778	1·0711450	1·0785347	1·0863604	60
1	1·0515617	1·0577267	1·0642905	1·0712647	1·0786616	1·0864946	59
2	1·0516612	1·0578329	1·0644033	1·0713844	1·0787885	1·0866289	58
3	1·0517608	1·0579390	1·0645163	1·0715013	1·0789156	1·0867634	57
4	1·0518606	1·0580453	1·0646294	1·0716244	1·0790427	1·0868979	56
5	1·0519605	1·0581517	1·0647425	1·0717445	1·0791700	1·0870326	55
6	1·0520604	1·0582583	1·0648558	1·0718647	1·0792975	1·0871675	54
7	1·0521605	1·0583649	1·0649693	1·0719851	1·0794250	1·0873024	53
8	1·0522607	1·0584717	1·0650828	1·0721056	1·0795527	1·0874375	52
9	1·0523610	1·0585786	1·0651964	1·0722262	1·0796805	1·0875727	51
10	1·0524614	1·0586855	1·0653102	1·0723469	1·0798084	1·0877080	50
11	1·0525619	1·0587926	1·0654240	1·0724678	1·0799364	1·0878435	49
12	1·0526625	1·0588999	1·0655380	1·0725887	1·0800646	1·0879791	48
13	1·0527633	1·0590072	1·0656521	1·0727098	1·0801928	1·0881148	47
14	1·0528641	1·0591146	1·0657663	1·0728310	1·0803212	1·0882506	46
15	1·0529651	1·0592221	1·0658807	1·0729523	1·0804497	1·0883866	45
16	1·0530661	1·0593298	1·0659951	1·0730737	1·0805784	1·0885226	44
17	1·0531673	1·0594376	1·0661097	1·0731953	1·0807071	1·0886589	43
18	1·0532686	1·0595454	1·0662243	1·0733170	1·0808360	1·0887952	42
19	1·0533699	1·0596534	1·0663391	1·0734388	1·0809650	1·0889317	41
20	1·0534714	1·0597615	1·0664540	1·0735607	1·0810942	1·0890682	40
21	1·0535730	1·0598697	1·0665690	1·0736827	1·0812234	1·0892050	39
22	1·0536747	1·0599781	1·0666842	1·0738048	1·0813528	1·0893418	38
23	1·0537765	1·0600865	1·0667994	1·0739271	1·0814823	1·0894788	37
24	1·0538785	1·0601951	1·0669148	1·0740495	1·0816119	1·0896159	36
25	1·0539805	1·0603037	1·0670303	1·0741720	1·0817417	1·0897531	35
26	1·0540826	1·0604125	1·0671458	1·0742946	1·0818715	1·0898904	34
27	1·0541849	1·0605214	1·0672615	1·0744173	1·0820015	1·0900279	33
28	1·0542873	1·0606304	1·0673774	1·0745403	1·0821316	1·0901655	32
29	1·0543897	1·0607395	1·0674933	1·0746631	1·0822618	1·0903032	31
30	1·0544923	1·0608487	1·0676094	1·0747862	1·0823923	1·0904411	30
31	1·0545950	1·0609580	1·0677255	1·0749095	1·0825227	1·0905791	29
32	1·0546978	1·0610675	1·0678418	1·0750328	1·0826533	1·0907172	28
33	1·0548007	1·0611770	1·0679582	1·0751562	1·0827840	1·0908554	27
34	1·0549037	1·0612867	1·0680747	1·0752798	1·0829149	1·0909938	26
35	1·0550068	1·0613965	1·0681914	1·0754035	1·0830458	1·0911323	25
36	1·0551101	1·0615064	1·0683081	1·0755273	1·0831769	1·0912709	24
37	1·0552134	1·0616164	1·0684250	1·0756512	1·0833081	1·0914097	23
38	1·0553169	1·0617265	1·0685420	1·0757753	1·0834395	1·0915485	22
39	1·0554204	1·0618367	1·0686591	1·0758995	1·0835709	1·0916876	21
40	1·0555241	1·0619471	1·0687763	1·0760237	1·0837025	1·0918267	20
41	1·0556279	1·0620575	1·0688936	1·0761481	1·0838342	1·0919659	19
42	1·0557318	1·0621681	1·0690110	1·0762727	1·0839661	1·0921053	18
43	1·0558358	1·0622788	1·0691286	1·0763973	1·0840980	1·0922448	17
44	1·0559399	1·0623896	1·0692463	1·0765221	1·0842301	1·0923845	16
45	1·0560441	1·0625005	1·0693641	1·0766470	1·0843623	1·0925243	15
46	1·0561485	1·0626115	1·0694820	1·0767720	1·0844947	1·0926642	14
47	1·0562529	1·0627227	1·0696000	1·0768971	1·0846271	1·0928042	13
48	1·0563575	1·0628339	1·0697182	1·0770224	1·0847597	1·0929444	12
49	1·0564621	1·0629453	1·0698364	1·0771477	1·0848924	1·0930846	11
50	1·0565669	1·0630568	1·0699548	1·0772732	1·0850252	1·0932251	10
51	1·0566718	1·0631684	1·0700733	1·0773988	1·0851582	1·0933656	9
52	1·0567768	1·0632801	1·0701919	1·0775246	1·0852913	1·0935063	8
53	1·0568819	1·0633919	1·0703105	1·0776504	1·0854245	1·0936471	7
54	1·0569871	1·0635038	1·0704295	1·0777764	1·0855578	1·0937880	6
55	1·0570924	1·0636158	1·0705484	1·0779025	1·0856912	1·0939291	5
56	1·0571978	1·0637280	1·0706675	1·0780287	1·0858248	1·0940702	4
57	1·0573034	1·0638403	1·0707867	1·0781550	1·0859585	1·0942116	3
58	1·0574090	1·0639527	1·0709060	1·0782815	1·0860924	1·0943530	2
59	1·0575148	1·0640652	1·0710254	1·0784080	1·0862263	1·0944946	1
60	1·0576207	1·0641778	1·0711450	1·0785347	1·0863604	1·0946363	0
′	71°	70°	69°	68°	67°	66°	′

COSECANTS.

′							′
4	1·0952644	1·1039777	1·1132315	1·1229925	1·1332719	1·1440827	55
5	1·0953467	1·1041279	1·1133929	1·1231598	1·1334478	1·1442778	55
6	1·0954892	1·1042783	1·1135546	1·1233269	1·1336238	1·1444630	54
7	1·0956318	1·1044289	1·1137103	1·1231942	1·1337999	1·1446484	53
8	1·0957746	1·1045795	1·1138692	1·1236616	1·1339762	1·1448339	52
9	1·0959174	1·1047303	1·1140282	1·1238292	1·1341527	1·1450196	51
10	1·0960604	1·1048813	1·1141874	1·1239969	1·1343293	1·1452055	50
11	1·0962036	1·1050324	1·1143467	1·1241618	1·1345060	1·1453915	49
12	1·0963468	1·1051836	1·1145062	1·1243328	1·1346829	1·1455776	48
13	1·0964902	1·1053349	1·1146658	1·1245010	1·1348600	1·1457639	47
14	1·0966337	1·1054864	1·1148255	1·1246693	1·1350372	1·1459504	46
15	1·0967774	1·1056380	1·1149854	1·1248377	1·1352146	1·1461371	45
16	1·0969212	1·1057898	1·1151454	1·1250063	1·1353921	1·1463238	44
17	1·0970651	1·1059417	1·1153056	1·1251750	1·1355697	1·1465108	43
18	1·0972091	1·1060937	1·1154659	1·1253439	1·1357476	1·1466979	42
19	1·0973533	1·1062458	1·1156263	1·1255130	1·1359255	1·1468852	41
20	1·0974976	1·1063981	1·1157869	1·1256821	1·1361036	1·1470726	40
21	1·0976420	1·1065506	1·1159476	1·1258514	1·1362819	1·1472602	39
22	1·0977866	1·1067031	1·1161084	1·1260209	1·1364603	1·1474479	38
23	1·0979313	1·1068558	1·1162694	1·1261905	1·1366389	1·1476358	37
24	1·0980761	1·1070087	1·1164306	1·1263603	1·1368176	1·1478239	36
25	1·0982211	1·1071616	1·1165919	1·1265302	1·1369965	1·1480121	35
26	1·0983662	1·1073147	1·1167533	1·1267003	1·1371755	1·1482005	34
27	1·0985114	1·1074680	1·1169148	1·1268705	1·1373547	1·1483890	33
28	1·0986568	1·1076214	1·1170766	1·1270408	1·1375341	1·1485777	32
29	1·0988023	1·1077749	1·1172384	1·1272113	1·1377135	1·1487665	31
30	1·0989479	1·1079285	1·1174004	1·1273819	1·1378932	1·1489555	30
31	1·0990936	1·1080823	1·1175625	1·1275527	1·1380730	1·1491447	29
32	1·0992395	1·1082363	1·1177248	1·1277237	1·1382529	1·1493340	28
33	1·0993855	1·1083903	1·1178872	1·1278948	1·1384330	1·1495235	27
34	1·0995317	1·1085445	1·1180498	1·1280660	1·1386133	1·1497132	26
35	1·0996779	1·1086989	1·1182124	1·1282374	1·1387937	1·1499030	25
36	1·0998243	1·1088533	1·1183753	1·1284089	1·1389742	1·1500930	24
37	1·0999709	1·1090079	1·1185383	1·1285806	1·1391550	1·1502831	23
38	1·1001175	1·1091627	1·1187014	1·1287524	1·1393358	1·1504734	22
39	1·1002644	1·1093176	1·1188647	1·1289244	1·1395169	1·1506638	21
40	1·1004113	1·1094726	1·1190281	1·1290965	1·1396980	1·1508544	20
41	1·1005584	1·1096277	1·1191916	1·1292687	1·1398794	1·1510452	19
42	1·1007056	1·1097830	1·1193553	1·1294412	1·1400608	1·1512361	18
43	1·1008529	1·1099385	1·1195191	1·1296137	1·1402425	1·1514272	17
44	1·1010004	1·1100940	1·1196831	1·1297864	1·1404243	1·1516185	16
45	1·1011480	1·1102498	1·1198472	1·1299593	1·1406062	1·1518099	15
46	1·1012957	1·1104056	1·1200115	1·1301323	1·1407883	1·1520015	14
47	1·1014436	1·1105616	1·1201759	1·1303055	1·1409706	1·1521932	13
48	1·1015916	1·1107177	1·1203405	1·1304788	1·1411530	1·1523851	12
49	1·1017397	1·1108740	1·1205051	1·1306522	1·1413356	1·1525772	11
50	1·1018879	1·1110304	1·1206700	1·1308258	1·1415183	1·1527694	10
51	1·1020363	1·1111869	1·1208350	1·1309996	1·1417012	1·1529618	9
52	1·1021849	1·1113436	1·1210001	1·1311735	1·1418842	1·1531543	8
53	1·1023335	1·1115004	1·1211653	1·1313475	1·1420674	1·1533470	7
54	1·1024823	1·1116573	1·1213308	1·1315217	1·1422507	1·1535399	6
55	1·1026313	1·1118144	1·1214963	1·1316961	1·1424342	1·1537329	5
56	1·1027803	1·1119716	1·1216620	1·1318706	1·1426179	1·1539261	4
57	1·1029295	1·1121290	1·1218278	1·1320452	1·1428017	1·1541195	3
58	1·1030789	1·1122865	1·1219938	1·1322200	1·1429857	1·1543130	2
59	1·1032283	1·1124442	1·1221600	1·1323950	1·1431698	1·1545067	1
60	1·1033779	1·1126019	1·1223262	1·1325701	1·1433541	1·1547005	0
′	65°	64°	63°	62°	61°	60°	′

COSECANTS.

TABLE XI.—SECANTS AND COSECANTS.

SECANTS.

′	30°	31°	32°	33°	34°	35°	′
0	1·1547005	1·1666334	1·1791784	1·1923833	1·2062179	1·2207746	60
1	1·1548945	1·1668374	1·1793928	1·1925886	1·2064547	1·2210233	59
2	1·1550887	1·1670416	1·1796074	1·1928142	1·2066917	1·2212723	58
3	1·1552830	1·1672459	1·1798222	1·1930399	1·2069288	1·2215215	57
4	1·1554775	1·1674504	1·1800372	1·1932658	1·2071662	1·2217708	56
5	1·1556722	1·1676551	1·1802523	1·1934918	1·2074037	1·2220204	55
6	1·1558670	1·1678599	1·1804676	1·1937181	1·2076415	1·2222702	54
7	1·1560620	1·1680649	1·1806831	1·1939446	1·2078794	1·2225202	53
8	1·1562572	1·1682701	1·1808988	1·1941712	1·2081175	1·2227703	52
9	1·1564525	1·1684755	1·1811146	1·1943980	1·2083559	1·2230207	51
10	1·1566480	1·1686810	1·1813307	1·1946251	1·2085944	1·2232713	50
11	1·1568436	1·1688867	1·1815469	1·1948523	1·2088331	1·2235222	49
12	1·1570394	1·1690926	1·1817633	1·1950796	1·2090720	1·2237732	48
13	1·1572354	1·1692986	1·1819798	1·1953072	1·2093112	1·2240244	47
14	1·1574315	1·1695048	1·1821966	1·1955350	1·2095505	1·2242758	46
15	1·1576278	1·1697112	1·1824135	1·1957629	1·2097900	1·2245274	45
16	1·1578243	1·1699178	1·1826306	1·1959911	1·2100297	1·2247793	44
17	1·1580209	1·1701245	1·1828479	1·1962194	1·2102696	1·2250313	43
18	1·1582177	1·1703314	1·1830654	1·1964479	1·2105097	1·2252836	42
19	1·1584146	1·1705385	1·1832830	1·1966767	1·2107500	1·2255361	41
20	1·1586118	1·1707457	1·1835008	1·1969056	1·2109905	1·2257887	40
21	1·1588091	1·1709531	1·1837188	1·1971346	1·2112312	1·2260416	39
22	1·1590065	1·1711607	1·1839370	1·1973639	1·2114721	1·2262947	38
23	1·1592041	1·1713685	1·1841554	1·1975934	1·2117132	1·2265480	37
24	1·1594019	1·1715764	1·1843739	1·1978230	1·2119545	1·2268015	36
25	1·1595999	1·1717845	1·1845927	1·1980529	1·2121960	1·2270552	35
26	1·1597980	1·1719928	1·1848116	1·1982829	1·2124377	1·2273091	34
27	1·1599963	1·1722013	1·1850307	1·1985131	1·2126795	1·2275633	33
28	1·1601947	1·1724099	1·1852500	1·1987435	1·2129216	1·2278176	32
29	1·1603933	1·1726187	1·1854694	1·1989741	1·2131639	1·2280722	31
30	1·1605921	1·1728277	1·1856890	1·1992049	1·2134064	1·2283269	30
31	1·1607911	1·1730368	1·1859089	1·1994359	1·2136491	1·2285819	29
32	1·1609902	1·1732462	1·1861289	1·1996671	1·2138920	1·2288371	28
33	1·1611894	1·1734557	1·1863490	1·1998985	1·2141351	1·2290924	27
34	1·1613889	1·1736653	1·1865694	1·2001300	1·2143784	1·2293480	26
35	1·1615885	1·1738752	1·1867900	1·2003618	1·2146218	1·2296039	25
36	1·1617883	1·1740852	1·1870107	1·2005937	1·2148655	1·2298599	24
37	1·1619882	1·1742954	1·1872316	1·2008258	1·2151094	1·2301161	23
38	1·1621883	1·1745058	1·1874527	1·2010582	1·2153535	1·2303725	22
39	1·1623886	1·1747163	1·1876740	1·2012907	1·2155978	1·2306292	21
40	1·1625891	1·1749270	1·1878954	1·2015234	1·2158423	1·2308861	20
41	1·1627897	1·1751379	1·1881171	1·2017563	1·2160870	1·2311432	19
42	1·1629905	1·1753490	1·1883389	1·2019894	1·2163319	1·2314004	18
43	1·1631914	1·1755603	1·1885609	1·2022226	1·2165770	1·2316579	17
44	1·1633925	1·1757717	1·1887831	1·2024561	1·2168223	1·2319156	16
45	1·1635938	1·1759833	1·1890055	1·2026898	1·2170678	1·2321736	15
46	1·1637953	1·1761951	1·1892280	1·2029236	1·2173135	1·2324317	14
47	1·1639969	1·1764070	1·1894508	1·2031577	1·2175594	1·2326900	13
48	1·1641987	1·1766191	1·1896737	1·2033919	1·2178055	1·2329486	12
49	1·1644007	1·1768314	1·1898968	1·2036264	1·2180518	1·2332074	11
50	1·1646028	1·1770439	1·1901201	1·2038610	1·2182983	1·2334664	10
51	1·1648051	1·1772566	1·1903436	1·2040958	1·2185450	1·2337256	9
52	1·1650076	1·1774694	1·1905673	1·2043308	1·2187919	1·2339850	8
53	1·1652102	1·1776824	1·1907911	1·2045660	1·2190390	1·2342446	7
54	1·1654130	1·1778956	1·1910152	1·2048014	1·2192864	1·2345044	6
55	1·1656160	1·1781089	1·1912394	1·2050370	1·2195339	1·2347645	5
56	1·1658191	1·1783225	1·1914638	1·2052728	1·2197816	1·2350248	4
57	1·1660224	1·1785362	1·1916884	1·2055088	1·2200296	1·2352853	3
58	1·1662259	1·1787501	1·1919132	1·2057450	1·2202777	1·2355459	2
59	1·1664296	1·1789642	1·1921381	1·2059814	1·2205260	1·2358069	1
60	1·1666334	1·1791784	1·1923633	1·2062179	1·2207746	1·2360680	0
′	59°	58°	57°	56°	55°	54°	′

COSECANTS.

TABLE XI.—SECANTS AND COSECANTS.

SECANTS.

′	36°	37°	38°	39°	40°	41°	′
0	1·2360680	1·2521357	1·2690182	1·2867596	1·3054073	1·3250130	60
1	1·2363293	1·2524102	1·2693067	1·2870628	1·3057261	1·3253482	59
2	1·2365909	1·2526850	1·2695955	1·2873663	1·3060451	1·3256837	58
3	1·2368526	1·2529601	1·2698845	1·2876700	1·3063644	1·3260194	57
4	1·2371146	1·2532353	1·2701737	1·2879740	1·3066839	1·3263554	56
5	1·2373768	1·2535108	1·2704632	1·2882782	1·3070038	1·3266918	55
6	1·2376393	1·2537865	1·2707529	1·2885827	1·3073239	1·3270284	54
7	1·2379019	1·2540625	1·2710429	1·2888875	1·3076442	1·3273653	53
8	1·2381647	1·2543387	1·2713331	1·2891925	1·3079649	1·3277024	52
9	1·2384278	1·2546151	1·2716235	1·2894977	1·3082858	1·3280399	51
10	1·2386911	1·2548917	1·2719142	1·2898032	1·3086069	1·3283776	50
11	1·2389546	1·2551685	1·2722052	1·2901090	1·3089284	1·3287156	49
12	1·2392183	1·2554456	1·2724963	1·2904150	1·3092501	1·3290539	48
13	1·2394823	1·2557229	1·2727877	1·2907213	1·3095720	1·3293925	47
14	1·2397464	1·2560005	1·2730794	1·2910278	1·3098943	1·3297314	46
15	1·2400108	1·2562783	1·2733712	1·2913346	1·3102168	1·3300706	45
16	1·2402754	1·2565563	1·2736634	1·2916416	1·3105396	1·3304100	44
17	1·2405402	1·2568345	1·2739557	1·2919489	1·3108696	1·3307497	43
18	1·2408052	1·2571129	1·2742484	1·2922564	1·3111860	1·3310897	42
19	1·2410704	1·2573916	1·2745412	1·2925642	1·3115095	1·3314301	41
20	1·2413359	1·2576705	1·2748343	1·2928723	1·3118334	1·3317707	40
21	1·2416016	1·2579497	1·2751276	1·2931806	1·3121575	1·3321115	39
22	1·2418675	1·2582291	1·2754212	1·2934892	1·3124820	1·3324527	38
23	1·2421336	1·2585087	1·2757151	1·2937980	1·3128066	1·3327942	37
24	1·2423999	1·2587885	1·2760091	1·2941071	1·3131316	1·3331359	36
25	1·2426665	1·2590686	1·2763034	1·2944164	1·3134568	1·3334779	35
26	1·2429333	1·2593489	1·2765980	1·2947260	1·3137823	1·3338203	34
27	1·2432003	1·2596294	1·2768928	1·2950359	1·3141081	1·3341629	33
28	1·2434675	1·2599102	1·2771878	1·2953460	1·3144341	1·3345058	32
29	1·2437349	1·2601912	1·2774831	1·2956564	1·3147604	1·3348489	31
30	1·2440026	1·2604724	1·2777787	1·2959670	1·3150870	1·3351924	30
31	1·2442704	1·2607539	1·2780744	1·2962779	1·3154139	1·3355362	29
32	1·2445385	1·2610356	1·2783705	1·2965890	1·3157410	1·3358802	28
33	1·2448069	1·2613175	1·2786667	1·2969004	1·3160684	1·3362246	27
34	1·2450754	1·2615997	1·2789633	1·2972121	1·3163961	1·3365692	26
35	1·2453442	1·2618820	1·2792600	1·2975240	1·3167240	1·3369141	25
36	1·2456131	1·2621647	1·2795570	1·2978362	1·3170523	1·3372594	24
37	1·2458823	1·2624475	1·2798543	1·2981487	1·3173808	1·3376049	23
38	1·2461518	1·2627306	1·2801518	1·2984614	1·3177096	1·3379507	22
39	1·2464214	1·2630140	1·2804495	1·2987743	1·3180386	1·3382968	21
40	1·2466913	1·2632975	1·2807475	1·2990876	1·3183680	1·3386432	20
41	1·2469614	1·2635813	1·2810457	1·2994011	1·3186976	1·3389898	19
42	1·2472317	1·2638653	1·2813442	1·2997148	1·3190274	1·3393368	18
43	1·2475022	1·2641496	1·2816430	1·3000288	1·3193576	1·3396841	17
44	1·2477730	1·2644341	1·2819419	1·3003431	1·3196881	1·3400316	16
45	1·2480440	1·2647188	1·2822412	1·3006576	1·3200188	1·3403795	15
46	1·2483152	1·2650038	1·2825407	1·3009724	1·3203498	1·3407276	14
47	1·2485866	1·2652890	1·2828404	1·3012875	1·3206810	1·3410761	13
48	1·2488583	1·2655745	1·2831404	1·3016028	1·3210126	1·3414248	12
49	1·2491302	1·2658601	1·2834406	1·3019184	1·3213444	1·3417738	11
50	1·2494023	1·2661460	1·2837411	1·3022343	1·3216765	1·3421232	10
51	1·2496746	1·2664322	1·2840418	1·3025504	1·3220089	1·3424728	9
52	1·2499471	1·2667186	1·2843428	1·3028667	1·3223416	1·3428227	8
53	1·2502199	1·2670052	1·2846440	1·3031834	1·3226745	1·3431729	7
54	1·2504929	1·2672921	1·2849455	1·3035003	1·3230078	1·3435234	6
55	1·2507661	1·2675792	1·2852472	1·3038175	1·3233413	1·3438742	5
56	1·2510396	1·2678665	1·2855492	1·3041349	1·3236750	1·3442253	4
57	1·2513133	1·2681541	1·2858514	1·3044526	1·3240091	1·3445767	3
58	1·2515872	1·2684419	1·2861539	1·3047706	1·3243435	1·3449284	2
59	1·2518613	1·2687299	1·2864566	1·3050888	1·3246781	1·3452804	1
60	1·2521357	1·2690182	1·2867596	1·3054073	1·3250130	1·3456327	0
′	53°	52°	51°	50°	49°	48°	′

COSECANTS.

TABLE XI.—SECANTS AND COSECANTS.

Secants.

′	42°	43°	44°	45°	46°	47°	′
0	1·3456327	1·3673275	1·3901636	1·4142136	1·4395565	1·4662792	60
1	1·3459853	1·3676985	1·3905543	1·4146251	1·4399904	1·4667368	59
2	1·3463382	1·3680699	1·3909453	1·4150370	1·4404246	1·4671948	58
3	1·3466914	1·3684416	1·3913366	1·4154493	1·4408592	1·4676532	57
4	1·3470449	1·3688136	1·3917283	1·4158619	1·4412941	1·4681120	56
5	1·3473987	1·3691859	1·3921203	1·4162749	1·4417295	1·4685713	55
6	1·3477528	1·3695586	1·3925127	1·4166883	1·4421652	1·4690309	54
7	1·3481072	1·3699315	1·3929054	1·4171020	1·4426013	1·4694910	53
8	1·3484619	1·3703048	1·3932985	1·4175161	1·4430379	1·4699514	52
9	1·3488168	1·3706784	1·3936918	1·4179306	1·4434748	1·4704123	51
10	1·3491721	1·3710523	1·3940856	1·4183454	1·4439120	1·4708736	50
11	1·3495277	1·3714266	1·3944796	1·4187605	1·4443497	1·4713354	49
12	1·3498836	1·3718011	1·3948740	1·4191761	1·4447878	1·4717975	48
13	1·3502398	1·3721760	1·3952688	1·4195920	1·4452262	1·4722600	47
14	1·3505963	1·3725512	1·3956639	1·4200083	1·4456651	1·4727230	46
15	1·3509531	1·3729268	1·3960593	1·4204248	1·4461043	1·4731864	45
16	1·3513102	1·3733026	1·3964551	1·4208418	1·4465439	1·4736502	44
17	1·3516677	1·3736788	1·3968512	1·4212592	1·4469839	1·4741144	43
18	1·3520254	1·3740553	1·3972477	1·4216769	1·4474243	1·4745790	42
19	1·3523834	1·3744321	1·3976445	1·4220950	1·4478651	1·4750440	41
20	1·3527417	1·3748092	1·3980416	1·4225134	1·4483063	1·4755095	40
21	1·3531003	1·3751867	1·3984391	1·4229323	1·4487478	1·4759754	39
22	1·3534593	1·3755645	1·3988369	1·4233514	1·4491898	1·4764417	38
23	1·3538185	1·3759426	1·3992351	1·4237710	1·4496322	1·4769084	37
24	1·3541780	1·3763210	1·3996336	1·4241909	1·4500749	1·4773755	36
25	1·3545379	1·3766998	1·4000325	1·4246112	1·4505181	1·4778431	35
26	1·3548980	1·3770789	1·4004317	1·4250319	1·4509616	1·4783111	34
27	1·3552585	1·3774583	1·4008313	1·4254529	1·4514055	1·4787795	33
28	1·3556193	1·3778380	1·4012312	1·4258743	1·4518498	1·4792483	32
29	1·3559803	1·3782181	1·4016315	1·4262961	1·4522946	1·4797176	31
30	1·3563417	1·3785985	1·4020321	1·4267182	1·4527397	1·4801872	30
31	1·3567034	1·3789792	1·4024330	1·4271407	1·4531852	1·4806573	29
32	1·3570654	1·3793602	1·4028343	1·4275636	1·4536311	1·4811278	28
33	1·3574277	1·3797416	1·4032360	1·4279868	1·4540774	1·4815988	27
34	1·3577903	1·3801233	1·4036380	1·4284105	1·4545241	1·4820702	26
35	1·3581532	1·3805053	1·4040403	1·4288345	1·4549712	1·4825420	25
36	1·3585164	1·3808877	1·4044430	1·4292588	1·4554187	1·4830142	24
37	1·3588800	1·3812704	1·4048461	1·4296836	1·4558666	1·4834868	23
38	1·3592438	1·3816534	1·4052494	1·4301087	1·4563149	1·4839599	22
39	1·3596080	1·3820367	1·4056532	1·4305342	1·4567636	1·4844334	21
40	1·3599725	1·3824204	1·4060573	1·4309600	1·4572127	1·4849073	20
41	1·3603372	1·3828044	1·4064617	1·4313863	1·4576621	1·4853817	19
42	1·3607023	1·3831887	1·4068665	1·4318129	1·4581120	1·4858565	18
43	1·3610677	1·3835734	1·4072717	1·4322399	1·4585623	1·4863317	17
44	1·3614334	1·3839584	1·4076772	1·4326672	1·4590130	1·4868073	16
45	1·3617995	1·3843437	1·4080831	1·4330950	1·4594641	1·4872834	15
46	1·3621658	1·3847294	1·4084893	1·4335231	1·4599156	1·4877599	14
47	1·3625324	1·3851153	1·4088958	1·4339516	1·4603675	1·4882369	13
48	1·3628994	1·3855017	1·4093028	1·4343805	1·4608198	1·4887142	12
49	1·3632667	1·3858883	1·4097100	1·4348097	1·4612726	1·4891920	11
50	1·3636343	1·3862753	1·4101177	1·4352393	1·4617257	1·4896703	10
51	1·3640022	1·3866626	1·4105257	1·4356693	1·4621792	1·4901489	9
52	1·3643704	1·3870503	1·4109340	1·4360997	1·4626331	1·4906280	8
53	1·3647389	1·3874383	1·4113427	1·4365305	1·4630875	1·4911076	7
54	1·3651078	1·3878266	1·4117517	1·4369616	1·4635422	1·4915876	6
55	1·3654770	1·3882153	1·4121612	1·4373933	1·4639973	1·4920680	5
56	1·3658464	1·3886043	1·4125709	1·4378251	1·4644529	1·4925488	4
57	1·3662162	1·3889936	1·4129810	1·4382574	1·4649099	1·4930301	3
58	1·3665863	1·3893832	1·4133915	1·4386900	1·4653652	1·4935118	2
59	1·3669567	1·3897733	1·4138024	1·4391231	1·4658220	1·4939940	1
60	1·3673275	1·3901636	1·4142136	1·4395565	1·4662792	1·4944765	0
′	47°	46°	45°	44°	43°	42°	′

Cosecants.

301

SECANTS.

′	48°	49°	50°	51°	52°	53°	′
0	1·4914765	1·5242531	1·5557238	1·5890157	1·6242692	1·6616401	60
1	1·4919396	1·5247634	1·5562634	1·5895868	1·6248743	1·6622819	59
2	1·4954431	1·5252741	1·5568035	1·5901584	1·6254799	1·6629243	58
3	1·4959270	1·5257854	1·5573441	1·5907306	1·6260861	1·6635673	57
4	1·4964113	1·5262971	1·5578852	1·5913033	1·6266929	1·6642110	56
5	1·4968961	1·5268093	1·5584268	1·5918766	1·6273003	1·6648553	55
6	1·4973813	1·5273219	1·5589689	1·5924504	1·6279083	1·6655002	54
7	1·4978670	1·5278351	1·5595115	1·5930247	1·6285169	1·6661458	53
8	1·4983531	1·5283487	1·5600546	1·5935996	1·6291261	1·6667920	52
9	1·4988397	1·5288627	1·5605982	1·5941751	1·6297359	1·6674389	51
10	1·4993267	1·5293773	1·5611424	1·5947511	1·6303462	1·6680864	50
11	1·4998141	1·5298923	1·5616871	1·5953276	1·6309572	1·6687345	49
12	1·5003020	1·5304078	1·5622322	1·5959048	1·6315688	1·6693833	48
13	1·5007903	1·5309238	1·5627779	1·5964824	1·6321809	1·6700328	47
14	1·5012791	1·5314403	1·5633241	1·5970606	1·6327937	1·6706828	46
15	1·5017683	1·5319572	1·5638708	1·5976394	1·6334070	1·6713336	45
16	1·5022580	1·5324746	1·5644181	1·5982187	1·6340210	1·6719850	44
17	1·5027481	1·5329925	1·5649658	1·5987986	1·6346355	1·6726370	43
18	1·5032387	1·5335109	1·5655141	1·5993790	1·6352507	1·6732897	42
19	1·5037297	1·5340297	1·5660628	1·5999600	1·6358664	1·6739430	41
20	1·5042211	1·5345491	1·5666121	1·6005416	1·6364828	1·6745970	40
21	1·5047131	1·5350689	1·5671619	1·6011237	1·6370997	1·6752517	39
22	1·5052054	1·5355893	1·5677123	1·6017064	1·6377173	1·6759070	38
23	1·5056982	1·5361100	1·5682631	1·6022896	1·6383355	1·6765629	37
24	1·5061915	1·5366313	1·5688145	1·6028734	1·6389542	1·6772195	36
25	1·5066852	1·5371530	1·5693664	1·6034577	1·6395736	1·6778768	35
26	1·5071793	1·5376752	1·5699188	1·6040426	1·6401936	1·6785347	34
27	1·5076739	1·5381980	1·5704717	1·6046281	1·6408142	1·6791933	33
28	1·5081690	1·5387212	1·5710252	1·6052142	1·6414354	1·6798525	32
29	1·5086645	1·5392449	1·5715792	1·6058008	1·6420572	1·6805124	31
30	1·5091605	1·5397690	1·5721337	1·6063870	1·6426706	1·6811730	30
31	1·5096569	1·5402937	1·5726887	1·6069757	1·6433027	1·6818342	29
32	1·5101538	1·5408189	1·5732443	1·6075640	1·6439263	1·6824961	28
33	1·5106511	1·5413445	1·5738004	1·6081528	1·6445506	1·6831586	27
34	1·5111489	1·5418706	1·5743570	1·6087423	1·6451754	1·6838219	26
35	1·5116472	1·5423973	1·5749141	1·6093323	1·6458009	1·6844857	25
36	1·5121459	1·5429244	1·5754718	1·6099228	1·6464270	1·6851503	24
37	1·5126450	1·5434520	1·5760300	1·6105140	1·6470537	1·6858155	23
38	1·5131446	1·5439801	1·5765887	1·6111057	1·6476811	1·6864814	22
39	1·5136447	1·5445087	1·5771479	1·6116980	1·6483090	1·6871479	21
40	1·5141452	1·5450378	1·5777077	1·6122908	1·6489376	1·6878151	20
41	1·5146462	1·5455673	1·5782680	1·6128843	1·6495668	1·6884830	19
42	1·5151477	1·5460974	1·5788289	1·6134783	1·6501966	1·6891516	18
43	1·5156496	1·5466280	1·5793902	1·6140728	1·6508270	1·6898208	17
44	1·5161520	1·5471590	1·5799521	1·6146680	1·6514581	1·6904907	16
45	1·5166548	1·5476906	1·5805146	1·6152637	1·6520898	1·6911613	15
46	1·5171581	1·5482226	1·5810776	1·6158600	1·6527221	1·6918326	14
47	1·5176619	1·5487552	1·5816411	1·6164569	1·6533550	1·6925045	13
48	1·5181661	1·5492882	1·5822051	1·6170544	1·6539885	1·6931771	12
49	1·5186708	1·5498218	1·5827697	1·6176524	1·6546227	1·6938504	11
50	1·5191759	1·5503558	1·5833348	1·6182510	1·6552575	1·6945244	10
51	1·5196815	1·5508904	1·5839005	1·6188502	1·6558929	1·6951990	9
52	1·5201876	1·5514254	1·5844667	1·6194500	1·6565290	1·6958744	8
53	1·5206942	1·5519610	1·5850334	1·6200504	1·6571657	1·6965504	7
54	1·5212012	1·5524970	1·5856007	1·6206513	1·6578030	1·6972271	6
55	1·5217087	1·5530335	1·5861685	1·6212528	1·6584409	1·6979044	5
56	1·5222166	1·5535706	1·5867369	1·6218549	1·6590795	1·6985825	4
57	1·5227250	1·5541081	1·5873058	1·6224576	1·6597187	1·6992612	3
58	1·5232339	1·5546462	1·5878752	1·6230609	1·6603586	1·6999407	2
59	1·5237433	1·5551848	1·5884452	1·6236648	1·6609990	1·7006208	1
60	1·5242531	1·5557238	1·5890157	1·6242692	1·6616401	1·7013016	0
′	41°	40°	39°	38°	37°	36°	′

COSECANTS.

SECANTS.

′	54°	55°	56°	57°	58°	59°	′
0	1·7013016	1·7434468	1·7882916	1·8360785	1·8870799	1·9416040	60
1	1·7019831	1·7441715	1·7890633	1·8369013	1·8879589	1·9425445	59
2	1·7026653	1·7448969	1·7898357	1·8377251	1·8888388	1·9434861	58
3	1·7033482	1·7456230	1·7906090	1·8385498	1·8897197	1·9444288	57
4	1·7040318	1·7463499	1·7913831	1·8393753	1·8906016	1·9453725	56
5	1·7047160	1·7470776	1·7921580	1·8402018	1·8914845	1·9463173	55
6	1·7054010	1·7478060	1·7929337	1·8410292	1·8923684	1·9472632	54
7	1·7060867	1·7485352	1·7937102	1·8418574	1·8932532	1·9482102	53
8	1·7067730	1·7492651	1·7944876	1·8426866	1·8941391	1·9491583	52
9	1·7074601	1·7499958	1·7952658	1·8435166	1·8950259	1·9501075	51
10	1·7081478	1·7507273	1·7960449	1·8443476	1·8959138	1·9510577	50
11	1·7088362	1·7514595	1·7968247	1·8451795	1·8968026	1·9520091	49
12	1·7095254	1·7521924	1·7976054	1·8460123	1·8976924	1·9529615	48
13	1·7102152	1·7529262	1·7983869	1·8468460	1·8985832	1·9539150	47
14	1·7109058	1·7536607	1·7991693	1·8476806	1·8994750	1·9548697	46
15	1·7115970	1·7543959	1·7999524	1·8485161	1·9003678	1·9558254	45
16	1·7122890	1·7551320	1·8007365	1·8493525	1·9012616	1·9567822	44
17	1·7129817	1·7558687	1·8015213	1·8501898	1·9021564	1·9577402	43
18	1·7136750	1·7566063	1·8023070	1·8510281	1·9030522	1·9586992	42
19	1·7143691	1·7573446	1·8030935	1·8518672	1·9039491	1·9596593	41
20	1·7150639	1·7580837	1·8038809	1·8527073	1·9048469	1·9606206	40
21	1·7157594	1·7588236	1·8046691	1·8535483	1·9057457	1·9615829	39
22	1·7164556	1·7595642	1·8054582	1·8543903	1·9066456	1·9625464	38
23	1·7171525	1·7603057	1·8062481	1·8552331	1·9075464	1·9635110	37
24	1·7178501	1·7610478	1·8070388	1·8560769	1·9084483	1·9644767	36
25	1·7185484	1·7617908	1·8078304	1·8569216	1·9093512	1·9654435	35
26	1·7192475	1·7625345	1·8086228	1·8577672	1·9102551	1·9664114	34
27	1·7199472	1·7632791	1·8094161	1·8586138	1·9111600	1·9673805	33
28	1·7206477	1·7640244	1·8102102	1·8594612	1·9120659	1·9683507	32
29	1·7213489	1·7647704	1·8110052	1·8603097	1·9129729	1·9693220	31
30	1·7220508	1·7655173	1·8118010	1·8611590	1·9138809	1·9702944	30
31	1·7227534	1·7662649	1·8125977	1·8620093	1·9147899	1·9712680	29
32	1·7234568	1·7670133	1·8133953	1·8628605	1·9156999	1·9722427	28
33	1·7241609	1·7677625	1·8141937	1·8637126	1·9166110	1·9732185	27
34	1·7248657	1·7685125	1·8149929	1·8645657	1·9175230	1·9741954	26
35	1·7255712	1·7692633	1·8157930	1·8654197	1·9184362	1·9751735	25
36	1·7262774	1·7700149	1·8165940	1·8662747	1·9193503	1·9761527	24
37	1·7269844	1·7707672	1·8173958	1·8671306	1·9202655	1·9771331	23
38	1·7276921	1·7715204	1·8181985	1·8679875	1·9211817	1·9781146	22
39	1·7284005	1·7722743	1·8190021	1·8688453	1·9220990	1·9790972	21
40	1·7291096	1·7730290	1·8198065	1·8697040	1·9230173	1·9800810	20
41	1·7298195	1·7737845	1·8206118	1·8705637	1·9239366	1·9810659	19
42	1·7305301	1·7745409	1·8214244	1·8714244	1·9248570	1·9820520	18
43	1·7312414	1·7752980	1·8222249	1·8722859	1·9257784	1·9830393	17
44	1·7319535	1·7760559	1·8230328	1·8731485	1·9267009	1·9840276	16
45	1·7326663	1·7768146	1·8238416	1·8740120	1·9276244	1·9850172	15
46	1·7333798	1·7775741	1·8246512	1·8748764	1·9285490	1·9860080	14
47	1·7340941	1·7783344	1·8254617	1·8757419	1·9294746	1·9869997	13
48	1·7348091	1·7790955	1·8262731	1·8766082	1·9304013	1·9879927	12
49	1·7355248	1·7798574	1·8270854	1·8774755	1·9313290	1·9889869	11
50	1·7362413	1·7806201	1·8278985	1·8783438	1·9322578	1·9899822	10
51	1·7369585	1·7813836	1·8287125	1·8792131	1·9331876	1·9909787	9
52	1·7376764	1·7821479	1·8295274	1·8800833	1·9341185	1·9919764	8
53	1·7383951	1·7829131	1·8303432	1·8809545	1·9350505	1·9929752	7
54	1·7391145	1·7836790	1·8311599	1·8818266	1·9359835	1·9939753	6
55	1·7398347	1·7844457	1·8319774	1·8826998	1·9369176	1·9949764	5
56	1·7405556	1·7852133	1·8327959	1·8835738	1·9378527	1·9959788	4
57	1·7412773	1·7859817	1·8336152	1·8844489	1·9387889	1·9969823	3
58	1·7419997	1·7867508	1·8344354	1·8853249	1·9397262	1·9979870	2
59	1·7427229	1·7875208	1·8352565	1·8862019	1·9406646	1·9989929	1
60	1·7434468	1·7882916	1·8360785	1·8870799	1·9416040	2·0000000	0
′	35°	34°	33°	32°	31°	30°	′

COSECANTS.

′							′
5	2·0050533	2·0680940	2·1358993	2·2089972	2·2879974	2·3736075	55
6	2·0060674	2·0691836	2·1370726	2·2102637	2·2893679	2·3750949	54
7	2·0070823	2·0702746	2·1382475	2·2115318	2·2907403	2·3765843	53
8	2·0080994	2·0713670	2·1394238	2·2128016	2·2921145	2·3780758	52
9	2·0091172	2·0724606	2·1406015	2·2140730	2·2934906	2·3795694	51
10	2·0101362	2·0735556	2·1417808	2·2153460	2·2948685	2·3810650	50
11	2·0111564	2·0746519	2·1429615	2·2166208	2·2962483	2·3825627	49
12	2·0121779	2·0757496	2·1441438	2·2178971	2·2976299	2·3840625	48
13	2·0132005	2·0768486	2·1453275	2·2191752	2·2990134	2·3855645	47
14	2·0142243	2·0779489	2·1465127	2·2204548	2·3003988	2·3870685	46
15	2·0152494	2·0790506	2·1476993	2·2217362	2·3017860	2·3885746	45
16	2·0162756	2·0801536	2·1488875	2·2230192	2·3031751	2·3900828	44
17	2·0173031	2·0812580	2·1500772	2·2243039	2·3045660	2·3915931	43
18	2·0183318	2·0823637	2·1512684	2·2255903	2·3059588	2·3931055	42
19	2·0193618	2·0834708	2·1524611	2·2268783	2·3073536	2·3946201	41
20	2·0203929	2·0845792	2·1536553	2·2281681	2·3087501	2·3961367	40
21	2·0214253	2·0856890	2·1548510	2·2294595	2·3101486	2·3976555	39
22	2·0224589	2·0868002	2·1560482	2·2307526	2·3115490	2·3991764	38
23	2·0234937	2·0879127	2·1572469	2·2320474	2·3129513	2·4006995	37
24	2·0245297	2·0890265	2·1584471	2·2333438	2·3143554	2·4022247	36
25	2·0255670	2·0901418	2·1596489	2·2346420	2·3157615	2·4037520	35
26	2·0266056	2·0912584	2·1608522	2·2359419	2·3171695	2·4052815	34
27	2·0276453	2·0923764	2·1620570	2·2372435	2·3185794	2·4068132	33
28	2·0286863	2·0934957	2·1632633	2·2385468	2·3199912	2·4083469	32
29	2·0297286	2·0946164	2·1644712	2·2398517	2·3214049	2·4098829	31
30	2·0307720	2·0957385	2·1656806	2·2411585	2·3228205	2·4114210	30
31	2·0318168	2·0968620	2·1668915	2·2424669	2·3242381	2·4129613	29
32	2·0328628	2·0979869	2·1681040	2·2437770	2·3256575	2·4145038	28
33	2·0339100	2·0991131	2·1693180	2·2450889	2·3270790	2·4160484	27
34	2·0349585	2·1002408	2·1705335	2·2464025	2·3285023	2·4175952	26
35	2·0360082	2·1013698	2·1717506	2·2477178	2·3299276	2·4191442	25
36	2·0370592	2·1025002	2·1729693	2·2490348	2·3313548	2·4206954	24
37	2·0381114	2·1036320	2·1741895	2·2503536	2·3327840	2·4222488	23
38	2·0391649	2·1047652	2·1754113	2·2516741	2·3342152	2·4238044	22
39	2·0402197	2·1058998	2·1766346	2·2529964	2·3356482	2·4253622	21
40	2·0412757	2·1070359	2·1778595	2·2543204	2·3370833	2·4269222	20
41	2·0423330	2·1081733	2·1790859	2·2556461	2·3385203	2·4284844	19
42	2·0433916	2·1093121	2·1803139	2·2569736	2·3399593	2·4300489	18
43	2·0444515	2·1104523	2·1815435	2·2583029	2·3414002	2·4316155	17
44	2·0455126	2·1115940	2·1827746	2·2596339	2·3428432	2·4331844	16
45	2·0465750	2·1127371	2·1840074	2·2609667	2·3442881	2·4347555	15
46	2·0476386	2·1138815	2·1852417	2·2623012	2·3457349	2·4363289	14
47	2·0487036	2·1150274	2·1864775	2·2636376	2·3471838	2·4379045	13
48	2·0497698	2·1161748	2·1877150	2·2649756	2·3486347	2·4394823	12
49	2·0508373	2·1173235	2·1889541	2·2663155	2·3500875	2·4410624	11
50	2·0519061	2·1184737	2·1901947	2·2676571	2·3515424	2·4426448	10
51	2·0529762	2·1196253	2·1914370	2·2690005	2·3529992	2·4442294	9
52	2·0540476	2·1207783	2·1926808	2·2703457	2·3544581	2·4458163	8
53	2·0551203	2·1219328	2·1939263	2·2716927	2·3559189	2·4474054	7
54	2·0561942	2·1230887	2·1951733	2·2730415	2·3573818	2·4489968	6
55	2·0572695	2·1242460	2·1964219	2·2743921	2·3588467	2·4505905	5
56	2·0583460	2·1254048	2·1976721	2·2757445	2·3603136	2·4521865	4
57	2·0594239	2·1265651	2·1989240	2·2770987	2·3617826	2·4537848	3
58	2·0605031	2·1277267	2·2001775	2·2784546	2·3632535	2·4553855	2
59	2·0615836	2·1288899	2·2014326	2·2798124	2·3647265	2·4569882	1
60	2·0626653	2·1300545	2·2026893	2·2811720	2·3662016	2·4585933	0
′	29°	28°	27°	26°	25°	24°	′

COSECANTS.

TABLE XI.—SECANTS AND COSECANTS.

Secants.

′	66°	67°	68°	69°	70°	71°	′
0	2·4585933	2·5593047	2·6094672	2·7904281	2·9238044	3·0715535	60
1	2·4602008	2·5610599	2·6713906	2·7925144	2·9261431	3·0741307	59
2	2·4618106	2·5628176	2·6733171	2·7946641	2·9284858	3·0767525	58
3	2·4634227	2·5645781	2·6752465	2·7967873	2·9308326	3·0793590	57
4	2·4650371	2·5663412	2·6771790	2·7989140	2·9331833	3·0819702	56
5	2·4666538	2·5681069	2·6791145	2·8010441	2·9355380	3·0845860	55
6	2·4682729	2·5698752	2·6810530	2·8031777	2·9378968	3·0872066	54
7	2·4698943	2·5716462	2·6829945	2·8053148	2·9402597	3·0898319	53
8	2·4715181	2·5734199	2·6849391	2·8074554	2·9426265	3·0924620	52
9	2·4731442	2·5751963	2·6868867	2·8095995	2·9449975	3·0950967	51
10	2·4747726	2·5769753	2·6888374	2·8117471	2·9473725	3·0977363	50
11	2·4764034	2·5787570	2·6907912	2·8138982	2·9497516	3·1003805	49
12	2·4780366	2·5805414	2·6927480	2·8160529	2·9521348	3·1030296	48
13	2·4796721	2·5823284	2·6947079	2·8182111	2·9545221	3·1056835	47
14	2·4813100	2·5841182	2·6966709	2·8203729	2·9569135	3·1083422	46
15	2·4829503	2·5859107	2·6986370	2·8225382	2·9593090	3·1110057	45
16	2·4845929	2·5877058	2·7006061	2·8247071	2·9617087	3·1136740	44
17	2·4862380	2·5895037	2·7025784	2·8268796	2·9641125	3·1163473	43
18	2·4878854	2·5913043	2·7045538	2·8290556	2·9665205	3·1190252	42
19	2·4895352	2·5931077	2·7065323	2·8312353	2·9689327	3·1217081	41
20	2·4911874	2·5949137	2·7085139	2·8334185	2·9713490	3·1243959	40
21	2·4928421	2·5967225	2·7104987	2·8356054	2·9737695	3·1270886	39
22	2·4944991	2·5985341	2·7124866	2·8377958	2·9761942	3·1297862	38
23	2·4961586	2·6003484	2·7144777	2·8399899	2·9786231	3·1324887	37
24	2·4978204	2·6021654	2·7164719	2·8421877	2·9810563	3·1351962	36
25	2·4994848	2·6039852	2·7184693	2·8443891	2·9834936	3·1379086	35
26	2·5011515	2·6058078	2·7204698	2·8465941	2·9859352	3·1406259	34
27	2·5028207	2·6076332	2·7224735	2·8488028	2·9883811	3·1433483	33
28	2·5044923	2·6094613	2·7244804	2·8510152	2·9908312	3·1460756	32
29	2·5061663	2·6112922	2·7264905	2·8532312	2·9932856	3·1488079	31
30	2·5078428	2·6131259	2·7285038	2·8554510	2·9957443	3·1515453	30
31	2·5095218	2·6149624	2·7305202	2·8576744	2·9982073	3·1542877	29
32	2·5112032	2·6168018	2·7325400	2·8599015	3·0006746	3·1570351	28
33	2·5128871	2·6186439	2·7345630	2·8621324	3·0031462	3·1597876	27
34	2·5145735	2·6204888	2·7365892	2·8643670	3·0056221	3·1625452	26
35	2·5162624	2·6223366	2·7386186	2·8666053	3·0081024	3·1653078	25
36	2·5179537	2·6241872	2·7406512	2·8688474	3·0105870	3·1680756	24
37	2·5196475	2·6260406	2·7426871	2·8710932	3·0130760	3·1708484	23
38	2·5213438	2·6278969	2·7447263	2·8733428	3·0155694	3·1736264	22
39	2·5230426	2·6297560	2·7467687	2·8755961	3·0180672	3·1764095	21
40	2·5247440	2·6316180	2·7488144	2·8778532	3·0205693	3·1791978	20
41	2·5264478	2·6334828	2·7508634	2·8801142	3·0230759	3·1819913	19
42	2·5281541	2·6353506	2·7529157	2·8823789	3·0255868	3·1847899	18
43	2·5298630	2·6372211	2·7549712	2·8846474	3·0281023	3·1875937	17
44	2·5315744	2·6390946	2·7570301	2·8869198	3·0306221	3·1904028	16
45	2·5332883	2·6409710	2·7590923	2·8891960	3·0331464	3·1932170	15
46	2·5350048	2·6428502	2·7611578	2·8914760	3·0356752	3·1960365	14
47	2·5367238	2·6447323	2·7632267	2·8937598	3·0382084	3·1988613	13
48	2·5384453	2·6466174	2·7652988	2·8960475	3·0407462	3·2016913	12
49	2·5401694	2·6485054	2·7673744	2·8983391	3·0432884	3·2045266	11
50	2·5418961	2·6503962	2·7694532	2·9006346	3·0458352	3·2073673	10
51	2·5436253	2·6522901	2·7715355	2·9029339	3·0483864	3·2102132	9
52	2·5453571	2·6541868	2·7736211	2·9052372	3·0509423	3·2130644	8
53	2·5470915	2·6560865	2·7757100	2·9075443	3·0535026	3·2159210	7
54	2·5488284	2·6579891	2·7778024	2·9098553	3·0560675	3·2187830	6
55	2·5505680	2·6598947	2·7798982	2·9121703	3·0586370	3·2216503	5
56	2·5523101	2·6618033	2·7819973	2·9144892	3·0612111	3·2245230	4
57	2·5540548	2·6637148	2·7840999	2·9168121	3·0637898	3·2274011	3
58	2·5558022	2·6656292	2·7862059	2·9191389	3·0663731	3·2302846	2
59	2·5575521	2·6675467	2·7883153	2·9214697	3·0689610	3·2331736	1
60	2·5593047	2·6694672	2·7904281	2·9238044	3·0715535	3·2360680	0
′	23°	22°	21°	20°	19°	18°	′

Cosecants.

305

TABLE XI.—SECANTS AND COSECANTS.

	|	SECANTS.						
		72°	73°	74°	75°	76°	77°	
0		3·2360680	3·4203036	3·6279558	3·8637033	4·1335655	4·4454115	60
1		3·2389078	3·4235611	3·6316395	3·8679025	4·1383939	4·4510198	59
2		3·2418732	3·4268251	3·6353316	3·8721112	4·1432339	4·4566428	58
3		3·2447840	3·4300956	3·6390315	3·8763293	4·1480856	4·4622803	57
4		3·2477003	3·4333727	3·6427392	3·8805570	4·1529491	4·4679324	56
5		3·2506222	3·4366563	3·6464548	3·8847943	4·1578243	4·4735993	55
6		3·2535496	3·4399465	3·6501783	3·8890411	4·1627114	4·4792810	54
7		3·2564825	3·4432433	3·6539097	3·8932976	4·1676102	4·4849775	53
8		3·2594211	3·4465467	3·6576491	3·8975637	4·1725210	4·4906889	52
9		3·2623652	3·4498568	3·6613964	3·9018395	4·1774438	4·4964152	51
10		3·2653149	3·4531735	3·6651518	3·9061250	4·1823785	4·5021565	50
11		3·2682702	3·4564969	3·6689151	3·9104203	4·1873252	4·5079129	49
12		3·2712311	3·4598269	3·6726865	3·9147254	4·1922840	4·5136844	48
13		3·2741977	3·4631637	3·6764660	3·9190403	4·1972549	4·5194711	47
14		3·2771700	3·4665073	3·6802536	3·9233651	4·2022380	4·5252730	46
15		3·2801479	3·4698576	3·6840493	3·9276997	4·2072333	4·5310903	45
16		3·2831316	3·4732146	3·6878532	3·9320443	4·2122408	4·5369229	44
17		3·2861209	3·4765785	3·6916652	3·9363988	4·2172606	4·5427709	43
18		3·2891160	3·4799492	3·6954854	3·9407633	4·2222928	4·5486344	42
19		3·2921168	3·4833267	3·6993139	3·9451379	4·2273373	4·5545134	41
20		3·2951234	3·4867110	3·7031506	3·9495224	4·2323943	4·5604080	40
21		3·2981357	3·4901023	3·7069956	3·9539171	4·2374637	4·5663183	39
22		3·3011539	3·4935004	3·7108489	3·9583219	4·2425457	4·5722444	38
23		3·3041778	3·4969055	3·7147105	3·9627369	4·2476403	4·5781862	37
24		3·3072076	3·5003175	3·7185805	3·9671621	4·2527474	4·5841439	36
25		3·3102432	3·5037365	3·7224589	3·9715975	4·2578671	4·5901174	35
26		3·3132847	3·5071625	3·7263457	3·9760431	4·2629996	4·5961070	34
27		3·3163320	3·5105954	3·7302409	3·9804991	4·2681449	4·6021126	33
28		3·3193853	3·5140354	3·7341446	3·9849654	4·2733029	4·6081343	32
29		3·3224444	3·5174824	3·7380568	3·9894421	4·2784738	4·6141722	31
30		3·3255095	3·5209365	3·7419775	3·9939292	4·2836576	4·6202263	30
31		3·3285805	3·5243977	3·7459068	3·9984267	4·2888543	4·6262967	29
32		3·3316575	3·5278660	3·7498447	4·0029347	4·2940640	4·6323835	28
33		3·3347405	3·5313414	3·7537911	4·0074532	4·2992867	4·6384867	27
34		3·3378294	3·5348240	3·7577462	4·0119823	4·3045225	4·6446064	26
35		3·3409244	3·5383138	3·7617100	4·0165219	4·3097715	4·6507427	25
36		3·3440254	3·5418107	3·7656824	4·0210722	4·3150336	4·6568956	24
37		3·3471324	3·5453149	3·7696636	4·0256332	4·3203090	4·6630652	23
38		3·3502455	3·5488263	3·7736535	4·0302048	4·3255977	4·6692516	22
39		3·3533647	3·5523450	3·7776522	4·0347872	4·3308996	4·6754548	21
40		3·3564900	3·5558710	3·7816596	4·0393804	4·3362150	4·6816748	20
41		3·3596214	3·5594042	3·7856760	4·0439844	4·3415438	4·6879119	19
42		3·3627589	3·5629448	3·7897011	4·0485992	4·3468861	4·6941660	18
43		3·3659026	3·5664928	3·7937352	4·0532249	4·3522419	4·7004372	17
44		3·3690524	3·5700481	3·7977782	4·0578615	4·3576113	4·7067256	16
45		3·3722084	3·5736108	3·8018301	4·0625091	4·3629943	4·7130313	15
46		3·3753707	3·5771810	3·8058911	4·0671677	4·3683910	4·7193542	14
47		3·3785391	3·5807586	3·8099610	4·0718374	4·3738015	4·7256945	13
48		3·3817138	3·5843437	3·8140399	4·0765181	4·3792257	4·7320524	12
49		3·3848948	3·5879362	3·8181280	4·0812100	4·3846638	4·7384277	11
50		3·3880820	3·5915363	3·8222251	4·0859160	4·3901158	4·7448206	10
51		3·3912755	3·5951439	3·8263313	4·0906272	4·3955817	4·7512312	9
52		3·3944754	3·5987590	3·8304467	4·0953526	4·4010616	4·7576596	8
53		3·3976816	3·6023818	3·8345713	4·1000893	4·4065556	4·7641058	7
54		3·4008941	3·6060121	3·8387052	4·1048374	4·4120637	4·7705699	6
55		3·4041130	3·6096501	3·8428482	4·1095967	4·4175859	4·7770519	5

Secants.

′	78°	79°	80°	81°	82°	83°	′
0	4·8097343	5·2408431	5·7587705	6·3924532	7·1852965	8·2055090	60
1	4·8163258	5·2486979	5·7682867	6·4042154	7·2001996	8·2249952	59
2	4·8229357	5·2565768	5·7778350	6·4160216	7·2151653	8·2445748	58
3	4·8295643	5·2644798	5·7874153	6·4278719	7·2301940	8·2642485	57
4	4·8362114	5·2724070	5·7970280	6·4397666	7·2452859	8·2840171	56
5	4·8428774	5·2803587	5·8066732	6·4517059	7·2604417	8·3038812	55
6	4·8495621	5·2883347	5·8163510	6·4636901	7·2756616	8·3238415	54
7	4·8562657	5·2963354	5·8260617	6·4757195	7·2909460	8·3438986	53
8	4·8629883	5·3043608	5·8358053	6·4877944	7·3062954	8·3640534	52
9	4·8697299	5·3124109	5·8455820	6·4999148	7·3217102	8·3843065	51
10	4·8764907	5·3204860	5·8553921	6·5120812	7·3371909	8·4046586	50
11	4·8832707	5·3285861	5·8652356	6·5242938	7·3527377	8·4251105	49
12	4·8900700	5·3367114	5·8751128	6·5365528	7·3683512	8·4456629	48
13	4·8968886	5·3448620	5·8850233	6·5488586	7·3840318	8·4663165	47
14	4·9037267	5·3530379	5·8949688	6·5612113	7·3997798	8·4870721	46
15	4·9105844	5·3612393	5·9049479	6·5736112	7·4155959	8·5079304	45
16	4·9174616	5·3694664	5·9149614	6·5860587	7·4314803	8·5288923	44
17	4·9243586	5·3777192	5·9250095	6·5985540	7·4474335	8·5499584	43
18	4·9312754	5·3859979	5·9350922	6·6110973	7·4634560	8·5711295	42
19	4·9382120	5·3943026	5·9452098	6·6236890	7·4795482	8·5924065	41
20	4·9451687	5·4026333	5·9553625	6·6363293	7·4957106	8·6137901	40
21	4·9521453	5·4109903	5·9655504	6·6490184	7·5119437	8·6352812	39
22	4·9591421	5·4193737	5·9757737	6·6617568	7·5282478	8·6568805	38
23	4·9661591	5·4277835	5·9860326	6·6745446	7·5446236	8·6785889	37
24	4·9731964	5·4362199	5·9963274	6·6873822	7·5610713	8·7004071	36
25	4·9802541	5·4446831	6·0066581	6·7002699	7·5775916	8·7223361	35
26	4·9873323	5·4531731	6·0170250	6·7132079	7·5941849	8·7443766	34
27	4·9944311	5·4616901	6·0274282	6·7261965	7·6108516	8·7665295	33
28	5·0015505	5·4702342	6·0378680	6·7392360	7·6275923	8·7887957	32
29	5·0086907	5·4788056	6·0483445	6·7523268	7·6444075	8·8111761	31
30	5·0158517	5·4874043	6·0588580	6·7654691	7·6612976	8·8336715	30
31	5·0230337	5·4960305	6·0694085	6·7786632	7·6782631	8·8562828	29
32	5·0302367	5·5046843	6·0799964	6·7919095	7·6953047	8·8790109	28
33	5·0374607	5·5133659	6·0906219	6·8052082	7·7124227	8·9018567	27
34	5·0447060	5·5220754	6·1012850	6·8185597	7·7296176	8·9248211	26
35	5·0519726	5·5308129	6·1119861	6·8319642	7·7468901	8·9479051	25
36	5·0592606	5·5395786	6·1227253	6·8454222	7·7642406	8·9711095	24
37	5·0665701	5·5483726	6·1335028	6·8589338	7·7816697	8·9944354	23
38	5·0739012	5·5571951	6·1443189	6·8724995	7·7991778	9·0178837	22
39	5·0812539	5·5660460	6·1551736	6·8861195	7·8167656	9·0414553	21
40	5·0886284	5·5749258	6·1660674	6·8997942	7·8344335	9·0651512	20
41	5·0960248	5·5838343	6·1770003	6·9135239	7·8521821	9·0889725	19
42	5·1034431	5·5927719	6·1879725	6·9273089	7·8700120	9·1129200	18
43	5·1108835	5·6017386	6·1989843	6·9411496	7·8879238	9·1369949	17
44	5·1183461	5·6107345	6·2100359	6·9550464	7·9059179	9·1611980	16
45	5·1258309	5·6197599	6·2211275	6·9689994	7·9239950	9·1855305	15
46	5·1333381	5·6288148	6·2322594	6·9830092	7·9421556	9·2099934	14
47	5·1408677	5·6378995	6·2434316	6·9970760	7·9604003	9·2345877	13
48	5·1484199	5·6470140	6·2546446	7·0112001	7·9787298	9·2593145	12
49	5·1559948	5·6561584	6·2658984	7·0253820	7·9971445	9·2841749	11
50	5·1635924	5·6653331	6·2771933	7·0396220	8·0156450	9·3091699	10
51	5·1712128	5·6745380	6·2885295	7·0539205	8·0342321	9·3343006	9
52	5·1788563	5·6837734	6·2999073	7·0682777	8·0529062	9·3595682	8
53	5·1865228	5·6930393	6·3113269	7·0826941	8·0716681	9·3849738	7
54	5·1942125	5·7023360	6·3227884	7·0971700	8·0905182	9·4105184	6
55	5·2019254	5·7116636	6·3342923	7·1117059	8·1094573	9·4362033	5
56	5·2096618	5·7210223	6·3458386	7·1263019	8·1284860	9·4620296	4
57	5·2174216	5·7304121	6·3574276	7·1409587	8·1476048	9·4879984	3
58	5·2252050	5·7398333	6·3690595	7·1556764	8·1668145	9·5141110	2
59	5·2330121	5·7492861	6·3807347	7·1704556	8·1861157	9·5403686	1
60	5·2408431	5·7587705	6·3924532	7·1852965	8·2055090	9·5667722	0
′	11°	10°	9°	8°	7°	6°	′

Cosecants.

TABLE XI.—SECANTS AND COSECANTS.

SECANTS.

′	84°	85°	86°	87°	88°	89°	′
0	9·5667722	11·473713	14·335587	19·107323	28·653708	57·298688	60
1	9·5933233	11·511990	14·395471	19·213970	28·894398	58·269755	59
2	9·6200229	11·550523	14·455859	19·321816	29·139169	59·274308	58
3	9·6468724	11·589316	14·516757	19·430682	29·388064	60·314110	57
4	9·6738730	11·628372	14·578172	19·541187	29·641373	61·391050	56
5	9·7010260	11·667693	14·640109	19·652754	29·899026	62·507153	55
6	9·7283327	11·707282	14·702576	19·765604	30·161201	63·664595	54
7	9·7557944	11·747141	14·765580	19·879758	30·428017	64·865716	53
8	9·7834124	11·787274	14·829128	19·995241	30·699598	66·113036	52
9	9·8111880	11·827683	14·893226	20·112075	30·976074	67·409272	51
10	9·8391227	11·868370	14·957882	20·230284	31·257577	68·757360	50
11	9·8672176	11·909340	15·023103	20·349893	31·544246	70·160474	49
12	9·8954744	11·950595	15·088896	20·470926	31·836225	71·622052	48
13	9·9238943	11·992137	15·155270	20·593409	32·133663	73·145827	47
14	9·9524787	12·033970	15·222231	20·717368	32·436713	74·735856	46
15	9·9812291	12·076098	15·289788	20·842830	32·745537	76·396554	45
16	10·010147	12·118523	15·357949	20·969824	23·060300	78·132742	44
17	10·039234	12·161246	15·426721	21·098376	33·381176	79·949684	43
18	10·068491	12·204274	15·496114	21·228515	33·708345	81·853150	42
19	10·097920	12·247606	15·566135	21·360272	34·041094	83·849470	41
20	10·127522	12·291252	15·636793	21·493676	34·382316	85·945609	40
21	10·157300	12·335210	15·708096	21·628759	34·729515	88·149244	39
22	10·187254	12·379484	15·780054	21·765553	35·083800	90·468863	38
23	10·217386	12·424078	15·852676	21·904090	35·445391	92·913869	37
24	10·247697	12·468995	15·925971	22·044403	35·814517	95·494711	36
25	10·278190	12·514240	15·999948	22·186528	36·191414	98·223033	35
26	10·308865	12·559815	16·074617	22·330499	36·576332	101·11185	34
27	10·339726	12·605724	16·149987	22·476353	36·969528	104·17574	33
28	10·370772	12·651971	16·226069	22·624126	37·371273	107·43114	32
29	10·402007	12·698560	16·302873	22·773857	37·781849	110·89656	31
30	10·433431	12·745495	16·380408	22·925586	38·201550	114·59301	30
31	10·465046	12·792779	16·458686	23·079351	38·630683	118·54440	29
32	10·496854	12·840416	16·537717	23·235196	39·069571	122·77803	28
33	10·528857	12·888410	16·617512	23·393161	39·518549	127·32526	27
34	10·561057	12·936765	16·698082	23·553291	39·977969	132·22219	26
35	10·593455	12·985486	16·779439	23·715630	40·448201	137·51108	25
36	10·626054	13·034576	16·861594	23·880224	40·929630	143·24061	24
37	10·658854	13·084040	16·944559	24·047121	41·422660	149·46837	23
38	10·691859	13·133882	17·028346	24·216370	41·927717	156·26128	22
39	10·725070	13·184106	17·112966	24·388020	42·445245	163·70325	21
40	10·758488	13·234717	17·198434	24·562123	42·975718	171·88831	20
41	10·792117	13·285719	17·284761	24·738731	43·519612	180·93496	19
42	10·825957	13·337116	17·371960	24·917900	44·077458	190·98680	18
43	10·860011	13·388914	17·460046	25·099685	44·649795	202·22122	17
44	10·894281	13·441118	17·549030	25·284144	45·237195	214·85995	16
45	10·928768	13·493731	17·638928	25·471337	45·840260	229·18385	15
46	10·963476	13·546758	17·729753	25·661324	46·459625	245·55402	14
47	10·998406	13·600205	17·821520	25·854169	47·095961	264·44269	13
48	11·033560	13·654077	17·914243	26·049937	47·749974	286·47948	12
49	11·068940	13·708379	18·007937	26·248694	48·422411	312·52297	11
50	11·104549	13·763115	18·102619	26·450510	49·114002	343·77516	10
51	11·140389	13·818291	18·198303	26·655455	49·825762	381·97230	9
52	11·176462	13·873913	18·295005	26·863603	50·558396	429·71873	8
53	11·212770	13·929985	18·392742	27·075030	51·312902	491·10702	7
54	11·249316	13·986514	18·491530	27·289814	52·090272	572·95809	6
55	11·286101	14·043504	18·591387	27·508035	52·891564	687·54960	5
56	11·323129	14·100963	18·692330	27·729777	53·717896	859·43689	4
57	11·360402	14·158894	18·794377	27·955125	54·570464	1145·9157	3
58	11·397922	14·217304	18·897545	28·184168	55·450534	1718·8735	2
59	11·435692	14·276200	19·001854	28·416997	56·359462	3437·7458	1
60	11·473713	14·335587	19·107323	28·653708	57·298668	Infinite.	0
′	5°	4°	3°	2°	1°	0°	′

COSECANTS.

TABLE XII.—TANGENTS AND COTANGENTS.

′	0°		1°		2°		3°		′
	Tang	Cotang	Tang	Cotang	Tang	Cotang	Tang	Cotang	
0	.00000	Infinite.	.01746	57.2900	.03492	28.6363	.05241	19.0811	60
1	.00029	3437.75	.01775	56.3506	.03521	28.3994	.05270	18.9755	59
2	.00058	1718.87	.01804	55.4415	.03550	28.1664	.05299	18.8711	58
3	.00087	1145.92	.01833	54.5613	.03579	27.9372	.05328	18.7678	57
4	.00116	859.436	.01862	53.7086	.03609	27.7117	.05357	18.6656	56
5	.00145	687.549	.01891	52.8821	.03638	27.4899	.05387	18.5645	55
6	.00175	572.957	.01920	52.0807	.03667	27.2715	.05416	18.4645	54
7	.00204	491.106	.01949	51.3032	.03696	27.0566	.05445	18.3655	53
8	.00233	429.718	.01978	50.5485	.03725	26.8450	.05474	18.2677	52
9	.00262	381.971	.02007	49.8157	.03754	26.6367	.05503	18.1708	51
10	.00291	343.774	.02036	49.1039	.03783	26.4316	.05533	18.0750	50
11	.00320	312.521	.02066	48.4121	.03812	26.2296	.05562	17.9802	49
12	.00349	286.478	.02095	47.7395	.03842	26.0307	.05591	17.8863	48
13	.00378	264.441	.02124	47.0853	.03871	25.8348	.05620	17.7934	47
14	.00407	245.552	.02153	46.4489	.03900	25.6418	.05649	17.7015	46
15	.00436	229.182	.02182	45.8294	.03929	25.4517	.05678	17.6106	45
16	.00465	214.858	.02211	45.2261	.03958	25.2644	.05708	17.5205	44
17	.00495	202.219	.02240	44.6386	.03987	25.0798	.05737	17.4314	43
18	.00524	190.984	.02269	44.0661	.04016	24.8978	.05766	17.3432	42
19	.00553	180.932	.02298	43.5081	.04046	24.7185	.05795	17.2558	41
20	.00582	171.885	.02328	42.9641	.04075	24.5418	.05824	17.1693	40
21	.00611	163.700	.02357	42.4335	.04104	24.3675	.05854	17.0837	39
22	.00640	156.259	.02386	41.9158	.04133	24.1957	.05883	16.9990	38
23	.00669	149.465	.02415	41.4106	.04162	24.0263	.05912	16.9150	37
24	.00698	143.237	.02444	40.9174	.04191	23.8593	.05941	16.8319	36
25	.00727	137.507	.02473	40.4358	.04220	23.6945	.05970	16.7496	35
26	.00756	132.219	.02502	39.9655	.04250	23.5321	.05999	16.6681	34
27	.00785	127.321	.02531	39.5059	.04279	23.3718	.06029	16.5874	33
28	.00815	122.774	.02560	39.0568	.04308	23.2137	.06058	16.5075	32
29	.00844	118.540	.02589	38.6177	.04337	23.0577	.06087	16.4283	31
30	.00873	114.589	.02619	38.1885	.04366	22.9038	.06116	16.3499	30
31	.00902	110.892	.02648	37.7686	.04395	22.7519	.06145	16.2722	29
32	.00931	107.426	.02677	37.3579	.04424	22.6020	.06175	16.1952	28
33	.00960	104.171	.02706	36.9560	.04454	22.4541	.06204	16.1190	27
34	.00989	101.107	.02735	36.5627	.04483	22.3081	.06233	16.0435	26
35	.01018	98.2179	.02764	36.1776	.04512	22.1640	.06262	15.9687	25
36	.01047	95.4895	.02793	35.8006	.04541	22.0217	.06291	15.8945	24
37	.01076	92.9085	.02822	35.4313	.04570	21.8813	.06321	15.8211	23
38	.01105	90.4633	.02851	35.0695	.04599	21.7426	.06350	15.7483	22
39	.01135	88.1436	.02881	34.7151	.04628	21.6056	.06379	15.6762	21
40	.01164	85.9398	.02910	34.3678	.04658	21.4704	.06408	15.6048	20
41	.01193	83.8435	.02939	34.0273	.04687	21.3369	.06437	15.5340	19
42	.01222	81.8470	.02968	33.6935	.04716	21.2049	.06467	15.4638	18
43	.01251	79.9434	.02997	33.3662	.04745	21.0747	.06496	15.3943	17
44	.01280	78.1263	.03026	33.0452	.04774	20.9460	.06525	15.3254	16
45	.01309	76.3900	.03055	32.7303	.04803	20.8188	.06554	15.2571	15
46	.01338	74.7292	.03084	32.4213	.04833	20.6932	.06584	15.1893	14
47	.01367	73.1390	.03114	32.1181	.04862	20.5691	.06613	15.1222	13
48	.01396	71.6151	.03143	31.8205	.04891	20.4465	.06642	15.0557	12
49	.01425	70.1533	.03172	31.5284	.04920	20.3253	.06671	14.9898	11
50	.01455	68.7501	.03201	31.2416	.04949	20.2056	.06700	14.9244	10
51	.01484	67.4019	.03230	30.9599	.04978	20.0872	.06730	14.8596	9
52	.01513	66.1055	.03259	30.6833	.05007	19.9702	.06759	14.7954	8
53	.01542	64.8580	.03288	30.4116	.05037	19.8546	.06788	14.7317	7
54	.01571	63.6567	.03317	30.1446	.05066	19.7403	.06817	14.6685	6
55	.01600	62.4992	.03346	29.8823	.05095	19.6273	.06847	14.6059	5
56	.01629	61.3829	.03376	29.6245	.05124	19.5156	.06876	14.5438	4
57	.01658	60.3058	.03405	29.3711	.05153	19.4051	.06905	14.4823	3
58	.01687	59.2659	.03434	29.1220	.05182	19.2959	.06934	14.4212	2
59	.01716	58.2612	.03463	28.8771	.05212	19.1879	.06963	14.3607	1
60	.01746	57.2900	.03492	28.6363	.05241	19.0811	.06993	14.3007	0
′	Cotang	Tang	Cotang	Tang	Cotang	Tang	Cotang	Tang	′
	89°		88°		87°		86°		

TABLE XII.—TANGENTS AND COTANGENTS.

′	4°		5°		6°		7°		′
	Tang	Cotang	Tang	Cotang	Tang	Cotang	Tang	Cotang	
0	.06993	14.3007	.08749	11.4301	.10510	9.51436	.12278	8.14435	60
1	.07022	14.2411	.08778	11.3919	.10540	9.48781	.12308	8.12481	59
2	.07051	14.1821	.08807	11.3540	.10569	9.46141	.12338	8.10536	58
3	.07080	14.1235	.08837	11.3163	.10599	9.43515	.12367	8.08600	57
4	.07110	14.0655	.08866	11.2789	.10628	9.40904	.12397	8.06674	56
5	.07139	14.0079	.08895	11.2417	.10657	9.38307	.12426	8.04756	55
6	.07168	13.9507	.08925	11.2048	.10687	9.35724	.12456	8.02848	54
7	.07197	13.8940	.08954	11.1681	.10716	9.33155	.12485	8.00948	53
8	.07227	13.8378	.08983	11.1316	.10746	9.30599	.12515	7.99058	52
9	.07256	13.7821	.09013	11.0954	.10775	9.28058	.12544	7.97176	51
10	.07285	13.7267	.09042	11.0594	.10805	9.25530	.12574	7.95302	50
11	.07314	13.6719	.09071	11.0237	.10834	9.23016	.12603	7.93438	49
12	.07344	13.6174	.09101	10.9882	.10863	9.20516	.12633	7.91582	48
13	.07373	13.5634	.09130	10.9529	.10893	9.18028	.12662	7.89734	47
14	.07402	13.5098	.09159	10.9178	.10922	9.15554	.12692	7.87895	46
15	.07431	13.4566	.09189	10.8829	.10952	9.13093	.12722	7.86064	45
16	.07461	13.4039	.09218	10.8483	.10981	9.10646	.12751	7.84242	44
17	.07490	13.3515	.09247	10.8139	.11011	9.08211	.12781	7.82428	43
18	.07519	13.2996	.09277	10.7797	.11040	9.05789	.12810	7.80622	42
19	.07548	13.2480	.09306	10.7457	.11070	9.03379	.12840	7.78825	41
20	.07578	13.1969	.09335	10.7119	.11099	9.00983	.12869	7.77035	40
21	.07607	13.1461	.09365	10.6783	.11128	8.98598	.12899	7.75254	39
22	.07636	13.0958	.09394	10.6450	.11158	8.96227	.12929	7.73480	38
23	.07665	13.0458	.09423	10.6118	.11187	8.93867	.12958	7.71715	37
24	.07695	12.9962	.09453	10.5789	.11217	8.91520	.12988	7.69957	36
25	.07724	12.9469	.09482	10.5462	.11246	8.89185	.13017	7.68208	35
26	.07753	12.8981	.09511	10.5136	.11276	8.86862	.13047	7.66466	34
27	.07782	12.8496	.09541	10.4813	.11305	8.84551	.13076	7.64732	33
28	.07812	12.8014	.09570	10.4491	.11335	8.82252	.13106	7.63005	32
29	.07841	12.7536	.09600	10.4172	.11364	8.79964	.13136	7.61287	31
30	.07870	12.7062	.09629	10.3854	.11394	8.77689	.13165	7.59575	30
31	.07899	12.6591	.09658	10.3538	.11423	8.75425	.13195	7.57872	29
32	.07929	12.6124	.09688	10.3224	.11452	8.73172	.13224	7.56176	28
33	.07958	12.5660	.09717	10.2913	.11482	8.70931	.13254	7.54487	27
34	.07987	12.5199	.09746	10.2602	.11511	8.68701	.13284	7.52806	26
35	.08017	12.4742	.09776	10.2294	.11541	8.66482	.13313	7.51132	25
36	.08046	12.4288	.09805	10.1988	.11570	8.64275	.13343	7.49465	24
37	.08075	12.3838	.09834	10.1683	.11600	8.62078	.13372	7.47806	23
38	.08104	12.3390	.09864	10.1381	.11629	8.59893	.13402	7.46154	22
39	.08134	12.2946	.09893	10.1080	.11659	8.57718	.13432	7.44509	21
40	.08163	12.2505	.09923	10.0780	.11688	8.55555	.13461	7.42871	20
41	.08192	12.2067	.09952	10.0483	.11718	8.53402	.13491	7.41240	19
42	.08221	12.1632	.09981	10.0187	.11747	8.51259	.13521	7.39616	18
43	.08251	12.1201	.10011	9.98931	.11777	8.49128	.13550	7.37999	17
44	.08280	12.0772	.10040	9.96007	.11806	8.47007	.13580	7.36389	16
45	.08309	12.0346	.10069	9.93101	.11836	8.44896	.13609	7.34786	15
46	.08339	11.9923	.10099	9.90211	.11865	8.42795	.13639	7.33190	14
47	.08368	11.9504	.10128	9.87338	.11895	8.40705	.13669	7.31600	13
48	.08397	11.9087	.10158	9.84482	.11924	8.38625	.13698	7.30018	12
49	.08427	11.8673	.10187	9.81641	.11954	8.36555	.13728	7.28442	11
50	.08456	11.8262	.10216	9.78817	.11983	8.34496	.13758	7.26873	10
51	.08485	11.7853	.10246	9.76009	.12013	8.32446	.13787	7.25310	9
52	.08514	11.7448	.10275	9.73217	.12042	8.30406	.13817	7.23754	8
53	.08544	11.7045	.10305	9.70441	.12072	8.28376	.13846	7.22204	7
54	.08573	11.6645	.10334	9.67680	.12101	8.26355	.13876	7.20661	6
55	.08602	11.6248	.10363	9.64935	.12131	8.24345	.13906	7.19125	5
56	.08632	11.5853	.10393	9.62205	.12160	8.22344	.13935	7.17594	4
57	.08661	11.5461	.10422	9.59490	.12190	8.20352	.13965	7.16071	3
58	.08690	11.5072	.10452	9.56791	.12219	8.18370	.13995	7.14553	2
59	.08720	11.4685	.10481	9.54106	.12249	8.16398	.14024	7.13042	1
60	.08749	11.4301	.10510	9.51436	.12278	8.14435	.14054	7.11537	0
′	Cotang	Tang	Cotang	Tang	Cotang	Tang	Cotang	Tang	′
	85°		84°		83°		82°		

TABLE XII.—TANGENTS AND COTANGENTS.

′	8° Tang	8° Cotang	9° Tang	9° Cotang	10° Tang	10° Cotang	11° Tang	11° Cotang	′
0	.14054	7.11537	.15838	6.31375	.17633	5.67128	.19438	5.14455	60
1	.14084	7.10038	.15868	6.30189	.17663	5.66165	.19468	5.13658	59
2	.14113	7.08546	.15898	6.29007	.17693	5.65205	.19498	5.12862	58
3	.14143	7.07059	.15928	6.27829	.17723	5.64248	.19529	5.12069	57
4	.14173	7.05579	.15958	6.26655	.17753	5.63295	.19559	5.11279	56
5	.14202	7.04105	.15988	6.25486	.17783	5.62344	.19589	5.10490	55
6	.14232	7.02637	.16017	6.24321	.17813	5.61397	.19619	5.09704	54
7	.14262	6.91174	.16047	6.23160	.17843	5.60452	.19649	5.08921	53
8	.14291	6.99718	.16077	6.22003	.17873	5.59511	.19680	5.08139	52
9	.14321	6.98268	.16107	6.20851	.17903	5.58573	.19710	5.07360	51
10	.14351	6.96823	.16137	6.19703	.17933	5.57638	.19740	5.06584	50
11	.14381	6.95385	.16167	6.18559	.17963	5.56706	.19770	5.05809	49
12	.14410	6.93952	.16196	6.17419	.17993	5.55777	.19801	5.05037	48
13	.14440	6.92525	.16226	6.16283	.18023	5.54851	.19831	5.04267	47
14	.14470	6.91104	.16256	6.15151	.18053	5.53927	.19861	5.03499	46
15	.14499	6.89688	.16286	6.14023	.18083	5.53007	.19891	5.02734	45
16	.14529	6.88278	.16316	6.12899	.18113	5.52090	.19921	5.01971	44
17	.14559	6.86874	.16346	6.11779	.18143	5.51176	.19952	5.01210	43
18	.14588	6.85475	.16376	6.10664	.18173	5.50264	.19982	5.00451	42
19	.14618	6.84082	.16405	6.09552	.18203	5.49356	.20012	4.99695	41
20	.14648	6.82694	.16435	6.08444	.18233	5.48451	.20042	4.98940	40
21	.14678	6.81312	.16465	6.07340	.18263	5.47548	.20073	4.98188	39
22	.14707	6.79936	.16495	6.06240	.18293	5.46648	.20103	4.97438	38
23	.14737	6.78564	.16525	6.05143	.18323	5.45751	.20133	4.96690	37
24	.14767	6.77199	.16555	6.04051	.18353	5.44857	.20164	4.95945	36
25	.14796	6.75838	.16585	6.02962	.18384	5.43966	.20194	4.95201	35
26	.14826	6.74483	.16615	6.01878	.18414	5.43077	.20224	4.94460	34
27	.14856	6.73133	.16645	6.00797	.18444	5.42192	.20254	4.93721	33
28	.14886	6.71789	.16674	5.99720	.18474	5.41309	.20285	4.92984	32
29	.14915	6.70450	.16704	5.98646	.18504	5.40429	.20315	4.92249	31
30	.14945	6.69116	.16734	5.97576	.18534	5.39552	.20345	4.91516	30
31	.14975	6.67787	.16764	5.96510	.18564	5.38677	.20376	4.90785	29
32	.15005	6.66463	.16794	5.95448	.18594	5.37805	.20406	4.90056	28
33	.15034	6.65144	.16824	5.94390	.18624	5.36936	.20436	4.89330	27
34	.15064	6.63831	.16854	5.93335	.18654	5.36070	.20466	4.88605	26
35	.15094	6.62523	.16884	5.92283	.18684	5.35206	.20497	4.87882	25
36	.15124	6.61219	.16914	5.91236	.18714	5.34345	.20527	4.87162	24
37	.15153	6.59921	.16944	5.90191	.18745	5.33487	.20557	4.86444	23
38	.15183	6.58627	.16974	5.89151	.18775	5.32631	.20588	4.85727	22
39	.15213	6.57339	.17004	5.88114	.18805	5.31778	.20618	4.85013	21
40	.15243	6.56055	.17033	5.87080	.18835	5.30928	.20648	4.84300	20
41	.15272	6.54777	.17063	5.86051	.18865	5.30080	.20679	4.83590	19
42	.15302	6.53503	.17093	5.85024	.18895	5.29235	.20709	4.82882	18
43	.15332	6.52234	.17123	5.84001	.18925	5.28393	.20739	4.82175	17
44	.15362	6.50970	.17153	5.82982	.18955	5.27553	.20770	4.81471	16
45	.15391	6.49710	.17183	5.81966	.18986	5.26715	.20800	4.80769	15
46	.15421	6.48456	.17213	5.80953	.19016	5.25880	.20830	4.80068	14
47	.15451	6.47206	.17243	5.79944	.19046	5.25048	.20861	4.79370	13
48	.15481	6.45961	.17273	5.78938	.19076	5.24218	.20891	4.78673	12
49	.15511	6.44720	.17303	5.77936	.19106	5.23391	.20921	4.77978	11
50	.15540	6.43484	.17333	5.76937	.19136	5.22566	.20952	4.77286	10
51	.15570	6.42253	.17363	5.75941	.19166	5.21744	.20982	4.76595	9
52	.15600	6.41026	.17393	5.74949	.19197	5.20925	.21013	4.75906	8
53	.15630	6.39804	.17423	5.73960	.19227	5.20107	.21043	4.75219	7
54	.15660	6.38587	.17453	5.72974	.19257	5.19293	.21073	4.74534	6
55	.15689	6.37374	.17483	5.71992	.19287	5.18480	.21104	4.73851	5
56	.15719	6.36165	.17513	5.71013	.19317	5.17671	.21134	4.73170	4
57	.15749	6.34961	.17543	5.70037	.19347	5.16863	.21164	4.72490	3
58	.15779	6.33761	.17573	5.69064	.19378	5.16058	.21195	4.71813	2
59	.15909	6.32566	.17603	5.68094	.19408	5.15256	.21225	4.71137	1
60	.15838	6.31375	.17633	5.67128	.19438	5.14455	.21256	4.70463	0
′	Cotang	Tang	Cotang	Tang	Cotang	Tang	Cotang	Tang	′
	81°		80°		79°		78°		

TABLE XII.—TANGENTS AND COTANGENTS.

′	12° Tang	Cotang	13° Tang	Cotang	14° Tang	Cotang	15° Tang	Cotang	′
0	.21256	4.70463	.23087	4.33148	.24933	4.01078	.26795	3.73205	60
1	.21286	4.69791	.23117	4.32573	.24964	4.00582	.26826	3.72771	59
2	.21316	4.69121	.23148	4.32001	.24995	4.00086	.26857	3.72338	58
3	.21347	4.68452	.23179	4.31430	.25026	3.99592	.26888	3.71907	57
4	.21377	4.67786	.23209	4.30860	.25056	3.99099	.26920	3.71476	56
5	.21408	4.67121	.23240	4.30291	.25087	3.98607	.26951	3.71046	55
6	.21438	4.66458	.23271	4.29724	.25118	3.98117	.26982	3.70616	54
7	.21469	4.65797	.23301	4.29159	.25149	3.97627	.27013	3.70188	53
8	.21499	4.65138	.23332	4.28595	.25180	3.97139	.27044	3.69761	52
9	.21529	4.64480	.23363	4.28032	.25211	3.96651	.27076	3.69335	51
10	.21560	4.63825	.23393	4.27471	.25242	3.96165	.27107	3.68909	50
11	.21590	4.63171	.23424	4.26911	.25273	3.95680	.27138	3.68485	49
12	.21621	4.62518	.23455	4.26352	.25304	3.95196	.27169	3.68061	48
13	.21651	4.61868	.23485	4.25795	.25335	3.94713	.27201	3.67638	47
14	.21682	4.61219	.23516	4.25239	.25366	3.94232	.27232	3.67217	46
15	.21712	4.60572	.23547	4.24685	.25397	3.93751	.27263	3.66796	45
16	.21743	4.59927	.23578	4.24132	.25428	3.93271	.27294	3.66376	44
17	.21773	4.59283	.23608	4.23580	.25459	3.92793	.27326	3.65957	43
18	.21804	4.58641	.23639	4.23030	.25490	3.92316	.27357	3.65538	42
19	.21834	4.58001	.23670	4.22481	.25521	3.91839	.27388	3.65121	41
20	.21864	4.57363	.23700	4.21933	.25552	3.91364	.27419	3.64705	40
21	.21895	4.56726	.23731	4.21387	.25583	3.90890	.27451	3.64289	39
22	.21925	4.56091	.23762	4.20842	.25614	3.90417	.27482	3.63874	38
23	.21956	4.55458	.23793	4.20298	.25645	3.89945	.27513	3.63461	37
24	.21986	4.54826	.23823	4.19756	.25676	3.89474	.27545	3.63048	36
25	.22017	4.54196	.23854	4.19215	.25707	3.89004	.27576	3.62636	35
26	.22047	4.53568	.23885	4.18675	.25738	3.88536	.27607	3.62224	34
27	.22078	4.52941	.23916	4.18137	.25769	3.88068	.27638	3.61814	33
28	.22108	4.52316	.23946	4.17600	.25800	3.87601	.27670	3.61405	32
29	.22139	4.51693	.23977	4.17064	.25831	3.87136	.27701	3.60996	31
30	.22169	4.51071	.24008	4.16530	.25862	3.86671	.27732	3.60588	30
31	.22200	4.50451	.24039	4.15997	.25893	3.86208	.27764	3.60181	29
32	.22231	4.49832	.24069	4.15465	.25924	3.85745	.27795	3.59775	28
33	.22261	4.49215	.24100	4.14934	.25955	3.85284	.27826	3.59370	27
34	.22292	4.48600	.24131	4.14405	.25986	3.84824	.27858	3.58966	26
35	.22322	4.47986	.24162	4.13877	.26017	3.84364	.27889	3.58562	25
36	.22353	4.47374	.24193	4.13350	.26048	3.83906	.27921	3.58160	24
37	.22383	4.46764	.24223	4.12825	.26079	3.83449	.27952	3.57758	23
38	.22414	4.46155	.24254	4.12301	.26110	3.82992	.27983	3.57357	22
39	.22444	4.45548	.24285	4.11778	.26141	3.82537	.28015	3.56957	21
40	.22475	4.44942	.24316	4.11256	.26172	3.82083	.28046	3.56557	20
41	.22505	4.44338	.24347	4.10736	.26203	3.81630	.28077	3.56159	19
42	.22536	4.43735	.24377	4.10216	.26235	3.81177	.28109	3.55761	18
43	.22567	4.43134	.24408	4.09699	.26266	3.80726	.28140	3.55364	17
44	.22597	4.42534	.24439	4.09182	.26297	3.80276	.28172	3.54968	16
45	.22628	4.41936	.24470	4.08666	.26328	3.79827	.28203	3.54573	15
46	.22658	4.41340	.24501	4.08152	.26359	3.79378	.28234	3.54179	14
47	.22689	4.40745	.24532	4.07639	.26390	3.78931	.28266	3.53785	13
48	.22719	4.40152	.24562	4.07127	.26421	3.78485	.28297	3.53393	12
49	.22750	4.39560	.24593	4.06616	.26452	3.78040	.28329	3.53001	11
50	.22781	4.38969	.24624	4.06107	.26483	3.77595	.28360	3.52609	10
51	.22811	4.38381	.24655	4.05599	.26515	3.77152	.28391	3.52219	9
52	.22842	4.37793	.24686	4.05092	.26546	3.76709	.28423	3.51829	8
53	.22872	4.37207	.24717	4.04586	.26577	3.76268	.28454	3.51441	7
54	.22903	4.36623	.24747	4.04081	.26608	3.75828	.28486	3.51053	6
55	.22934	4.36040	.24778	4.03578	.26639	3.75388	.28517	3.50666	5
56	.22964	4.35459	.24809	4.03076	.26670	3.74950	.28549	3.50279	4
57	.22995	4.34879	.24840	4.02574	.26701	3.74512	.28580	3.49894	3
58	.23026	4.34300	.24871	4.02074	.26733	3.74075	.28612	3.49509	2
59	.23056	4.33723	.24902	4.01576	.26764	3.73640	.28643	3.49125	1
60	.23087	4.33148	.24933	4.01078	.26795	3.73205	.28675	3.48741	0
′	Cotang	Tang	Cotang	Tang	Cotang	Tang	Cotang	Tang	′
	77°		76°		75°		74°		

TABLE XII.—TANGENTS AND COTANGENTS.

′	16° Tang	16° Cotang	17° Tang	17° Cotang	18° Tang	18° Cotang	19° Tang	19° Cotang	′
0	.28675	3.48741	.30573	3.27085	.32492	3.07768	.34433	2.90421	60
1	.28706	3.48359	.30605	3.26745	.32524	3.07464	.34465	2.90147	59
2	.28738	3.47977	.30637	3.26406	.32556	3.07160	.34498	2.89873	58
3	.28769	3.47596	.30669	3.26067	.32588	3.06857	.34530	2.89600	57
4	.28800	3.47216	.30700	3.25729	.32621	3.06554	.34563	2.89327	56
5	.28832	3.46837	.30732	3.25392	.32653	3.06252	.34596	2.89055	55
6	.28864	3.46458	.30764	3.25055	.32685	3.05950	.34628	2.88783	54
7	.28895	3.46080	.30796	3.24719	.32717	3.05649	.34661	2.88511	53
8	.28927	3.45703	.30828	3.24383	.32749	3.05349	.34693	2.88240	52
9	.28958	3.45327	.30860	3.24049	.32782	3.05049	.34726	2.87970	51
10	.28990	3.44951	.30891	3.23714	.32814	3.04749	.34758	2.87700	50
11	.29021	3.44576	.30923	3.23381	.32846	3.04450	.34791	2.87430	49
12	.29053	3.44202	.30955	3.23048	.32878	3.04152	.34824	2.87161	48
13	.29084	3.43829	.30987	3.22715	.32911	3.03854	.34856	2.86892	47
14	.29116	3.43456	.31019	3.22384	.32943	3.03556	.34889	2.86624	46
15	.29147	3.43084	.31051	3.22053	.32975	3.03260	.34922	2.86356	45
16	.29179	3.42713	.31083	3.21722	.33007	3.02963	.34954	2.86089	44
17	.29210	3.42343	.31115	3.21392	.33040	3.02667	.34987	2.85822	43
18	.29242	3.41973	.31147	3.21063	.33072	3.02372	.35020	2.85555	42
19	.29274	3.41604	.31178	3.20734	.33104	3.02077	.35052	2.85289	41
20	.29305	3.41236	.31210	3.20406	.33136	3.01783	.35085	2.85023	40
21	.29337	3.40869	.31242	3.20079	.33169	3.01489	.35118	2.84758	39
22	.29368	3.40502	.31274	3.19752	.33201	3.01196	.35150	2.84494	38
23	.29400	3.40136	.31306	3.19426	.33233	3.00903	.35183	2.84229	37
24	.29432	3.39771	.31338	3.19100	.33266	3.00611	.35216	2.83965	36
25	.29463	3.39406	.31370	3.18775	.33298	3.00319	.35248	2.83702	35
26	.29495	3.39042	.31402	3.18451	.33330	3.00028	.35281	2.83439	34
27	.29526	3.38679	.31434	3.18127	.33363	2.99738	.35314	2.83176	33
28	.29558	3.38317	.31466	3.17804	.33395	2.99447	.35346	2.82914	32
29	.29590	3.37955	.31498	3.17481	.33427	2.99158	.35379	2.82653	31
30	.29621	3.37594	.31530	3.17159	.33460	2.98868	.35412	2.82391	30
31	.29653	3.37234	.31562	3.16838	.33492	2.98580	.35445	2.82130	29
32	.29685	3.36875	.31594	3.16517	.33524	2.98292	.35477	2.81870	28
33	.29716	3.36516	.31626	3.16197	.33557	2.98004	.35510	2.81610	27
34	.29748	3.36158	.31658	3.15877	.33589	2.97717	.35543	2.81350	26
35	.29780	3.35800	.31690	3.15558	.33621	2.97430	.35576	2.81091	25
36	.29811	3.35443	.31722	3.15240	.33654	2.97144	.35608	2.80833	24
37	.29843	3.35087	.31754	3.14922	.33686	2.96858	.35641	2.80574	23
38	.29875	3.34732	.31786	3.14605	.33718	2.96573	.35674	2.80316	22
39	.29906	3.34377	.31818	3.14288	.33751	2.96288	.35707	2.80059	21
40	.29938	3.34023	.31850	3.13972	.33783	2.96004	.35740	2.79802	20
41	.29970	3.33670	.31882	3.13656	.33816	2.95721	.35772	2.79545	19
42	.30001	3.33317	.31914	3.13341	.33848	2.95437	.35805	2.79289	18
43	.30033	3.32965	.31946	3.13027	.33881	2.95155	.35838	2.79033	17
44	.30065	3.32614	.31978	3.12713	.33913	2.94872	.35871	2.78778	16
45	.30097	3.32264	.32010	3.12400	.33945	2.94591	.35904	2.78523	15
46	.30128	3.31914	.32042	3.12087	.33978	2.94309	.35937	2.78269	14
47	.30160	3.31565	.32074	3.11775	.34010	2.94028	.35969	2.78014	13
48	.30192	3.31216	.32106	3.11464	.34043	2.93748	.36002	2.77761	12
49	.30224	3.30868	.32139	3.11153	.34075	2.93468	.36035	2.77507	11
50	.30255	3.30521	.32171	3.10842	.34108	2.93189	.36068	2.77254	10
51	.30287	3.30174	.32203	3.10532	.34140	2.92910	.36101	2.77002	9
52	.30319	3.29829	.32235	3.10223	.34173	2.92632	.36134	2.76750	8
53	.30351	3.29483	.32267	3.09914	.34205	2.92354	.36167	2.76498	7
54	.30382	3.29139	.32299	3.09606	.34238	2.92076	.36199	2.76247	6
55	.30414	3.28795	.32331	3.09298	.34270	2.91799	.36232	2.75996	5
56	.30446	3.28452	.32363	3.08991	.34303	2.91523	.36265	2.75746	4
57	.30478	3.28109	.32396	3.08685	.34335	2.91246	.36298	2.75496	3
58	.30509	3.27767	.32428	3.08379	.34368	2.90971	.36331	2.75246	2
59	.30541	3.27426	.32460	3.08073	.34400	2.90696	.36364	2.74997	1
60	.30573	3.27085	.32492	3.07768	.34433	2.90421	.36397	2.74748	0
	Cotang	Tang	Cotang	Tang	Cotang	Tang	Cotang	Tang	′
	73°		72°		71°		70°		

TABLE XII.—TANGENTS AND COTANGENTS.

′	20° Tang	Cotang	21° Tang	Cotang	22° Tang	Cotang	23° Tang	Cotang	′
0	.36397	2.74748	.38386	2.60509	.40403	2.47509	.42447	2.35585	60
1	.36430	2.74499	.38420	2.60283	.40436	2.47302	.42482	2.35395	59
2	.36463	2.74251	.38453	2.60057	.40470	2.47095	.42516	2.35205	58
3	.36496	2.74004	.38487	2.59831	.40504	2.46888	.42551	2.35015	57
4	.36529	2.73756	.38520	2.59606	.40538	2.46682	.42585	2.34825	56
5	.36562	2.73509	.38553	2.59381	.40572	2.46476	.42619	2.34636	55
6	.36595	2.73263	.38587	2.59156	.40606	2.46270	.42654	2.34447	54
7	.36628	2.73017	.38620	2.58932	.40640	2.46065	.42688	2.34258	53
8	.36661	2.72771	.38654	2.58708	.40674	2.45860	.42722	2.34069	52
9	.36694	2.72526	.38687	2.58484	.40707	2.45655	.42757	2.33881	51
10	.36727	2.72281	.38721	2.58261	.40741	2.45451	.42791	2.33693	50
11	.36760	2.72036	.38754	2.58038	.40775	2.45246	.42826	2.33505	49
12	.36793	2.71792	.38787	2.57815	.40809	2.45043	.42860	2.33317	48
13	.36826	2.71548	.38821	2.57593	.40843	2.44839	.42894	2.33130	47
14	.36859	2.71305	.38854	2.57371	.40877	2.44636	.42929	2.32943	46
15	.36892	2.71062	.38888	2.57150	.40911	2.44433	.42963	2.32756	45
16	.36925	2.70819	.38921	2.56928	.40945	2.44230	.42998	2.32570	44
17	.36958	2.70577	.38955	2.56707	.40979	2.44027	.43032	2.32383	43
18	.36991	2.70335	.38988	2.56487	.41013	2.43825	.43067	2.32197	42
19	.37024	2.70094	.39022	2.56266	.41047	2.43623	.43101	2.32012	41
20	.37057	2.69853	.39055	2.56046	.41081	2.43422	.43136	2.31826	40
21	.37090	2.69612	.39089	2.55827	.41115	2.43220	.43170	2.31641	39
22	.37123	2.69371	.39122	2.55608	.41149	2.43019	.43205	2.31456	38
23	.37157	2.69131	.39156	2.55389	.41183	2.42819	.43239	2.31271	37
24	.37190	2.68892	.39190	2.55170	.41217	2.42618	.43274	2.31086	36
25	.37223	2.68653	.39223	2.54952	.41251	2.42418	.43308	2.30902	35
26	.37256	2.68414	.39257	2.54734	.41285	2.42218	.43343	2.30718	34
27	.37289	2.68175	.39290	2.54516	.41319	2.42019	.43378	2.30534	33
28	.37322	2.67937	.39324	2.54299	.41353	2.41819	.43412	2.30351	32
29	.37355	2.67700	.39357	2.54082	.41387	2.41620	.43447	2.30167	31
30	.37388	2.67462	.39391	2.53865	.41421	2.41421	.43481	2.29984	30
31	.37422	2.67225	.39425	2.53648	.41455	2.41223	.43516	2.29801	29
32	.37455	2.66989	.39458	2.53432	.41490	2.41025	.43550	2.29619	28
33	.37488	2.66752	.39492	2.53217	.41524	2.40827	.43585	2.29437	27
34	.37521	2.66516	.39526	2.53001	.41558	2.40629	.43620	2.29254	26
35	.37554	2.66281	.39559	2.52786	.41592	2.40432	.43654	2.29073	25
36	.37588	2.66046	.39593	2.52571	.41626	2.40235	.43689	2.28891	24
37	.37621	2.65811	.39626	2.52357	.41660	2.40038	.43724	2.28710	23
38	.37654	2.65576	.39660	2.52142	.41694	2.39841	.43758	2.28528	22
39	.37687	2.65342	.39694	2.51929	.41728	2.39645	.43793	2.28348	21
40	.37720	2.65109	.39727	2.51715	.41763	2.39449	.43828	2.28167	20
41	.37754	2.64875	.39761	2.51502	.41797	2.39253	.43862	2.27987	19
42	.37787	2.64642	.39795	2.51289	.41831	2.39058	.43897	2.27806	18
43	.37820	2.64410	.39829	2.51076	.41865	2.38863	.43932	2.27626	17
44	.37853	2.64177	.39862	2.50864	.41899	2.38668	.43966	2.27447	16
45	.37887	2.63945	.39896	2.50652	.41933	2.38473	.44001	2.27267	15
46	.37920	2.63714	.39930	2.50440	.41968	2.38279	.44036	2.27088	14
47	.37953	2.63483	.39963	2.50229	.42002	2.38084	.44071	2.26909	13
48	.37986	2.63252	.39997	2.50018	.42036	2.37891	.44105	2.26730	12
49	.38020	2.63021	.40031	2.49807	.42070	2.37697	.44140	2.26552	11
50	.38053	2.62791	.40065	2.49597	.42105	2.37504	.44175	2.26374	10
51	.38086	2.62561	.40098	2.49386	.42139	2.37311	.44210	2.26196	9
52	.38120	2.62332	.40132	2.49177	.42173	2.37118	.44244	2.26018	8
53	.38153	2.62103	.40166	2.48967	.42207	2.36925	.44279	2.25840	7
54	.38186	2.61874	.40200	2.48758	.42242	2.36733	.44314	2.25663	6
55	.38220	2.61646	.40234	2.48549	.42276	2.36541	.44349	2.25486	5
56	.38253	2.61418	.40267	2.48340	.42310	2.36349	.44384	2.25309	4
57	.38286	2.61190	.40301	2.48132	.42345	2.36158	.44418	2.25132	3
58	.38320	2.60963	.40335	2.47924	.42379	2.35967	.44453	2.24956	2
59	.38353	2.60736	.40369	2.47716	.42413	2.35776	.44488	2.24780	1
60	.38386	2.60509	.40403	2.47509	.42447	2.35585	.44523	2.24604	0
	Cotang	Tang	Cotang	Tang	Cotang	Tang	Cotang	Tang	′
	69°		68°		67°		66°		

314

TABLE XII.—TANGENTS AND COTANGENTS.

′	24°		25°		26°		27°		′
	Tang	Cotang	Tang	Cotang	Tang	Cotang	Tang	Cotang	
0	.44523	2.24604	.46631	2.14451	.48773	2.05030	.50953	1.96261	60
1	.44558	2.24428	.46666	2.14288	.48809	2.04879	.50989	1.96120	59
2	.44593	2.24252	.46702	2.14125	.48845	2.04728	.51026	1.95979	58
3	.44627	2.24077	.46737	2.13963	.48881	2.04577	.51063	1.95838	57
4	.44662	2.23902	.46772	2.13801	.48917	2.04426	.51099	1.95698	56
5	.44697	2.23727	.46808	2.13639	.48953	2.04276	.51136	1.95557	55
6	.44732	2.23553	.46843	2.13477	.48989	2.04125	.51173	1.95417	54
7	.44767	2.23378	.46879	2.13316	.49026	2.03975	.51209	1.95277	53
8	.44802	2.23204	.46914	2.13154	.49062	2.03825	.51246	1.95137	52
9	.44837	2.23030	.46950	2.12993	.49098	2.03675	.51283	1.94997	51
10	.44872	2.22857	.46985	2.12832	.49134	2.03526	.51319	1.94858	50
11	.44907	2.22683	.47021	2.12671	.49170	2.03376	.51356	1.94718	49
12	.44942	2.22510	.47056	2.12511	.49206	2.03227	.51393	1.94579	48
13	.44977	2.22337	.47092	2.12350	.49242	2.03078	.51430	1.94440	47
14	.45012	2.22164	.47128	2.12190	.49278	2.02929	.51467	1.94301	46
15	.45047	2.21992	.47163	2.12030	.49315	2.02780	.51503	1.94162	45
16	.45082	2.21819	.47199	2.11871	.49351	2.02631	.51540	1.94023	44
17	.45117	2.21647	.47234	2.11711	.49387	2.02483	.51577	1.93885	43
18	.45152	2.21475	.47270	2.11552	.49423	2.02335	.51614	1.93746	42
19	.45187	2.21304	.47305	2.11392	.49459	2.02187	.51651	1.93608	41
20	.45222	2.21132	.47341	2.11233	.49495	2.02039	.51688	1.93470	40
21	.45257	2.20961	.47377	2.11075	.49532	2.01891	.51724	1.93332	39
22	.45292	2.20790	.47412	2.10916	.49568	2.01743	.51761	1.93195	38
23	.45327	2.20619	.47448	2.10758	.49604	2.01596	.51798	1.93057	37
24	.45362	2.20449	.47483	2.10600	.49640	2.01449	.51835	1.92920	36
25	.45397	2.20278	.47519	2.10442	.49677	2.01302	.51872	1.92782	35
26	.45432	2.20108	.47555	2.10284	.49713	2.01155	.51909	1.92645	34
27	.45467	2.19938	.47590	2.10126	.49749	2.01008	.51946	1.92508	33
28	.45502	2.19769	.47626	2.09969	.49786	2.00862	.51983	1.92371	32
29	.45538	2.19599	.47662	2.09811	.49822	2.00715	.52020	1.92235	31
30	.45573	2.19430	.47698	2.09654	.49858	2.00569	.52057	1.92098	30
31	.45608	2.19261	.47733	2.09498	.49894	2.00423	.52094	1.91962	29
32	.45643	2.19092	.47769	2.09341	.49931	2.00277	.52131	1.91826	28
33	.45678	2.18923	.47805	2.09184	.49967	2.00131	.52168	1.91690	27
34	.45713	2.18755	.47840	2.09028	.50004	1.99986	.52205	1.91554	26
35	.45748	2.18587	.47876	2.08872	.50040	1.99841	.52242	1.91418	25
36	.45784	2.18419	.47912	2.08716	.50076	1.99695	.52279	1.91282	24
37	.45819	2.18251	.47948	2.08560	.50113	1.99550	.52316	1.91147	23
38	.45854	2.18084	.47984	2.08405	.50149	1.99406	.52353	1.91012	22
39	.45889	2.17916	.48019	2.08250	.50185	1.99261	.52390	1.90876	21
40	.45924	2.17749	.48055	2.08094	.50222	1.99116	.52427	1.90741	20
41	.45960	2.17582	.48091	2.07939	.50258	1.98972	.52464	1.90607	19
42	.45995	2.17416	.48127	2.07785	.50295	1.98828	.52501	1.90472	18
43	.46030	2.17249	.48163	2.07630	.50331	1.98684	.52538	1.90337	17
44	.46065	2.17083	.48198	2.07476	.50368	1.98540	.52575	1.90203	16
45	.46101	2.16917	.48234	2.07321	.50404	1.98396	.52613	1.90069	15
46	.46136	2.16751	.48270	2.07167	.50441	1.98253	.52650	1.89935	14
47	.46171	2.16585	.48306	2.07014	.50477	1.98110	.52687	1.89801	13
48	.46206	2.16420	.48342	2.06860	.50514	1.97966	.52724	1.89667	12
49	.46242	2.16255	.48378	2.06706	.50550	1.97823	.52761	1.89533	11
50	.46277	2.16090	.48414	2.06553	.50587	1.97681	.52798	1.89400	10
51	.46312	2.15925	.48450	2.06400	.50623	1.97538	.52836	1.89266	9
52	.46348	2.15760	.48486	2.06247	.50660	1.97395	.52873	1.89133	8
53	.46383	2.15596	.48521	2.06094	.50696	1.97253	.52910	1.89000	7
54	.46418	2.15432	.48557	2.05942	.50733	1.97111	.52947	1.88867	6
55	.46454	2.15268	.48593	2.05790	.50769	1.96969	.52985	1.88734	5
56	.46489	2.15104	.48629	2.05637	.50806	1.96827	.53022	1.88602	4
57	.46525	2.14940	.48665	2.05485	.50843	1.96685	.53059	1.88469	3
58	.46560	2.14777	.48701	2.05333	.50879	1.96544	.53096	1.88337	2
59	.46595	2.14614	.48737	2.05182	.50916	1.96402	.53134	1.88205	1
60	.46631	2.14451	.48773	2.05030	.50953	1.96261	.53171	1.88073	0
	Cotang	Tang	Cotang	Tang	Cotang	Tang	Cotang	Tang	′
	65°		64°		63°		62°		

TABLE XII.—TANGENTS AND COTANGENTS.

′	28°		29°		30°		31°		′
	Tang	Cotang	Tang	Cotang	Tang	Cotang	Tang	Cotang	
0	.53171	1.88073	.55431	1.80405	.57735	1.73205	.60086	1.66428	60
1	.53208	1.87941	.55469	1.80281	.57774	1.73089	.60126	1.66318	59
2	.53246	1.87809	.55507	1.80158	.57813	1.72973	.60165	1.66209	58
3	.53283	1.87677	.55545	1.80034	.57851	1.72857	.60205	1.66099	57
4	.53320	1.87546	.55583	1.79911	.57890	1.72741	.60245	1.65990	56
5	.53358	1.87415	.55621	1.79788	.57929	1.72625	.60284	1.65881	55
6	.53395	1.87283	.55659	1.79665	.57968	1.72509	.60324	1.65772	54
7	.53432	1.87152	.55697	1.79542	.58007	1.72393	.60364	1.65663	53
8	.53470	1.87021	.55736	1.79419	.58046	1.72278	.60403	1.65554	52
9	.53507	1.86891	.55774	1.79296	.58085	1.72163	.60443	1.65445	51
10	.53545	1.86760	.55812	1.79174	.58124	1.72047	.60483	1.65337	50
11	.53582	1.86630	.55850	1.79051	.58162	1.71932	.60522	1.65228	49
12	.53620	1.86499	.55888	1.78929	.58201	1.71817	.60562	1.65120	48
13	.53657	1.86369	.55926	1.78807	.58240	1.71702	.60602	1.65011	47
14	.53694	1.86239	.55964	1.78685	.58279	1.71588	.60642	1.64903	46
15	.53732	1.86109	.56003	1.78563	.58318	1.71473	.60681	1.64795	45
16	.53769	1.85979	.56041	1.78441	.58357	1.71358	.60721	1.64687	44
17	.53807	1.85850	.56079	1.78319	.58396	1.71244	.60761	1.64579	43
18	.53844	1.85720	.56117	1.78198	.58435	1.71129	.60801	1.64471	42
19	.53882	1.85591	.56156	1.78077	.58474	1.71015	.60841	1.64363	41
20	.53920	1.85462	.56194	1.77955	.58513	1.70901	.60881	1.64256	40
21	.53957	1.85333	.56232	1.77834	.58552	1.70787	.60921	1.64148	39
22	.53995	1.85204	.56270	1.77713	.58591	1.70673	.60960	1.64041	38
23	.54032	1.85075	.56309	1.77592	.58631	1.70560	.61000	1.63934	37
24	.54070	1.84946	.56347	1.77471	.58670	1.70446	.61040	1.63826	36
25	.54107	1.84818	.56385	1.77351	.58709	1.70332	.61080	1.63719	35
26	.54145	1.84689	.56424	1.77230	.58748	1.70219	.61120	1.63612	34
27	.54183	1.84561	.56462	1.77110	.58787	1.70106	.61160	1.63505	33
28	.54220	1.84433	.56501	1.76990	.58826	1.69992	.61200	1.63398	32
29	.54258	1.84305	.56539	1.76869	.58865	1.69879	.61240	1.63292	31
30	.54296	1.84177	.56577	1.76749	.58905	1.69766	.61280	1.63185	30
31	.54333	1.84049	.56616	1.76629	.58944	1.69653	.61320	1.63079	29
32	.54371	1.83922	.56654	1.76510	.58983	1.69541	.61360	1.62972	28
33	.54409	1.83794	.56693	1.76390	.59022	1.69428	.61400	1.62866	27
34	.54446	1.83667	.56731	1.76271	.59061	1.69316	.61440	1.62760	26
35	.54484	1.83540	.56769	1.76151	.59101	1.69203	.61480	1.62654	25
36	.54522	1.83413	.56808	1.76032	.59140	1.69091	.61520	1.62548	24
37	.54560	1.83286	.56846	1.75913	.59179	1.68979	.61561	1.62442	23
38	.54597	1.83159	.56885	1.75794	.59218	1.68866	.61601	1.62336	22
39	.54635	1.83033	.56923	1.75675	.59258	1.68754	.61641	1.62230	21
40	.54673	1.82906	.56962	1.75556	.59297	1.68643	.61681	1.62125	20
41	.54711	1.82780	.57000	1.75437	.59336	1.68531	.61721	1.62019	19
42	.54748	1.82654	.57039	1.75319	.59376	1.68419	.61761	1.61914	18
43	.54786	1.82528	.57078	1.75200	.59415	1.68308	.61801	1.61808	17
44	.54824	1.82402	.57116	1.75082	.59454	1.68196	.61842	1.61703	16
45	.54862	1.82276	.57155	1.74964	.59494	1.68085	.61882	1.61598	15
46	.54900	1.82150	.57193	1.74846	.59533	1.67974	.61922	1.61493	14
47	.54938	1.82025	.57232	1.74728	.59573	1.67863	.61962	1.61388	13
48	.54975	1.81899	.57271	1.74610	.59612	1.67752	.62003	1.61283	12
49	.55013	1.81774	.57309	1.74492	.59651	1.67641	.62043	1.61179	11
50	.55051	1.81649	.57348	1.74375	.59691	1.67530	.62083	1.61074	10
51	.55089	1.81524	.57386	1.74257	.59730	1.67419	.62124	1.60970	9
52	.55127	1.81399	.57425	1.74140	.59770	1.67309	.62164	1.60865	8
53	.55165	1.81274	.57464	1.74022	.59809	1.67198	.62204	1.60761	7
54	.55203	1.81150	.57503	1.73905	.59849	1.67088	.62245	1.60657	6
55	.55241	1.81025	.57541	1.73788	.59888	1.66978	.62285	1.60553	5
56	.55279	1.80901	.57580	1.73671	.59928	1.66867	.62325	1.60449	4
57	.55317	1.80777	.57619	1.73555	.59967	1.66757	.62366	1.60345	3
58	.55355	1.80653	.57657	1.73438	.60007	1.66647	.62406	1.60241	2
59	.55393	1.80529	.57696	1.73321	.60046	1.66538	.62446	1.60137	1
60	.55431	1.80405	.57735	1.73205	.60086	1.66428	.62487	1.60033	0
′	Cotang	Tang	Cotang	Tang	Cotang	Tang	Cotang	Tang	′
	61°		60°		59°		58°		

TABLE XII.—TANGENTS AND COTANGENTS.

′	32° Tang	32° Cotang	33° Tang	33° Cotang	34° Tang	34° Cotang	35° Tang	35° Cotang	′
0	.62487	1.60033	.64941	1.53986	.67451	1.48256	.70021	1.42815	60
1	.62527	1.59930	.64982	1.53888	.67493	1.48163	.70064	1.42726	59
2	.62568	1.59826	.65024	1.53791	.67536	1.48070	.70107	1.42638	58
3	.62608	1.59723	.65065	1.53693	.67578	1.47977	.70151	1.42550	57
4	.62649	1.59620	.65106	1.53595	.67620	1.47885	.70194	1.42462	56
5	.62689	1.59517	.65148	1.53497	.67663	1.47792	.70238	1.42374	55
6	.62730	1.59414	.65189	1.53400	.67705	1.47699	.70281	1.42286	54
7	.62770	1.59311	.65231	1.53302	.67748	1.47607	.70325	1.42198	53
8	.62811	1.59208	.65272	1.53205	.67790	1.47514	.70368	1.42110	52
9	.62852	1.59105	.65314	1.53107	.67832	1.47422	.70412	1.42022	51
10	.62892	1.59002	.65355	1.53010	.67875	1.47330	.70455	1.41934	50
11	.62933	1.58900	.65397	1.52913	.67917	1.47238	.70499	1.41847	49
12	.62973	1.58797	.65438	1.52816	.67960	1.47146	.70542	1.41759	48
13	.63014	1.58695	.65480	1.52719	.68002	1.47053	.70586	1.41672	47
14	.63055	1.58593	.65521	1.52622	.68045	1.46962	.70629	1.41584	46
15	.63095	1.58490	.65563	1.52525	.68088	1.46870	.70673	1.41497	45
16	.63136	1.58388	.65604	1.52429	.68130	1.46778	.70717	1.41409	44
17	.63177	1.58286	.65646	1.52332	.68173	1.46686	.70760	1.41322	43
18	.63217	1.58184	.65688	1.52235	.68215	1.46595	.70804	1.41235	42
19	.63258	1.58083	.65729	1.52139	.68258	1.46503	.70848	1.41148	41
20	.63299	1.57981	.65771	1.52043	.68301	1.46411	.70891	1.41061	40
21	.63340	1.57879	.65813	1.51946	.68343	1.46320	.70935	1.40974	39
22	.63380	1.57778	.65854	1.51850	.68396	1.46229	.70979	1.40887	38
23	.63421	1.57676	.65896	1.51754	.68429	1.46137	.71023	1.40800	37
24	.63462	1.57575	.65938	1.51658	.68471	1.46046	.71066	1.40714	36
25	.63503	1.57474	.65980	1.51562	.68514	1.45955	.71110	1.40627	35
26	.63544	1.57372	.66021	1.51466	.68557	1.45864	.71154	1.40540	34
27	.63584	1.57271	.66063	1.51370	.68600	1.45773	.71198	1.40454	33
28	.63625	1.57170	.66105	1.51275	.68642	1.45682	.71242	1.40367	32
29	.63666	1.57069	.66147	1.51179	.68685	1.45592	.71285	1.40281	31
30	.63707	1.56969	.66189	1.51084	.68728	1.45501	.71329	1.40195	30
31	.63748	1.56868	.66230	1.50988	.68771	1.45410	.71373	1.40109	29
32	.63789	1.56767	.66272	1.50893	.68814	1.45320	.71417	1.40022	28
33	.63830	1.56667	.66314	1.50797	.68857	1.45229	.71461	1.39936	27
34	.63871	1.56566	.66356	1.50702	.68900	1.45139	.71505	1.39850	26
35	.63912	1.56466	.66398	1.50607	.68942	1.45049	.71549	1.39764	25
36	.63953	1.56366	.66440	1.50512	.68985	1.44958	.71593	1.39679	24
37	.63994	1.56265	.66482	1.50417	.69028	1.44868	.71637	1.39593	23
38	.64035	1.56165	.66524	1.50322	.69071	1.44778	.71681	1.39507	22
39	.64076	1.56065	.66566	1.50228	.69114	1.44688	.71725	1.39421	21
40	.64117	1.55966	.66608	1.50133	.69157	1.44598	.71769	1.39336	20
41	.64158	1.55866	.66650	1.50038	.69200	1.44508	.71813	1.39250	19
42	.64199	1.55766	.66692	1.49944	.69243	1.44418	.71857	1.39165	18
43	.64240	1.55666	.66734	1.49849	.69286	1.44329	.71901	1.39079	17
44	.64281	1.55567	.66776	1.49755	.69329	1.44239	.71946	1.38994	16
45	.64322	1.55467	.66818	1.49661	.69372	1.44149	.71990	1.38909	15
46	.64363	1.55368	.66860	1.49566	.69416	1.44060	.72034	1.38824	14
47	.64404	1.55269	.66902	1.49472	.69459	1.43970	.72078	1.38738	13
48	.64446	1.55170	.66944	1.49378	.69502	1.43881	.72122	1.38653	12
49	.64487	1.55071	.66986	1.49284	.69545	1.43792	.72167	1.38568	11
50	.64528	1.54972	.67028	1.49190	.69588	1.43703	.72211	1.38484	10
51	.64569	1.54873	.67071	1.49097	.69631	1.43614	.72255	1.38399	9
52	.64610	1.54774	.67113	1.49003	.69675	1.43525	.72299	1.38314	8
53	.64652	1.54675	.67155	1.48909	.69718	1.43436	.72344	1.38229	7
54	.64693	1.54576	.67197	1.48816	.69761	1.43347	.72388	1.38145	6
55	.64734	1.54478	.67239	1.48722	.69804	1.43258	.72432	1.38060	5
56	.64775	1.54379	.67282	1.48629	.69847	1.43169	.72477	1.37976	4
57	.64817	1.54281	.67324	1.48536	.69891	1.43080	.72521	1.37891	3
58	.64858	1.54183	.67366	1.48442	.69934	1.42992	.72565	1.37807	2
59	.64899	1.54085	.67409	1.48349	.69977	1.42903	.72610	1.37722	1
60	.64941	1.53986	.67451	1.48256	.70021	1.42815	.72654	1.37638	0
′	Cotang	Tang	Cotang	Tang	Cotang	Tang	Cotang	Tang	′
	57°		56°		55°		54°		

,	36°		37°		38°		39°		,
	Tang	Cotang	Tang	Cotang	Tang	Cotang	Tang	Cotang	
0	.72654	1.37638	.75355	1.32704	.78129	1.27994	.80978	1.23490	60
1	.72699	1.37554	.75401	1.32624	.78175	1.27917	.81027	1.23416	59
2	.72743	1.37470	.75447	1.32544	.78222	1.27841	.81075	1.23343	58
3	.72788	1.37386	.75492	1.32464	.78269	1.27764	.81123	1.23270	57
4	.72832	1.37302	.75538	1.32384	.78316	1.27688	.81171	1.23196	56
5	.72877	1.37218	.75584	1.32304	.78363	1.27611	.81220	1.23123	55
6	.72921	1.37134	.75629	1.32224	.78410	1.27535	.81268	1.23050	54
7	.72966	1.37050	.75675	1.32144	.78457	1.27458	.81316	1.22977	53
8	.73010	1.36967	.75721	1.32064	.78504	1.27382	.81364	1.22904	52
9	.73055	1.36883	.75767	1.31984	.78551	1.27306	.81413	1.22831	51
10	.73100	1.36800	.75812	1.31904	.78598	1.27230	.81461	1.22758	50
11	.73144	1.36716	.75858	1.31825	.78645	1.27153	.81510	1.22685	49
12	.73189	1.36633	.75904	1.31745	.78692	1.27077	.81558	1.22612	48
13	.73234	1.36549	.75950	1.31666	.78739	1.27001	.81606	1.22539	47
14	.73278	1.36466	.75996	1.31586	.78786	1.26925	.81655	1.22467	46
15	.73323	1.36383	.76042	1.31507	.78834	1.26849	.81703	1.22394	45
16	.73368	1.36300	.76088	1.31427	.78881	1.26774	.81752	1.22321	44
17	.73413	1.36217	.76134	1.31348	.78928	1.26698	.81800	1.22249	43
18	.73457	1.36134	.76180	1.31269	.78975	1.26622	.81849	1.22176	42
19	.73502	1.36051	.76226	1.31190	.79022	1.26546	.81898	1.22104	41
20	.73547	1.35968	.76272	1.31110	.79070	1.26471	.81946	1.22031	40
21	.73592	1.35885	.76318	1.31031	.79117	1.26395	.81995	1.21959	39
22	.73637	1.35802	.76364	1.30952	.79164	1.26319	.82044	1.21886	38
23	.73681	1.35719	.76410	1.30873	.79212	1.26244	.82092	1.21814	37
24	.73726	1.35637	.76456	1.30795	.79259	1.26169	.82141	1.21742	36
25	.73771	1.35554	.76502	1.30716	.79306	1.26093	.82190	1.21670	35
26	.73816	1.35472	.76548	1.30637	.79354	1.26018	.82238	1.21598	34
27	.73861	1.35389	.76594	1.30558	.79401	1.25943	.82287	1.21526	33
28	.73906	1.35307	.76640	1.30480	.79449	1.25867	.82336	1.21454	32
29	.73951	1.35224	.76686	1.30401	.79496	1.25792	.82385	1.21382	31
30	.73996	1.35142	.76733	1.30323	.79544	1.25717	.82434	1.21310	30
31	.74041	1.35060	.76779	1.30244	.79591	1.25642	.82483	1.21238	29
32	.74086	1.34978	.76825	1.30166	.79639	1.25567	.82531	1.21166	28
33	.74131	1.34896	.76871	1.30087	.79686	1.25492	.82580	1.21094	27
34	.74176	1.34814	.76918	1.30009	.79734	1.25417	.82629	1.21023	26
35	.74221	1.34732	.76964	1.29931	.79781	1.25343	.82678	1.20951	25
36	.74267	1.34650	.77010	1.29853	.79829	1.25268	.82727	1.20879	24
37	.74312	1.34568	.77057	1.29775	.79877	1.25193	.82776	1.20808	23
38	.74357	1.34487	.77103	1.29696	.79924	1.25118	.82825	1.20736	22
39	.74402	1.34405	.77149	1.29618	.79972	1.25044	.82874	1.20665	21
40	.74447	1.34323	.77196	1.29541	.80020	1.24969	.82923	1.20593	20
41	.74492	1.34242	.77242	1.29463	.80067	1.24895	.82972	1.20522	19
42	.74538	1.34160	.77289	1.29385	.80115	1.24820	.83022	1.20451	18
43	.74583	1.34079	.77335	1.29307	.80163	1.24746	.83071	1.20379	17
44	.74628	1.33998	.77382	1.29229	.80211	1.24672	.83120	1.20308	16
45	.74674	1.33916	.77428	1.29152	.80258	1.24597	.83169	1.20237	15
46	.74719	1.33835	.77475	1.29074	.80306	1.24523	.83218	1.20166	14
47	.74764	1.33754	.77521	1.28997	.80354	1.24449	.83268	1.20095	13
48	.74810	1.33673	.77568	1.28919	.80402	1.24375	.83317	1.20024	12
49	.74855	1.33592	.77615	1.28842	.80450	1.24301	.83366	1.19953	11
50	.74900	1.33511	.77661	1.28764	.80498	1.24227	.83415	1.19882	10
51	.74946	1.33430	.77708	1.28687	.80546	1.24153	.83465	1.19811	9
52	.74991	1.33349	.77754	1.28610	.80594	1.24079	.83514	1.19740	8
53	.75037	1.33268	.77801	1.28533	.80642	1.24005	.83564	1.19669	7
54	.75082	1.33187	.77848	1.28456	.80690	1.23931	.83613	1.19599	6
55	.75128	1.33107	.77895	1.28379	.80738	1.23858	.83662	1.19528	5
56	.75173	1.33026	.77941	1.28302	.80786	1.23784	.83712	1.19457	4
57	.75219	1.32946	.77988	1.28225	.80834	1.23710	.83761	1.19387	3
58	.75264	1.32865	.78035	1.28148	.80882	1.23637	.83811	1.19316	2
59	.75310	1.32785	.78082	1.28071	.80930	1.23563	.83860	1.19246	1
60	.75355	1.32704	.78129	1.27994	.80978	1.23490	.83910	1.19175	0
,	Cotang	Tang	Cotang	Tang	Cotang	Tang	Cotang	Tang	,
	53°		52°		51°		50°		

TABLE XII.—TANGENTS AND COTANGENTS.

′	40° Tang	40° Cotang	41° Tang	41° Cotang	42° Tang	42° Cotang	43° Tang	43° Cotang	′
0	.83910	1.19175	.86929	1.15037	.90040	1.11061	.93252	1.07237	60
1	.83960	1.19105	.86980	1.14969	.90093	1.10996	.93306	1.07174	59
2	.84009	1.19035	.87031	1.14902	.90146	1.10931	.93360	1.07112	58
3	.84059	1.18964	.87082	1.14834	.90199	1.10867	.93415	1.07049	57
4	.84108	1.18894	.87133	1.14767	.90251	1.10802	.93469	1.06987	56
5	.84158	1.18824	.87184	1.14699	.90304	1.10737	.93524	1.06925	55
6	.84208	1.18754	.87236	1.14632	.90357	1.10672	.93578	1.06862	54
7	.84258	1.18684	.87287	1.14565	.90410	1.10607	.93633	1.06800	53
8	.84307	1.18614	.87338	1.14498	.90463	1.10543	.93688	1.06738	52
9	.84357	1.18544	.87389	1.14430	.90516	1.10478	.93742	1.06676	51
10	.84407	1.18474	.87441	1.14363	.90569	1.10414	.93797	1.06613	50
11	.84457	1.18404	.87492	1.14296	.90621	1.10349	.93852	1.06551	49
12	.84507	1.18334	.87543	1.14229	.90674	1.10285	.93906	1.06489	48
13	.84556	1.18264	.87595	1.14162	.90727	1.10220	.93961	1.06427	47
14	.84606	1.18194	.87646	1.14095	.90781	1.10156	.94016	1.06365	46
15	.84656	1.18125	.87698	1.14028	.90834	1.10091	.94071	1.06303	45
16	.84706	1.18055	.87749	1.13961	.90887	1.10027	.94125	1.06241	44
17	.84756	1.17986	.87801	1.13894	.90940	1.09963	.94180	1.06179	43
18	.84806	1.17916	.87852	1.13828	.90993	1.09899	.94235	1.06117	42
19	.84856	1.17846	.87904	1.13761	.91046	1.09834	.94290	1.06056	41
20	.84906	1.17777	.87955	1.13694	.91099	1.09770	.94345	1.05994	40
21	.84956	1.17708	.88007	1.13627	.91153	1.09706	.94400	1.05932	39
22	.85006	1.17638	.88059	1.13561	.91206	1.09642	.94455	1.05870	38
23	.85057	1.17569	.88110	1.13494	.91259	1.09578	.94510	1.05809	37
24	.85107	1.17500	.88162	1.13428	.91313	1.09514	.94565	1.05747	36
25	.85157	1.17430	.88214	1.13361	.91366	1.09450	.94620	1.05685	35
26	.85207	1.17361	.88265	1.13295	.91419	1.09386	.94676	1.05624	34
27	.85257	1.17292	.88317	1.13228	.91473	1.09322	.94731	1.05562	33
28	.85308	1.17223	.88369	1.13162	.91526	1.09258	.94786	1.05501	32
29	.85358	1.17154	.88421	1.13096	.91580	1.09195	.94841	1.05439	31
30	.85408	1.17085	.88473	1.13029	.91633	1.09131	.94896	1.05378	30
31	.85458	1.17016	.88524	1.12963	.91687	1.09067	.94952	1.05317	29
32	.85509	1.16947	.88576	1.12897	.91740	1.09003	.95007	1.05255	28
33	.85559	1.16878	.88628	1.12831	.91794	1.08940	.95062	1.05194	27
34	.85609	1.16809	.88680	1.12765	.91847	1.08876	.95118	1.05133	26
35	.85660	1.16741	.88732	1.12699	.91901	1.08813	.95173	1.05072	25
36	.85710	1.16672	.88784	1.12633	.91955	1.08749	.95229	1.05010	24
37	.85761	1.16603	.88836	1.12567	.92008	1.08686	.95284	1.04949	23
38	.85811	1.16535	.88888	1.12501	.92062	1.08622	.95340	1.04888	22
39	.85862	1.16466	.88940	1.12435	.92116	1.08559	.95395	1.04827	21
40	.85912	1.16398	.88992	1.12369	.92170	1.08496	.95451	1.04766	20
41	.85963	1.16329	.89045	1.12303	.92224	1.08432	.95506	1.04705	19
42	.86014	1.16261	.89097	1.12238	.92277	1.08369	.95562	1.04644	18
43	.86064	1.16192	.89149	1.12172	.92331	1.08306	.95618	1.04583	17
44	.86115	1.16124	.89201	1.12106	.92385	1.08243	.95673	1.04522	16
45	.86166	1.16056	.89253	1.12041	.92439	1.08179	.95729	1.04461	15
46	.86216	1.15987	.89306	1.11975	.92493	1.08116	.95785	1.04401	14
47	.86267	1.15919	.89358	1.11909	.92547	1.08053	.95841	1.04340	13
48	.86318	1.15851	.89410	1.11844	.92601	1.07990	.95897	1.04279	12
49	.86368	1.15783	.89463	1.11778	.92655	1.07927	.95952	1.04218	11
50	.86419	1.15715	.89515	1.11713	.92709	1.07864	.96008	1.04158	10
51	.86470	1.15647	.89567	1.11648	.92763	1.07801	.96064	1.04097	9
52	.86521	1.15579	.89620	1.11582	.92817	1.07738	.96120	1.04036	8
53	.86572	1.15511	.89672	1.11517	.92872	1.07676	.96176	1.03976	7
54	.86623	1.15443	.89725	1.11452	.92926	1.07613	.96232	1.03915	6
55	.86674	1.15375	.89777	1.11387	.92980	1.07550	.96288	1.03855	5
56	.86725	1.15308	.89830	1.11321	.93034	1.07487	.96344	1.03794	4
57	.86776	1.15240	.89883	1.11256	.93088	1.07425	.96400	1.03734	3
58	.86827	1.15172	.89935	1.11191	.93143	1.07362	.96457	1.03674	2
59	.86878	1.15104	.89988	1.11126	.93197	1.07299	.96513	1.03613	1
60	.86929	1.15037	.90040	1.11061	.93252	1.07237	.96569	1.03553	0
′	Cotang	Tang	Cotang	Tang	Cotang	Tang	Cotang	Tang	′
	49°		48°		47°		46°		

TABLE XII.—TANGENTS AND COTANGENTS.

′	44°		′	′	44°		′	′	44°		′
	Tang	Cotang			Tang	Cotang			Tang	Cotang	
0	.96569	1.03553	60	20	.97700	1.02355	40	40	.98843	1.01170	20
1	.96625	1.03493	59	21	.97756	1.02295	39	41	.98901	1.01112	19
2	.96681	1.03433	58	22	.97813	1.02236	38	42	.98958	1.01053	18
3	.96738	1.03372	57	23	.97870	1.02176	37	43	.99016	1.00994	17
4	.96794	1.03312	56	24	.97927	1.02117	36	44	.99073	1.00935	16
5	.96850	1.03252	55	25	.97984	1.02057	35	45	.99131	1.00876	15
6	.96907	1.03192	54	26	.98041	1.01998	34	46	.99189	1.00818	14
7	.96963	1.03132	53	27	.98098	1.01939	33	47	.99247	1.00759	13
8	.97020	1.03072	52	28	.98155	1.01879	32	48	.99304	1.00701	12
9	.97076	1.03012	51	29	.98213	1.01820	31	49	.99362	1.00642	11
10	.97133	1.02952	50	30	.98270	1.01761	30	50	.99420	1.00583	10
11	.97189	1.02892	49	31	.98327	1.01702	29	51	.99478	1.00525	9
12	.97246	1.02832	48	32	.98384	1.01642	28	52	.99536	1.00467	8
13	.97302	1.02772	47	33	.98441	1.01583	27	53	.99594	1.00408	7
14	.97359	1.02713	46	34	.98499	1.01524	26	54	.99652	1.00350	6
15	.97416	1.02653	45	35	.98556	1.01465	25	55	.99710	1.00291	5
16	.97472	1.02593	44	36	.98613	1.01406	24	56	.99768	1.00233	4
17	.97529	1.02533	43	37	.98671	1.01347	23	57	.99826	1.00175	3
18	.97586	1.02474	42	38	.98728	1.01288	22	58	.99884	1.00116	2
19	.97643	1.02414	41	39	.98786	1.01229	21	59	.99942	1.00058	1
20	.97700	1.02355	40	40	.98843	1.01170	20	60	1.00000	1.00000	0
	Cotang	Tang			Cotang	Tang			Cotang	Tang	
′	45°		′	′	45°		′	′	45°		′

TABLE XIII.—VERSINES AND EXSECANTS.

′	0°		1°		2°		3°		′
	Vers.	Exsec.	Vers.	Exsec.	Vers.	Exsec.	Vers.	Exsec.	
0	.00000	.00000	.00015	.00015	.00061	.00061	.00137	.00137	0
1	.00000	.00000	.00016	.00016	.00062	.00062	.00139	.00139	1
2	.00000	.00000	.00016	.00016	.00063	.00063	.00140	.00140	2
3	.00000	.00000	.00017	.00017	.00064	.00064	.00142	.00142	3
4	.00000	.00000	.00017	.00017	.00065	.00065	.00143	.00143	4
5	.00000	.00000	.00018	.00018	.00066	.00066	.00145	.00145	5
6	.00000	.00000	.00018	.00018	.00067	.00067	.00146	.00147	6
7	.00000	.00000	.00019	.00019	.00068	.00068	.00148	.00148	7
8	.00000	.00000	.00020	.00020	.00069	.00069	.00150	.00150	8
9	.00000	.00000	.00020	.00020	.00070	.00070	.00151	.00151	9
10	.00000	.00000	.00021	.00021	.00071	.00072	.00153	.00153	10
11	.00001	.00001	.00021	.00021	.00073	.00073	.00154	.00155	11
12	.00001	.00001	.00022	.00022	.00074	.00074	.00156	.00156	12
13	.00001	.00001	.00023	.00023	.00075	.00075	.00158	.00158	13
14	.00001	.00001	.00023	.00023	.00076	.00076	.00159	.00159	14
15	.00001	.00001	.00024	.00024	.00077	.00077	.00161	.00161	15
16	.00001	.00001	.00024	.00024	.00078	.00078	.00162	.00163	16
17	.00001	.00001	.00025	.00025	.00079	.00079	.00164	.00164	17
18	.00001	.00001	.00026	.00026	.00081	.00081	.00166	.00166	18
19	.00002	.00002	.00026	.00026	.00082	.00082	.00168	.00168	19
20	.00002	.00002	.00027	.00027	.00083	.00083	.00169	.00169	20
21	.00002	.00002	.00028	.00028	.00084	.00084	.00171	.00171	21
22	.00002	.00002	.00028	.00028	.00085	.00085	.00173	.00173	22
23	.00002	.00002	.00029	.00029	.00087	.00087	.00174	.00175	23
24	.00002	.00002	.00030	.00030	.00088	.00088	.00176	.00176	24
25	.00003	.00003	.00031	.00031	.00089	.00089	.00178	.00178	25
26	.00003	.00003	.00031	.00031	.00090	.00090	.00179	.00180	26
27	.00003	.00003	.00032	.00032	.00091	.00091	.00181	.00182	27
28	.00003	.00003	.00033	.00033	.00093	.00093	.00183	.00183	28
29	.00004	.00004	.00034	.00034	.00094	.00094	.00185	.00185	29
30	.00004	.00004	.00034	.00034	.00095	.00095	.00187	.00187	30
31	.00004	.00004	.00035	.00035	.00096	.00097	.00188	.00189	31
32	.00004	.00004	.00036	.00036	.00098	.00098	.00190	.00190	32
33	.00005	.00005	.00037	.00037	.00099	.00099	.00192	.00192	33
34	.00005	.00005	.00037	.00037	.00100	.00100	.00194	.00194	34
35	.00005	.00005	.00038	.00038	.00102	.00102	.00196	.00196	35
36	.00005	.00005	.00039	.00039	.00103	.00103	.00197	.00198	36
37	.00006	.00006	.00040	.00040	.00104	.00104	.00199	.00200	37
38	.00006	.00006	.00041	.00041	.00106	.00106	.00201	.00201	38
39	.00006	.00006	.00041	.00041	.00107	.00107	.00203	.00203	39
40	.00007	.00007	.00042	.00042	.00108	.00108	.00205	.00205	40
41	.00007	.00007	.00043	.00043	.00110	.00110	.00207	.00207	41
42	.00007	.00007	.00044	.00044	.00111	.00111	.00208	.00209	42
43	.00008	.00008	.00045	.00045	.00112	.00113	.00210	.00211	43
44	.00008	.00008	.00046	.00046	.00114	.00114	.00212	.00213	44
45	.00009	.00009	.00047	.00047	.00115	.00115	.00214	.00215	45
46	.00009	.00009	.00048	.00048	.00117	.00117	.00216	.00216	46
47	.00009	.00009	.00048	.00048	.00118	.00118	.00218	.00218	47
48	.00010	.00010	.00049	.00049	.00119	.00120	.00220	.00220	48
49	.00010	.00010	.00050	.00050	.00121	.00121	.00222	.00222	49
50	.00011	.00011	.00051	.00051	.00122	.00122	.00224	.00224	50
51	.00011	.00011	.00052	.00052	.00124	.00124	.00226	.00226	51
52	.00011	.00011	.00053	.00053	.00125	.00125	.00228	.00228	52

TABLE XIII.—VERSINES AND EXSECANTS.

′	4°		5°		6°		7°		′
	Vers.	Exsec.	Vers.	Exsec.	Vers.	Exsec.	Vers.	Exsec.	
0	.00244	.00244	.00381	.00382	.00548	.00551	.00745	.00751	0
1	.00246	.00246	.00383	.00385	.00551	.00554	.00749	.00755	1
2	.00248	.00248	.00386	.00387	.00554	.00557	.00752	.00758	2
3	.00250	.00250	.00388	.00390	.00557	.00560	.00756	.00762	3
4	.00252	.00252	.00391	.00392	.00560	.00563	.00760	.00765	4
5	.00254	.00254	.00393	.00395	.00563	.00566	.00763	.00769	5
6	.00256	.00257	.00396	.00397	.00566	.00569	.00767	.00773	6
7	.00258	.00259	.00398	.00400	.00569	.00573	.00770	.00776	7
8	.00260	.00261	.00401	.00403	.00572	.00576	.00774	.00780	8
9	.00262	.00263	.00404	.00405	.00576	.00579	.00778	.00784	9
10	.00264	.00265	.00406	.00408	.00579	.00582	.00781	.00787	10
11	.00266	.00267	.00409	.00411	.00582	.00585	.00785	.00791	11
12	.00269	.00269	.00412	.00413	.00585	.00588	.00789	.00795	12
13	.00271	.00271	.00414	.00416	.00588	.00592	.00792	.00799	13
14	.00273	.00274	.00417	.00419	.00591	.00595	.00796	.00802	14
15	.00275	.00276	.00420	.00421	.00594	.00598	.00800	.00806	15
16	.00277	.00278	.00422	.00424	.00598	.00601	.00803	.00810	16
17	.00279	.00280	.00425	.00427	.00601	.00604	.00807	.00813	17
18	.00281	.00282	.00428	.00429	.00604	.00608	.00811	.00817	18
19	.00284	.00284	.00430	.00432	.00607	.00611	.00814	.00821	19
20	.00286	.00287	.00433	.00435	.00610	.00614	.00818	.00825	20
21	.00288	.00289	.00436	.00438	.00614	.00617	.00822	.00828	21
22	.00290	.00291	.00438	.00440	.00617	.00621	.00825	.00832	22
23	.00293	.00293	.00441	.00443	.00620	.00624	.00829	.00836	23
24	.00295	.00296	.00444	.00446	.00623	.00627	.00833	.00840	24
25	.00297	.00298	.00447	.00449	.00626	.00630	.00837	.00844	25
26	.00299	.00300	.00449	.00451	.00630	.00634	.00840	.00848	26
27	.00301	.00302	.00452	.00454	.00633	.00637	.00844	.00851	27
28	.00304	.00305	.00455	.00457	.00636	.00640	.00848	.00855	28
29	.00306	.00307	.00458	.00460	.00640	.00644	.00852	.00859	29
30	.00308	.00309	.00460	.00463	.00643	.00647	.00856	.00863	30
31	.00311	.00312	.00463	.00465	.00646	.00650	.00859	.00867	31
32	.00313	.00314	.00466	.00468	.00649	.00654	.00863	.00871	32
33	.00315	.00316	.00469	.00471	.00653	.00657	.00867	.00875	33
34	.00317	.00318	.00472	.00474	.00656	.00660	.00871	.00878	34
35	.00320	.00321	.00474	.00477	.00659	.00664	.00875	.00882	35
36	.00322	.00323	.00477	.00480	.00663	.00667	.00878	.00886	36
37	.00324	.00326	.00480	.00482	.00666	.00671	.00882	.00890	37
38	.00327	.00328	.00483	.00485	.00669	.00674	.00886	.00894	38
39	.00329	.00330	.00486	.00488	.00673	.00677	.00890	.00898	39
40	.00332	.00333	.00489	.00491	.00676	.00681	.00894	.00902	40
41	.00334	.00335	.00492	.00494	.00680	.00684	.00898	.00906	41
42	.00336	.00337	.00494	.00497	.00683	.00688	.00902	.00910	42
43	.00339	.00340	.00497	.00500	.00686	.00691	.00906	.00914	43
44	.00341	.00342	.00500	.00503	.00690	.00695	.00909	.00918	44
45	.00343	.00345	.00503	.00506	.00693	.00698	.00913	.00922	45
46	.00346	.00347	.00506	.00509	.00697	.00701	.00917	.00926	46
47	.00348	.00350	.00509	.00512	.00700	.00705	.00921	.00930	47
48	.00351	.00352	.00512	.00515	.00703	.00708	.00925	.00934	48
49	.00353	.00354	.00515	.00518	.00707	.00712	.00929	.00938	49
50	.00356	.00357	.00518	.00521	.00710	.00715	.00933	.00942	50
51	.00358	.00359	.00521	.00524	.00714	.00719	.00937	.00946	51
52	.00361	.00362	.00524	.00527	.00717	.00722	.00941	.00950	52
53	.00363	.00364	.00527	.00530	.00721	.00726	.00945	.00954	53
54	.00365	.00367	.00530	.00533	.00724	.00730	.00949	.00958	54
55	.00368	.00369	.00533	.00536	.00728	.00733	.00953	.00962	55
56	.00370	.00372	.00536	.00539	.00731	.00737	.00957	.00966	56
57	.00373	.00374	.00539	.00542	.00735	.00740	.00961	.00970	57
58	.00375	.00377	.00542	.00545	.00738	.00744	.00965	.00975	58
59	.00378	.00379	.00545	.00548	.00742	.00747	.00969	.00979	59
60	.00381	.00382	.00548	.00551	.00745	.00751	.00973	.00983	60

TABLE XIII.—VERSINES AND EXSECANTS.

′	8°		9°		10°		11°		′
	Vers.	Exsec.	Vers.	Exsec.	Vers.	Exsec.	Vers.	Exsec.	
0	.00973	.00983	.01231	.01247	.01519	.01543	.01837	.01872	0
1	.00977	.00987	.01236	.01251	.01524	.01548	.01843	.01877	1
2	.00981	.00991	.01240	.01256	.01529	.01553	.01848	.01883	2
3	.00985	.00995	.01245	.01261	.01534	.01558	.01854	.01889	3
4	.00989	.00999	.01249	.01265	.01540	.01564	.01860	.01895	4
5	.00994	.01004	.01254	.01270	.01545	.01569	.01865	.01901	5
6	.00998	.01008	.01259	.01275	.01550	.01574	.01871	.01906	6
7	.01002	.01012	.01263	.01279	.01555	.01579	.01876	.01912	7
8	.01006	.01016	.01268	.01284	.01560	.01585	.01882	.01918	8
9	.01010	.01020	.01272	.01289	.01565	.01590	.01888	.01924	9
10	.01014	.01024	.01277	.01294	.01570	.01595	.01893	.01930	10
11	.01018	.01029	.01282	.01298	.01575	.01601	.01899	.01936	11
12	.01022	.01033	.01286	.01303	.01580	.01606	.01904	.01941	12
13	.01027	.01037	.01291	.01308	.01586	.01611	.01910	.01947	13
14	.01031	.01041	.01296	.01313	.01591	.01616	.01916	.01953	14
15	.01035	.01046	.01300	.01318	.01596	.01622	.01921	.01959	15
16	.01039	.01050	.01305	.01322	.01601	.01627	.01927	.01965	16
17	.01043	.01054	.01310	.01327	.01606	.01633	.01933	.01971	17
18	.01047	.01059	.01314	.01332	.01612	.01638	.01939	.01977	18
19	.01052	.01063	.01319	.01337	.01617	.01643	.01944	.01983	19
20	.01056	.01067	.01324	.01342	.01622	.01649	.01950	.01989	20
21	.01060	.01071	.01329	.01346	.01627	.01654	.01956	.01995	21
22	.01064	.01076	.01333	.01351	.01632	.01659	.01961	.02001	22
23	.01069	.01080	.01338	.01356	.01638	.01665	.01967	.02007	23
24	.01073	.01084	.01343	.01361	.01643	.01670	.01973	.02013	24
25	.01077	.01089	.01348	.01366	.01648	.01676	.01979	.02019	25
26	.01081	.01093	.01352	.01371	.01653	.01681	.01984	.02025	26
27	.01085	.01097	.01357	.01376	.01659	.01687	.01990	.02031	27
28	.01090	.01102	.01362	.01381	.01664	.01692	.01996	.02037	28
29	.01094	.01106	.01367	.01386	.01669	.01698	.02002	.02043	29
30	.01098	.01111	.01371	.01391	.01675	.01703	.02008	.02049	30
31	.01103	.01115	.01376	.01395	.01680	.01709	.02013	.02055	31
32	.01107	.01119	.01381	.01400	.01685	.01714	.02019	.02061	32
33	.01111	.01124	.01386	.01405	.01690	.01720	.02025	.02067	33
34	.01116	.01128	.01391	.01410	.01696	.01725	.02031	.02073	34
35	.01120	.01133	.01396	.01415	.01701	.01731	.02037	.02079	35
36	.01124	.01137	.01400	.01420	.01706	.01736	.02042	.02085	36
37	.01129	.01142	.01405	.01425	.01712	.01742	.02048	.02091	37
38	.01133	.01146	.01410	.01430	.01717	.01747	.02054	.02097	38
39	.01137	.01151	.01415	.01435	.01723	.01753	.02060	.02103	39
40	.01142	.01155	.01420	.01440	.01728	.01758	.02066	.02110	40
41	.01146	.01160	.01425	.01445	.01733	.01764	.02072	.02116	41
42	.01151	.01164	.01430	.01450	.01739	.01769	.02078	.02122	42
43	.01155	.01169	.01435	.01455	.01744	.01775	.02084	.02128	43
44	.01159	.01173	.01439	.01461	.01750	.01781	.02090	.02134	44
45	.01164	.01178	.01444	.01466	.01755	.01786	.02095	.02140	45
46	.01168	.01182	.01449	.01471	.01760	.01792	.02101	.02146	46
47	.01173	.01187	.01454	.01476	.01766	.01798	.02107	.02153	47
48	.01177	.01191	.01459	.01481	.01771	.01803	.02113	.02159	48
49	.01182	.01196	.01464	.01486	.01777	.01809	.02119	.02165	49
50	.01186	.01200	.01469	.01491	.01782	.01815	.02125	.02171	50
51	.01191	.01205	.01474	.01496	.01788	.01820	.02131	.02178	51
52	.01195	.01209	.01479	.01501	.01793	.01826	.02137	.02184	52
53	.01200	.01214	.01484	.01506	.01799	.01832	.02143	.02190	53
54	.01204	.01219	.01489	.01512	.01804	.01837	.02149	.02196	54
55	.01209	.01223	.01494	.01517	.01810	.01843	.02155	.02203	55
56	.01213	.01228	.01499	.01522	.01815	.01849	.02161	.02209	56
57	.01218	.01233	.01504	.01527	.01821	.01854	.02167	.02215	57
58	.01222	.01237	.01509	.01532	.01826	.01860	.02173	.02221	58
59	.01227	.01242	.01514	.01537	.01832	.01866	.02179	.02228	59
60	.01231	.01247	.01519	.01543	.01837	.01872	.02185	.02234	60

TABLE XIII.—VERSINES AND EXSECANTS.

′	12° Vers.	12° Exsec.	13° Vers.	13° Exsec.	14° Vers.	14° Exsec.	15° Vers.	15° Exsec.	′
0	.02185	.02234	.02563	.02630	.02970	.03061	.03407	.03528	0
1	.02191	.02240	.02570	.02637	.02977	.03069	.03415	.03536	1
2	.02197	.02247	.02576	.02644	.02985	.03076	.03422	.03544	2
3	.02203	.02253	.02583	.02651	.02992	.03084	.03430	.03552	3
4	.02210	.02259	.02589	.02658	.02999	.03091	.03438	.03560	4
5	.02216	.02266	.02596	.02665	.03006	.03099	.03445	.03568	5
6	.02222	.02272	.02602	.02672	.03013	.03106	.03453	.03576	6
7	.02228	.02279	.02609	.02679	.03020	.03114	.03460	.03584	7
8	.02234	.02285	.02616	.02686	.03027	.03121	.03468	.03592	8
9	.02240	.02291	.02622	.02693	.03034	.03129	.03476	.03601	9
10	.02246	.02298	.02629	.02700	.03041	.03137	.03483	.03609	10
11	.02252	.02304	.02635	.02707	.03048	.03144	.03491	.03617	11
12	.02258	.02311	.02642	.02714	.03055	.03152	.03498	.03625	12
13	.02265	.02317	.02649	.02721	.03063	.03159	.03506	.03633	13
14	.02271	.02323	.02655	.02728	.03070	.03167	.03514	.03642	14
15	.02277	.02330	.02662	.02735	.03077	.03175	.03521	.03650	15
16	.02283	.02336	.02669	.02742	.03084	.03182	.03529	.03658	16
17	.02289	.02343	.02675	.02749	.03091	.03190	.03537	.03666	17
18	.02295	.02349	.02682	.02756	.03098	.03198	.03544	.03674	18
19	.02302	.02356	.02689	.02763	.03106	.03205	.03552	.03683	19
20	.02308	.02362	.02696	.02770	.03113	.03213	.03560	.03691	20
21	.02314	.02369	.02702	.02777	.03120	.03221	.03567	.03699	21
22	.02320	.02375	.02709	.02784	.03127	.03228	.03575	.03708	22
23	.02327	.02382	.02716	.02791	.03134	.03236	.03583	.03716	23
24	.02333	.02388	.02722	.02799	.03142	.03244	.03590	.03724	24
25	.02339	.02395	.02729	.02806	.03149	.03251	.03598	.03732	25
26	.02345	.02402	.02736	.02813	.03156	.03259	.03606	.03741	26
27	.02352	.02408	.02743	.02820	.03163	.03267	.03614	.03749	27
28	.02358	.02415	.02749	.02827	.03171	.03275	.03621	.03758	28
29	.02364	.02421	.02756	.02834	.03178	.03282	.03629	.03766	29
30	.02370	.02428	.02763	.02842	.03185	.03290	.03637	.03774	30
31	.02377	.02435	.02770	.02849	.03193	.03298	.03645	.03783	31
32	.02383	.02441	.02777	.02856	.03200	.03306	.03653	.03791	32
33	.02389	.02448	.02783	.02863	.03207	.03313	.03660	.03799	33
34	.02396	.02454	.02790	.02870	.03214	.03321	.03668	.03808	34
35	.02402	.02461	.02797	.02878	.03222	.03329	.03676	.03816	35
36	.02408	.02468	.02804	.02885	.03229	.03337	.03684	.03825	36
37	.02415	.02474	.02811	.02892	.03236	.03345	.03692	.03833	37
38	.02421	.02481	.02818	.02899	.03244	.03353	.03699	.03842	38
39	.02427	.02488	.02824	.02907	.03251	.03360	.03707	.03850	39
40	.02434	.02494	.02831	.02914	.03258	.03368	.03715	.03858	40
41	.02440	.02501	.02838	.02921	.03266	.03376	.03723	.03867	41
42	.02447	.02508	.02845	.02928	.03273	.03384	.03731	.03875	42
43	.02453	.02515	.02852	.02936	.03281	.03392	.03739	.03884	43
44	.02459	.02521	.02859	.02943	.03288	.03400	.03747	.03892	44
45	.02466	.02528	.02866	.02950	.03295	.03408	.03754	.03901	45
46	.02472	.02535	.02873	.02958	.03303	.03416	.03762	.03909	46
47	.02479	.02542	.02880	.02965	.03310	.03424	.03770	.03918	47
48	.02485	.02548	.02887	.02972	.03318	.03432	.03778	.03927	48
49	.02492	.02555	.02894	.02980	.03325	.03439	.03786	.03935	49
50	.02498	.02562	.02900	.02987	.03333	.03447	.03794	.03944	50
51	.02504	.02569	.02907	.02994	.03340	.03455	.03802	.03952	51
52	.02511	.02576	.02914	.03002	.03347	.03463	.03810	.03961	52
53	.02517	.02582	.02921	.03009	.03355	.03471	.03818	.03969	53
54	.02524	.02589	.02928	.03017	.03362	.03479	.03826	.03978	54
55	.02530	.02596	.02935	.03024	.03370	.03487	.03834	.03987	55
56	.02537	.02603	.02942	.03032	.03377	.03495	.03842	.03995	56
57	.02543	.02610	.02949	.03039	.03385	.03503	.03850	.04004	57
58	.02550	.02617	.02956	.03046	.03392	.03512	.03858	.04013	58
59	.02556	.02624	.02963	.03054	.03400	.03520	.03866	.04021	59
60	.02563	.02630	.02970	.03061	.03407	.03528	.03874	.04030	60

TABLE XIII.—VERSINES AND EXSECANTS.

′	16°		17°		18°		19°		′
	Vers.	Exsec.	Vers.	Exsec.	Vers.	Exsec.	Vers.	Exsec.	
0	.03874	.04030	.04370	.04569	.04894	.05146	.05448	.05762	0
1	.03882	.04039	.04378	.04578	.04903	.05156	.05458	.05773	1
2	.03890	.04047	.04387	.04588	.04912	.05166	.05467	.05783	2
3	.03898	.04056	.04395	.04597	.04921	.05176	.05477	.05794	3
4	.03906	.04065	.04404	.04606	.04930	.05186	.05486	.05805	4
5	.03914	.04073	.04412	.04616	.04939	.05196	.05496	.05815	5
6	.03922	.04082	.04421	.04625	.04948	.05206	.05505	.05826	6
7	.03930	.04091	.04429	.04635	.04957	.05216	.05515	.05836	7
8	.03938	.04100	.04438	.04644	.04967	.05226	.05524	.05847	8
9	.03946	.04108	.04446	.04653	.04976	.05236	.05534	.05858	9
10	.03954	.04117	.04455	.04663	.04985	.05246	.05543	.05869	10
11	.03963	.04126	.04464	.04672	.04994	.05256	.05553	.05879	11
12	.03971	.04135	.04472	.04682	.05003	.05266	.05562	.05890	12
13	.03979	.04144	.04481	.04691	.05012	.05276	.05572	.05901	13
14	.03987	.04152	.04489	.04700	.05021	.05286	.05582	.05911	14
15	.03995	.04161	.04498	.04710	.05030	.05297	.05591	.05922	15
16	.04003	.04170	.04507	.04719	.05059	.05307	.05601	.05933	16
17	.04011	.04179	.04515	.04729	.05048	.05317	.05610	.05944	17
18	.04019	.04188	.04524	.04738	.05057	.05327	.05620	.05955	18
19	.04028	.04197	.04533	.04748	.05067	.05337	.05630	.05965	19
20	.04036	.04206	.04541	.04757	.05076	.05347	.05639	.05976	20
21	.04044	.04214	.04550	.04767	.05085	.05357	.05649	.05987	21
22	.04052	.04223	.04559	.04776	.05094	.05367	.05658	.05998	22
23	.04060	.04232	.04567	.04786	.05103	.05378	.05668	.06009	23
24	.04069	.04241	.04576	.04795	.05112	.05388	.05678	.06020	24
25	.04077	.04250	.04585	.04805	.05122	.05398	.05687	.06030	25
26	.04085	.04259	.04593	.04815	.05131	.05408	.05697	.06041	26
27	.04093	.04268	.04602	.04824	.05140	.05418	.05707	.06052	27
28	.04102	.04277	.04611	.04834	.05149	.05429	.05716	.06063	28
29	.04110	.04286	.04620	.04843	.05158	.05439	.05726	.06074	29
30	.04118	.04295	.04628	.04853	.05168	.05449	.05736	.06085	30
31	.04126	.04304	.04637	.04863	.05177	.05460	.05746	.06096	31
32	.04135	.04313	.04646	.04872	.05186	.05470	.05755	.06107	32
33	.04143	.04322	.04655	.04882	.05195	.05480	.05765	.06118	33
34	.04151	.04331	.04663	.04891	.05205	.05490	.05775	.06129	34
35	.04159	.04340	.04672	.04901	.05214	.05501	.05785	.06140	35
36	.04168	.04349	.04681	.04911	.05223	.05511	.05794	.06151	36
37	.04176	.04358	.04690	.04920	.05232	.05521	.05804	.06162	37
38	.04184	.04367	.04699	.04930	.05242	.05532	.05814	.06173	38
39	.04193	.04376	.04707	.04940	.05251	.05542	.05824	.06184	39
40	.04201	.04385	.04716	.04950	.05260	.05552	.05833	.06195	40
41	.04209	.04394	.04725	.04959	.05270	.05563	.05843	.06206	41
42	.04218	.04403	.04734	.04969	.05279	.05573	.05853	.06217	42
43	.04226	.04413	.04743	.04979	.05288	.05584	.05863	.06228	43
44	.04234	.04422	.04752	.04989	.05298	.05594	.05873	.06239	44
45	.04243	.04431	.04760	.04998	.05307	.05604	.05882	.06250	45
46	.04251	.04440	.04769	.05008	.05316	.05615	.05892	.06261	46
47	.04260	.04449	.04778	.05018	.05326	.05625	.05902	.06272	47
48	.04268	.04458	.04787	.05028	.05335	.05636	.05912	.06283	48
49	.04276	.04468	.04796	.05038	.05344	.05646	.05922	.06295	49
50	.04285	.04477	.04805	.05047	.05354	.05657	.05932	.06306	50
51	.04293	.04486	.04814	.05057	.05363	.05667	.05942	.06317	51
52	.04302	.04495	.04823	.05067	.05373	.05678	.05951	.06328	52
53	.04310	.04504	.04832	.05077	.05382	.05688	.05961	.06339	53
54	.04319	.04514	.04841	.05087	.05391	.05699	.05971	.06350	54
55	.04327	.04523	.04850	.05097	.05401	.05709	.05981	.06362	55
56	.04336	.04532	.04858	.05107	.05410	.05720	.05991	.06373	56
57	.04344	.04541	.04867	.05116	.05420	.05730	.06001	.06384	57
58	.04353	.04551	.04876	.05126	.05429	.05741	.06011	.06395	58
59	.04361	.04560	.04885	.05136	.05439	.05751	.06021	.06407	59
60	.04370	.04569	.04894	.05146	.05448	.05762	.06031	.06418	60

TABLE XIII.—VERSINES AND EXSECANTS.

′	20°		21°		22°		23°		′
	Vers.	Exsec.	Vers.	Exsec.	Vers.	Exsec.	Vers.	Exsec.	
0	.06031	.06418	.06642	.07115	.07282	.07853	.07950	.08636	0
1	.06041	.06429	.06652	.07126	.07293	.07866	.07961	.08649	1
2	.06051	.06440	.06663	.07138	.07303	.07879	.07972	.08663	2
3	.06061	.06452	.06673	.07150	.07314	.07892	.07984	.08676	3
4	.06071	.06463	.06684	.07162	.07325	.07904	.07995	.08690	4
5	.06081	.06474	.06694	.07174	.07336	.07917	.08006	.08703	5
6	.06091	.06486	.06705	.07186	.07347	.07930	.08018	.08717	6
7	.06101	.06497	.06715	.07199	.07358	.07943	.08029	.08730	7
8	.06111	.06508	.06726	.07211	.07369	.07955	.08041	.08744	8
9	.06121	.06520	.06736	.07223	.07380	.07968	.08052	.08757	9
10	.06131	.06531	.06747	.07235	.07391	.07981	.08064	.08771	10
11	.06141	.06542	.06757	.07247	.07402	.07994	.08075	.08784	11
12	.06151	.06554	.06768	.07259	.07413	.08006	.08086	.08798	12
13	.06161	.06565	.06778	.07271	.07424	.08019	.08098	.08811	13
14	.06171	.06577	.06789	.07283	.07435	.08032	.08109	.08825	14
15	.06181	.06588	.06799	.07295	.07446	.08045	.08121	.08839	15
16	.06191	.06600	.06810	.07307	.07457	.08058	.08132	.08852	16
17	.06201	.06611	.06820	.07320	.07468	.08071	.08144	.08866	17
18	.06211	.06622	.06831	.07332	.07479	.08084	.08155	.08880	18
19	.06221	.06634	.06841	.07344	.07490	.08097	.08167	.08893	19
20	.06231	.06645	.06852	.07356	.07501	.08109	.08178	.08907	20
21	.06241	.06657	.06863	.07368	.07512	.08122	.08190	.08921	21
22	.06252	.06668	.06873	.07380	.07523	.08135	.08201	.08934	22
23	.06262	.06680	.06884	.07393	.07534	.08148	.08213	.08948	23
24	.06272	.06691	.06894	.07405	.07545	.08161	.08225	.08962	24
25	.06282	.06703	.06905	.07417	.07556	.08174	.08236	.08975	25
26	.06292	.06715	.06916	.07429	.07568	.08187	.08248	.08989	26
27	.06302	.06726	.06926	.07442	.07579	.08200	.08259	.09003	27
28	.06312	.06738	.06937	.07454	.07590	.08213	.08271	.09017	28
29	.06323	.06749	.06948	.07466	.07601	.08226	.08282	.09030	29
30	.06333	.06761	.06958	.07479	.07612	.08239	.08294	.09044	30
31	.06343	.06773	.06969	.07491	.07623	.08252	.08306	.09058	31
32	.06353	.06784	.06980	.07503	.07634	.08265	.08317	.09072	32
33	.06363	.06796	.06990	.07516	.07645	.08278	.08329	.09086	33
34	.06374	.06807	.07001	.07528	.07657	.08291	.08340	.09099	34
35	.06384	.06819	.07012	.07540	.07668	.08305	.08352	.09113	35
36	.06394	.06831	.07022	.07553	.07679	.08318	.08364	.09127	36
37	.06404	.06843	.07033	.07565	.07690	.08331	.08375	.09141	37
38	.06415	.06854	.07044	.07578	.07701	.08344	.08387	.09155	38
39	.06425	.06866	.07055	.07590	.07713	.08357	.08399	.09169	39
40	.06435	.06878	.07065	.07602	.07724	.08370	.08410	.09183	40
41	.06445	.06889	.07076	.07615	.07735	.08383	.08422	.09197	41
42	.06456	.06901	.07087	.07627	.07746	.08397	.08434	.09211	42
43	.06466	.06913	.07098	.07640	.07757	.08410	.08445	.09224	43
44	.06476	.06925	.07108	.07652	.07769	.08423	.08457	.09238	44
45	.06486	.06936	.07119	.07665	.07780	.08436	.08469	.09252	45
46	.06497	.06948	.07130	.07677	.07791	.08449	.08481	.09266	46
47	.06507	.06960	.07141	.07690	.07802	.08463	.08492	.09280	47
48	.06517	.06972	.07151	.07702	.07814	.08476	.08504	.09294	48
49	.06528	.06984	.07162	.07715	.07825	.08489	.08516	.09308	49
50	.06538	.06995	.07173	.07727	.07836	.08503	.08528	.09323	50
51	.06548	.07007	.07184	.07740	.07848	.08516	.08539	.09337	51
52	.06559	.07019	.07195	.07752	.07859	.08529	.08551	.09351	52
53	.06569	.07031	.07206	.07765	.07870	.08542	.08563	.09365	53
54	.06580	.07043	.07216	.07778	.07881	.08556	.08575	.09379	54
55	.06590	.07055	.07227	.07790	.07893	.08569	.08586	.09393	55
56	.06600	.07067	.07238	.07803	.07904	.08582	.08598	.09407	56
57	.06611	.07079	.07249	.07816	.07915	.08596	.08610	.09421	57
58	.06621	.07091	.07260	.07828	.07927	.08609	.08622	.09435	58
59	.06632	.07103	.07271	.07841	.07938	.08623	.08634	.09449	59
60	.06642	.07115	.07282	.07853	.07950	.08636	.08645	.09464	60

TABLE XIII.—VERSINES AND EXSECANTS.

′	24°		25°		26°		27°		′
	Vers.	Exsec.	Vers.	Exsec.	Vers.	Exsec.	Vers.	Exsec.	
0	.08645	.09464	.09369	.10338	.10121	.11260	.10899	.12233	0
1	.08657	.09478	.09382	.10353	.10133	.11276	.10913	.12249	1
2	.08669	.09492	.09394	.10368	.10146	.11292	.10926	.12266	2
3	.08681	.09506	.09406	.10383	.10159	.11308	.10939	.12283	3
4	.08693	.09520	.09418	.10398	.10172	.11323	.10952	.12299	4
5	.08705	.09535	.09431	.10413	.10184	.11339	.10965	.12316	5
6	.08717	.09549	.09443	.10428	.10197	.11355	.10979	.12333	6
7	.08728	.09563	.09455	.10443	.10210	.11371	.10992	.12349	7
8	.08740	.09577	.09468	.10458	.10223	.11387	.11005	.12366	8
9	.08752	.09592	.09480	.10473	.10236	.11403	.11019	.12383	9
10	.08764	.09606	.09493	.10488	.10248	.11419	.11032	.12400	10
11	.08776	.09620	.09505	.10503	.10261	.11435	.11045	.12416	11
12	.08788	.09635	.09517	.10518	.10274	.11451	.11058	.12433	12
13	.08800	.09649	.09530	.10533	.10287	.11467	.11072	.12450	13
14	.08812	.09663	.09542	.10549	.10300	.11483	.11085	.12467	14
15	.08824	.09678	.09554	.10564	.10313	.11499	.11098	.12484	15
16	.08836	.09692	.09567	.10579	.10326	.11515	.11112	.12501	16
17	.08848	.09707	.09579	.10594	.10338	.11531	.11125	.12518	17
18	.08860	.09721	.09592	.10609	.10351	.11547	.11138	.12534	18
19	.08872	.09735	.09604	.10625	.10364	.11563	.11152	.12551	19
20	.08884	.09750	.09617	.10640	.10377	.11579	.11165	.12568	20
21	.08896	.09764	.09629	.10655	.10390	.11595	.11178	.12585	21
22	.08908	.09779	.09642	.10670	.10403	.11611	.11192	.12602	22
23	.08920	.09793	.09654	.10686	.10416	.11627	.11205	.12619	23
24	.08932	.09808	.09666	.10701	.10429	.11643	.11218	.12636	24
25	.08944	.09822	.09679	.10716	.10442	.11659	.11232	.12653	25
26	.08956	.09837	.09691	.10731	.10455	.11675	.11245	.12670	26
27	.08968	.09851	.09704	.10747	.10468	.11691	.11258	.12687	27
28	.08980	.09866	.09716	.10762	.10481	.11708	.11272	.12704	28
29	.08992	.09880	.09729	.10777	.10494	.11724	.11285	.12721	29
30	.09004	.09895	.09741	.10793	.10507	.11740	.11299	.12738	30
31	.09016	.09909	.09754	.10808	.10520	.11756	.11312	.12755	31
32	.09028	.09924	.09767	.10824	.10533	.11772	.11326	.12772	32
33	.09040	.09939	.09779	.10839	.10546	.11789	.11339	.12789	33
34	.09052	.09953	.09792	.10854	.10559	.11805	.11353	.12807	34
35	.09064	.09968	.09804	.10870	.10572	.11821	.11366	.12824	35
36	.09076	.09982	.09817	.10885	.10585	.11838	.11380	.12841	36
37	.09089	.09997	.09829	.10901	.10598	.11854	.11393	.12858	37
38	.09101	.10012	.09842	.10916	.10611	.11870	.11407	.12875	38
39	.09113	.10026	.09854	.10932	.10624	.11886	.11420	.12892	39
40	.09125	.10041	.09867	.10947	.10637	.11903	.11434	.12910	40
41	.09137	.10055	.09880	.10963	.10650	.11919	.11447	.12927	41
42	.09149	.10071	.09892	.10978	.10663	.11936	.11461	.12944	42
43	.09161	.10085	.09905	.10994	.10676	.11952	.11474	.12961	43
44	.09174	.10100	.09918	.11009	.10689	.11968	.11488	.12979	44
45	.09186	.10115	.09930	.11025	.10702	.11985	.11501	.12996	45
46	.09198	.10130	.09943	.11041	.10715	.12001	.11515	.13013	46
47	.09210	.10144	.09955	.11056	.10728	.12018	.11528	.13031	47
48	.09222	.10159	.09968	.11072	.10741	.12034	.11542	.13048	48
49	.09234	.10174	.09981	.11087	.10755	.12051	.11555	.13065	49
50	.09247	.10189	.09993	.11103	.10768	.12067	.11569	.13083	50
51	.09259	.10204	.10006	.11119	.10781	.12084	.11583	.13100	51
52	.09271	.10218	.10019	.11134	.10794	.12100	.11596	.13117	52
53	.09283	.10233	.10032	.11150	.10807	.12117	.11610	.13135	53
54	.09296	.10248	.10044	.11166	.10820	.12133	.11623	.13152	54
55	.09308	.10263	.10057	.11181	.10833	.12150	.11637	.13170	55
56	.09320	.10278	.10070	.11197	.10847	.12166	.11651	.13187	56
57	.09332	.10293	.10082	.11213	.10860	.12183	.11664	.13205	57
58	.09345	.10308	.10095	.11229	.10873	.12199	.11678	.13222	58
59	.09357	.10323	.10108	.11244	.10886	.12216	.11692	.13240	59
60	.09369	.10338	.10121	.11260	.10899	.12233	.11705	.13257	60

TABLE XIII.—VERSINES AND EXSECANTS.

′	28°		29°		30°		31°		′
	Vers.	Exsec.	Vers.	Exsec.	Vers.	Exsec.	Vers.	Exsec.	
0	.11705	.13257	.12538	.14335	.13397	.15470	.14283	.16663	0
1	.11719	.13275	.12552	.14354	.13412	.15489	.14298	.16684	1
2	.11733	.13292	.12566	.14372	.13427	.15509	.14313	.16704	2
3	.11746	.13310	.12580	.14391	.13441	.15528	.14328	.16725	3
4	.11760	.13327	.12595	.14409	.13456	.15548	.14343	.16745	4
5	.11774	.13345	.12609	.14428	.13470	.15567	.14358	.16766	5
6	.11787	.13362	.12623	.14446	.13485	.15587	.14373	.16786	6
7	.11801	.13380	.12637	.14465	.13499	.15606	.14388	.16806	7
8	.11815	.13398	.12651	.14483	.13514	.15626	.14403	.16827	8
9	.11828	.13415	.12665	.14502	.13529	.15645	.14418	.16848	9
10	.11842	.13433	.12679	.14521	.13543	.15665	.14433	.16868	10
11	.11856	.13451	.12694	.14539	.13558	.15684	.14449	.16889	11
12	.11870	.13468	.12708	.14558	.13573	.15704	.14464	.16909	12
13	.11883	.13486	.12722	.14576	.13587	.15724	.14479	.16930	13
14	.11897	.13504	.12736	.14595	.13602	.15743	.14494	.16950	14
15	.11911	.13521	.12750	.14614	.13616	.15763	.14509	.16971	15
16	.11925	.13539	.12765	.14632	.13631	.15782	.14524	.16992	16
17	.11938	.13557	.12779	.14651	.13646	.15802	.14539	.17012	17
18	.11952	.13575	.12793	.14670	.13660	.15822	.14554	.17033	18
19	.11966	.13593	.12807	.14689	.13675	.15841	.14569	.17054	19
20	.11980	.13610	.12822	.14707	.13690	.15861	.14584	.17075	20
21	.11994	.13628	.12836	.14726	.13705	.15881	.14599	.17095	21
22	.12007	.13646	.12850	.14745	.13719	.15901	.14615	.17116	22
23	.12021	.13664	.12864	.14764	.13734	.15920	.14630	.17137	23
24	.12035	.13682	.12879	.14782	.13749	.15940	.14645	.17158	24
25	.12049	.13700	.12893	.14801	.13763	.15960	.14660	.17178	25
26	.12063	.13718	.12907	.14820	.13778	.15980	.14675	.17199	26
27	.12077	.13735	.12921	.14839	.13793	.16000	.14690	.17220	27
28	.12091	.13753	.12936	.14858	.13808	.16019	.14706	.17241	28
29	.12104	.13771	.12950	.14877	.13822	.16039	.14721	.17262	29
30	.12118	.13789	.12964	.14896	.13837	.16059	.14736	.17283	30
31	.12132	.13807	.12979	.14914	.13852	.16079	.14751	.17304	31
32	.12146	.13825	.12993	.14933	.13867	.16099	.14766	.17325	32
33	.12160	.13843	.13007	.14952	.13881	.16119	.14782	.17346	33
34	.12174	.13861	.13022	.14971	.13896	.16139	.14797	.17367	34
35	.12188	.13879	.13036	.14990	.13911	.16159	.14812	.17388	35
36	.12202	.13897	.13051	.15009	.13926	.16179	.14827	.17409	36
37	.12216	.13916	.13065	.15028	.13941	.16199	.14843	.17430	37
38	.12230	.13934	.13079	.15047	.13955	.16219	.14858	.17451	38
39	.12244	.13952	.13094	.15066	.13970	.16239	.14873	.17472	39
40	.12257	.13970	.13108	.15085	.13985	.16259	.14888	.17493	40
41	.12271	.13988	.13122	.15105	.14000	.16279	.14904	.17514	41
42	.12285	.14006	.13137	.15124	.14015	.16299	.14919	.17535	42
43	.12299	.14024	.13151	.15143	.14030	.16319	.14934	.17556	43
44	.12313	.14042	.13166	.15162	.14044	.16339	.14949	.17577	44
45	.12327	.14061	.13180	.15181	.14059	.16359	.14965	.17598	45
46	.12341	.14079	.13195	.15200	.14074	.16380	.14980	.17620	46
47	.12355	.14097	.13209	.15219	.14089	.16400	.14995	.17641	47
48	.12369	.14115	.13223	.15239	.14104	.16420	.15011	.17662	48
49	.12383	.14134	.13238	.15258	.14119	.16440	.15026	.17683	49
50	.12397	.14152	.13252	.15277	.14134	.16460	.15041	.17704	50
51	.12411	.14170	.13267	.15296	.14149	.16481	.15057	.17726	51
52	.12425	.14188	.13281	.15315	.14164	.16501	.15072	.17747	52
53	.12439	.14207	.13296	.15335	.14179	.16521	.15087	.17768	53
54	.12454	.14225	.13310	.15354	.14194	.16541	.15103	.17790	54
55	.12468	.14243	.13325	.15373	.14208	.16562	.15118	.17811	55
56	.12482	.14262	.13339	.15393	.14223	.16582	.15134	.17832	56
57	.12496	.14280	.13354	.15412	.14238	.16602	.15149	.17854	57
58	.12510	.14299	.13368	.15431	.14253	.16623	.15164	.17875	58
59	.12524	.14317	.13383	.15451	.14268	.16643	.15180	.17896	59
60	.12538	.14335	.13397	.15470	.14283	.16663	.15195	.17918	60

TABLE XIII.—VERSINES AND EXSECANTS.

′	32°		33°		34°		35°		′
	Vers.	Exsec.	Vers.	Exsec.	Vers.	Exsec.	Vers.	Exsec.	
0	.15195	.17918	.16133	.19236	.17096	.20622	.18085	.22077	0
1	.15211	.17939	.16149	.19259	.17113	.20645	.18101	.22102	1
2	.15226	.17961	.16165	.19281	.17129	.20669	.18118	.22127	2
3	.15241	.17982	.16181	.19304	.17145	.20693	.18135	.22152	3
4	.15257	.18004	.16196	.19327	.17161	.20717	.18152	.22177	4
5	.15272	.18025	.16212	.19349	.17178	.20740	.18168	.22202	5
6	.15288	.18047	.16228	.19372	.17194	.20764	.18185	.22227	6
7	.15303	.18068	.16244	.19394	.17210	.20788	.18202	.22252	7
8	.15319	.18090	.16260	.19417	.17227	.20812	.18218	.22277	8
9	.15334	.18111	.16276	.19440	.17243	.20836	.18235	.22302	9
10	.15350	.18133	.16292	.19463	.17259	.20859	.18252	.22327	10
11	.15365	.18155	.16308	.19485	.17276	.20883	.18269	.22352	11
12	.15381	.18176	.16324	.19508	.17292	.20907	.18286	.22377	12
13	.15396	.18198	.16340	.19531	.17308	.20931	.18302	.22402	13
14	.15412	.18220	.16355	.19554	.17325	.20955	.18319	.22428	14
15	.15427	.18241	.16371	.19576	.17341	.20979	.18336	.22453	15
16	.15443	.18263	.16387	.19599	.17357	.21003	.18353	.22478	16
17	.15458	.18285	.16403	.19622	.17374	.21027	.18369	.22503	17
18	.15474	.18307	.16419	.19645	.17390	.21051	.18386	.22528	18
19	.15489	.18328	.16435	.19668	.17407	.21075	.18403	.22554	19
20	.15505	.18350	.16451	.19691	.17423	.21099	.18420	.22579	20
21	.15520	.18372	.16467	.19713	.17439	.21123	.18437	.22604	21
22	.15536	.18394	.16483	.19736	.17456	.21147	.18454	.22629	22
23	.15552	.18416	.16499	.19759	.17472	.21171	.18470	.22655	23
24	.15567	.18437	.16515	.19782	.17489	.21195	.18487	.22680	24
25	.15583	.18459	.16531	.19805	.17505	.21220	.18504	.22706	25
26	.15598	.18481	.16547	.19828	.17522	.21244	.18521	.22731	26
27	.15614	.18503	.16563	.19851	.17538	.21268	.18538	.22756	27
28	.15630	.18525	.16579	.19874	.17554	.21292	.18555	.22782	28
29	.15645	.18547	.16595	.19897	.17571	.21316	.18572	.22807	29
30	.15661	.18569	.16611	.19920	.17587	.21341	.18588	.22833	30
31	.15676	.18591	.16627	.19944	.17604	.21365	.18605	.22858	31
32	.15692	.18613	.16644	.19967	.17620	.21389	.18622	.22884	32
33	.15708	.18635	.16660	.19990	.17637	.21414	.18639	.22909	33
34	.15723	.18657	.16676	.20013	.17653	.21438	.18656	.22935	34
35	.15739	.18679	.16692	.20036	.17670	.21462	.18673	.22960	35
36	.15755	.18701	.16708	.20059	.17686	.21487	.18690	.22986	36
37	.15770	.18723	.16724	.20083	.17703	.21511	.18707	.23012	37
38	.15786	.18745	.16740	.20106	.17719	.21535	.18724	.23037	38
39	.15802	.18767	.16756	.20129	.17736	.21560	.18741	.23063	39
40	.15818	.18790	.16772	.20152	.17752	.21584	.18758	.23089	40
41	.15833	.18812	.16788	.20176	.17769	.21609	.18775	.23114	41
42	.15849	.18834	.16805	.20199	.17786	.21633	.18792	.23140	42
43	.15865	.18856	.16821	.20222	.17802	.21658	.18809	.23166	43
44	.15880	.18878	.16837	.20246	.17819	.21682	.18826	.23192	44
45	.15896	.18901	.16853	.20269	.17835	.21707	.18843	.23217	45
46	.15912	.18923	.16869	.20292	.17852	.21731	.18860	.23243	46
47	.15928	.18945	.16885	.20316	.17868	.21756	.18877	.23269	47
48	.15943	.18967	.16902	.20339	.17885	.21781	.18894	.23295	48
49	.15959	.18990	.16918	.20363	.17902	.21805	.18911	.23321	49
50	.15975	.19012	.16934	.20386	.17918	.21830	.18928	.23347	50
51	.15991	.19034	.16950	.20410	.17935	.21855	.18945	.23373	51
52	.16006	.19057	.16966	.20433	.17952	.21879	.18962	.23399	52
53	.16022	.19079	.16983	.20457	.17968	.21904	.18979	.23424	53
54	.16038	.19102	.16999	.20480	.17985	.21929	.18996	.23450	54
55	.16054	.19124	.17015	.20504	.18001	.21953	.19013	.23476	55
56	.16070	.19146	.17031	.20527	.18018	.21978	.19030	.23502	56
57	.16085	.19169	.17047	.20551	.18035	.22003	.19047	.23529	57
58	.16101	.19191	.17064	.20575	.18051	.22028	.19064	.23555	58
59	.16117	.19214	.17080	.20598	.18068	.22053	.19081	.23581	59
60	.16133	.19236	.17096	.20622	.18085	.22077	.19098	.23607	60

TABLE XIII.—VERSINES AND EXSECANTS.

′	36°		37°		38°		39°		′
	Vers.	Exsec.	Vers.	Exsec.	Vers.	Exsec.	Vers.	Exsec.	
0	.19098	.23607	.20136	.25214	.21199	.26902	.22285	.28676	0
1	.19115	.23633	.20154	.25241	.21217	.26931	.22304	.28706	1
2	.19133	.23659	.20171	.25269	.21235	.26960	.22322	.28737	2
3	.19150	.23685	.20189	.25296	.21253	.26988	.22340	.28767	3
4	.19167	.23711	.20207	.25324	.21271	.27017	.22359	.28797	4
5	.19184	.23738	.20224	.25351	.21289	.27046	.22377	.28828	5
6	.19201	.23764	.20242	.25379	.21307	.27075	.22395	.28858	6
7	.19218	.23790	.20259	.25406	.21324	.27104	.22414	.28889	7
8	.19235	.23816	.20277	.25434	.21342	.27133	.22432	.28919	8
9	.19252	.23843	.20294	.25462	.21360	.27162	.22450	.28950	9
10	.19270	.23869	.20312	.25489	.21378	.27191	.22469	.28980	10
11	.19287	.23895	.20329	.25517	.21396	.27221	.22487	.29011	11
12	.19304	.23922	.20347	.25545	.21414	.27250	.22506	.29042	12
13	.19321	.23948	.20365	.25572	.21432	.27279	.22524	.29072	13
14	.19338	.23975	.20382	.25600	.21450	.27308	.22543	.29103	14
15	.19356	.24001	.20400	.25628	.21468	.27337	.22561	.29133	15
16	.19373	.24028	.20417	.25656	.21486	.27366	.22579	.29164	16
17	.19390	.24054	.20435	.25683	.21504	.27396	.22598	.29195	17
18	.19407	.24081	.20453	.25711	.21522	.27425	.22616	.29226	18
19	.19424	.24107	.20470	.25739	.21540	.27454	.22634	.29256	19
20	.19442	.24134	.20488	.25767	.21558	.27483	.22653	.29287	20
21	.19459	.24160	.20506	.25795	.21576	.27513	.22671	.29318	21
22	.19476	.24187	.20523	.25823	.21595	.27542	.22690	.29349	22
23	.19493	.24213	.20541	.25851	.21613	.27572	.22708	.29380	23
24	.19511	.24240	.20559	.25879	.21631	.27601	.22727	.29411	24
25	.19528	.24267	.20576	.25907	.21649	.27630	.22745	.29442	25
26	.19545	.24293	.20594	.25935	.21667	.27660	.22764	.29473	26
27	.19562	.24320	.20612	.25963	.21685	.27689	.22782	.29504	27
28	.19580	.24347	.20629	.25991	.21703	.27719	.22801	.29535	28
29	.19597	.24373	.20647	.26019	.21721	.27748	.22819	.29566	29
30	.19614	.24400	.20665	.26047	.21739	.27778	.22838	.29597	30
31	.19632	.24427	.20682	.26075	.21757	.27807	.22856	.29628	31
32	.19649	.24454	.20700	.26104	.21775	.27837	.22875	.29659	32
33	.19666	.24481	.20718	.26132	.21794	.27867	.22893	.29690	33
34	.19684	.24508	.20736	.26160	.21812	.27896	.22912	.29721	34
35	.19701	.24534	.20753	.26188	.21830	.27926	.22930	.29752	35
36	.19718	.24561	.20771	.26216	.21848	.27956	.22949	.29784	36
37	.19736	.24588	.20789	.26245	.21866	.27985	.22967	.29815	37
38	.19753	.24615	.20807	.26273	.21884	.28015	.22986	.29846	38
39	.19770	.24642	.20824	.26301	.21902	.28045	.23004	.29877	39
40	.19788	.24669	.20842	.26330	.21921	.28075	.23023	.29909	40
41	.19805	.24696	.20860	.26358	.21939	.28105	.23041	.29940	41
42	.19822	.24723	.20878	.26387	.21957	.28134	.23060	.29971	42
43	.19840	.24750	.20895	.26415	.21975	.28164	.23079	.30003	43
44	.19857	.24777	.20913	.26443	.21993	.28194	.23097	.30034	44
45	.19875	.24804	.20931	.26472	.22012	.28224	.23116	.30065	45
46	.19892	.24832	.20949	.26500	.22030	.28254	.23134	.30097	46
47	.19909	.24859	.20967	.26529	.22048	.28284	.23153	.30129	47
48	.19927	.24886	.20985	.26557	.22066	.28314	.23172	.30160	48
49	.19944	.24913	.21002	.26586	.22084	.28344	.23190	.30192	49
50	.19962	.24940	.21020	.26615	.22103	.28374	.23209	.30223	50
51	.19979	.24967	.21038	.26643	.22121	.28404	.23228	.30255	51
52	.19997	.24995	.21056	.26672	.22139	.28434	.23246	.30287	52
53	.20014	.25023	.21074	.26701	.22157	.28464	.23265	.30318	53
54	.20032	.25049	.21092	.26729	.22176	.28495	.23283	.30350	54
55	.20049	.25077	.21109	.26758	.22194	.28525	.23302	.30382	55
56	.20066	.25104	.21127	.26787	.22212	.28555	.23321	.30413	56
57	.20084	.25131	.21145	.26815	.22231	.28585	.23339	.30445	57
58	.20101	.25159	.21163	.26844	.22249	.28615	.23358	.30477	58
59	.20119	.25186	.21181	.26873	.22267	.28646	.23377	.30509	59
60	.20136	.25214	.21199	.26902	.22285	.28676	.23396	.30541	60

TABLE XIII.—VERSINES AND EXSECANTS.

′	40°		41°		42°		43°		′
	Vers.	Exsec.	Vers.	Exsec.	Vers.	Exsec.	Vers.	Exsec.	
0	.23396	.30541	.24529	.32501	.25686	.34563	.26865	.36733	0
1	.23414	.30573	.24548	.32535	.25705	.34599	.26884	.36770	1
2	.23433	.30605	.24567	.32568	.25724	.34634	.26904	.36807	2
3	.23452	.30636	.24586	.32602	.25744	.34669	.26924	.36844	3
4	.23470	.30668	.24605	.32636	.25763	.34704	.26944	.36881	4
5	.23489	.30700	.24625	.32669	.25783	.34740	.26964	.36919	5
6	.23508	.30732	.24644	.32703	.25802	.34775	.26984	.36956	6
7	.23527	.30764	.24663	.32737	.25822	.34811	.27004	.36993	7
8	.23545	.30796	.24682	.32770	.25841	.34846	.27024	.37030	8
9	.23564	.30829	.24701	.32804	.25861	.34882	.27043	.37068	9
10	.23583	.30861	.24720	.32838	.25880	.34917	.27063	.37105	10
11	.23602	.30893	.24739	.32872	.25900	.34953	.27083	.37143	11
12	.23620	.30925	.24759	.32905	.25920	.34988	.27103	.37180	12
13	.23639	.30957	.24778	.32939	.25939	.35024	.27123	.37218	13
14	.23658	.30989	.24797	.32973	.25959	.35060	.27143	.37255	14
15	.23677	.31022	.24816	.33007	.25978	.35095	.27163	.37293	15
16	.23696	.31054	.24835	.33041	.25998	.35131	.27183	.37330	16
17	.23714	.31086	.24854	.33075	.26017	.35167	.27203	.37368	17
18	.23733	.31119	.24874	.33109	.26037	.35203	.27223	.37406	18
19	.23752	.31151	.24893	.33143	.26056	.35238	.27243	.37443	19
20	.23771	.31183	.24912	.33177	.26076	.35274	.27263	.37481	20
21	.23790	.31216	.24931	.33211	.26096	.35310	.27283	.37519	21
22	.23808	.31248	.24950	.33245	.26115	.35346	.27303	.37556	22
23	.23827	.31281	.24970	.33279	.26135	.35382	.27323	.37594	23
24	.23846	.31313	.24989	.33314	.26154	.35418	.27343	.37632	24
25	.23865	.31346	.25008	.33348	.26174	.35454	.27363	.37670	25
26	.23884	.31378	.25027	.33382	.26194	.35490	.27383	.37708	26
27	.23903	.31411	.25047	.33416	.26213	.35526	.27403	.37746	27
28	.23922	.31443	.25066	.33451	.26233	.35562	.27423	.37784	28
29	.23941	.31476	.25085	.33485	.26253	.35598	.27443	.37822	29
30	.23959	.31509	.25104	.33519	.26272	.35634	.27463	.37860	30
31	.23978	.31541	.25124	.33554	.26292	.35670	.27483	.37898	31
32	.23997	.31574	.25143	.33588	.26312	.35707	.27503	.37936	32
33	.24016	.31607	.25162	.33622	.26331	.35743	.27523	.37974	33
34	.24035	.31640	.25182	.33657	.26351	.35779	.27543	.38012	34
35	.24054	.31672	.25201	.33691	.26371	.35815	.27563	.38051	35
36	.24073	.31705	.25220	.33726	.26390	.35852	.27583	.38089	36
37	.24092	.31738	.25240	.33760	.26410	.35888	.27603	.38127	37
38	.24111	.31771	.25259	.33795	.26430	.35924	.27623	.38165	38
39	.24130	.31804	.25278	.33830	.26449	.35961	.27643	.38204	39
40	.24149	.31837	.25297	.33864	.26469	.35997	.27663	.38242	40
41	.24168	.31870	.25317	.33899	.26489	.36034	.27683	.38280	41
42	.24187	.31903	.25336	.33934	.26509	.36070	.27703	.38319	42
43	.24206	.31936	.25356	.33968	.26528	.36107	.27723	.38357	43
44	.24225	.31969	.25375	.34003	.26548	.36143	.27743	.38396	44
45	.24244	.32002	.25394	.34038	.26568	.36180	.27764	.38434	45
46	.24262	.32035	.25414	.34073	.26588	.36217	.27784	.38473	46
47	.24281	.32068	.25433	.34108	.26607	.36253	.27804	.38512	47
48	.24300	.32101	.25452	.34142	.26627	.36290	.27824	.38550	48
49	.24320	.32134	.25472	.34177	.26647	.36327	.27844	.38589	49
50	.24339	.32168	.25491	.34212	.26667	.36363	.27864	.38628	50
51	.24358	.32201	.25511	.34247	.26686	.36400	.27884	.38666	51
52	.24377	.32234	.25530	.34282	.26706	.36437	.27905	.38705	52
53	.24396	.32267	.25549	.34317	.26726	.36474	.27925	.38744	53
54	.24415	.32301	.25569	.34352	.26746	.36511	.27945	.38783	54
55	.24434	.32334	.25588	.34387	.26766	.36548	.27965	.38822	55
56	.24453	.32368	.25608	.34423	.26785	.36585	.27985	.38860	56
57	.24472	.32401	.25627	.34458	.26805	.36622	.28005	.38899	57
58	.24491	.32434	.25647	.34493	.26825	.36659	.28026	.38938	58
59	.24510	.32468	.25666	.34528	.26845	.36696	.28046	.38977	59
60	.24529	.32501	.25686	.34563	.26865	.36733	.28066	.39016	60

TABLE XIII.—VERSINES AND EXSECANTS.

′	44°		45°		46°		47°		′
	Vers.	Exsec.	Vers.	Exsec.	Vers.	Exsec.	Vers.	Exsec.	
0	.28066	.39016	.29289	.41421	.30534	.43956	.31800	.46628	0
1	.28086	.39055	.29310	.41463	.30555	.43999	.31821	.46674	1
2	.28106	.39095	.29330	.41504	.30576	.44042	.31843	.46719	2
3	.28127	.39134	.29351	.41545	.30597	.44086	.31864	.46765	3
4	.28147	.39173	.29372	.41586	.30618	.44129	.31885	.46811	4
5	.28167	.39212	.29392	.41627	.30639	.44173	.31907	.46857	5
6	.28187	.39251	.29413	.41669	.30660	.44217	.31928	.46903	6
7	.28208	.39291	.29433	.41710	.30681	.44260	.31949	.46949	7
8	.28228	.39330	.29454	.41752	.30702	.44304	.31971	.46995	8
9	.28248	.39369	.29475	.41793	.30723	.44347	.31992	.47041	9
10	.28268	.39409	.29495	.41835	.30744	.44391	.32013	.47087	10
11	.28289	.39448	.29516	.41876	.30765	.44435	.32035	.47134	11
12	.28309	.39487	.29537	.41918	.30786	.44479	.32056	.47180	12
13	.28329	.39527	.29557	.41959	.30807	.44523	.32077	.47226	13
14	.28350	.39566	.29578	.42001	.30828	.44567	.32099	.47272	14
15	.28370	.39606	.29599	.42042	.30849	.44610	.32120	.47319	15
16	.28390	.39646	.29619	.42084	.30870	.44654	.32141	.47365	16
17	.28410	.39685	.29640	.42126	.30891	.44698	.32163	.47411	17
18	.28431	.39725	.29661	.42168	.30912	.44742	.32184	.47458	18
19	.28451	.39764	.29681	.42210	.30933	.44787	.32205	.47504	19
20	.28471	.39804	.29702	.42251	.30954	.44831	.32227	.47551	20
21	.28492	.39844	.29723	.42293	.30975	.44875	.32248	.47598	21
22	.28512	.39884	.29743	.42335	.30996	.44919	.32270	.47644	22
23	.28532	.39924	.29764	.42377	.31017	.44963	.32291	.47691	23
24	.28553	.39963	.29785	.42419	.31038	.45007	.32312	.47738	24
25	.28573	.40003	.29805	.42461	.31059	.45052	.32334	.47784	25
26	.28593	.40043	.29826	.42503	.31080	.45096	.32355	.47831	26
27	.28614	.40083	.29847	.42545	.31101	.45141	.32377	.47878	27
28	.28634	.40123	.29868	.42587	.31122	.45185	.32398	.47925	28
29	.28655	.40163	.29888	.42630	.31143	.45229	.32420	.47972	29
30	.28675	.40203	.29909	.42672	.31165	.45274	.32441	.48019	30
31	.28695	.40243	.29930	.42714	.31186	.45319	.32462	.48066	31
32	.28716	.40283	.29951	.42756	.31207	.45363	.32484	.48113	32
33	.28736	.40324	.29971	.42799	.31228	.45408	.32505	.48160	33
34	.28757	.40364	.29992	.42841	.31249	.45452	.32527	.48207	34
35	.28777	.40404	.30013	.42883	.31270	.45497	.32548	.48254	35
36	.28797	.40444	.30034	.42926	.31291	.45542	.32570	.48301	36
37	.28818	.40485	.30054	.42968	.31312	.45587	.32591	.48349	37
38	.28838	.40525	.30075	.43011	.31334	.45631	.32613	.48396	38
39	.28859	.40565	.30096	.43053	.31355	.45676	.32634	.48443	39
40	.28879	.40606	.30117	.43096	.31376	.45721	.32656	.48491	40
41	.28900	.40646	.30138	.43139	.31397	.45766	.32677	.48538	41
42	.28920	.40687	.30158	.43181	.31418	.45811	.32699	.48586	42
43	.28941	.40727	.30179	.43224	.31439	.45856	.32720	.48633	43
44	.28961	.40768	.30200	.43267	.31461	.45901	.32742	.48681	44
45	.28981	.40808	.30221	.43310	.31482	.45946	.32763	.48728	45
46	.29002	.40849	.30242	.43352	.31503	.45992	.32785	.48776	46
47	.29022	.40890	.30263	.43395	.31524	.46037	.32806	.48824	47
48	.29043	.40930	.30283	.43438	.31545	.46082	.32828	.48871	48
49	.29063	.40971	.30304	.43481	.31567	.46127	.32849	.48919	49
50	.29084	.41012	.30325	.43524	.31588	.46173	.32871	.48967	50
51	.29104	.41053	.30346	.43567	.31609	.46218	.32893	.49015	51
52	.29125	.41093	.30367	.43610	.31630	.46263	.32914	.49063	52
53	.29145	.41134	.30388	.43653	.31651	.46309	.32936	.49111	53
54	.29166	.41175	.30409	.43696	.31673	.46354	.32957	.49159	54
55	.29187	.41216	.30430	.43739	.31694	.46400	.32979	.49207	55
56	.29207	.41257	.30451	.43783	.31715	.46445	.33001	.49255	56
57	.29228	.41298	.30471	.43826	.31736	.46491	.33022	.49303	57
58	.29248	.41339	.30492	.43869	.31758	.46537	.33044	.49351	58
59	.29269	.41380	.30513	.43912	.31779	.46582	.33065	.49399	59
60	.29289	.41421	.30534	.43956	.31800	.46628	.33087	.49448	60

TABLE XIII.—VERSINES AND EXSECANTS.

,	48°		49°		50°		51°		,
	Vers.	Exsec.	Vers.	Exsec.	Vers.	Exsec.	Vers.	Exsec.	
0	.33087	.49448	.34394	.52425	.35721	.55572	.37068	.58902	0
1	.33109	.49496	.34416	.52476	.35744	.55626	.37091	.58959	1
2	.33130	.49544	.34438	.52527	.35766	.55680	.37113	.59016	2
3	.33152	.49593	.34460	.52579	.35788	.55734	.37136	.59073	3
4	.33173	.49641	.34482	.52630	.35810	.55789	.37158	.59130	4
5	.33195	.49690	.34504	.52681	.35833	.55843	.37181	.59188	5
6	.33217	.49738	.34526	.52732	.35855	.55897	.37204	.59245	6
7	.33238	.49787	.34548	.52784	.35877	.55951	.37226	.59302	7
8	.33260	.49835	.34570	.52835	.35900	.56005	.37249	.59360	8
9	.33282	.49884	.34592	.52886	.35922	.56060	.37272	.59418	9
10	.33303	.49933	.34614	.52938	.35944	.56114	.37294	.59475	10
11	.33325	.49981	.34636	.52989	.35967	.56169	.37317	.59533	11
12	.33347	.50030	.34658	.53041	.35989	.56223	.37340	.59590	12
13	.33368	.50079	.34680	.53092	.36011	.56278	.37362	.59648	13
14	.33390	.50128	.34702	.53144	.36034	.56332	.37385	.59706	14
15	.33412	.50177	.34724	.53196	.36056	.56387	.37408	.59764	15
16	.33434	.50226	.34746	.53247	.36078	.56442	.37430	.59822	16
17	.33455	.50275	.34768	.53299	.36101	.56497	.37453	.59880	17
18	.33477	.50324	.34790	.53351	.36123	.56551	.37476	.59938	18
19	.33499	.50373	.34812	.53403	.36146	.56606	.37498	.59996	19
20	.33520	.50422	.34834	.53455	.36168	.56661	.37521	.60054	20
21	.33542	.50471	.34856	.53507	.36190	.56716	.37544	.60112	21
22	.33564	.50521	.34878	.53559	.36213	.56771	.37567	.60171	22
23	.33586	.50570	.34900	.53611	.36235	.56826	.37589	.60229	23
24	.33607	.50619	.34923	.53663	.36258	.56881	.37612	.60287	24
25	.33629	.50669	.34945	.53715	.36280	.56937	.37635	.60346	25
26	.33651	.50718	.34967	.53768	.36302	.56992	.37658	.60404	26
27	.33673	.50767	.34989	.53820	.36325	.57047	.37680	.60463	27
28	.33694	.50817	.35011	.53872	.36347	.57103	.37703	.60521	28
29	.33716	.50866	.35033	.53924	.36370	.57158	.37726	.60580	29
30	.33738	.50916	.35055	.53977	.36392	.57213	.37749	.60689	30
31	.33760	.50966	.35077	.54029	.36415	.57269	.37771	.60698	31
32	.33782	.51015	.35099	.54082	.36437	.57324	.37794	.60736	32
33	.33803	.51065	.35122	.54134	.36460	.57380	.37817	.60815	33
34	.33825	.51115	.35144	.54187	.36482	.57436	.37840	.60874	34
35	.33847	.51165	.35166	.54240	.36504	.57491	.37862	.60933	35
36	.33869	.51215	.35188	.54292	.36527	.57547	.37885	.60992	36
37	.33891	.51265	.35210	.54345	.36549	.57603	.37908	.61051	37
38	.33912	.51314	.35232	.54398	.36572	.57659	.37931	.61111	38
39	.33934	.51364	.35254	.54451	.36594	.57715	.37954	.61170	39
40	.33956	.51415	.35277	.54504	.36617	.57771	.37976	.61229	40
41	.33978	.51465	.35299	.54557	.36639	.57827	.37999	.61288	41
42	.34000	.51515	.35321	.54610	.36662	.57883	.38022	.61348	42
43	.34022	.51565	.35343	.54663	.36684	.57939	.38045	.61407	43
44	.34044	.51615	.35365	.54716	.36707	.57995	.38068	.61467	44
45	.34065	.51665	.35388	.54769	.36729	.58051	.38091	.61526	45
46	.34087	.51716	.35410	.54822	.36752	.58108	.38113	.61586	46
47	.34109	.51766	.35432	.54876	.36775	.58164	.38136	.61646	47
48	.34131	.51817	.35454	.54929	.36797	.58221	.38159	.61705	48
49	.34153	.51867	.35476	.54982	.36820	.58277	.38182	.61765	49
50	.34175	.51918	.35499	.55036	.36842	.58333	.38205	.61825	50
51	.34197	.51968	.35521	.55089	.36865	.58390	.38228	.61885	51
52	.34219	.52019	.35543	.55143	.36887	.58447	.38251	.61945	52
53	.34241	.52069	.35565	.55196	.36910	.58503	.38274	.62005	53
54	.34262	.52120	.35588	.55250	.36932	.58560	.38296	.62065	54
55	.34284	.52171	.35610	.55303	.36955	.58617	.38319	.62125	55
56	.34306	.52222	.35632	.55357	.36978	.58674	.38342	.62185	56
57	.34328	.52273	.35654	.55411	.37000	.58731	.38365	.62246	57
58	.34350	.52323	.35677	.55465	.37023	.58788	.38388	.62306	58
59	.34372	.52374	.35699	.55518	.37045	.58845	.38411	.62366	59
60	.34394	.52425	.35721	.55572	.37068	.58902	.38434	.62427	60

TABLE XIII.—VERSINES AND EXSECANTS.

′	52°		53°		54°		55°		′
	Vers.	Exsec.	Vers.	Exsec.	Vers.	Exsec.	Vers.	Exsec.	
0	.38434	.62427	.39819	.66164	.41221	.70130	.42642	.74345	0
1	.38457	.62487	.39842	.66228	.41245	.70198	.42666	.74417	1
2	.38480	.62548	.39865	.66292	.41269	.70267	.42690	.74490	2
3	.38503	.62609	.39888	.66357	.41292	.70335	.42714	.74562	3
4	.38526	.62669	.39911	.66421	.41316	.70403	.42738	.74635	4
5	.38549	.62730	.39935	.66486	.41339	.70472	.42762	.74708	5
6	.38571	.62791	.39958	.66550	.41363	.70540	.42785	.74781	6
7	.38594	.62852	.39981	.66615	.41386	.70609	.42809	.74854	7
8	.38617	.62913	.40005	.66679	.41410	.70677	.42833	.74927	8
9	.38640	.62974	.40028	.66744	.41433	.70746	.42857	.75000	9
10	.38663	.63035	.40051	.66809	.41457	.70815	.42881	.75073	10
11	.38686	.63096	.40074	.66873	.41481	.70884	.42905	.75146	11
12	.38709	.63157	.40098	.66938	.41504	.70953	.42929	.75219	12
13	.38732	.63218	.40121	.67003	.41528	.71022	.42953	.75293	13
14	.38755	.63279	.40144	.67068	.41551	.71091	.42976	.75366	14
15	.38778	.63341	.40168	.67133	.41575	.71160	.43000	.75440	15
16	.38801	.63402	.40191	.67199	.41599	.71229	.43024	.75513	16
17	.38824	.63464	.40214	.67264	.41622	.71298	.43048	.75587	17
18	.38847	.63525	.40237	.67329	.41646	.71368	.43072	.75661	18
19	.38870	.63587	.40261	.67394	.41670	.71437	.43096	.75734	19
20	.38893	.63648	.40284	.67460	.41693	.71506	.43120	.75808	20
21	.38916	.63710	.40307	.67525	.41717	.71576	.43144	.75882	21
22	.38939	.63772	.40331	.67591	.41740	.71646	.43168	.75956	22
23	.38962	.63834	.40354	.67656	.41764	.71715	.43192	.76031	23
24	.38985	.63895	.40378	.67722	.41788	.71785	.43216	.76105	24
25	.39009	.63957	.40401	.67788	.41811	.71855	.43240	.76179	25
26	.39032	.64019	.40424	.67853	.41835	.71925	.43264	.76253	26
27	.39055	.64081	.40448	.67919	.41859	.71995	.43287	.76328	27
28	.39078	.64144	.40471	.67985	.41882	.72065	.43311	.76402	28
29	.39101	.64206	.40494	.68051	.41906	.72135	.43335	.76477	29
30	.39124	.64268	.40518	.68117	.41930	.72205	.43359	.76552	30
31	.39147	.64330	.40541	.68183	.41953	.72275	.43383	.76626	31
32	.39170	.64393	.40565	.68250	.41977	.72346	.43407	.76701	32
33	.39193	.64455	.40588	.68316	.42001	.72416	.43431	.76776	33
34	.39216	.64518	.40611	.68382	.42024	.72487	.43455	.76851	34
35	.39239	.64580	.40635	.68449	.42048	.72557	.43479	.76926	35
36	.39262	.64643	.40658	.68515	.42072	.72628	.43503	.77001	36
37	.39286	.64705	.40682	.68582	.42096	.72698	.43527	.77077	37
38	.39309	.64768	.40705	.68648	.42119	.72769	.43551	.77152	38
39	.39332	.64831	.40728	.68715	.42143	.72840	.43575	.77227	39
40	.39355	.64894	.40752	.68782	.42167	.72911	.43599	.77303	40
41	.39378	.64957	.40775	.68848	.42191	.72982	.43623	.77378	41
42	.39401	.65020	.40799	.68915	.42214	.73053	.43647	.77454	42
43	.39424	.65083	.40822	.68982	.42238	.73124	.43671	.77530	43
44	.39447	.65146	.40846	.69049	.42262	.73195	.43695	.77606	44
45	.39471	.65209	.40869	.69116	.42285	.73267	.43720	.77681	45
46	.39494	.65272	.40893	.69183	.42309	.73338	.43744	.77757	46
47	.39517	.65336	.40916	.69250	.42333	.73409	.43768	.77833	47
48	.39540	.65399	.40939	.69318	.42357	.73481	.43792	.77910	48
49	.39563	.65462	.40963	.69385	.42381	.73552	.43816	.77986	49
50	.39586	.65526	.40986	.69452	.42404	.73624	.43840	.78062	50
51	.39610	.65589	.41010	.69520	.42428	.73696	.43864	.78138	51
52	.39633	.65653	.41033	.69587	.42452	.73768	.43888	.78215	52
53	.39656	.65717	.41057	.69655	.42476	.73840	.43912	.78291	53
54	.39679	.65780	.41080	.69723	.42499	.73911	.43936	.78368	54
55	.39702	.65844	.41104	.69790	.42523	.73983	.43960	.78445	55
56	.39726	.65908	.41127	.69858	.42547	.74056	.43984	.78521	56
57	.39749	.65972	.41151	.69926	.42571	.74128	.44008	.78598	57
58	.39772	.66036	.41174	.69994	.42595	.74200	.44032	.78675	58
59	.39795	.66100	.41198	.70062	.42619	.74272	.44057	.78752	59
60	.39819	.66164	.41221	.70130	.42642	.74345	.44081	.78829	60

TABLE XIII.—VERSINES AND EXSECANTS.

′	56°		57°		58°		59°		′
	Vers.	Exsec.	Vers.	Exsec.	Vers.	Exsec.	Vers.	Exsec.	
0	.44081	.78829	.45536	.83608	.47008	.88708	.48496	.94160	0
1	.44105	.78906	.45560	.83690	.47033	.88796	.48521	.94254	1
2	.44129	.78984	.45585	.83773	.47057	.88884	.48546	.94340	2
3	.44153	.79061	.45609	.83855	.47082	.88972	.48571	.94443	3
4	.44177	.79138	.45634	.83938	.47107	.89060	.48596	.94537	4
5	.44201	.79216	.45658	.84020	.47131	.89148	.48621	.94632	5
6	.44225	.79293	.45683	.84103	.47156	.89237	.48646	.94726	6
7	.44250	.79371	.45707	.84186	.47181	.89325	.48671	.94821	7
8	.44274	.79449	.45731	.84269	.47206	.89414	.48696	.94916	8
9	.44298	.79527	.45756	.84352	.47230	.89503	.48721	.95011	9
10	.44322	.79604	.45780	.84435	.47255	.89591	.48746	.95106	10
11	.44346	.79682	.45805	.84518	.47280	.89680	.48771	.95201	11
12	.44370	.79761	.45829	.84601	.47304	.89769	.48796	.95296	12
13	.44395	.79839	.45854	.84685	.47329	.89858	.48821	.95392	13
14	.44419	.79917	.45878	.84768	.47354	.89948	.48846	.95487	14
15	.44443	.79995	.45903	.84852	.47379	.90037	.48871	.95583	15
16	.44467	.80074	.45927	.84935	.47403	.90126	.48896	.95678	16
17	.44491	.80152	.45951	.85019	.47428	.90216	.48921	.95774	17
18	.44516	.80231	.45976	.85103	.47453	.90305	.48946	.95870	18
19	.44540	.80309	.46000	.85187	.47478	.90395	.48971	.95966	19
20	.44564	.80388	.46025	.85271	.47502	.90485	.48996	.96062	20
21	.44588	.80467	.46049	.85355	.47527	.90575	.49021	.96158	21
22	.44612	.80546	.46074	.85439	.47552	.90665	.49046	.96255	22
23	.44637	.80625	.46098	.85523	.47577	.90755	.49071	.96351	23
24	.44661	.80704	.46123	.85608	.47601	.90845	.49096	.96448	24
25	.44685	.80783	.46147	.85692	.47626	.90935	.49121	.96544	25
26	.44709	.80862	.46172	.85777	.47651	.91026	.49146	.96641	26
27	.44734	.80942	.46196	.85861	.47676	.91116	.49171	.96738	27
28	.44758	.81021	.46221	.85946	.47701	.91207	.49196	.96835	28
29	.44782	.81101	.46246	.86031	.47725	.91297	.49221	.96932	29
30	.44806	.81180	.46270	.86116	.47750	.91388	.49246	.97029	30
31	.44831	.81260	.46295	.86201	.47775	.91479	.49271	.97127	31
32	.44855	.81340	.46319	.86286	.47800	.91570	.49296	.97224	32
33	.44879	.81419	.46344	.86371	.47825	.91661	.49321	.97322	33
34	.44903	.81499	.46368	.86457	.47849	.91752	.49346	.97420	34
35	.44928	.81579	.46393	.86542	.47874	.91844	.49372	.97517	35
36	.44952	.81659	.46417	.86627	.47899	.91935	.49397	.97615	36
37	.44976	.81740	.46442	.86713	.47924	.92027	.49422	.97713	37
38	.45001	.81820	.46466	.86799	.47949	.92118	.49447	.97811	38
39	.45025	.81900	.46491	.86885	.47974	.92210	.49472	.97910	39
40	.45049	.81981	.46516	.86970	.47998	.92302	.49497	.98008	40
41	.45073	.82061	.46540	.87056	.48023	.92394	.49522	.98107	41
42	.45098	.82142	.46565	.87142	.48048	.92486	.49547	.98205	42
43	.45122	.82222	.46589	.87229	.48073	.92578	.49572	.98304	43
44	.45146	.82303	.46614	.87315	.48098	.92670	.49597	.98403	44
45	.45171	.82384	.46639	.87401	.48123	.92762	.49623	.98502	45
46	.45195	.82465	.46663	.87488	.48148	.92855	.49648	.98601	46
47	.45219	.82546	.46688	.87574	.48172	.92947	.49673	.98700	47
48	.45244	.82627	.46712	.87661	.48197	.93040	.49698	.98799	48
49	.45268	.82709	.46737	.87748	.48222	.93133	.49723	.98899	49
50	.45292	.82790	.46762	.87834	.48247	.93226	.49748	.98998	50
51	.45317	.82871	.46786	.87921	.48272	.93319	.49773	.99098	51
52	.45341	.82953	.46811	.88008	.48297	.93412	.49799	.99198	52
53	.45365	.83034	.46836	.88095	.48322	.93505	.49824	.99298	53
54	.45390	.83116	.46860	.88183	.48347	.93598	.49849	.99398	54
55	.45414	.83198	.46885	.88270	.48372	.93692	.49874	.99498	55
56	.45439	.83280	.46909	.88357	.48396	.93785	.49899	.99598	56
57	.45463	.83362	.46934	.88445	.48421	.93879	.49924	.99698	57
58	.45487	.83444	.46959	.88532	.48446	.93973	.49950	.99799	58
59	.45512	.83526	.46983	.88620	.48471	.94066	.49975	.99899	59
60	.45536	.83608	.47008	.88708	.48496	.94160	.50000	1.00000	60

TABLE XIII.—VERSINES AND EXSECANTS.

′	60°		61°		62°		63°		′
	Vers.	Exsec.	Vers.	Exsec.	Vers.	Exsec.	Vers.	Exsec.	
0	.50000	1.00000	.51519	1.06267	.53053	1.13005	.54601	1.20269	0
1	.50025	1.00101	.51544	1.06375	.53079	1.13122	.54627	1.20395	1
2	.50050	1.00202	.51570	1.06483	.53104	1.13239	.54653	1.20521	2
3	.50076	1.00303	.51595	1.06592	.53130	1.13356	.54679	1.20647	3
4	.50101	1.00404	.51621	1.06701	.53156	1.13473	.54705	1.20773	4
5	.50126	1.00505	.51646	1.06809	.53181	1.13590	.54731	1.20900	5
6	.50151	1.00607	.51672	1.06918	.53207	1.13707	.54757	1.21026	6
7	.50176	1.00708	.51697	1.07027	.53233	1.13825	.54782	1.21153	7
8	.50202	1.00810	.51723	1.07137	.53258	1.13942	.54808	1.21280	8
9	.50227	1.00912	.51748	1.07246	.53284	1.14060	.54834	1.21407	9
10	.50252	1.01014	.51774	1.07356	.53310	1.14178	.54860	1.21535	10
11	.50277	1.01116	.51799	1.07465	.53336	1.14296	.54886	1.21662	11
12	.50303	1.01218	.51825	1.07575	.53361	1.14414	.54912	1.21790	12
13	.50328	1.01320	.51850	1.07685	.53387	1.14533	.54938	1.21918	13
14	.50353	1.01422	.51876	1.07795	.53413	1.14651	.54964	1.22045	14
15	.50378	1.01525	.51901	1.07905	.53439	1.14770	.54990	1.22174	15
16	.50404	1.01628	.51927	1.08015	.53464	1.14889	.55016	1.22302	16
17	.50429	1.01730	.51952	1.08126	.53490	1.15008	.55042	1.22430	17
18	.50454	1.01833	.51978	1.08236	.53516	1.15127	.55068	1.22559	18
19	.50479	1.01936	.52003	1.08347	.53542	1.15246	.55094	1.22688	19
20	.50505	1.02039	.52029	1.08458	.53567	1.15366	.55120	1.22817	20
21	.50530	1.02143	.52054	1.08569	.53593	1.15485	.55146	1.22946	21
22	.50555	1.02246	.52080	1.08680	.53619	1.15605	.55172	1.23075	22
23	.50581	1.02349	.52105	1.08791	.53645	1.15725	.55198	1.23205	23
24	.50606	1.02453	.52131	1.08903	.53670	1.15845	.55224	1.23334	24
25	.50631	1.02557	.52156	1.09014	.53696	1.15965	.55250	1.23464	25
26	.50656	1.02661	.52182	1.09126	.53722	1.16085	.55276	1.23594	26
27	.50682	1.02765	.52207	1.09238	.53748	1.16206	.55302	1.23724	27
28	.50707	1.02869	.52233	1.09350	.53774	1.16326	.55328	1.23855	28
29	.50732	1.02973	.52259	1.09462	.53799	1.16447	.55354	1.23985	29
30	.50758	1.03077	.52284	1.09574	.53825	1.16568	.55380	1.24116	30
31	.50783	1.03182	.52310	1.09686	.53851	1.16689	.55406	1.24247	31
32	.50808	1.03286	.52335	1.09799	.53877	1.16810	.55432	1.24378	32
33	.50834	1.03391	.52361	1.09911	.53903	1.16932	.55458	1.24509	33
34	.50859	1.03496	.52386	1.10024	.53928	1.17053	.55484	1.24640	34
35	.50884	1.03601	.52412	1.10137	.53954	1.17175	.55510	1.24772	35
36	.50910	1.03706	.52438	1.10250	.53980	1.17297	.55536	1.24903	36
37	.50935	1.03811	.52463	1.10363	.54006	1.17419	.55563	1.25035	37
38	.50960	1.03916	.52489	1.10477	.54032	1.17541	.55589	1.25167	38
39	.50986	1.04022	.52514	1.10590	.54058	1.17663	.55615	1.25300	39
40	.51011	1.04128	.52540	1.10704	.54083	1.17786	.55641	1.25432	40
41	.51036	1.04233	.52566	1.10817	.54109	1.17909	.55667	1.25565	41
42	.51062	1.04339	.52591	1.10931	.54135	1.18031	.55693	1.25697	42
43	.51087	1.04445	.52617	1.11045	.54161	1.18154	.55719	1.25830	43
44	.51113	1.04551	.52642	1.11159	.54187	1.18277	.55745	1.25963	44
45	.51138	1.04658	.52668	1.11274	.54213	1.18401	.55771	1.26097	45
46	.51163	1.04764	.52694	1.11388	.54238	1.18524	.55797	1.26230	46
47	.51189	1.04870	.52719	1.11503	.54264	1.18648	.55823	1.26364	47
48	.51214	1.04977	.52745	1.11617	.54290	1.18772	.55849	1.26498	48
49	.51239	1.05084	.52771	1.11732	.54316	1.18895	.55876	1.26632	49
50	.51265	1.05191	.52796	1.11847	.54342	1.19019	.55902	1.26766	50
51	.51290	1.05298	.52822	1.11963	.54368	1.19144	.55928	1.26900	51
52	.51316	1.05405	.52848	1.12078	.54394	1.19268	.55954	1.27035	52
53	.51341	1.05512	.52873	1.12193	.54420	1.19393	.55980	1.27169	53
54	.51366	1.05619	.52899	1.12309	.54446	1.19517	.56006	1.27304	54
55	.51392	1.05727	.52924	1.12425	.54471	1.19642	.56032	1.27439	55
56	.51417	1.05835	.52950	1.12540	.54497	1.19767	.56058	1.27574	56
57	.51443	1.05942	.52976	1.12657	.54523	1.19892	.56084	1.27710	57
58	.51468	1.06050	.53001	1.12773	.54549	1.20018	.56111	1.27845	58
59	.51494	1.06158	.53027	1.12889	.54575	1.20143	.56137	1.27981	59
60	.51519	1.06267	.53053	1.13005	.54601	1.20269	.56163	1.28117	60

TABLE XIII.—VERSINES AND EXSECANTS.

′	64°		65°		66°		67°		
	Vers.	Exsec.	Vers.	Exsec.	Vers.	Exsec.	Vers.	Exsec.	
0	.56163	1.28117	.57738	1.36620	.59326	1.45859	.60927	1.55930	0
1	.56189	1.28253	.57765	1.36768	.59353	1.46020	.60954	1.56106	1
2	.56215	1.28390	.57791	1.36916	.59379	1.46181	.60980	1.56282	2
3	.56241	1.28526	.57817	1.37064	.59406	1.46342	.61007	1.56458	3
4	.56267	1.28663	.57844	1.37212	.59433	1.46504	.61034	1.56634	4
5	.56294	1.28800	.57870	1.37361	.59459	1.46665	.61061	1.56811	5
6	.56320	1.28937	.57896	1.37509	.59486	1.46827	.61088	1.56988	6
7	.56346	1.29074	.57923	1.37658	.59512	1.46989	.61114	1.57165	7
8	.56372	1.29211	.57949	1.37808	.59539	1.47152	.61141	1.57342	8
9	.56398	1.29349	.57976	1.37957	.59566	1.47314	.61168	1.57520	9
10	.56425	1.29487	.58002	1.38107	.59592	1.47477	.61195	1.57698	10
11	.56451	1.29625	.58028	1.38256	.59619	1.47640	.61222	1.57876	11
12	.56477	1.29763	.58055	1.38406	.59645	1.47804	.61248	1.58054	12
13	.56503	1.29901	.58081	1.38556	.59672	1.47967	.61275	1.58233	13
14	.56529	1.30040	.58108	1.38707	.59699	1.48131	.61302	1.58412	14
15	.56555	1.30179	.58134	1.38857	.59725	1.48295	.61329	1.58591	15
16	.56582	1.30318	.58160	1.39008	.59752	1.48459	.61356	1.58771	16
17	.56608	1.30457	.58187	1.39159	.59779	1.48624	.61383	1.58950	17
18	.56634	1.30596	.58213	1.39311	.59805	1.48789	.61409	1.59130	18
19	.56660	1.30735	.58240	1.39462	.59832	1.48954	.61436	1.59311	19
20	.56687	1.30875	.58266	1.39614	.59859	1.49119	.61463	1.59491	20
21	.56713	1.31015	.58293	1.39766	.59885	1.49284	.61490	1.59672	21
22	.56739	1.31155	.58319	1.39918	.59912	1.49450	.61517	1.59853	22
23	.56765	1.31295	.58345	1.40070	.59938	1.49616	.61544	1.60035	23
24	.56791	1.31436	.58372	1.40222	.59965	1.49782	.61570	1.60217	24
25	.56818	1.31576	.58398	1.40375	.59992	1.49948	.61597	1.60399	25
26	.56844	1.31717	.58425	1.40528	.60018	1.50115	.61624	1.60581	26
27	.56870	1.31858	.58451	1.40681	.60045	1.50282	.61651	1.60763	27
28	.56896	1.31999	.58478	1.40835	.60072	1.50449	.61678	1.60946	28
29	.56923	1.32140	.58504	1.40988	.60098	1.50617	.61705	1.61129	29
30	.56949	1.32282	.58531	1.41142	.60125	1.50784	.61732	1.61313	30
31	.56975	1.32424	.58557	1.41296	.60152	1.50952	.61759	1.61496	31
32	.57001	1.32566	.58584	1.41450	.60178	1.51120	.61785	1.61680	32
33	.57028	1.32708	.58610	1.41605	.60205	1.51289	.61812	1.61864	33
34	.57054	1.32850	.58637	1.41760	.60232	1.51457	.61839	1.62049	34
35	.57080	1.32993	.58663	1.41914	.60259	1.51626	.61866	1.62234	35
36	.57106	1.33135	.58690	1.42070	.60285	1.51795	.61893	1.62419	36
37	.57133	1.33278	.58716	1.42225	.60312	1.51965	.61920	1.62604	37
38	.57159	1.33422	.58743	1.42380	.60339	1.52134	.61947	1.62790	38
39	.57185	1.33565	.58769	1.42536	.60365	1.52304	.61974	1.62976	39
40	.57212	1.33708	.58796	1.42692	.60392	1.52474	.62001	1.63162	40
41	.57238	1.33852	.58822	1.42848	.60419	1.52645	.62027	1.63348	41
42	.57264	1.33996	.58849	1.43005	.60445	1.52815	.62054	1.63535	42
43	.57291	1.34140	.58875	1.43162	.60472	1.52986	.62081	1.63722	43
44	.57317	1.34284	.58902	1.43318	.60499	1.53157	.62108	1.63909	44
45	.57343	1.34429	.58928	1.43476	.60526	1.53329	.62135	1.64097	45
46	.57369	1.34573	.58955	1.43633	.60552	1.53500	.62162	1.64285	46
47	.57396	1.34718	.58981	1.43790	.60579	1.53672	.62189	1.64473	47
48	.57422	1.34863	.59008	1.43948	.60606	1.53845	.62216	1.64662	48
49	.57448	1.35009	.59034	1.44106	.60633	1.54017	.62243	1.64851	49
50	.57475	1.35154	.59061	1.44264	.60659	1.54190	.62270	1.65040	50
51	.57501	1.35300	.59087	1.44423	.60686	1.54363	.62297	1.65229	51
52	.57527	1.35446	.59114	1.44582	.60713	1.54536	.62324	1.65419	52
53	.57554	1.35592	.59140	1.44741	.60740	1.54709	.62351	1.65609	53
54	.57580	1.35738	.59167	1.44900	.60766	1.54883	.62378	1.65799	54
55	.57606	1.35885	.59194	1.45059	.60793	1.55057	.62405	1.65989	55
56	.57633	1.36031	.59220	1.45219	.60820	1.55231	.62431	1.66180	56
57	.57659	1.36178	.59247	1.45378	.60847	1.55405	.62458	1.66371	57
58	.57685	1.36325	.59273	1.45539	.60873	1.55580	.62485	1.66563	58
59	.57712	1.36473	.59300	1.45699	.60900	1.55755	.62512	1.66755	59
60	.57738	1.36620	.59326	1.45859	.60927	1.55930	.62539	1.66947	60

TABLE XIII.—VERSINES AND EXSECANTS.

′	68°		69°		70°		71°	
	Vers.	Exsec.	Vers.	Exsec.	Vers.	Exsec.	Vers.	Ex
0	.62539	1.66947	.64163	1.79043	.65798	1.92380	.67443	2.0
1	.62566	1.67139	.64190	1.79254	.65825	1.92614	.67471	2.0
2	.62593	1.67332	.64218	1.79466	.65853	1.92849	.67498	2.0
3	.62620	1.67525	.64245	1.79679	.65880	1.93083	.67526	2.0
4	.62647	1.67718	.64272	1.79891	.65907	1.93318	.67553	2.0
5	.62674	1.67911	.64299	1.80104	.65935	1.93554	.67581	2.0
6	.62701	1.68105	.64326	1.80318	.65962	1.93790	.67608	2.0
7	.62728	1.68299	.64353	1.80531	.65989	1.94026	.67636	2.0
8	.62755	1.68494	.64381	1.80746	.66017	1.94263	.67663	2.0
9	.62782	1.68689	.64408	1.80960	.66044	1.94500	.67691	2.0
10	.62809	1.68884	.64435	1.81175	.66071	1.94737	.67718	2.0
11	.62836	1.69079	.64462	1.81390	.66099	1.94975	.67746	2.1
12	.62863	1.69275	.64489	1.81605	.66126	1.95213	.67773	2.1
13	.62890	1.69471	.64517	1.81821	.66154	1.95452	.67801	2.1
14	.62917	1.69667	.64544	1.82037	.66181	1.95691	.67829	2.1
15	.62944	1.69864	.64571	1.82254	.66208	1.95931	.67856	2.1
16	.62971	1.70061	.64598	1.82471	.66236	1.96171	.67884	2.1
17	.62998	1.70258	.64625	1.82688	.66263	1.96411	.67911	2.1
18	.63025	1.70455	.64653	1.82906	.66290	1.96652	.67939	2.1
19	.63052	1.70653	.64680	1.83124	.66318	1.96893	.67966	2.1
20	.63079	1.70851	.64707	1.83342	.66345	1.97135	.67994	2.1
21	.63106	1.71050	.64734	1.83561	.66373	1.97377	.68021	2.1
22	.63133	1.71249	.64761	1.83780	.66400	1.97619	.68049	2.1
23	.63161	1.71448	.64789	1.83999	.66427	1.97862	.68077	2.1
24	.63188	1.71647	.64816	1.84219	.66455	1.98106	.68104	2.1
25	.63215	1.71847	.64843	1.84439	.66482	1.98349	.68132	2.1
26	.63242	1.72047	.64870	1.84659	.66510	1.98594	.68159	2.1
27	.63269	1.72247	.64898	1.84880	.66537	1.98838	.68187	2.1
28	.63296	1.72448	.64925	1.85102	.66564	1.99083	.68214	2.1
29	.63323	1.72649	.64952	1.85323	.66592	1.99329	.68242	2.1
30	.63350	1.72850	.64979	1.85545	.66619	1.99574	.68270	2.1
31	.63377	1.73052	.65007	1.85767	.66647	1.99821	.68297	2.1
32	.63404	1.73254	.65034	1.85990	.66674	2.00067	.68325	2.1
33	.63431	1.73456	.65061	1.86213	.66702	2.00315	.68352	2.1
34	.63458	1.73659	.65088	1.86437	.66729	2.00562	.68380	2.1
35	.63485	1.73862	.65116	1.86661	.66756	2.00810	.68408	2.1
36	.63512	1.74065	.65143	1.86885	.66784	2.01059	.68435	2.1
37	.63539	1.74269	.65170	1.87109	.66811	2.01308	.68463	2.1
38	.63566	1.74473	.65197	1.87334	.66839	2.01557	.68490	2.1
39	.63594	1.74677	.65225	1.87560	.66866	2.01807	.68518	2.1
40	.63621	1.74881	.65252	1.87785	.66894	2.02057	.68546	2.1
41	.63648	1.75086	.65279	1.88011	.66921	2.02308	.68573	2.1
42	.63675	1.75292	.65306	1.88238	.66949	2.02559	.68601	2.1
43	.63702	1.75497	.65334	1.88465	.66976	2.02810	.68628	2.1
44	.63729	1.75703	.65361	1.88692	.67003	2.03062	.68656	2.1
45	.63756	1.75909	.65388	1.88920	.67031	2.03315	.68684	2.1
46	.63783	1.76116	.65416	1.89148	.67058	2.03568	.68711	2.1
47	.63810	1.76323	.65443	1.89376	.67086	2.03821	.68739	2.1
48	.63838	1.76530	.65470	1.89605	.67113	2.04075	.68767	2.2
49	.63865	1.76737	.65497	1.89834	.67141	2.04329	.68794	2.2
50	.63892	1.76945	.65525	1.90063	.67168	2.04584	.68822	2.2
51	.63919	1.77154	.65552	1.90293	.67196	2.04839	.68849	2.2
52	.63946	1.77362	.65579	1.90524	.67223	2.05094	.68877	2.2
53	.63973	1.77571	.65607	1.90754	.67251	2.05350	.68905	2.2
54	.64000	1.77780	.65634	1.90986	.67278	2.05607	.68932	2.2
55	.64027	1.77990	.65661	1.91217	.67306	2.05864	.68960	2.2
56	.64055	1.78200	.65689	1.91449	.67333	2.06121	.68988	2.2
57	.64082	1.78410	.65716	1.91681	.67361	2.06379	.69015	2.2
58	.64109	1.78621	.65743	1.91914	.67388	2.06637	.69043	2.2
59	.64136	1.78832	.65771	1.92147	.67416	2.06896	.69071	2.2
60	.64163	1.79043	.65798	1.92380	.67443	2.07155	.69098	2.2

TABLE XIII.—VERSINES AND EXSECANTS.

′	72°		73°		74°		75°		′
	Vers.	Exsec.	Vers.	Exsec.	Vers.	Exsec.	Vers.	Exsec.	
0	.69098	2.23607	.70763	2.42030	.72436	2.62796	.74118	2.86370	0
1	.69126	2.23897	.70791	2.42356	.72464	2.63164	.74146	2.86790	1
2	.69154	2.24187	.70818	2.42683	.72492	2.63533	.74174	2.87211	2
3	.69181	2.24478	.70846	2.43010	.72520	2.63903	.74202	2.87633	3
4	.69209	2.24770	.70874	2.43337	.72548	2.64274	.74231	2.88056	4
5	.69237	2.25062	.70902	2.43666	.72576	2.64645	.74259	2.88479	5
6	.69264	2.25355	.70930	2.43995	.72604	2.65018	.74287	2.88904	6
7	.69292	2.25648	.70958	2.44324	.72632	2.65391	.74315	2.89330	7
8	.69320	2.25942	.70985	2.44655	.72660	2.65765	.74343	2.89756	8
9	.69347	2.26237	.71013	2.44986	.72688	2.66140	.74371	2.90184	9
10	.69375	2.26531	.71041	2.45317	.72716	2.66515	.74399	2.90613	10
11	.69403	2.26827	.71069	2.45650	.72744	2.66892	.74427	2.91042	11
12	.69430	2.27123	.71097	2.45983	.72772	2.67269	.74455	2.91473	12
13	.69458	2.27420	.71125	2.46316	.72800	2.67647	.74484	2.91904	13
14	.69486	2.27717	.71153	2.46651	.72828	2.68025	.74512	2.92337	14
15	.69514	2.28015	.71180	2.46986	.72856	2.68405	.74540	2.92770	15
16	.69541	2.28313	.71208	2.47321	.72884	2.68785	.74568	2.93204	16
17	.69569	2.28612	.71236	2.47658	.72912	2.69167	.74596	2.93640	17
18	.69597	2.28912	.71264	2.47995	.72940	2.69549	.74624	2.94076	18
19	.69624	2.29212	.71292	2.48333	.72968	2.69931	.74652	2.94514	19
20	.69652	2.29512	.71320	2.48671	.72996	2.70315	.74680	2.94952	20
21	.69680	2.29814	.71348	2.49010	.73024	2.70700	.74709	2.95392	21
22	.69708	2.30115	.71375	2.49350	.73052	2.71085	.74737	2.95832	22
23	.69735	2.30418	.71403	2.49691	.73080	2.71471	.74765	2.96274	23
24	.69763	2.30721	.71431	2.50032	.73108	2.71858	.74793	2.96716	24
25	.69791	2.31024	.71459	2.50374	.73136	2.72246	.74821	2.97160	25
26	.69818	2.31328	.71487	2.50716	.73164	2.72635	.74849	2.97604	26
27	.69846	2.31633	.71515	2.51060	.73192	2.73024	.74878	2.98050	27
28	.69874	2.31939	.71543	2.51404	.73220	2.73414	.74906	2.98497	28
29	.69902	2.32244	.71571	2.51748	.73248	2.73806	.74934	2.98944	29
30	.69929	2.32551	.71598	2.52094	.73276	2.74198	.74962	2.99393	30
31	.69957	2.32858	.71626	2.52440	.73304	2.74591	.74990	2.99843	31
32	.69985	2.33166	.71654	2.52787	.73332	2.74984	.75018	3.00293	32
33	.70013	2.33474	.71682	2.53134	.73360	2.75379	.75047	3.00745	33
34	.70040	2.33783	.71710	2.53482	.73388	2.75775	.75075	3.01198	34
35	.70068	2.34092	.71738	2.53831	.73416	2.76171	.75103	3.01652	35
36	.70096	2.34403	.71766	2.54181	.73444	2.76568	.75131	3.02107	36
37	.70124	2.34713	.71794	2.54531	.73472	2.76966	.75159	3.02563	37
38	.70151	2.35025	.71822	2.54883	.73500	2.77365	.75187	3.03020	38
39	.70179	2.35336	.71850	2.55235	.73529	2.77765	.75216	3.03479	39
40	.70207	2.35649	.71877	2.55587	.73557	2.78166	.75244	3.03938	40
41	.70235	2.35962	.71905	2.55940	.73585	2.78568	.75272	3.04398	41
42	.70263	2.36276	.71933	2.56294	.73613	2.78970	.75300	3.04860	42
43	.70290	2.36590	.71961	2.56649	.73641	2.79374	.75328	3.05322	43
44	.70318	2.36905	.71989	2.57005	.73669	2.79778	.75356	3.05786	44
45	.70346	2.37221	.72017	2.57361	.73697	2.80183	.75385	3.06251	45
46	.70374	2.37537	.72045	2.57718	.73725	2.80589	.75413	3.06717	46
47	.70401	2.37854	.72073	2.58076	.73753	2.80996	.75441	3.07184	47
48	.70429	2.38171	.72101	2.58434	.73781	2.81404	.75469	3.07652	48
49	.70457	2.38489	.72129	2.58794	.73809	2.81813	.75497	3.08121	49
50	.70485	2.38808	.72157	2.59154	.73837	2.82223	.75526	3.08591	50
51	.70513	2.39128	.72185	2.59514	.73865	2.82633	.75554	3.09063	51
52	.70540	2.39448	.72213	2.59876	.73893	2.83045	.75582	3.09535	52
53	.70568	2.39768	.72241	2.60238	.73921	2.83457	.75610	3.10009	53
54	.70596	2.40089	.72269	2.60601	.73950	2.83871	.75639	3.10484	54
55	.70624	2.40411	.72296	2.60965	.73978	2.84285	.75667	3.10960	55
56	.70652	2.40734	.72324	2.61330	.74006	2.84700	.75695	3.11437	56
57	.70679	2.41057	.72352	2.61695	.74034	2.85116	.75723	3.11915	57
58	.70707	2.41381	.72380	2.62061	.74062	2.85533	.75751	3.12394	58
59	.70735	2.41705	.72408	2.62428	.74090	2.85951	.75780	3.12875	59
60	.70763	2.42030	.72436	2.62796	.74118	2.86370	.75808	3.13357	60

TABLE XIII.—VERSINES AND EXSECANTS.

′	76°		77°		78°		79°		′
	Vers.	Exsec.	Vers.	Exsec.	Vers.	Exsec.	Vers.	Exsec.	
0	.75808	3.13357	.77505	3.44541	.79209	3.80973	.80919	4.24084	0
1	.75836	3.13839	.77533	3.45102	.79237	3.81633	.80948	4.24870	1
2	.75864	3.14323	.77562	3.45664	.79266	3.82294	.80976	4.25658	2
3	.75892	3.14809	.77590	3.46228	.79294	3.82956	.81005	4.26448	3
4	.75921	3.15295	.77618	3.46793	.79323	3.83621	.81033	4.27241	4
5	.75949	3.15782	.77647	3.47360	.79351	3.84288	.81062	4.28036	5
6	.75977	3.16271	.77675	3.47928	.79380	3.84956	.81090	4.28833	6
7	.76005	3.16761	.77703	3.48498	.79408	3.85627	.81119	4.29634	7
8	.76034	3.17252	.77732	3.49069	.79437	3.86299	.81148	4.30436	8
9	.76062	3.17744	.77760	3.49642	.79465	3.86973	.81176	4.31241	9
10	.76090	3.18238	.77788	3.50216	.79493	3.87649	.81205	4.32049	10
11	.76118	3.18733	.77817	3.50791	.79522	3.88327	.81233	4.32859	11
12	.76147	3.19228	.77845	3.51368	.79550	3.89007	.81262	4.33671	12
13	.76175	3.19725	.77874	3.51947	.79579	3.89689	.81290	4.34486	13
14	.76203	3.20224	.77902	3.52527	.79607	3.90373	.81319	4.35304	14
15	.76231	3.20723	.77930	3.53109	.79636	3.91058	.81348	4.36124	15
16	.76260	3.21224	.77959	3.53692	.79664	3.91746	.81376	4.36947	16
17	.76288	3.21726	.77987	3.54277	.79693	3.92436	.81405	4.37772	17
18	.76316	3.22229	.78015	3.54863	.79721	3.93128	.81433	4.38600	18
19	.76344	3.22734	.78044	3.55451	.79750	3.93821	.81462	4.39430	19
20	.76373	3.23239	.78072	3.56041	.79778	3.94517	.81491	4.40263	20
21	.76401	3.23746	.78101	3.56632	.79807	3.95215	.81519	4.41099	21
22	.76429	3.24255	.78129	3.57224	.79835	3.95914	.81548	4.41937	22
23	.76458	3.24764	.78157	3.57819	.79864	3.96616	.81576	4.42778	23
24	.76486	3.25275	.78186	3.58414	.79892	3.97320	.81605	4.43622	24
25	.76514	3.25787	.78214	3.59012	.79921	3.98025	.81633	4.44468	25
26	.76542	3.26300	.78242	3.59611	.79949	3.98733	.81662	4.45317	26
27	.76571	3.26814	.78271	3.60211	.79978	3.99443	.81691	4.46169	27
28	.76599	3.27330	.78299	3.60813	.80006	4.00155	.81719	4.47023	28
29	.76627	3.27847	.78328	3.61417	.80035	4.00869	.81748	4.47881	29
30	.76655	3.28366	.78356	3.62023	.80063	4.01585	.81776	4.48740	30
31	.76684	3.28885	.78384	3.62630	.80092	4.02303	.81805	4.49603	31
32	.76712	3.29406	.78413	3.63238	.80120	4.03024	.81834	4.50468	32
33	.76740	3.29929	.78441	3.63849	.80149	4.03746	.81862	4.51337	33
34	.76769	3.30452	.78470	3.64461	.80177	4.04471	.81891	4.52208	34
35	.76797	3.30977	.78498	3.65074	.80206	4.05197	.81919	4.53081	35
36	.76825	3.31503	.78526	3.65690	.80234	4.05926	.81948	4.53958	36
37	.76854	3.32031	.78555	3.66307	.80263	4.06657	.81977	4.54837	37
38	.76882	3.32560	.78583	3.66925	.80291	4.07390	.82005	4.55720	38
39	.76910	3.33090	.78612	3.67545	.80320	4.08125	.82034	4.56605	39
40	.76938	3.33622	.78640	3.68167	.80348	4.08863	.82063	4.57493	40
41	.76967	3.34154	.78669	3.68791	.80377	4.09602	.82091	4.58383	41
42	.76995	3.34689	.78697	3.69417	.80405	4.10344	.82120	4.59277	42
43	.77023	3.35224	.78725	3.70044	.80434	4.11088	.82148	4.60174	43
44	.77052	3.35761	.78754	3.70673	.80462	4.11835	.82177	4.61073	44
45	.77080	3.36299	.78782	3.71303	.80491	4.12583	.82206	4.61976	45
46	.77108	3.36839	.78811	3.71935	.80520	4.13334	.82234	4.62881	46
47	.77137	3.37380	.78839	3.72569	.80548	4.14087	.82263	4.63790	47
48	.77165	3.37923	.78868	3.73205	.80577	4.14842	.82292	4.64701	48
49	.77193	3.38466	.78896	3.73843	.80605	4.15599	.82320	4.65616	49
50	.77222	3.39012	.78924	3.74482	.80634	4.16359	.82349	4.66533	50
51	.77250	3.39558	.78953	3.75123	.80662	4.17121	.82377	4.67454	51
52	.77278	3.40106	.78981	3.75766	.80691	4.17886	.82406	4.68377	52
53	.77307	3.40656	.79010	3.76411	.80719	4.18652	.82435	4.69304	53
54	.77335	3.41206	.79038	3.77057	.80748	4.19421	.82463	4.70234	54
55	.77363	3.41759	.79067	3.77705	.80776	4.20193	.82492	4.71166	55
56	.77392	3.42312	.79095	3.78355	.80805	4.20966	.82521	4.72102	56
57	.77420	3.42867	.79123	3.79007	.80833	4.21742	.82549	4.73041	57
58	.77448	3.43424	.79152	3.79661	.80862	4.22521	.82578	4.73983	58
59	.77477	3.43982	.79180	3.80316	.80891	4.23301	.82607	4.74929	59
60	.77505	3.44541	.79209	3.80973	.80919	4.24084	.82635	4.75877	60

TABLE XIII.—VERSINES AND EXSECANTS.

′	80°		81°		82°		83°		′
	Vers.	Exsec.	Vers.	Exsec.	Vers.	Exsec.	Vers.	Exsec.	
0	.82635	4.75877	.84357	5.39245	.86083	6.18530	.87813	7.20551	0
1	.82664	4.76829	.84385	5.40422	.86112	6.20020	.87842	7.22500	1
2	.82692	4.77784	.84414	5.41002	.86140	6.21517	.87871	7.24457	2
3	.82721	4.78742	.84443	5.42787	.86169	6.23019	.87900	7.26425	3
4	.82750	4.79703	.84471	5.43977	.86198	6.24529	.87929	7.28402	4
5	.82778	4.80667	.84500	5.45171	.86227	6.26044	.87957	7.30388	5
6	.82807	4.81635	.84529	5.46369	.86256	6.27566	.87986	7.32384	6
7	.82836	4.82606	.84558	5.47572	.86284	6.29095	.88015	7.34390	7
8	.82864	4.83581	.84586	5.48779	.86313	6.30630	.88044	7.36405	8
9	.82893	4.84558	.84615	5.49991	.86342	6.32171	.88073	7.38431	9
10	.82922	4.85539	.84644	5.51208	.86371	6.33719	.88102	7.40466	10
11	.82950	4.86524	.84673	5.52429	.86400	6.35274	.88131	7.42511	11
12	.82979	4.87511	.84701	5.53655	.86428	6.36835	.88160	7.44566	12
13	.83008	4.88502	.84730	5.54886	.86457	6.38403	.88188	7.46632	13
14	.83036	4.89497	.84759	5.56121	.86486	6.39978	.88217	7.48707	14
15	.83065	4.90495	.84788	5.57361	.86515	6.41560	.88246	7.50793	15
16	.83094	4.91496	.84816	5.58606	.86544	6.43148	.88275	7.52889	16
17	.83122	4.92501	.84845	5.59855	.86573	6.44743	.88304	7.54996	17
18	.83151	4.93509	.84874	5.61110	.86601	6.46346	.88333	7.57113	18
19	.83180	4.94521	.84903	5.62369	.86630	6.47955	.88362	7.59241	19
20	.83208	4.95536	.84931	5.63633	.86659	6.49571	.88391	7.61379	20
21	.83237	4.96555	.84960	5.64902	.86688	6.51194	.88420	7.63528	21
22	.83266	4.97577	.84989	5.66176	.86717	6.52825	.88448	7.65688	22
23	.83294	4.98603	.85018	5.67454	.86746	6.54462	.88477	7.67859	23
24	.83323	4.99633	.85046	5.68738	.86774	6.56107	.88506	7.70041	24
25	.83352	5.00666	.85075	5.70027	.86803	6.57759	.88535	7.72234	25
26	.83380	5.01703	.85104	5.71321	.86832	6.59418	.88564	7.74438	26
27	.83409	5.02743	.85133	5.72620	.86861	6.61085	.88593	7.76653	27
28	.83438	5.03787	.85162	5.73924	.86890	6.62759	.88622	7.78880	28
29	.83467	5.04834	.85190	5.75233	.86919	6.64441	.88651	7.81118	29
30	.83495	5.05886	.85219	5.76547	.86947	6.66130	.88680	7.83367	30
31	.83524	5.06941	.85248	5.77866	.86976	6.67826	.88709	7.85628	31
32	.83553	5.08000	.85277	5.79191	.87005	6.69530	.88737	7.87901	32
33	.83581	5.09062	.85305	5.80521	.87034	6.71242	.88766	7.90186	33
34	.83610	5.10129	.85334	5.81856	.87063	6.72962	.88795	7.92482	34
35	.83639	5.11199	.85363	5.83196	.87092	6.74689	.88824	7.94791	35
36	.83667	5.12273	.85392	5.84542	.87120	6.76424	.88853	7.97111	36
37	.83696	5.13350	.85420	5.85893	.87149	6.78167	.88882	7.99444	37
38	.83725	5.14432	.85449	5.87250	.87178	6.79918	.88911	8.01788	38
39	.83754	5.15517	.85478	5.88612	.87207	6.81677	.88940	8.04146	39
40	.83782	5.16607	.85507	5.89979	.87236	6.83443	.88969	8.06515	40
41	.83811	5.17700	.85536	5.91352	.87265	6.85218	.88998	8.08897	41
42	.83840	5.18797	.85564	5.92731	.87294	6.87001	.89027	8.11292	42
43	.83868	5.19898	.85593	5.94115	.87322	6.88792	.89055	8.13699	43
44	.83897	5.21004	.85622	5.95505	.87351	6.90592	.89084	8.16120	44
45	.83926	5.22113	.85651	5.96900	.87380	6.92400	.89113	8.18553	45
46	.83954	5.23226	.85680	5.98301	.87409	6.94216	.89142	8.20999	46
47	.83983	5.24343	.85708	5.99708	.87438	6.96040	.89171	8.23459	47
48	.84012	5.25464	.85737	6.01120	.87467	6.97873	.89200	8.25931	48
49	.84041	5.26590	.85766	6.02538	.87496	6.99714	.89229	8.28417	49
50	.84069	5.27719	.85795	6.03962	.87524	7.01563	.89258	8.30917	50
51	.84098	5.28853	.85823	6.05392	.87553	7.03423	.89287	8.33430	51
52	.84127	5.29991	.85852	6.06828	.87582	7.05291	.89316	8.35957	52
53	.84155	5.31133	.85881	6.08269	.87611	7.07167	.89345	8.38497	53
54	.84184	5.32279	.85910	6.09717	.87640	7.09052	.89374	8.41052	54
55	.84213	5.33429	.85939	6.11171	.87669	7.10946	.89403	8.43620	55
56	.84242	5.34584	.85967	6.12630	.87698	7.12849	.89431	8.46203	56
57	.84270	5.35743	.85996	6.14096	.87726	7.14760	.89460	8.48800	57
58	.84299	5.36906	.86025	6.15568	.87755	7.16681	.89489	8.51411	58
59	.84328	5.38073	.86054	6.17046	.87784	7.18612	.89518	8.54037	59
60	.84357	5.39245	.86083	6.18530	.87813	7.20551	.89547	8.56677	60

TABLE XIII.—VERSINES AND EXSECANTS.

′	84°		85°		86°		′
	Vers.	Exsec.	Vers.	Exsec.	Vers.	Exsec.	
0	.89547	8.56677	.91284	10.47371	.93024	13.33559	0
1	.89576	8.59332	.91313	10.51199	.93053	13.39547	1
2	.89605	8.62002	.91342	10.55052	.93082	13.45586	2
3	.89634	8.64687	.91371	10.58932	.93111	13.51676	3
4	.89663	8.67387	.91400	10.62837	.93140	13.57817	4
5	.89692	8.70103	.91429	10.66769	.93169	13.64011	5
6	.89721	8.72833	.91458	10.70728	.93198	13.70258	6
7	.89750	8.75579	.91487	10.74714	.93227	13.76558	7
8	.89779	8.78341	.91516	10.78727	.93257	13.82913	8
9	.89808	8.81119	.91545	10.82768	.93286	13.89323	9
10	.89836	8.83912	.91574	10.86837	.93315	13.95788	10
11	.89865	8.86722	.91603	10.90934	.93344	14.02310	11
12	.89894	8.89547	.91632	10.95060	.93373	14.08890	12
13	.89923	8.92389	.91661	10.99214	.93402	14.15527	13
14	.89952	8.95248	.91690	11.03397	.93431	14.22223	14
15	.89981	8.98123	.91719	11.07610	.93460	14.28979	15
16	.90010	9.01015	.91748	11.11852	.93489	14.35795	16
17	.90039	9.03923	.91777	11.16125	.93518	14.42672	17
18	.90068	9.06849	.91806	11.20427	.93547	14.49611	18
19	.90097	9.09792	.91835	11.24761	.93576	14.56614	19
20	.90126	9.12752	.91864	11.29125	.93605	14.63679	20
21	.90155	9.15730	.91893	11.33521	.93634	14.70810	21
22	.90184	9.18725	.91922	11.37948	.93663	14.78005	22
23	.90213	9.21739	.91951	11.42408	.93692	14.85268	23
24	.90242	9.24770	.91980	11.46900	.93721	14.92597	24
25	.90271	9.27819	.92009	11.51424	.93750	14.99995	25
26	.90300	9.30887	.92038	11.55982	.93779	15.07462	26
27	.90329	9.33973	.92067	11.60572	.93808	15.14999	27
28	.90358	9.37077	.92096	11.65197	.93837	15.22607	28
29	.90386	9.40201	.92125	11.69856	.93866	15.30287	29
30	.90415	9.43343	.92154	11.74550	.93895	15.38041	30
31	.90444	9.46505	.92183	11.79278	.93924	15.45869	31
32	.90473	9.49685	.92212	11.84042	.93953	15.53772	32
33	.90502	9.52886	.92241	11.88841	.93982	15.61751	33
34	.90531	9.56106	.92270	11.93677	.94011	15.69808	34
35	.90560	9.59346	.92299	11.98549	.94040	15.77944	35
36	.90589	9.62605	.92328	12.03458	.94069	15.86159	36
37	.90618	9.65885	.92357	12.08404	.94098	15.94456	37
38	.90647	9.69186	.92386	12.13388	.94127	16.02835	38
39	.90676	9.72507	.92415	12.18411	.94156	16.11297	39
40	.90705	9.75849	.92444	12.23472	.94186	16.19843	40
41	.90734	9.79212	.92473	12.28572	.94215	16.28476	41
42	.90763	9.82596	.92502	12.33712	.94244	16.37196	42
43	.90792	9.86001	.92531	12.38891	.94273	16.46005	43
44	.90821	9.89428	.92560	12.44112	.94302	16.54903	44
45	.90850	9.92877	.92589	12.49373	.94331	16.63893	45
46	.90879	9.96348	.92618	12.54676	.94360	16.72975	46
47	.90908	9.99841	.92647	12.60021	.94389	16.82152	47
48	.90937	10.03356	.92676	12.65408	.94418	16.91424	48
49	.90966	10.06894	.92705	12.70838	.94447	17.00794	49
50	.90995	10.10455	.92734	12.76312	.94476	17.10262	50
51	.91024	10.14039	.92763	12.81829	.94505	17.19830	51
52	.91053	10.17646	.92792	12.87391	.94534	17.29501	52
53	.91082	10.21277	.92821	12.92999	.94563	17.39274	53
54	.91111	10.24932	.92850	12.98651	.94592	17.49153	54
55	.91140	10.28610	.92879	13.04350	.94621	17.59139	55
56	.91169	10.32313	.92908	13.10096	.94650	17.69233	56
57	.91197	10.36040	.92937	13.15889	.94679	17.79438	57
58	.91226	10.39792	.92966	13.21730	.94708	17.89755	58
59	.91255	10.43569	.92995	13.27620	.94737	18.00185	59
60	.91284	10.47371	.93024	13.33559	.94766	18.10732	60

TABLE XIII.—VERSINES AND EXSECANTS.

′	87°		88°		89°		′
	Vers.	Exsec.	Vers.	Exsec.	Vers.	Exsec.	
0	.94766	18.10732	.96510	27.65371	.98255	56.29869	0
1	.94795	18.21397	.96539	27.89440	.98284	57.26976	1
2	.94825	18.32182	.96568	28.13917	.98313	58.27431	2
3	.94854	18.43088	.96597	28.38812	.98342	59.31411	3
4	.94883	18.54119	.96626	28.64137	.98371	60.39105	4
5	.94912	18.65275	.96655	28.89903	.98400	61.50715	5
6	.94941	18.76560	.96684	29.16120	.98429	62.66460	6
7	.94970	18.87976	.96714	29.42802	.98458	63.86572	7
8	.94999	18.99524	.96743	29.69960	.98487	65.11304	8
9	.95028	19.11208	.96772	29.97607	.98517	66.40927	9
10	.95057	19.23028	.96801	30.25758	.98546	67.75736	10
11	.95086	19.34989	.96830	30.54425	.98575	69.16047	11
12	.95115	19.47033	.96859	30.83623	.98604	70.62285	12
13	.95144	19.59341	.96888	31.13366	.98633	72.14583	13
14	.95173	19.71737	.96917	31.43671	.98662	73.73586	14
15	.95202	19.84283	.96946	31.74554	.98691	75.39655	15
16	.95231	19.96982	.96975	32.06030	.98720	77.13274	16
17	.95260	20.09838	.97004	32.38118	.98749	78.94968	17
18	.95289	20.22852	.97033	32.70835	.98778	80.85315	18
19	.95318	20.36027	.97062	33.04199	.98807	82.84947	19
20	.95347	20.49368	.97092	33.38232	.98836	84.94561	20
21	.95377	20.62876	.97121	33.72952	.98866	87.14924	21
22	.95406	20.76555	.97150	34.08380	.98895	89.46886	22
23	.95435	20.90403	.97179	34.44539	.98924	91.91387	23
24	.95464	21.04440	.97208	34.81452	.98953	94.49471	24
25	.95493	21.18658	.97237	35.19141	.98982	97.22303	25
26	.95522	21.33050	.97266	35.57633	.99011	100.1119	26
27	.95551	21.47635	.97295	35.96953	.99040	103.1757	27
28	.95580	21.62413	.97324	36.37127	.99069	106.4311	28
29	.95609	21.77386	.97353	36.78185	.99098	109.8966	29
30	.95638	21.92559	.97382	37.20155	.99127	113.5930	30
31	.95667	22.07935	.97411	37.63068	.99156	117.5444	31
32	.95696	22.23520	.97440	38.06957	.99186	121.7780	32
33	.95725	22.39316	.97470	38.51855	.99215	126.3253	33
34	.95754	22.55329	.97499	38.97797	.99244	131.2223	34
35	.95783	22.71563	.97528	39.44820	.99273	136.5111	35
36	.95812	22.88022	.97557	39.92963	.99302	142.2406	36
37	.95842	23.04712	.97586	40.42266	.99331	148.4684	37
38	.95871	23.21637	.97615	40.92772	.99360	155.2623	38
39	.95900	23.38802	.97644	41.44525	.99389	162.7033	39
40	.95929	23.56212	.97673	41.97571	.99418	170.8883	40
41	.95958	23.73873	.97702	42.51961	.99447	179.9350	41
42	.95987	23.91790	.97731	43.07746	.99476	189.9868	42
43	.96016	24.09969	.97760	43.64980	.99505	201.2212	43
44	.96045	24.28414	.97789	44.23720	.99535	213.8600	44
45	.96074	24.47134	.97819	44.84026	.99564	228.1839	45
46	.96103	24.66132	.97848	45.45963	.99593	244.5540	46
47	.96132	24.85417	.97877	46.09596	.99622	263.4427	47
48	.96161	25.04994	.97906	46.74997	.99651	285.4795	48
49	.96190	25.24869	.97935	47.42241	.99680	311.5230	49
50	.96219	25.45051	.97964	48.11406	.99709	342.7752	50
51	.96248	25.65546	.97993	48.82576	.99738	380.9723	51
52	.96277	25.86360	.98022	49.55840	.99767	428.7187	52
53	.96307	26.07508	.98051	50.31290	.99796	490.1070	53
54	.96336	26.28981	.98080	51.09027	.99825	571.9581	54
55	.96365	26.50804	.98109	51.89156	.99855	686.5496	55
56	.96394	26.72978	.98138	52.71790	.99884	858.4369	56
57	.96423	26.95518	.98168	53.57046	.99913	1144.916	57
58	.96452	27.18417	.98197	54.45053	.99942	1717.874	58
59	.96481	27.41700	.98226	55.35946	.99971	3436.747	59
60	.96510	27.65371	.98255	56.29869	1.00000	Infinite	60

TABLE XIV.—CUBIC YARDS PER 100 FEET. SLOPES ¼ : 1.

Depth	Base 12	Base 14	Base 16	Base 18	Base 22	Base 24	Base 26	Base 28
1	45	53	60	68	82	90	97	105
2	93	107	122	137	167	181	196	211
3	142	163	186	208	253	275	297	319
4	193	222	252	281	341	370	400	430
5	245	282	319	356	431	468	505	542
6	300	344	389	433	522	567	611	656
7	356	408	460	512	616	668	719	771
8	415	474	533	593	711	770	830	889
9	475	542	608	675	808	875	942	1008
10	537	611	685	759	907	981	1056	1130
11	601	682	764	845	1008	1090	1171	1253
12	667	756	844	933	1111	1200	1289	1378
13	734	831	926	1023	1216	1312	1408	1505
14	804	907	1010	1115	1322	1426	1530	1633
15	875	986	1096	1208	1431	1542	1653	1764
16	948	1067	1184	1304	1541	1659	1778	1896
17	1023	1149	1274	1401	1653	1779	1905	2031
18	1100	1233	1366	1500	1767	1900	2033	2167
19	1179	1319	1460	1601	1882	2023	2164	2305
20	1259	1407	1555	1704	2000	2148	2296	2444
21	1342	1497	1653	1808	2119	2275	2431	2586
22	1426	1589	1752	1915	2241	2404	2567	2730
23	1512	1682	1853	2023	2364	2534	2705	2875
24	1600	1778	1955	2133	2489	2667	2844	3022
25	1690	1875	2060	2245	2616	2801	2986	3171
26	1781	1974	2166	2359	2744	2937	3130	3322
27	1875	2075	2274	2475	2875	3075	3275	3475
28	1970	2178	2384	2593	3007	3215	3422	3630
29	2068	2282	2496	2712	3142	3356	3571	3786
30	2167	2389	2610	2833	3278	3500	3722	3944
31	2268	2497	2726	2956	3416	3645	3875	4105
32	2370	2607	2844	3081	3556	3793	4030	4267
33	2475	2719	2964	3208	3697	3942	4186	4431
34	2581	2833	3085	3337	3841	4093	4344	4596
35	2690	2940	3208	3468	3986	4245	4505	4764
36	2800	3067	3333	3600	4133	4400	4667	4933
37	2912	3186	3460	3734	4282	4556	4831	5105
38	3026	3307	3589	3870	4433	4715	4996	5278
39	3142	3431	3719	4008	4586	4875	5164	5453
40	3259	3556	3852	4148	4741	5037	5333	5630
41	3379	3682	3986	4290	4897	5201	5505	5808
42	3500	3811	4122	4433	5056	5367	5678	5989
43	3623	3942	4260	4579	5216	5534	5853	6171
44	3748	4074	4400	4726	5378	5704	6030	6356
45	3875	4208	4541	4875	5542	5875	6208	6542
46	4004	4344	4684	5026	5707	6048	6389	6730
47	4134	4482	4830	5179	5875	6223	6571	6919
48	4267	4622	4978	5333	6044	6400	6756	7111
49	4401	4764	5127	5490	6216	6579	6942	7305
50	4537	4907	5278	5648	6389	6759	7130	7500
51	4675	5053	5430	5808	6564	6942	7319	7697
52	4815	5200	5584	5970	6741	7126	7511	7896
53	4956	5349	5741	6134	6919	7312	7705	8097
54	5100	5500	5900	6300	7100	7500	7900	8300
55	5245	5653	6060	6468	7282	7690	8097	8505
56	5393	5807	6222	6637	7467	7881	8296	8711
57	5542	5964	6386	6808	7653	8075	8497	8919
58	5693	6122	6552	6981	7841	8270	8700	9130
59	5845	6282	6719	7156	8031	8468	8905	9342
60	6000	6444	6889	7333	8222	8667	9111	9556

TABLE XIV.—CUBIC YARDS PER 100 FEET. SLOPES ¼ : 1.

Depth	Base 12	Base 14	Base 16	Base 18	Base 22	Base 24	Base 26	Base 28
1	46	54	61	69	83	91	98	106
2	96	111	126	141	170	185	200	215
3	150	172	194	217	261	283	306	328
4	207	237	267	296	356	385	415	444
5	269	306	343	380	454	491	528	565
6	333	378	422	467	556	600	644	689
7	402	454	506	557	661	713	765	817
8	474	533	593	652	770	830	889	948
9	550	617	683	750	883	950	1017	1083
10	630	704	778	852	1000	1074	1148	1222
11	713	794	876	957	1120	1202	1283	1365
12	800	889	978	1067	1244	1333	1422	1511
13	891	987	1083	1180	1372	1469	1565	1661
14	985	1089	1193	1296	1504	1607	1711	1815
15	1083	1194	1306	1417	1639	1750	1861	1972
16	1185	1304	1422	1541	1779	1896	2015	2133
17	1291	1417	1543	1669	1920	2046	2172	2298
18	1400	1533	1667	1800	2067	2200	2333	2467
19	1513	1654	1794	1935	2217	2357	2498	2639
20	1630	1778	1926	2074	2370	2519	2667	2815
21	1750	1906	2061	2217	2528	2683	2839	2994
22	1874	2037	2200	2363	2689	2852	3015	3178
23	2002	2172	2343	2513	2854	3024	3194	3365
24	2133	2311	2489	2667	3022	3200	3378	3556
25	2269	2454	2639	2824	3194	3380	3565	3750
26	2407	2600	2793	2985	3370	3563	3756	3948
27	2550	2750	2950	3150	3550	3750	3950	4151
28	2696	2904	3111	3319	3733	3941	4148	4356
29	2846	3061	3276	3491	3920	4135	4350	4565
30	3000	3222	3444	3667	4111	4333	4556	4778
31	3157	3387	3617	3846	4306	4535	4765	4994
32	3319	3556	3793	4030	4504	4741	4978	5215
33	3483	3728	3972	4217	4706	4950	5194	5439
34	3652	3904	4156	4407	4911	5163	5415	5667
35	3824	4083	4343	4602	5120	5380	5639	5898
36	4000	4267	4533	4800	5333	5600	5867	6133
37	4180	4454	4728	5002	5550	5824	6098	6372
38	4363	4644	4926	5207	5770	6052	6333	6615
39	4550	4839	5128	5417	5994	6283	6572	6861
40	4741	5037	5333	5630	6222	6519	6815	7111
41	4935	5239	5543	5846	6454	6757	7061	7365
42	5133	5444	5756	6067	6689	7000	7311	7622
43	5335	5654	5972	6291	6928	7246	7565	7883
44	5541	5867	6193	6519	7170	7496	7822	8148
45	5750	6083	6417	6750	7417	7750	8083	8417
46	5963	6304	6644	6985	7667	8007	8348	8689
47	6180	6528	6876	7224	7920	8269	8617	8965
48	6400	6756	7111	7467	8178	8533	8889	9244
49	6624	6987	7350	7713	8439	8802	9165	9528
50	6852	7222	7593	7963	8704	9074	9444	9815
51	7083	7461	7839	8217	8972	9350	9728	10106
52	7319	7704	8089	8474	9244	9630	10015	10400
53	7557	7950	8343	8735	9520	9913	10306	10698
54	7800	8200	8600	9000	9800	10200	10600	11000
55	8046	8454	8861	9269	10083	10491	10898	11306
56	8296	8711	9126	9541	10370	10785	11200	11615
57	8550	8972	9394	9817	10661	11083	11506	11928
58	8807	9237	9667	10096	10956	11385	11815	12244
59	9069	9506	9943	10380	11254	11691	12128	12565
60	9333	9778	10222	10667	11556	12000	12444	12889

TABLE XIV.—CUBIC YARDS PER 100 FEET. SLOPES 1 : 1.

Depth	Base 12	Base 14	Base 16	Base 18	Base 20	Base 28	Base 30	Base 32
1	48	56	63	70	78	107	115	122
2	104	119	133	148	163	222	237	252
3	167	189	211	233	256	344	367	389
4	237	267	296	326	356	474	504	533
5	315	352	389	426	463	611	648	685
6	400	444	489	533	578	756	800	844
7	493	544	596	648	700	907	959	1011
8	593	652	711	770	830	1067	1126	1185
9	700	767	833	900	967	1233	1300	1367
10	815	889	963	1037	1111	1407	1481	1556
11	937	1019	1100	1181	1263	1589	1670	1752
12	1067	1156	1244	1333	1422	1778	1867	1956
13	1204	1300	1396	1493	1589	1974	2070	2167
14	1348	1452	1556	1659	1763	2178	2281	2385
15	1500	1611	1722	1833	1944	2389	2500	2611
16	1659	1778	1896	2015	2133	2607	2726	2844
17	1826	1952	2078	2204	2330	2833	2959	3085
18	2000	2133	2267	2400	2533	3067	3200	3333
19	2181	2322	2463	2604	2744	3307	3448	3589
20	2370	2519	2667	2815	2963	3556	3704	3852
21	2567	2722	2878	3033	3189	3811	3967	4122
22	2770	2933	3096	3259	3422	4074	4237	4444
23	2981	3152	3322	3493	3663	4344	4515	4685
24	3200	3378	3556	3733	3911	4622	4800	4978
25	3426	3611	3796	3981	4167	4907	5093	5278
26	3659	3852	4044	4237	4430	5200	5393	5585
27	3900	4100	4300	4500	4700	5500	5700	5900
28	4148	4356	4563	4770	4978	5807	6015	6222
29	4404	4619	4833	5048	5263	6122	6337	6552
30	4667	4889	5111	5333	5556	6444	6667	6889
31	4937	5167	5396	5626	5856	6774	7004	7233
32	5215	5452	5689	5926	6163	7111	7348	7585
33	5500	5744	5989	6233	6478	7456	7700	7944
34	5793	6044	6296	6548	6800	7807	8059	8311
35	6093	6352	6611	6870	7130	8167	8426	8685
36	6400	6667	6933	7200	7467	8533	8800	9067
37	6715	6989	7263	7537	7811	8907	9181	9456
38	7037	7319	7600	7881	8163	9289	9570	9852
39	7367	7656	7944	8233	8522	9678	9967	10256
40	7704	8000	8296	8593	8889	10074	10370	10667
41	8048	8352	8656	8959	9263	10478	10781	11085
42	8400	8711	9022	9333	9644	10889	11200	11511
43	8759	9078	9396	9715	10033	11307	11626	11944
44	9126	9452	9778	10104	10430	11733	12059	12385
45	9500	9833	10167	10500	10833	12167	12500	12833
46	9881	10222	10563	10904	11244	12607	12948	13289
47	10270	10619	10967	11315	11663	13056	13404	13752
48	10667	11022	11378	11733	12089	13511	13867	14222
49	11070	11433	11796	12159	12522	13974	14337	14700
50	11481	11852	12222	12593	12963	14444	14815	15185
51	11900	12278	12656	13033	13411	14922	15300	15678
52	12326	12711	13096	13481	13867	15407	15793	16178
53	12759	13152	13544	13937	14330	15900	16293	16685
54	13200	13600	14000	14400	14800	16400	16800	17200
55	13648	14056	14463	14870	15278	16907	17315	17722
56	14104	14519	14933	15348	15763	17422	17837	18252
57	14567	14989	15411	15833	16256	17944	18367	18789
58	15037	15467	15896	16326	16756	18474	18904	19333
59	15515	15952	16389	16826	17263	19011	19448	19885
60	16000	16444	16889	17333	17778	19556	20000	20444

TABLE XIV.—CUBIC YARDS PER 100 FEET. SLOPES 1½ : 1.

Depth	Base 12	Base 14	Base 16	Base 18	Base 20	Base 28	Base 30	Base 32
1	50	57	65	72	80	109	117	124
2	111	126	141	156	170	230	244	259
3	183	206	228	250	272	361	383	406
4	267	296	326	356	385	504	533	563
5	361	398	435	472	509	657	694	731
6	467	511	556	600	644	822	867	911
7	583	635	687	739	791	998	1050	1102
8	711	770	830	889	948	1185	1244	1304
9	850	917	983	1050	1116	1383	1450	1517
10	1000	1074	1148	1222	1296	1593	1667	1741
11	1161	1243	1324	1406	1487	1813	1894	1976
12	1333	1422	1511	1600	1689	2044	2133	2222
13	1517	1613	1709	1806	1902	2287	2383	2480
14	1711	1815	1919	2022	2126	2541	2644	2748
15	1917	2028	2139	2250	2361	2806	2917	3028
16	2133	2252	2370	2489	2607	3081	3200	3319
17	2361	2487	2613	2739	2865	3369	3494	3620
18	2600	2733	2867	3000	3133	3667	3800	3933
19	2850	2991	3131	3272	3413	3976	4117	4257
20	3111	3259	3407	3556	3704	4296	4444	4592
21	3383	3539	3694	3850	4005	4628	4783	4939
22	3667	3830	3993	4156	4318	4970	5133	5296
23	3961	4131	4302	4472	4642	5324	5494	5665
24	4267	4444	4622	4800	4978	5689	5867	6044
25	4583	4769	4954	5139	5324	6065	6250	6435
26	4911	5104	5296	5489	5681	6452	6644	6837
27	5250	5450	5650	5850	6050	6850	7050	7250
28	5600	5807	6015	6222	6430	7259	7467	7674
29	5961	6176	6391	6606	6820	7680	7894	8109
30	6333	6556	6778	7000	7222	8111	8333	8555
31	6717	6946	7176	7406	7635	8554	8783	9013
32	7111	7348	7585	7822	8059	9007	9244	9482
33	7517	7761	8006	8250	8494	9472	9717	9962
34	7933	8185	8437	8689	8941	9948	10200	10452
35	8361	8620	8880	9139	9398	10435	10694	10954
36	8900	9067	9333	9600	9867	10933	11200	11467
37	9250	9524	9798	10072	10346	11443	11717	11991
38	9711	9993	10274	10556	10837	11963	12244	12526
39	10183	10472	10761	11050	11339	12494	12783	13072
40	10667	10963	11259	11556	11852	13037	13333	13630
41	11161	11465	11769	12072	12376	13591	13894	14198
42	11667	11978	12289	12600	12911	14156	14467	14778
43	12183	12502	12820	13139	13457	14731	15050	15369
44	12711	13037	13363	13689	14015	15319	15644	15970
45	13250	13583	13917	14250	14583	15917	16250	16583
46	13800	14141	14481	14822	15163	16526	16867	17207
47	14361	14709	15057	15406	15754	17146	17494	17843
48	14933	15289	15644	16000	16356	17778	18133	18489
49	15517	15880	16243	16606	16968	18420	18783	19146
50	16111	16481	16852	17222	17592	19074	19444	19815
51	16717	17094	17472	17850	18228	19739	20117	20494
52	17333	17719	18104	18489	18874	20415	20800	21185
53	17961	18354	18746	19139	19531	21102	21494	21887
54	18600	19000	19400	19800	20200	21800	22200	22600
55	19250	19657	20065	20472	20880	22509	22917	23324
56	19911	20326	20741	21156	21570	23230	23644	24059
57	20583	21006	21428	21850	22272	23961	24383	24805
58	21267	21696	22126	22556	22985	24704	25133	25563
59	21961	22398	22835	23272	23709	25457	25894	26332
60	22667	23111	23556	24000	24444	26222	26667	27111

TABLE XIV.—CUBIC YARDS PER 100 FEET. SLOPES 2 : 1.

Depth	Base 12	Base 14	Base 16	Base 18	Base 20	Base 28	Base 30	Base 32
1	52	59	67	74	81	111	119	126
2	119	133	148	163	178	237	252	267
3	200	222	244	267	289	378	400	422
4	296	326	356	385	415	533	563	593
5	407	444	481	519	556	704	741	778
6	533	578	622	667	711	889	933	978
7	674	726	778	830	881	1089	1141	1193
8	830	889	948	1007	1067	1304	1363	1422
9	1000	1067	1133	1200	1267	1533	1600	1667
10	1185	1259	1333	1407	1481	1778	1852	1926
11	1385	1467	1548	1630	1711	2037	2119	2200
12	1600	1689	1778	1867	1956	2311	2400	2489
13	1830	1926	2022	2119	2215	2600	2696	2793
14	2074	2178	2281	2385	2489	2904	3007	3111
15	2333	2444	2556	2667	2778	3222	3333	3444
16	2607	2726	2844	2963	3081	3556	3674	3793
17	2896	3022	3148	3274	3400	3904	4030	4156
18	3200	3333	3467	3600	3733	4267	4400	4533
19	3519	3659	3800	3941	4081	4644	4785	4926
20	3852	4000	4148	4296	4444	5037	5185	5333
21	4200	4356	4511	4667	4822	5444	5600	5756
22	4563	4730	4889	5052	5215	5867	6030	6193
23	4941	5111	5281	5452	5622	6304	6474	6644
24	5333	5511	5689	5867	6044	6756	6933	7111
25	5741	5926	6111	6296	6481	7222	7407	7593
26	6163	6356	6548	6741	6933	7704	7896	8089
27	6600	6800	7000	7200	7400	8200	8400	8600
28	7052	7259	7467	7674	7881	8711	8919	9126
29	7519	7733	7948	8163	8378	9237	9452	9667
30	8000	8222	8444	8667	8889	9778	10000	10222
31	8496	8726	8956	9185	9415	10333	10563	10793
32	9007	9244	9481	9719	9956	10904	11141	11378
33	9533	9778	10022	10267	10511	11489	11733	11978
34	10074	10326	10578	10830	11081	12089	12341	12593
35	10630	10889	11148	11407	11667	12704	12963	13222
36	11200	11467	11733	12000	12267	13333	13600	13867
37	11785	12059	12333	12607	12881	13978	14252	14526
38	12385	12667	12948	13230	13511	14637	14919	15200
39	13000	13289	13578	13867	14156	15311	15600	15889
40	13630	13926	14222	14519	14815	16000	16296	16593
41	14274	14578	14881	15185	15489	16704	17007	17311
42	14933	15244	15556	15867	16178	17422	17733	18044
43	15607	15926	16244	16563	16881	18156	18474	18793
44	16296	16622	16948	17274	17600	18904	19230	19556
45	17000	17333	17667	18000	18333	19667	20000	20333
46	17719	18059	18400	18741	19081	20444	20785	21126
47	18452	18800	19148	19496	19844	21237	21585	21933
48	19200	19556	19911	20267	20622	22044	22400	22756
49	19963	20326	20689	21052	21415	22867	23230	23593
50	20741	20711	21481	21852	22222	23704	24074	24444
51	21533	21911	22289	22667	23044	24556	24933	25311
52	22341	22726	23111	23496	23881	25422	25807	26193
53	23163	23556	23948	24341	24733	26304	26696	27089
54	24000	24400	24800	25200	25600	27200	27600	28000
55	24852	25259	25667	26074	26481	28111	28519	28926
56	25719	26133	26548	26963	27378	29037	29452	29867
57	26600	27022	27444	27867	28289	29978	30400	30822
58	27496	27926	28356	28785	29215	30933	31363	31793
59	28407	28844	29281	29719	30156	31904	32341	32778
60	29333	29778	30222	30667	31111	32889	33333	33778

TABLE XIV.—CUBIC YARDS PER 100 FEET. SLOPES 3 : 1.

Depth	Base 12	Base 14	Base 16	Base 18	Base 20	Base 28	Base 30	Base 32
1	56	63	70	78	85	115	122	130
2	133	148	163	178	193	252	267	281
3	233	256	278	300	322	411	433	456
4	356	385	415	444	474	593	622	652
5	500	537	574	611	648	796	833	870
6	667	711	756	800	844	1022	1067	1111
7	856	907	959	1011	1063	1270	1322	1374
8	1067	1126	1185	1244	1304	1541	1600	1659
9	1300	1367	1433	1500	1567	1833	1900	1967
10	1556	1630	1704	1778	1852	2148	2222	2296
11	1833	1915	1996	2078	2159	2485	2567	2648
12	2133	2222	2311	2400	2489	2844	2933	3022
13	2456	2552	2648	2744	2841	3226	3322	3419
14	2800	2904	3007	3111	3215	3630	3733	3837
15	3167	3278	3389	3500	3611	4056	4167	4278
16	3556	3674	3793	3911	4030	4504	4622	4741
17	3967	4093	4219	4344	4470	4974	5100	5226
18	4400	4533	4667	4800	4933	5467	5600	5733
19	4856	4996	5137	5278	5419	5981	6122	6263
20	5333	5481	5630	5778	5926	6519	6667	6815
21	5833	5989	6144	6300	6456	7078	7233	7389
22	6356	6519	6681	6844	7007	7659	7822	7985
23	6900	7070	7241	7411	7581	8263	8433	8504
24	7467	7644	7822	8000	8178	8889	9067	9144
25	8056	8241	8426	8611	8796	9537	9722	9807
26	8667	8859	9052	9244	9437	10207	10400	10593
27	9300	9500	9700	9900	10100	10900	11100	11300
28	9956	10163	10370	10578	10785	11615	11822	12030
29	10633	10848	11063	11278	11493	12352	12567	12781
30	11333	11556	11778	12000	12222	13111	13333	13556
31	12056	12285	12515	12744	12074	13893	14122	14352
32	12800	13037	13274	13511	13748	14696	14933	15170
33	13567	13811	14056	14300	14544	15522	15767	16011
34	14356	14607	14859	15111	15363	16370	16622	16874
35	15167	15426	15685	15944	16204	17241	17500	17759
36	16000	16267	16533	16800	17067	18133	18400	18667
37	16856	17130	17404	17678	17952	19048	19322	19596
38	17733	18015	18296	18578	18859	19985	20267	20548
39	18633	18922	19211	19500	19789	20944	21233	21522
40	19556	19852	20148	20444	20741	21926	22222	22516
41	20500	20804	21107	21411	21715	22980	23283	23537
42	21467	21778	22089	22400	22711	23956	24267	24578
43	22456	22774	23093	23411	23730	25004	25322	25641
44	23467	23793	24119	24444	24770	26074	26400	26726
45	24500	24833	25167	25500	25833	27167	27500	27833
46	25556	25896	26237	26578	26919	28281	28622	28063
47	26633	26981	27330	27678	28026	29419	29767	30115
48	27733	28089	28444	28800	29156	30578	30933	31289
49	28856	29219	29581	29944	30307	31759	32122	32485
50	30000	30370	30741	31111	31481	32963	33333	33704
51	31167	31544	31922	32300	32678	34189	34567	34944
52	32356	32741	33126	33511	33896	35437	35822	36207
53	33567	33959	34352	34744	35137	36707	37100	37493
54	34800	35200	35600	36000	36400	38000	38400	38800
55	36056	36463	36870	37278	37685	39315	39722	40130
56	37333	37748	38163	38578	38993	40652	41067	41481
57	38633	39056	39478	39900	40322	42011	42433	42856
58	39956	40385	40815	41244	41674	43393	43822	44252
59	41300	41737	42174	42611	43048	44796	45233	45670
60	42667	43111	43556	44000	44444	46222	46667	47111

TABLE XV.—CUBIC YARDS IN 100 FEET LENGTH.

Area. Sq. Ft.	Cubic Yards.	Area. Sq. Ft.	Cubic Yards.	Area. Sq. Ft.	Cubic Yards.	Area. Sq. Ft.	Cubic Yards.	Area. Sq. Ft.	Cubic Yards.
1	3.7	51	188.9	101	374.1	151	559.3	201	744.4
2	7.4	52	192.6	102	377.8	152	563.0	202	748.2
3	11.1	53	196.3	103	381.5	153	566.7	203	751.9
4	14.8	54	200.0	104	385.2	154	570.4	204	755.6
5	18.5	55	203.7	105	388.9	155	574.1	205	759.3
6	22.2	56	207.4	106	392.6	156	577.8	206	763.0
7	25.9	57	211.1	107	396.3	157	581.5	207	766.7
8	29.6	58	214.8	108	400.0	158	585.2	208	770.4
9	33.3	59	218.5	109	403.7	159	588.9	209	774.1
10	37.0	60	222.2	110	407.4	160	592.6	210	777.8
11	40.7	61	225.9	111	411.1	161	596.3	211	781.5
12	44.4	62	229.6	112	414.8	162	600.0	212	785.2
13	48.1	63	233.3	113	418.5	163	603.7	213	788.9
14	51.9	64	237.0	114	422.2	164	607.4	214	792.6
15	55.6	65	240.7	115	425.9	165	611.1	215	796.3
16	59.3	66	244.4	116	429.6	166	614.8	216	800.0
17	63.0	67	248.2	117	433.3	167	618.5	217	803.7
18	66.7	68	251.9	118	437.0	168	622.2	218	807.4
19	70.4	69	255.6	119	440.7	169	625.9	219	811.1
20	74.1	70	259.3	120	444.4	170	629.6	220	814.8
21	77.8	71	263.0	121	448.2	171	633.3	221	818.5
22	81.5	72	266.7	122	451.9	172	637.0	222	822.2
23	85.2	73	270.4	123	455.6	173	640.7	223	825.9
24	88.9	74	274.1	124	459.3	174	644.4	224	829.6
25	92.6	75	277.8	125	463.0	175	648.2	225	833.3
26	96.3	76	281.5	126	466.7	176	651.9	226	837.0
27	100.0	77	285.2	127	470.4	177	655.6	227	840.7
28	103.7	78	288.9	128	474.1	178	659.3	228	844.4
29	107.4	79	292.6	129	477.8	179	663.0	229	848.2
30	111.1	80	296.3	130	481.5	180	666.7	230	851.9
31	114.8	81	300.0	131	485.2	181	670.4	231	855.6
32	118.5	82	303.7	132	488.9	182	674.1	232	859.3
33	122.2	83	307.4	133	492.6	183	677.8	233	863.0
34	125.9	84	311.1	134	496.3	184	681.5	234	866.7
35	129.6	85	314.8	135	500.0	185	685.2	235	870.4
36	133.3	86	318.5	136	503.7	186	688.9	236	874.1
37	137.0	87	322.2	137	507.4	187	692.6	237	877.8
38	140.7	88	325.9	138	511.1	188	696.3	238	881.5
39	144.4	89	329.6	139	514.8	189	700.0	239	885.2
40	148.2	90	333.3	140	518.5	190	703.7	240	888.9
41	151.9	91	337.0	141	522.2	191	707.4	241	892.6
42	155.6	92	340.7	142	525.9	192	711.1	242	896.3
43	159.3	93	344.4	143	529.6	193	714.8	243	900.0
44	163.0	94	348.2	144	533.3	194	718.5	244	903.7
45	166.7	95	351.9	145	537.0	195	722.2	245	907.4
46	170.4	96	355.6	146	540.7	196	725.9	246	911.1
47	174.1	97	359.3	147	544.4	197	729.6	247	914.8
48	177.8	98	363.0	148	548.2	198	733.3	248	918.5
49	181.5	99	366.7	149	551.9	199	737.0	249	922.2
50	185.2	100	370.4	150	555.6	200	740.7	250	925.9

TABLE XV.—CUBIC YARDS IN 100 FEET LENGTH.

Area. Sq. Ft.	Cubic Yards.	Area. Sq. Ft.	Cubic Yards.	Area. Sq. Ft.	Cubic Yards.	Area. Sq. Ft.	Cubic Yards.	Area. Sq. Ft.	Cubic Yards.
251	929.6	301	1114.8	351	1300.0	401	1485.2	451	1670.4
252	933.3	302	1118.5	352	1303.7	402	1488.9	452	1674.1
253	937.0	303	1122.2	353	1307.4	403	1492.6	453	1677.8
254	940.7	304	1125.9	354	1311.1	404	1496.3	454	1681.5
255	944.4	305	1129.6	355	1314.8	405	1500.0	455	1685.2
256	948.2	306	1133.3	356	1318.5	406	1503.7	456	1688.9
257	951.9	307	1137.0	357	1322.2	407	1507.4	457	1692.6
258	955.6	308	1140.7	358	1325.9	408	1511.1	458	1696.3
259	959.3	309	1144.4	359	1329.6	409	1514.8	459	1700.0
260	963.0	310	1148.2	360	1333.3	410	1518.5	460	1703.7
261	966.7	311	1151.9	361	1337.0	411	1522.2	461	1707.4
262	970.4	312	1155.6	362	1340.7	412	1525.9	462	1711.1
263	974.1	313	1159.3	363	1344.4	413	1529.6	463	1714.8
264	977.8	314	1163.0	364	1348.2	414	1533.3	464	1718.5
265	981.5	315	1166.7	365	1351.9	415	1537.0	465	1722.2
266	985.2	316	1170.4	366	1355.6	416	1540.7	466	1725.9
267	988.9	317	1174.1	367	1359.3	417	1544.4	467	1729.6
268	992.6	318	1177.8	368	1363.0	418	1548.2	468	1733.3
269	996.3	319	1181.5	369	1366.7	419	1551.9	469	1737.0
270	1000.0	320	1185.2	370	1370.4	420	1555.6	470	1740.7
271	1003.7	321	1188.9	371	1374.1	421	1559.3	471	1744.4
272	1007.4	322	1192.6	372	1377.8	422	1563.0	472	1748.2
273	1011.1	323	1196.3	373	1381.5	423	1566.7	473	1751.9
274	1014.8	324	1200.0	374	1385.2	424	1570.4	474	1755.6
275	1018.5	325	1203.7	375	1388.9	425	1574.1	475	1759.3
276	1022.2	326	1207.4	376	1392.6	426	1577.8	476	1763.0
277	1025.9	327	1211.1	377	1396.3	427	1581.5	477	1766.7
278	1029.6	328	1214.8	378	1400.0	428	1585.2	478	1770.4
279	1033.3	329	1218.5	379	1403.7	429	1588.9	479	1774.1
280	1037.0	330	1222.2	380	1407.4	430	1592.6	480	1777.8
281	1040.7	331	1225.9	381	1411.1	431	1596.3	481	1781.5
282	1044.4	332	1229.6	382	1414.8	432	1600.0	482	1785.2
283	1048.2	333	1233.3	383	1418.5	433	1603.7	483	1788.9
284	1051.9	334	1237.0	384	1422.2	434	1607.4	484	1792.6
285	1055.6	335	1240.7	385	1425.9	435	1611.1	485	1796.3
286	1059.3	336	1244.4	386	1429.6	436	1614.8	486	1800.0
287	1063.0	337	1248.2	387	1433.3	437	1618.5	487	1803.7
288	1066.7	338	1251.9	388	1437.0	438	1622.2	488	1807.4
289	1070.4	339	1255.6	389	1440.7	439	1625.9	489	1811.1
290	1074.1	340	1259.3	390	1444.4	440	1629.6	490	1814.8
291	1077.8	341	1263.0	391	1448.2	441	1633.3	491	1818.5
292	1081.5	342	1266.7	392	1451.9	442	1637.0	492	1822.2
293	1085.2	343	1270.4	393	1455.6	443	1640.7	493	1825.9
294	1088.9	344	1274.1	394	1459.3	444	1644.4	494	1829.6
295	1092.6	345	1277.8	395	1463.0	445	1648.2	495	1833.3
296	1096.3	346	1281.5	396	1466.7	446	1651.9	496	1837.0
297	1100.0	347	1285.2	397	1470.4	447	1655.6	497	1840.7
298	1103.7	348	1288.9	398	1474.1	448	1659.3	498	1844.4
299	1107.4	349	1292.6	399	1477.8	449	1663.0	499	1848.2
300	1111.1	350	1296.3	400	1481.5	450	1666.7	500	1851.9

TABLE XV.—CUBIC YARDS IN 100 FEET LENGTH.

Area. Sq. Ft.	Cubic Yards.	Area. Sq. Ft.	Cubic Yards.	Area. Sq. Ft.	Cubic Yards.	Area. Sq. Ft.	Cubic Yards.	Area. Sq. Ft.	Cubic Yards.
501	1855.6	551	2040.7	601	2225.9	651	2411.1	701	2596.3
502	1859.3	552	2044.4	602	2229.6	652	2414.8	702	2600.0
503	1863.0	553	2048.2	603	2233.3	653	2418.5	703	2603.7
504	1866.7	554	2051.9	604	2237.0	654	2422.2	704	2607.4
505	1870.4	555	2055.6	605	2240.7	655	2425.9	705	2611.1
506	1874.1	556	2059.3	606	2244.4	656	2429.6	706	2614.8
507	1877.8	557	2063.0	607	2248.2	657	2433.3	707	2618.5
508	1881.5	558	2066.7	608	2251.9	658	2437.0	708	2622.2
509	1885.2	559	2070.4	609	2255.6	659	2440.7	709	2625.9
510	1888.9	560	2074.1	610	2259.3	660	2444.4	710	2629.6
511	1892.6	561	2077.8	611	2263.0	661	2448.2	711	2633.3
512	1896.3	562	2081.5	612	2266.7	662	2451.9	712	2637.0
513	1900.0	563	2085.2	613	2270.4	663	2455.6	713	2640.7
514	1903.7	564	2088.9	614	2274.1	664	2459.3	714	2644.4
515	1907.4	565	2092.6	615	2277.8	665	2463.0	715	2648.2
516	1911.1	566	2096.3	616	2281.5	666	2466.7	716	2651.9
517	1914.8	567	2100.0	617	2285.2	667	2470.4	717	2655.6
518	1918.5	568	2103.7	618	2288.9	668	2474.1	718	2659.3
519	1922.2	569	2107.4	619	2292.6	669	2477.8	719	2663.0
520	1925.9	570	2111.1	620	2296.3	670	2481.5	720	2666.7
521	1929.6	571	2114.8	621	2300.0	671	2485.2	721	2670.4
522	1933.3	572	2118.5	622	2303.7	672	2488.9	722	2674.1
523	1937.0	573	2122.2	623	2307.4	673	2492.6	723	2677.8
524	1940.7	574	2125.9	624	2311.1	674	2496.3	724	2681.5
525	1944.4	575	2129.6	625	2314.8	675	2500.0	725	2685.2
526	1948.2	576	2133.3	626	2318.5	676	2503.7	726	2688.9
527	1951.9	577	2137.0	627	2322.2	677	2507.4	727	2692.6
528	1955.6	578	2140.7	628	2325.9	678	2511.1	728	2696.3
529	1959.3	579	2144.4	629	2329.6	679	2514.8	729	2700.0
530	1963.0	580	2148.2	630	2333.3	680	2518.5	730	2703.7
531	1966.7	581	2151.9	631	2337.0	681	2522.2	731	2707.4
532	1970.4	582	2155.6	632	2340.7	682	2525.9	732	2711.1
533	1974.1	583	2159.3	633	2344.4	683	2529.6	733	2714.8
534	1977.8	584	2163.0	634	2348.2	684	2533.3	734	2718.5
535	1981.5	585	2166.7	635	2351.9	685	2537.0	735	2722.2
536	1985.2	586	2170.4	636	2355.6	686	2540.7	736	2725.9
537	1988.9	587	2174.1	637	2359.3	687	2544.4	737	2729.6
538	1992.6	588	2177.8	638	2363.0	688	2548.2	738	2733.3
539	1996.3	589	2181.5	639	2366.7	689	2551.9	739	2737.0
540	2000.0	590	2185.2	640	2370.4	690	2555.6	740	2740.7
541	2003.7	591	2188.9	641	2374.1	691	2559.3	741	2744.4
542	2007.4	592	2192.6	642	2377.8	692	2563.0	742	2748.2
543	2011.1	593	2196.3	643	2381.5	693	2566.7	743	2751.9
544	2014.8	594	2200.0	644	2385.2	694	2570.4	744	2755.6
545	2018.5	595	2203.7	645	2388.9	695	2574.1	745	2759.3
546	2022.2	596	2207.4	646	2392.6	696	2577.8	746	2763.0
547	2025.9	597	2211.1	647	2396.3	697	2581.5	747	2766.7
548	2029.6	598	2214.8	648	2400.0	698	2585.2	748	2770.4
549	2033.3	599	2218.5	649	2403.7	699	2588.9	749	2774.1
550	2037.0	600	2222.2	650	2407.4	700	2592.6	750	2777.8

TABLE XV.—CUBIC YARDS IN 100 FEET LENGTH.

Area. Sq. Ft.	Cubic Yards.	Area. Sq. Ft.	Cubic Yards.	Area. Sq. Ft.	Cubic Yards.	Area. Sq. Ft.	Cubic Yards.	Area. Sq. Ft.	Cubic Yards.
751	2781.5	801	2966.7	851	3151.9	901	3337.0	951	3522.2
752	2785.2	802	2970.4	852	3155.6	902	3340.7	952	3525.9
753	2788.9	803	2974.1	853	3159.3	903	3344.4	953	3529.6
754	2792.6	804	2977.8	854	3163.0	904	3348.2	954	3533.3
755	2796.3	805	2981.5	855	3166.7	905	3351.9	955	3537.0
756	2800.0	806	2985.2	856	3170.4	906	3355.6	956	3540.7
757	2803.7	807	2988.9	857	3174.1	907	3359.3	957	3544.4
758	2807.4	808	2992.6	858	3177.8	908	3363.0	958	3548.2
759	2811.1	809	2996.3	859	3181.5	909	3366.7	959	3551.9
760	2814.8	810	3000.0	860	3185.2	910	3370.4	960	3555.6
761	2818.5	811	3003.7	861	3188.9	911	3374.1	961	3559.3
762	2822.2	812	3007.4	862	3192.6	912	3377.8	962	3563.0
763	2825.9	813	3011.1	863	3196.3	913	3381.5	963	3566.7
764	2829.6	814	3014.8	864	3200.0	914	3385.2	964	3570.4
765	2833.3	815	3018.5	865	3203.7	915	3388.9	965	3574.1
766	2837.0	816	3022.2	866	3207.4	916	3392.6	966	3577.8
767	2840.7	817	3025.9	867	3211.1	917	3396.3	967	3581.5
768	2844.4	818	3029.6	868	3214.8	918	3400.0	968	3585.2
769	2848.2	819	3033.3	869	3218.5	919	3403.7	969	3588.9
770	2851.9	820	3037.0	870	3222.2	920	3407.4	970	3592.6
771	2855.6	821	3040.7	871	3225.9	921	3411.1	971	3596.3
772	2859.3	822	3044.4	872	3229.6	922	3414.8	972	3600.0
773	2863.0	823	3048.2	873	3233.3	923	3418.5	973	3603.7
774	2866.7	824	3051.9	874	3237.0	924	3422.2	974	3607.4
775	2870.4	825	3055.6	875	3240.7	925	3425.9	975	3611.1
776	2874.1	826	3059.3	876	3244.4	926	3429.6	976	3614.8
777	2877.8	827	3063.0	877	3248.2	927	3433.3	977	3618.5
778	2881.5	828	3066.7	878	3251.9	928	3437.0	978	3622.2
779	2885.2	829	3070.4	879	3255.6	929	3440.7	979	3625.9
780	2888.9	830	3074.1	880	3259.3	930	3444.4	980	3629.6
781	2892.6	831	3077.8	881	3263.0	931	3448.2	981	3633.3
782	2896.3	832	3081.5	882	3266.7	932	3451.9	982	3637.0
783	2900.0	833	3085.2	883	3270.4	933	3455.6	983	3640.7
784	2903.7	834	3088.9	884	3274.1	934	3459.3	984	3644.4
785	2907.4	835	3092.6	885	3277.8	935	3463.0	985	3648.2
786	2911.1	836	3096.3	886	3281.5	936	3466.7	986	3651.9
787	2914.8	837	3100.0	887	3285.2	937	3470.4	987	3655.6
788	2918.5	838	3103.7	888	3288.9	938	3474.1	988	3659.3
789	2922.2	839	3107.4	889	3292.6	939	3477.8	989	3663.0
790	2925.9	840	3111.1	890	3296.3	940	3481.5	990	3666.7
791	2929.6	841	3114.8	891	3300.0	941	3485.2	991	3670.4
792	2933.3	842	3118.5	892	3303.7	942	3488.9	992	3674.1
793	2937.0	843	3122.2	893	3307.4	943	3492.6	993	3677.8
794	2940.7	844	3125.9	894	3311.1	944	3496.3	994	3681.5
795	2944.4	845	3129.6	895	3314.8	945	3500.0	995	3685.2
796	2948.2	846	3133.3	896	3318.5	946	3503.7	996	3688.9
797	2951.9	847	3137.0	897	3322.2	947	3507.4	997	3692.6
798	2955.6	848	3140.7	898	3325.9	948	3511.1	998	3696.3
799	2959.3	849	3144.4	899	3329.6	949	3514.8	999	3700.0
800	2963.0	850	3148.2	900	3333.3	950	3518.5	1000	3703.7

TABLE XVI.

CONVERSION OF ENGLISH INCHES INTO CENTIMETRES.

Ins.	0	1	2	3	4	5	6	7	8	9
	Cm.	Cm.	Cm.	Cm.	Cm.	Cm.	Cm.	Cm.	Cm.	Cm.
0	0.000	2.540	5.080	7.620	10.16	12.70	15.24	17.78	20.32	22.86
10	25.40	27.94	30.48	33.02	35.56	38.10	40.64	43.18	45.72	48.26
20	50.80	53.34	55.88	58.42	60.96	63.50	66.04	68.58	71.12	73.66
30	76.20	78.74	81.28	83.82	86.36	88.90	91.44	93.98	96.52	99.06
40	101.60	104.14	106.68	109.22	111.76	114.30	116.84	119.38	121.92	124.46
50	127.00	129.54	132.08	134.62	137.16	139.70	142.24	144.78	147.32	149.86
60	152.40	154.94	157.48	160.02	162.56	165.10	167.64	170.18	172.72	175.26
70	177.80	180.34	182.88	185.42	187.96	190.50	193.04	195.58	198.12	200.96
80	203.20	205.74	208.28	210.82	213.36	215.90	218.44	220.98	223.52	226.06
90	228.60	231.14	233.68	236.22	238.76	241.30	243.84	246.38	248.92	251.46
100	254.00	256.54	259.08	261.62	264.16	266.70	269.24	271.78	274.32	276.86

CONVERSION OF CENTIMETRES INTO ENGLISH INCHES.

Cm.	0	1	2	3	4	5	6	7	8	9
	Ins.	Ins.	Ins.	Ins.	Ins.	Ins.	Ins.	Ins.	Ins.	Ins.
0	0.000	0.394	0.787	1.181	1.575	1.969	2.362	2.756	3.150	3.543
10	3.937	4.331	4.742	5.118	5.512	5.906	6.299	6.693	7.087	7.480
20	7.874	8.268	8.662	9.055	9.449	9.843	10.236	10.630	11.024	11.418
30	11.811	12.205	12.599	12.992	13.386	13.780	14.173	14.567	14.961	15.355
40	15.748	16.142	16.536	16.929	17.323	17.717	18.111	18.504	18.898	19.292
50	19.685	20.079	20.473	20.867	21.260	21.654	22.048	22.441	22.835	23.229
60	23.622	24.016	24.410	24.804	25.197	25.591	25.985	26.378	26.772	27.166
70	27.560	27.953	28.347	28.741	29.134	29.528	29.922	30.316	30.709	31.103
80	31.497	31.890	32.284	32.678	33.071	33.465	33.859	34.253	34.646	35.040
90	35.434	35.827	36.221	36.615	37.009	37.402	37.796	38.190	38.583	38.977
100	39.370	39.764	40.158	40.552	40.945	41.339	41.733	42.126	42.520	42.914

CONVERSION OF ENGLISH FEET INTO METRES.

Feet.	0	1	2	3	4	5	6	7	8	9
	Met.	Met.	Met.	Met.	Met.	Met.	Met.	Met.	Met.	Met.
0	0.000	0.3048	0.6096	0.9144	1.2192	1.5239	1.8287	2.1335	2.4383	2.7431
10	3.0479	3.3527	3.6575	3.9623	4.2671	4.5719	4.8767	5.1815	5.4863	5.7911
20	6.0359	6.4006	6.7055	7.0102	7.3150	7.6198	7.9246	8.2294	8.5342	8.8390
30	9.1438	9.4486	9.7534	10.058	10.363	10.668	10.972	11.277	11.582	11.887
40	12.192	12.496	12.801	13.106	13.411	13.716	14.020	14.325	14.630	14.935
50	15.239	15.544	15.849	16.154	16.459	16.763	17.068	17.373	17.678	17.983
60	18.287	18.592	18.897	19.202	19.507	19.811	20.116	20.421	20.726	21.031
70	21.335	21.640	21.945	22.250	22.555	22.859	23.164	23.469	23.774	24.079
80	24.383	24.688	24.993	25.298	25.602	25.907	26.212	26.517	26.822	27.126
90	27.431	27.736	28.041	28.346	28.651	28.955	29.260	29.565	29.870	30.174
100	30.479	30.784	31.089	31.394	31.698	32.003	32.308	32.613	32.918	33.222

CONVERSION OF METRES INTO ENGLISH FEET.

Met.	0	1	2	3	4	5	6	7	8	9
	Feet.	Feet.	Feet.	Feet.	Feet.	Feet.	Feet.	Feet.	Feet.	Feet.
0	0.000	3.2809	6.5618	9.8427	13.123	16.404	19.685	22.966	26.247	29.528
10	32.809	36.090	39.371	42.651	45.932	49.213	52.494	55.775	59.056	62.337
20	65.618	68.899	72.179	75.461	78.741	82.022	85.303	88.584	91.865	95.146
30	98.427	101.71	104.99	108.27	111.55	114.83	118.11	121.39	124.67	127.96
40	131.24	134.52	137.80	141.08	144.36	147.64	150.92	154.20	157.48	160.76
50	164.04	167.33	170.61	173.89	177.17	180.45	183.73	187.01	190.29	193.57
60	196.85	200.13	203.42	206.70	209.98	213.26	216.54	219.82	223.10	226.38
70	229.66	232.94	236.22	239.51	242.79	246.07	249.35	252.63	255.91	259.19
80	262.47	265.75	269.03	272.31	275.60	278.88	282.16	285.44	288.72	292.00
90	295.28	298.56	301.84	305.12	308.40	311.69	314.97	318.25	321.53	324.81
100	328.09	331.37	334.65	337.93	341.21	344.49	347.78	351.06	354.34	357.62

TABLE XVII.

CONVERSION OF ENGLISH STATUTE-MILES INTO KILOMETRES.

Miles.	0	1	2	3	4	5	6	7	8	9
	Kilo.	Kilo.	Kilo.	Kilo.	Kilo.	Kilo.	Kilo.	Kilo.	Kilo.	Kilo.
0	0.0000	1.6093	3.2186	4.8279	6.4372	8.0465	9.6558	11.2652	12.8745	14.4848
10	16.093	17.702	19.312	20.921	22.530	24.139	25.749	27.358	28.967	30.577
20	32.186	33.795	35.405	37.014	38.623	40.232	41.842	43.451	45.060	46.670
30	48.279	49.888	51.498	53.107	54.716	56.325	57.935	59.544	61.153	62.763
40	64.372	65.981	67.591	69.200	70.809	72.418	74.028	75.637	77.246	78.856
50	80.465	82.074	83.684	85.293	86.902	88.511	90.121	91.730	93.339	94.949
60	96.558	98.167	99.777	101.39	102.99	104.60	106.21	107.82	109.43	111.04
70	112.65	114.26	115.87	117.48	119.08	120.69	122.30	123.91	125.52	127.13
80	128.74	130.35	131.96	133.57	135.17	136.78	138.39	140.00	141.61	143.22
90	144.85	146.44	148.05	149.66	151.26	152.87	154.48	156.09	157.70	159.31
100	160.93	162.53	164.14	165.75	167.35	168.96	170.57	172.18	173.79	175.40

CONVERSION OF KILOMETRES INTO ENGLISH STATUTE-MILES.

Kilom.	0	1	2	3	4	5	6	7	8	9
	Miles.	Miles.	Miles.	Miles.	Miles.	Miles.	Miles.	Miles.	Miles.	Miles.
0	0.0000	0.6214	1.2427	1.8641	2.4855	3.1069	3.7282	4.3497	4.9711	5.5924
10	6.2138	6.8352	7.4565	8.0780	8.6994	9.3208	9.9421	10.562	11.185	11.805
20	12.427	13.049	13.670	14.292	14.913	15.534	16.156	16.776	17.399	18.019
30	18.641	19.263	19.884	20.506	21.127	21.748	22.370	22.990	23.613	24.233
40	24.855	25.477	26.098	26.720	27.341	27.962	28.584	29.204	29.827	30.447
50	31.069	31.690	32.311	32.933	33.554	34.175	34.797	35.417	36.040	36.660
60	37.282	37.904	38.525	39.147	39.768	40.389	41.011	41.631	42.254	42.874
70	43.497	44.118	44.739	45.361	45.982	46.603	47.225	47.845	48.468	49.088
80	49.711	50.332	50.953	51.575	52.196	52.817	53.439	54.059	54.682	55.302
90	55.924	56.545	57.166	57.788	58.409	59.030	59.652	60.272	60.895	61.515
100	62.138	62.759	63.380	64.002	64.623	65.244	65.866	66.486	67.109	67.729

TABLE XVIII.
LENGTH IN FEET OF 1' ARCS OF LATITUDE AND LONGITUDE.

Lat.	1' Lat.	1' Long.	Lat.	1' Lat.	1' Long.
1°	6045	6085	31°	6061	5222
2°	6045	6083	32°	6062	5166
3°	6045	6078	33°	6063	5109
4°	6045	6071	34°	6064	5051
5°	6045	6063	35°	6065	4991
6°	6045	6053	36°	6066	4930
7°	6046	6041	37°	6067	4867
8°	6046	6027	38°	6068	4802
9°	6046	6012	39°	6070	4736
10°	6047	5994	40°	6071	4669
11°	6047	5975	41°	6072	4600
12°	6048	5954	42°	6073	4530
13°	6048	5931	43°	6074	4458
14°	6049	5907	44°	6075	4385
15°	6049	5880	45°	6076	4311
16°	6050	5852	46°	6077	4235
17°	6050	5822	47°	6078	4158
18°	6051	5790	48°	6079	4080
19°	6052	5757	49°	6080	4001
20°	6052	5721	50°	6081	3920
21°	6053	5684	51°	6082	3838
22°	6054	5646	52°	6084	3755
23°	6054	5605	53°	6085	3671
24°	6055	5563	54°	6086	3586
25°	6056	5519	55°	6087	3499
26°	6057	5474	56°	6088	3413
27°	6058	5427	57°	6089	3323
28°	6059	5378	58°	6090	3233
29°	6060	5327	59°	6091	3142
30°	6061	5275	60°	6092	3051

TABLE XIX.—TO REDUCE MEAN TO SIDEREAL TIME.

Solar Hours.	Add Min. Sec.	Solar Min.	Add Sec.	Solar Min.	Add Sec.	Solar Sec.	Add Sec.	Solar Sec.	Add Sec.
1	0 9.86	1	0.16	31	5.09	1	0.00	31	0.08
2	0 19.71	2	0.33	32	5.26	2	0.01	32	0.09
3	0 29.57	3	0.49	33	5.42	3	0.01	33	0.09
4	0 39.43	4	0.66	34	5.59	4	0.01	34	0.09
5	0 49.28	5	0.82	35	5.75	5	0 01	35	0.10
6	0 59.14	6	0.99	36	5.92	6	0.02	36	0.10
7	1 9.00	7	1.15	37	6.08	7	0.02	37	0.10
8	1 18.85	8	1.31	38	6.24	8	0.02	38	0.10
9	1 28.71	9	1.48	39	6.41	9	0.03	39	0.11
10	1 38.57	10	1.64	40	6.57	10	0.03	40	0.11
11	1 48.42	11	1.81	41	6.74	11	0.03	41	0.11
12	1 58.28	12	1.97	42	6.90	12	0.03	42	0.12
13	2 8.13	13	2.14	43	7.07	13	0.04	43	0.12
14	2 17.99	14	2.30	44	7.23	14	0.04	44	0.12
15	2 27.85	15	2.46	45	7.39	15	0.04	45	0.12
16	2 37.70	16	2.63	46	7.56	16	0.04	46	0.13
17	2 47.56	17	2.79	47	7.72	17	0.05	47	0.13
18	2 57.42	18	2.96	48	7.89	18	0.05	48	0.13
19	3 7.27	19	3.12	49	8.05	19	0.05	49	0.13
20	3 17.13	20	3.29	50	8.22	20	0.06	50	0.14
21	3 26.99	21	3.45	51	8.38	21	0.06	51	0.14
22	3 36.84	22	3.61	52	8.54	22	0.06	52	0.14
23	3 46.70	23	3.78	53	8.71	23	0.06	53	0.15
24	3 56.56	24	3.94	54	8.87	24	0.07	54	0.15
25	4 6.40	25	4.11	55	9.04	25	0.07	55	0.15
26	4 16.26	26	4.27	56	9.20	26	0.07	56	0.15
27	4 26.13	27	4.44	57	9.37	27	0.08	57	0.16
28	4 36.00	28	4.60	58	9.53	28	0.08	58	0.16
29	4 45.86	29	4.76	59	9.69	29	0.08	59	0.16
30	4 55.71	30	4.93	60	9.86	30	0.08	60	0.16

TABLE XIX—Continued.—TO REDUCE SIDEREAL TO MEAN TIME.

Sid. Hours.	Subtract Min. Sec.	Sid. Min.	Subtract Sec.	Sid. Min.	Subtract Sec.	Sid. Sec.	Subtract Sec.	Sid. Sec.	Subtract Sec.
1	0 9.83	1	0.16	31	5.08	1	0.00	31	0.08
2	0 19.66	2	0.33	32	5.24	2	0.01	32	0.09
3	0 29.49	3	0.49	33	5.41	3	0.01	33	0.09
4	0 39.32	4	0.66	34	5.57	4	0.01	34	0.09
5	0 49.15	5	0.82	35	5.73	5	0.01	35	0.10
6	0 58.98	6	0.98	36	5.90	6	0.02	36	0.10
7	1 8.81	7	1.15	37	6.06	7	0.02	37	0.10
8	1 18.64	8	1.31	38	6.23	8	0.02	38	0.10
9	1 28.47	9	1.47	39	6.39	9	0.03	39	0.11
10	1 38.30	10	1.64	40	6.55	10	0.03	40	0.11
11	1 48.12	11	1.80	41	6.72	11	0.03	41	0.11
12	1 57.95	12	1.97	42	6.88	12	0.03	42	0.11
13	2 7.78	13	2.13	43	7.04	13	0.04	43	0.12
14	2 17.61	14	2.29	44	7.21	14	0.04	44	0.12
15	2 27.44	15	2.46	45	7.37	15	0.04	45	0.12
16	2 37.27	16	2.62	46	7.54	16	0.04	46	0.13
17	2 47.10	17	2.79	47	7.70	17	0.05	47	0.13
18	2 56.93	18	2.95	48	7.86	18	0.05	48	0.13
19	3 6.76	19	3.11	49	8.03	19	0.05	49	0.13
20	3 16.59	20	3.28	50	8.19	20	0.06	50	0.14
21	3 26.42	21	3.44	51	8.36	21	0.06	51	0.14
22	3 36.25	22	3.60	52	8.52	22	0.06	52	0.14
23	3 46.08	23	3.77	53	8.68	23	0.06	53	0.14
24	3 55.91	24	3.93	54	8.85	24	0.07	54	0.15
25	4 5.74	25	4.10	55	9.01	25	0.07	55	0.15
26	4 15.57	26	4.26	56	9.17	26	0.07	56	0.15
27	4 25.41	27	4.42	57	9.34	27	0.07	57	0.16
28	4 35.24	28	4.59	58	9.50	28	0.08	58	0.16
29	4 45.07	29	4.75	59	9.67	29	0.08	59	0.16
30	4 54.90	30	4.92	60	9.83	30	0.08	60	0.16

www.ingramcontent.com/pod-product-compliance
Lightning Source LLC
Chambersburg PA
CBHW020310240426
43673CB00039B/764